Asphaltenes

Chemical Transformation during Hydroprocessing of Heavy Oils

CHEMICAL INDUSTRIES

A Series of Reference Books and Textbooks

Founding Editor

HEINZ HEINEMANN
Berkeley, California

Series Editor

JAMES G. SPEIGHT
University of Trinidad and Tobago
O'Meara Campus, Trinidad

1. *Fluid Catalytic Cracking with Zeolite Catalysts,* Paul B. Venuto and E. Thomas Habib, Jr.
2. *Ethylene: Keystone to the Petrochemical Industry,* Ludwig Kniel, Olaf Winter, and Karl Stork
3. *The Chemistry and Technology of Petroleum,* James G. Speight
4. *The Desulfurization of Heavy Oils and Residua,* James G. Speight
5. *Catalysis of Organic Reactions,* edited by William R. Moser
6. *Acetylene-Based Chemicals from Coal and Other Natural Resources,* Robert J. Tedeschi
7. *Chemically Resistant Masonry,* Walter Lee Sheppard, Jr.
8. *Compressors and Expanders: Selection and Application for the Process Industry,* Heinz P. Bloch, Joseph A. Cameron, Frank M. Danowski, Jr., Ralph James, Jr., Judson S. Swearingen, and Marilyn E. Weightman
9. *Metering Pumps: Selection and Application,* James P. Poynton
10. *Hydrocarbons from Methanol,* Clarence D. Chang
11. *Form Flotation: Theory and Applications,* Ann N. Clarke and David J. Wilson
12. *The Chemistry and Technology of Coal,* James G. Speight
13. *Pneumatic and Hydraulic Conveying of Solids,* O. A. Williams

14. *Catalyst Manufacture: Laboratory and Commercial Preparations,* Alvin B. Stiles
15. *Characterization of Heterogeneous Catalysts,* edited by Francis Delannay
16. *BASIC Programs for Chemical Engineering Design,* James H. Weber
17. *Catalyst Poisoning,* L. Louis Hegedus and Robert W. McCabe
18. *Catalysis of Organic Reactions,* edited by John R. Kosak
19. *Adsorption Technology: A Step-by-Step Approach to Process Evaluation and Application,* edited by Frank L. Slejko
20. *Deactivation and Poisoning of Catalysts,* edited by Jacques Oudar and Henry Wise
21. *Catalysis and Surface Science: Developments in Chemicals from Methanol, Hydrotreating of Hydrocarbons, Catalyst Preparation, Monomers and Polymers, Photocatalysis and Photovoltaics,* edited by Heinz Heinemann and Gabor A. Somorjai
22. *Catalysis of Organic Reactions,* edited by Robert L. Augustine
23. *Modern Control Techniques for the Processing Industries,* T. H. Tsai, J. W. Lane, and C. S. Lin
24. *Temperature-Programmed Reduction for Solid Materials Characterization,* Alan Jones and Brian McNichol
25. *Catalytic Cracking: Catalysts, Chemistry, and Kinetics,* Bohdan W. Wojciechowski and Avelino Corma
26. *Chemical Reaction and Reactor Engineering,* edited by J. J. Carberry and A. Varma
27. *Filtration: Principles and Practices: Second Edition,* edited by Michael J. Matteson and Clyde Orr
28. *Corrosion Mechanisms,* edited by Florian Mansfeld
29. *Catalysis and Surface Properties of Liquid Metals and Alloys,* Yoshisada Ogino
30. *Catalyst Deactivation,* edited by Eugene E. Petersen and Alexis T. Bell
31. *Hydrogen Effects in Catalysis: Fundamentals and Practical Applications,* edited by Zoltán Paál and P. G. Menon
32. *Flow Management for Engineers and Scientists,* Nicholas P. Cheremisinoff and Paul N. Cheremisinoff
33. *Catalysis of Organic Reactions,* edited by Paul N. Rylander, Harold Greenfield, and Robert L. Augustine
34. *Powder and Bulk Solids Handling Processes: Instrumentation and Control,* Koichi Iinoya, Hiroaki Masuda, and Kinnosuke Watanabe
35. *Reverse Osmosis Technology: Applications for High-Purity-Water Production,* edited by Bipin S. Parekh
36. *Shape Selective Catalysis in Industrial Applications,* N. Y. Chen, William E. Garwood, and Frank G. Dwyer

37. *Alpha Olefins Applications Handbook,* edited by George R. Lappin and Joseph L. Sauer
38. *Process Modeling and Control in Chemical Industries,* edited by Kaddour Najim
39. *Clathrate Hydrates of Natural Gases,* E. Dendy Sloan, Jr.
40. *Catalysis of Organic Reactions,* edited by Dale W. Blackburn
41. *Fuel Science and Technology Handbook,* edited by James G. Speight
42. *Octane-Enhancing Zeolitic FCC Catalysts,* Julius Scherzer
43. Oxygen in Catalysis, Adam Bielanski and Jerzy Haber
44. *The Chemistry and Technology of Petroleum: Second Edition, Revised and Expanded,* James G. Speight
45. *Industrial Drying Equipment: Selection and Application,* C. M. van't Land
46. *Novel Production Methods for Ethylene, Light Hydrocarbons, and Aromatics*, edited by Lyle F. Albright, Billy L. Crynes, and Siegfried Nowak
47. *Catalysis of Organic Reactions*, edited by William E. Pascoe
48. *Synthetic Lubricants and High-Performance Functional Fluids,* edited by Ronald L. Shubkin
49. *Acetic Acid and Its Derivatives*, edited by Victor H. Agreda and Joseph R. Zoeller
50. *Properties and Applications of Perovskite-Type Oxides,* edited by L. G. Tejuca and J. L. G. Fierro
51. *Computer-Aided Design of Catalysts*, edited by E. Robert Becker and Carmo J. Pereira
52. *Models for Thermodynamic and Phase Equilibria Calculations,* edited by Stanley I. Sandler
53. *Catalysis of Organic Reactions*, edited by John R. Kosak and Thomas A. Johnson
54. *Composition and Analysis of Heavy Petroleum Fractions,* Klaus H. Altgelt and Mieczyslaw M. Boduszynski
55. *NMR Techniques in Catalysis,* edited by Alexis T. Bell and Alexander Pines
56. *Upgrading Petroleum Residues and Heavy Oils,* Murray R. Gray
57. *Methanol Production and Use*, edited by Wu-Hsun Cheng and Harold H. Kung
58. *Catalytic Hydroprocessing of Petroleum and Distillates,* edited by Michael C. Oballah and Stuart S. Shih
59. *The Chemistry and Technology of Coal: Second Edition, Revised and Expanded,* James G. Speight
60. *Lubricant Base Oil and Wax Processing*, Avilino Sequeira, Jr.
61. *Catalytic Naphtha Reforming: Science and Technology,* edited by George J. Antos, Abdullah M. Aitani, and José M. Parera

62. *Catalysis of Organic Reactions*, edited by Mike G. Scaros and Michael L. Prunier
63. *Catalyst Manufacture,* Alvin B. Stiles and Theodore A. Koch
64. *Handbook of Grignard Reagents,* edited by Gary S. Silverman and Philip E. Rakita
65. *Shape Selective Catalysis in Industrial Applications: Second Edition, Revised and Expanded,* N. Y. Chen, William E. Garwood, and Francis G. Dwyer
66. *Hydrocracking Science and Technology,* Julius Scherzer and A. J. Gruia
67. *Hydrotreating Technology for Pollution Control: Catalysts, Catalysis, and Processes,* edited by Mario L. Occelli and Russell Chianelli
68. *Catalysis of Organic Reactions,* edited by Russell E. Malz, Jr.
69. *Synthesis of Porous Materials: Zeolites, Clays, and Nanostructures,* edited by Mario L. Occelli and Henri Kessler
70. *Methane and Its Derivatives,* Sunggyu Lee
71. *Structured Catalysts and Reactors,* edited by Andrzej Cybulski and Jacob A. Moulijn
72. *Industrial Gases in Petrochemical Processing,* Harold Gunardson
73. *Clathrate Hydrates of Natural Gases: Second Edition, Revised and Expanded,* E. Dendy Sloan, Jr.
74. *Fluid Cracking Catalysts,* edited by Mario L. Occelli and Paul O'Connor
75. *Catalysis of Organic Reactions,* edited by Frank E. Herkes
76. *The Chemistry and Technology of Petroleum: Third Edition, Revised and Expanded,* James G. Speight
77. *Synthetic Lubricants and High-Performance Functional Fluids: Second Edition, Revised and Expanded,* Leslie R. Rudnick and Ronald L. Shubkin
78. *The Desulfurization of Heavy Oils and Residua, Second Edition, Revised and Expanded,* James G. Speight
79. *Reaction Kinetics and Reactor Design: Second Edition, Revised and Expanded,* John B. Butt
80. *Regulatory Chemicals Handbook,* Jennifer M. Spero, Bella Devito, and Louis Theodore
81. *Applied Parameter Estimation for Chemical Engineers,* Peter Englezos and Nicolas Kalogerakis
82. *Catalysis of Organic Reactions,* edited by Michael E. Ford
83. *The Chemical Process Industries Infrastructure: Function and Economics,* James R. Couper, O. Thomas Beasley, and W. Roy Penney
84. *Transport Phenomena Fundamentals,* Joel L. Plawsky

85. *Petroleum Refining Processes*, James G. Speight and Baki Özüm
86. *Health, Safety, and Accident Management in the Chemical Process Industries*, Ann Marie Flynn and Louis Theodore
87. *Plantwide Dynamic Simulators in Chemical Processing and Control*, William L. Luyben
88. *Chemical Reactor Design*, Peter Harriott
89. *Catalysis of Organic Reactions*, edited by Dennis G. Morrell
90. *Lubricant Additives: Chemistry and Applications*, edited by Leslie R. Rudnick
91. *Handbook of Fluidization and Fluid-Particle Systems*, edited by Wen-Ching Yang
92. *Conservation Equations and Modeling of Chemical and Biochemical Processes*, Said S. E. H. Elnashaie and Parag Garhyan
93. *Batch Fermentation: Modeling, Monitoring, and Control*, Ali Çinar, Gülnur Birol, Satish J. Parulekar, and Cenk Ündey
94. *Industrial Solvents Handbook, Second Edition*, Nicholas P. Cheremisinoff
95. *Petroleum and Gas Field Processing*, H. K. Abdel-Aal, Mohamed Aggour, and M. Fahim
96. *Chemical Process Engineering: Design and Economics*, Harry Silla
97. *Process Engineering Economics*, James R. Couper
98. *Re-Engineering the Chemical Processing Plant: Process Intensification*, edited by Andrzej Stankiewicz and Jacob A. Moulijn
99. *Thermodynamic Cycles: Computer-Aided Design and Optimization*, Chih Wu
100. *Catalytic Naphtha Reforming: Second Edition, Revised and Expanded*, edited by George T. Antos and Abdullah M. Aitani
101. *Handbook of MTBE and Other Gasoline Oxygenates*, edited by S. Halim Hamid and Mohammad Ashraf Ali
102. *Industrial Chemical Cresols and Downstream Derivatives*, Asim Kumar Mukhopadhyay
103. *Polymer Processing Instabilities: Control and Understanding*, edited by Savvas Hatzikiriakos and Kalman B. Migler
104. *Catalysis of Organic Reactions*, John Sowa
105. *Gasification Technologies: A Primer for Engineers and Scientists*, edited by John Rezaiyan and Nicholas P. Cheremisinoff
106. *Batch Processes*, edited by Ekaterini Korovessi and Andreas A. Linninger
107. *Introduction to Process Control*, Jose A. Romagnoli and Ahmet Palazoglu

108. *Metal Oxides: Chemistry and Applications*, edited by J. L. G. Fierro
109. *Molecular Modeling in Heavy Hydrocarbon Conversions*, Michael T. Klein, Ralph J. Bertolacini, Linda J. Broadbelt, Ankush Kumar and Gang Hou
110. *Structured Catalysts and Reactors, Second Edition*, edited by Andrzej Cybulski and Jacob A. Moulijn
111. *Synthetics, Mineral Oils, and Bio-Based Lubricants: Chemistry and Technology*, edited by Leslie R. Rudnick
112. *Alcoholic Fuels*, edited by Shelley Minteer
113. *Bubbles, Drops, and Particles in Non-Newtonian Fluids, Second Edition*, R. P. Chhabra
114. *The Chemistry and Technology of Petroleum, Fourth Edition*, James G. Speight
115. *Catalysis of Organic Reactions*, edited by Stephen R. Schmidt
116. *Process Chemistry of Lubricant Base Stocks*, Thomas R. Lynch
117. *Hydroprocessing of Heavy Oils and Residua*, edited by James G. Speight and Jorge Ancheyta
118. *Chemical Process Performance Evaluation*, Ali Cinar, Ahmet Palazoglu, and Ferhan Kayihan
119. *Clathrate Hydrates of Natural Gases, Third Edition*, E. Dendy Sloan and Carolyn Koh
120. *Interfacial Properties of Petroleum Products*, Lilianna Z. Pillon
121. *Process Chemistry of Petroleum Macromolecules*, Irwin A. Wiehe
122. *The Scientist or Engineer as an Expert Witness*, James G. Speight
123. *Catalysis of Organic Reactions*, edited by Michael L. Prunier
124. *Lubricant Additives: Chemistry and Applications, Second Edition*, edited by Leslie R. Rudnick
125. *Chemical Reaction Engineering and Reactor Technology*, Tapio O. Salmi, Jyri-Pekka Mikkola, and Johan P. Warna
126. *Asphaltenes: Chemical Transformation during Hydroprocessing of Heavy Oils*, Jorge Ancheyta, Fernando Trejo, and Mohan Singh Rana

Asphaltenes

Chemical Transformation during Hydroprocessing of Heavy Oils

Jorge Ancheyta

Mexican Institute of Petroleum
National Polytechnic Institute (ESIQIE-IPN)
Mexico City, Mexico

Fernando Trejo

Research Center for Applied Science and Advanced Technology
National Polytechnic Institute
Mexico City, Mexico

Mohan Singh Rana

Kuwait Institute for Scientific Research
Ahmadi, State of Kuwait

CRC Press
Taylor & Francis Group
Boca Raton London New York

CRC Press is an imprint of the
Taylor & Francis Group, an **informa** business

CRC Press
Taylor & Francis Group
6000 Broken Sound Parkway NW, Suite 300
Boca Raton, FL 33487-2742

Printed in the United States of America on acid-free paper
10 9 8 7 6 5 4 3 2 1

International Standard Book Number: 978-1-4200-6630-2 (Hardback)

Library of Congress Cataloging-in-Publication Data

Ancheyta, Jorge.
Asphaltenes : chemical transformation during hydroprocessing of heavy oils / Jorge Ancheyta, F. Trejo, and Mohan Singh Rana.
p. cm. -- (Chemical industries ; v. 126)
Includes bibliographical references and index.
ISBN 978-1-4200-6630-2
1. Petroleum--Refining. 2. Heavy oil. 3. Hydrogenation. 4. Hydrocracking. I. Trejo, F. (Fernando) II. Rana, Mohan Singh. III. Title. IV. Series.

TP690.45.A53 2009
665.5'3--dc22 2009006451

Visit the Taylor & Francis Web site at
http://www.taylorandfrancis.com

and the CRC Press Web site at
http://www.crcpress.com

Contents

Preface

In the past, as with many other complex molecules, details of asphaltene's characterization were completely unknown and only bulk properties were employed to elucidate its possible structure and composition. However, the development of sophisticated techniques along with the need to have a deeper understanding of asphaltene structural characterization have motivated researchers to study this macromolecule, with the main objective of solving particular problems associated to formation (reservoir) damage due to asphaltene precipitation, well and production facilities plugging, as well as asphaltene's precipitation during blending and transportation of crude oil, which are some of the most frequent inconveniences during different stages of petroleum production. In addition to this, the increasing production of heavy crude oils has played an important role for refiners and research centers to continue with the investigation of asphaltenes with such an intensity that nowadays asphaltene has become one of the most studied molecules, not only for academic/research centers personnel, but also for those working for petroleum companies.

The following definition of asphaltenes based on their solubility properties is accepted worldwide: "Asphaltene is the material present in petroleum that is insoluble in alkanes (*n*-pentane, *n*-hexane, *n*-heptane, or higher), but soluble in aromatic solvents (i.e., benzene, toluene)." The most commonly used solvent is *n*-heptane. The main constituents of asphaltenes are aromatic rings carrying alkyl chains up to 30 carbon atoms, sulfur as benzothiophene rings, nickel, and vanadium complexed with pyrrole nitrogen atoms in porphyrinic rings.

Presently refining processes are focused on the upgrading of heavy crude oils and residua, which is a consequence of the increasing demand for producing more gasoline and middle distillates, as well as the decrease of production of light crude oils. As the petroleum becomes heavier, the content and complexity of asphaltenes present in it increases considerably. The need to upgrade and convert heavy and extra-heavy crude oils in order to accomplish not only the increasing demand of automotive fuels, but also the strict legislations about fuel quality, i.e., the so-called clean fuels, have forced researchers to accelerate the comprehension of asphaltenes structures and also their changes suffered when processing heavy petroleum.

During upgrading of heavy petroleum, asphaltene is the most problematic impurity, since it is the main cause of catalyst deactivation and sediments formation. Asphaltenes and metals associated with them are deposited on the catalyst surface during heavy oils hydroprocessing with the consequent blockage of the pores, so that the evolution of asphaltenes properties and structure during hydroprocessing is very important to analyze for better design of catalyst and processes.

Asphaltenes are composed of aromatic rings and alkyl side chains, and when they are in a colloidal state in crude oil, they are surrounded by resins, which stabilize forming micelles. However, during hydroprocessing reactions, different compounds turn into lighter and valuable products, and asphaltenes experience significant changes in their structure because they are also exposed to reactions. These changes alter the "perfect" equilibrium between all molecules present in petroleum, and depending on reaction severity, high sludge and sediment formation can be encountered. Once sediment is formed, it is easily deposited on the catalyst surface and process equipments, provoking premature shutdown of commercial hydroprocessing units. Thus, the maximum conversion achieved is determined by the limit of sediment formation rather than by catalyst activity or other operating and equipment constrains. As can be observed, asphaltene is the molecule that strongly participates in catalyst deactivation and sediment formation during hydroprocessing of heavy oils, that is why a better knowledge of its changes and effects on product quality as the reaction proceeds is mandatory.

This book is devoted to highlighting various aspects relevant to the chemical transformations that asphaltenes undergo during hydroprocessing of heavy oils. The text is organized in seven chapters in order to have complete understanding of all aspects related to asphaltenes and their conversion during hydroprocessing. Because there is a great controversy about asphaltene structure, the different points of view about its definition and characterization are discussed in Chapter 1. Chapter 2 is designed to introduce those readers requiring an in-depth knowledge on topics related to hydroprocessing of heavy oils, such as the composition of petroleum, processes, and catalysts for upgrading of heavy oils. Special emphasis has been given to the effect of asphaltenes on catalyst stability and life. Characterization of asphaltenes after hydroprocessing and the effect of reaction conditions on their structures are treated with sufficient detail in Chapter 3. Deactivation and characterization of hydroprocessing the spent catalysts, as well as the role played by asphaltenes, are analyzed in Chapter 4. Methods for preventing coke deposition, regeneration, and rejuvenation of spent catalysts are also discussed in this chapter. An entire chapter (Chapter 5), covering sediments formation during hydroprocessing and the role of asphaltenes on it, has been included. Chapters 6 and 7 provide detailed studies of hydrocracking and kinetics of asphaltenes, and descriptions of fractionation of heavy crudes and asphaltenes.

We would like to acknowledge all our colleagues for their numerous and valuable suggestions to improve the quality of this book. We are also indebted to our friends, students, and technicians for all the support they provided.

Jorge Ancheyta

Fernando Trejo

Mohan S. Rana

About the Authors

Jorge Ancheyta, PhD, graduated with a bachelor's degree in petrochemical engineering (1989), master's degree in chemical engineering (1993), and master's degree in administration, planning, and economics of hydrocarbons (1997) from the National Polytechnic Institute (IPN) of Mexico. He split his PhD between the Metropolitan Autonomous University (UAM) of Mexico and the Imperial College London (1998), and was awarded a postdoctoral fellowship in the Laboratory of Catalytic Process Engineering of the CPE-CNRS in Lyon, France (1999). He has also been visiting professor at the Laboratoire de Catalyse et Spectrochimie (LCS), Université de Caen, France (2008, 2009), and Imperial College London (2009).

Dr. Ancheyta has worked for the Mexican Institute of Petroleum (IMP) since 1989 and his present position is project leader of Research and Development. He has also worked as professor at the undergraduate and postgraduate levels for the School of Chemical Engineering and Extractive Industries at the National Polytechnic Institute of Mexico (ESIQIE-IPN) since 1992 and for the IMP postgrade since 2003. He has been supervisor of more than 90 BSc, MSc, and PhD theses. Dr. Ancheyta has also been supervisor of a number of postdoctoral and sabbatical year professors.

Dr. Ancheyta has been working in the development and application of petroleum refining catalysts, kinetic, and reactor models, and process technologies, mainly in catalytic cracking, catalytic reforming, middle distillate hydrotreating, and heavy oils upgrading. He is the author and co-author of a number of patents, books, and scientific papers, and has been awarded the Highest Level (III) National Researcher Distinction by the Mexican government and is a member of the Mexican Academy of Science. He has also been guest editor of various international journals, e.g., *Catalysis Today*, *Petroleum Science and Technology*, *Industrial Engineering Chemistry Research*, *Chemical Engineering Communications*, and *Fuel*. Dr. Ancheyta has also chaired numerous international conferences.

Fernando Trejo, PhD, graduated with a bachelor's degree in chemical engineering (1999) and a master's degree in chemical engineering (2002) from the National Polytechnic Institute (IPN) in Mexico. He obtained a PhD degree in chemical engineering from National Autonomous University of Mexico (UNAM) in 2006 and completed a postdoctoral degree at Mexican Institute of Petroleum (IMP) in 2007. Since 2008, Dr. Trejo has worked as a researcher at the Center of Research in Applied Science and Advanced Technology (CICATA-Legaria), which is part of the National Polytechnic Institute (IPN).

Dr. Trejo's experience in petroleum fractions and hydroprocessing began in 2000 when he started his master's degree thesis at the School of Chemical Engineering and Extractive Industries (ESIQIE-IPN).

He is the author and co-author of a number of scientific papers and participated in various international congresses. The Mexican government has awarded him the distinction of National Researcher.

Mohan S. Rana, PhD, was born and raised in Uttranchal (North), India. He obtained the BSc and MSc degrees in chemistry from the University of HNB Grahwal, Srinagar, India, in 1990 and 1992, respectively. He later received his PhD in heterogeneous catalysis from the HNB Garhwal University (research center: Indian Institute of Petroleum, CSIR), India, in 2000. He then worked for a couple of years as a postdoctoral fellow at the University of Caen, CNRS, France, on the inhibition effect carried out by reaction intermediate in hydrotreating catalysts. Following that, Dr. Rana worked as a postdoctoral and research scientist at the Instituto Mexicano del Petroleo, Mexico, from 2002 to 2008, covering different areas, such as upgrading of crude oil by catalytic as well as noncatalytic methods. Presently, he is working as a research scientist for the Petroleum Refining Division, Kuwait Institute for Scientific Research in Kuwait.

Dr. Rana has more than 13 years of experience in areas associated with the heterogeneous catalysis and petroleum refining processes, mainly in hydroprocessing. His research has involved petroleum oil upgrading, improvement of middle distillates, heavy gas/oil and oil sands bitumen, including catalyst development, hydrotreating, mild hydrocracking, hydrocracking, catalytic cracking, and hydrogenation. He has published more than 50 papers in international journals.

He has also coedited a special issue of *Catalysis Today* (volume 109) "Hydroprocessing of Heavy Oil Fractions."

1 Definition and Structure of Asphaltenes

1.1 INTRODUCTION

Nowadays, the refining processes are focused on the conversion of heavy crude oils and residua as the consequence of the increasing demand to produce more gasoline and middle distillates and the decreasing production of light crude oils. Heavy fractions of petroleum can be defined as those molecules, which have more than 25 carbon atoms distributed in asphaltenes and resins, along with high boiling points (Merdrignac and Espinat, 2007).

Asphaltenes are known to be the major precursors of sludge or sediments. They are polyaromatic compounds that have boiling points higher than 500°C. Asphaltenes and metals associated with them are deposited on the catalyst surface during heavy oils hydroprocessing with the consequent blockage of the pores (Martínez et al., 1997).

Unstable crudes display larger amounts of saturates whereas stable crudes have higher aromatics content. Unstable crudes intrinsically have a potential for solids precipitation because they possess aromatic-condensed asphaltenes, which tend to separate from an alkane-rich phase so that selective precipitation of highly aromatic asphaltenes will occur when oils are paraffinic in nature (Carbognani et al., 1999).

Asphaltenes are constituted by the following components (Demirbaş, 2002):

- Aromatic rings carrying alkyl chains up to C_{30}
- Sulfur as benzothiophene rings and nitrogen is contained in pyrrol and pyridine
- Ketones, phenols, carboxylic acids
- Nickel and vanadium complexed with pyrrole nitrogen atoms in porphyrinic rings

Maltenes are conformed by:

- Resins, which are structures similar to asphaltenes but having lower molecular weightssssss
- O, N, and S are not always present in aromatic structures
- Naphtenes and other saturate hydrocarbons, such as straight or branched chains

The most common and widely used definition of asphaltenes is done in terms of their solubility properties. They are insoluble in alkanes, such as *n*-pentane, *n*-hexane, *n*-heptane, or higher, but soluble in aromatic solvents (i.e., benzene, toluene). When adding *n*-pentane for separating asphaltenes, higher content of precipitate is obtained compared with when *n*-heptane is used. Lower paraffinic solvents precipitate higher amounts of insoluble material because not only asphaltenes are precipitated, but also resins. Pillon (2001) obtained up to 30 wt% of material adsorbed onto asphaltenes when they were precipitated with *n*-pentane.

Asphaltenes are commonly precipitated from crudes by adding *n*-heptane because their properties do not exhibit significant changes when using *n*-heptane or higher carbon number alkanes (Andersen, 1994). A number of methods suggest the use of solvent-to-crude or bitumen ratio of 40 (v/w) with contact time between 16 to 24 h. Elemental composition of asphaltenes varies with the solvent used for precipitation. The use of *n*-heptane yields smaller amounts of asphaltenes, which, in some cases, comprised between 15 and 98% less than those obtained with *n*-pentane, depending on the sample. The difference is due to the presence of resins and low molecular weight asphaltenes (Strausz et al., 2002). A research group from Canada (Frakman et al., 1990; Peng et al., 1997; Strausz et al., 1999a, 1999b) has proposed acetone extraction using a Soxhlet extractor for one week to remove light components.

Individual compounds are easily identified in light fractions; however, there is a big diversity of compounds as the boiling point increases and heteroatomic species become more abundant. Thus, the extreme sample complexity of residues makes compositional analysis very difficult, and separation is carried out for compound types instead of individual compounds. When more insolubles are removed, the soluble fraction is rich in hydrogen. Sulfur, nitrogen, and oxygen of soluble fractions decrease as more insolubles are removed. According to Sharma et al. (2007), insoluble fraction contains 7 to 21 heteroatoms per molecule, and the soluble fraction has 0.8 to 1.7. The number of heteroatoms in a vacuum residue with API (American Petroleum Institute) gravity of 3.16 was calculated to be 5.

It is recognized that heavy fractions possess contaminants, such as metals (Ni, V, Ca, Na, Fe, etc.), sulfur, nitrogen, and asphaltenes. Sulfur is present in heavy fractions mainly as benzo-, dibenzo-, and napthobenzo-thiophenes-type of structures; however, sulfide derivatives have been also found.

The idea about asphaltene structure has changed with time. Large structures were initially proposed containing up to 10 fused rings bridged with sulfur and alkyl bonds constituting oligomers; thus, the average molecule of asphaltenes contained between 40 and 70 aromatic rings with heteroatoms, such as S, O, and N. More recent data obtained with nondestructive tests, such as nuclear magnetic resonance (NMR), Fourier transform infra-red (FT-IR), and electron paramagnetic resonance (EPR), have shown that the average asphaltene molecule can be well-represented by isolated clusters of polycondensed groups containing between 5 and 7 rings bridged with heteroatoms and aliphatic chains (Calemma et al., 1995). These authors have stated, according to their results obtained with asphaltenes from different sources, that the trend of some properties of asphaltenes can be

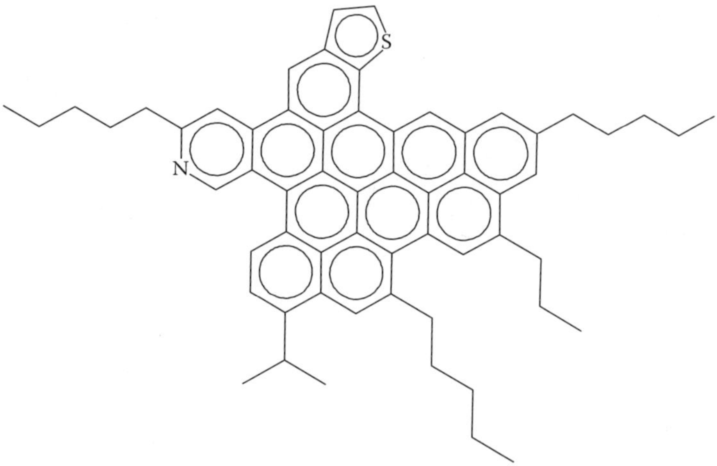

FIGURE 1.1 Hypothetical molecular structure of asphaltenes constituted by an aromatic core, alkyl chains, and heteroatoms.

summarized as: Aromaticity increases reducing the average length of alkyl side chains and the heteroatom content causing a diminution of the average molecular weight and, in consequence, the average aromatic core size increases. Other techniques, such as fluorescence depolarization, have demonstrated similar conclusions (Groenzin and Mullins, 1999; 2000). A typical schematic view of asphaltene structure is shown in Figure 1.1, where different constituents of the molecules are observed, i.e., aromatic core, alkyl chains, and heteroatoms.

Speight and Moschopedis (1981) have defined asphaltene as a structure with an aromatic core bearing aliphatic chains. The following empirical formula can be established from mass spectrometry: $C_{80}H_{80}N_2S_2O$, with molecular weight of 1,150 g/mol, which composition can be estimated as 83.58 wt% carbon, 7.01 wt% hydrogen, 2.44 wt% nitrogen, 5.58 wt% sulfur, and 1.39 wt% oxygen.

Recent techniques, such as fluorescence depolarization, have given smaller molecular weight of asphaltenes (~750 g/mol) and, consequently, the complex structure of asphaltenes believed in the past is now simplified. Asphaltenes are covalently coordinated to porphyrins, thiophene, sulfide, pyrrole, pyridine, alkyl chains, aromatics, etc. (Groenzin and Mullins, 2000). The variety of asphaltenes are huge because of the presence of linkages with sulfur, with nitrogen, or with other aromatic structures, so that taking into account the aromatic fraction of carbon, the $-CH_2$-to$-CH_3$ ratio, sulfur and nitrogen content, type of alkyl structures, the size of rings, and the number of fused rings, three idealized structures can be defined, which are represented in Figure 1.2. The number of fused rings is established to be between 4 and 10 rings. Figure 1.2 clearly shows the different constituents of asphaltenes, i.e., aliphatic chains, aromatic core, and the presence of heteroatoms, such as nitrogen and sulfur.

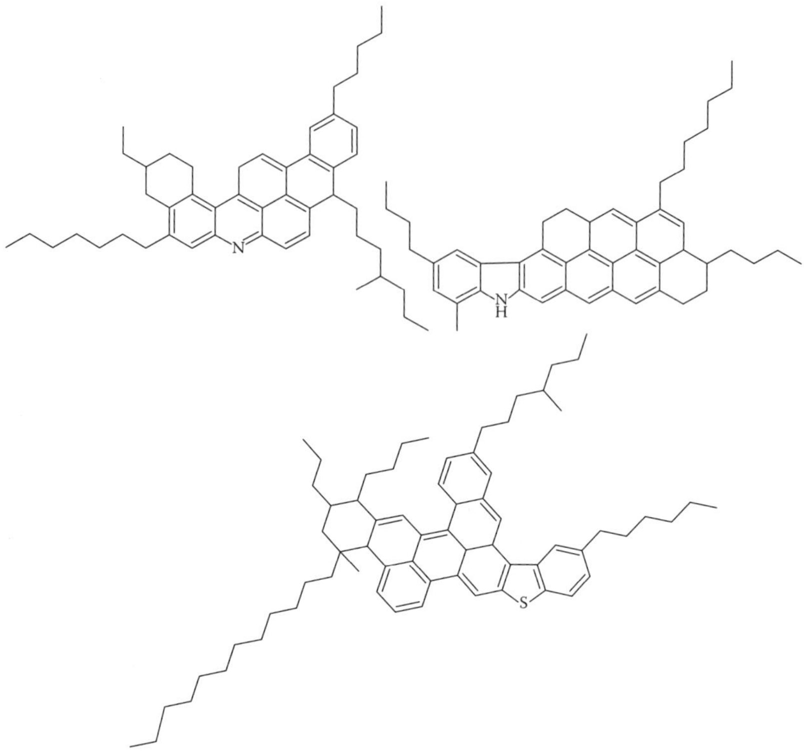

FIGURE 1.2 Idealized molecular structures for asphaltenes consistent with molecular size, aromatic ring systems, and chemical composition. (From Groenzin, H. and Mullins, O.C., 2000. *Energy Fuels* 14 (3): 677–684. With permission.)

Chemical characterization and molecular simulations are useful tools for representing some hypothetical structures of asphaltenes and resins. The use of a molecular model for simulating the aggregation of asphaltenes and resins in crude oil has been reported by Aguilera-Mercado et al. (2006). Asphaltenes can be modeled in two different ways, the *continental type* in which asphaltene cores are constituted by more than seven aromatic rings, and the *archipelago type* in which asphaltenes are represented by small aromatic cores linked to other cores by means of bridging alkanes.

Good examples of continental-type asphaltenes are proposed by Zhao et al. (2001), who used fractionation with supercritical pentane for extracting different cuts of residual oils. After characterization, the authors proposed the structure shown in Figure 1.3, which can be classified as a continental type that corresponds to pericondensed structures. Other examples are proposed by Rogel and Carbognani (2003), who worked with stable and unstable asphaltenes from Venezuelan crudes. Stable asphaltenes are characterized for having molecular weight of ~1,000 g/mol

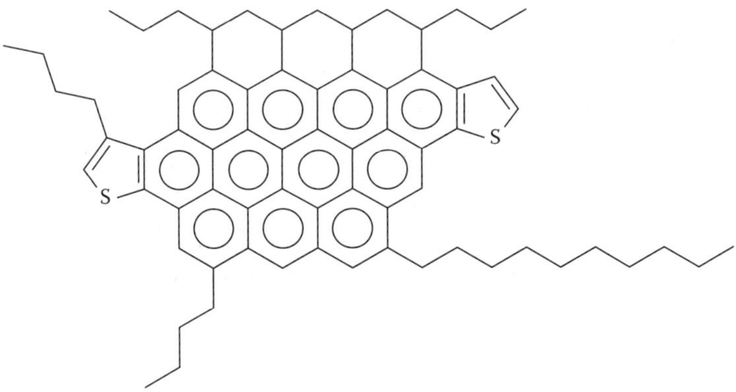

FIGURE 1.3 Average structure of continental-type asphaltenes from hydrocracked residue. (From Zhao et al., 2001. *Fuel* 80 (8): 1155–1163. With permission.)

($C_{76}H_{90}$), whereas unstable asphaltenes have molecular weight of ~1,200 g/mol ($C_{85}H_{92}NO_3S$) and higher with the presence of heteroatoms in higher amounts.

The other hypothetical representation of asphaltenes is the archipelago type. Some studies of characterization of asphaltenes have revealed distinct conformational representation of asphaltenes in which alkane linkages are present. Archipelago type has been supported by pyrolysis, oxidation, thermal degradation, and small-angle neutron scattering analysis as reported by Gawrys et al. (2002). Speight (1994) has also reported an archipelago-type asphaltene having molecular weight of 1,653 g/mol ($C_{116}H_{133}NOS_3$).

Sheremata et al. (2004) have obtained by Monte Carlo simulation a quantitative representation of asphaltenes to be archipelago type, indicating the presence of islands of small aromatic cores linked by alkyl and sulfur bonds. One proposed structure for archipelago-type asphaltenes is shown in Figure 1.4. The way in which both types of asphaltenes aggregate is different. Archipelago-type asphaltenes form planar aggregates in asphaltenes solutions, whereas continental-type asphaltenes are able to stack forming columns (Murgich et al., 1996).

Different techniques have been used to determine the number of aromatic rings. For example, Zajac et al. (1994; 1997) employed Scanning Tunneling Microscopy and stated that asymmetric structures have dimension averages of 10.4 ± 1.9 Å. Fluorescence Emission Spectroscopy supports the value of four-ring systems (Ralston et al., 1996). The authors stated that asphaltenes contain only small quantities of one-, two-, and even three-ring aromatic molecules. Intensity observed in asphaltene fluorescence spectra is almost the same as the emission from molecules in the range of 4 to 10 fused aromatic rings. The diameter of the aromatic core has been also confirmed with x-ray diffraction to be around 11 to 14 Å. These values were obtained for asphaltenes from four different sources and by using Gaussian distributions (Andersen et al., 2005). The structure of asphaltenes is dependent on the geographical source and the maturation processes of the crude oil from which

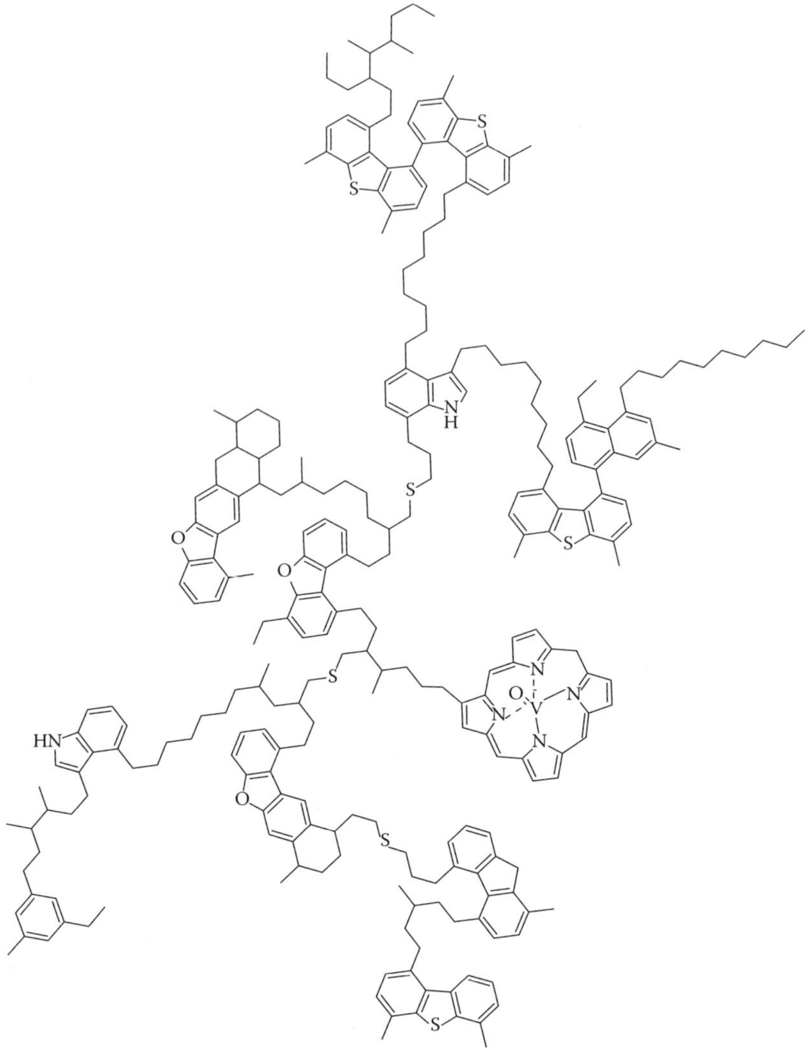

FIGURE 1.4 Archipelago-type asphaltenes generated by Monte Carlo simulation. Formula: $C_{318}H_{395}N_6O_6S_8V$; molecular weight: 4705 g/mol. (From Sheremata et al., 2004. *Energy Fuels* 18 (5): 1377–1384. With permission.)

they are precipitated. Taking this into account, some other structures of asphaltenes have been proposed recently based on experimental characterization reported in the literature. It is convenient to mention that asphaltenes have more pericondensed structures than resins and are composed by heteroatoms, such as N, S, and O, and some naphthenic structures bearing alkyl chains. It is then expected that asphaltenes have larger structures compared with resins and, in the case of more complex asphaltenes, resins with larger structures (more aromatic and naphthenic

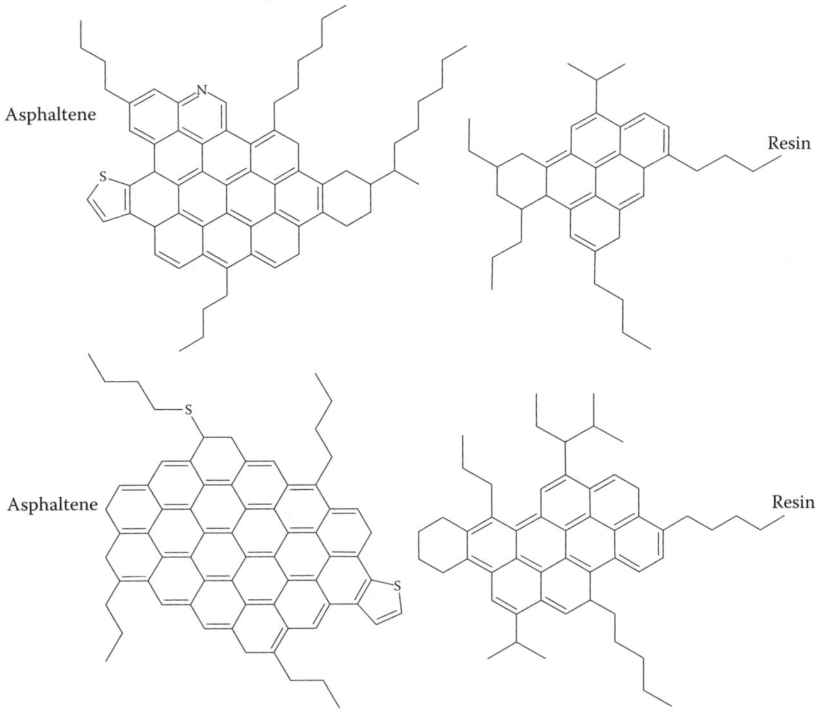

FIGURE 1.5 Hypothetical structural models for asphaltenes and resins.

rings) are necessary to stabilize them into the crude. A schematic representation of this idea is shown in Figure 1.5. From this figure, significant differences between asphaltenes and resins are clearly evident. Rogel et al. (1999) have stated that the aromatic core of resins is composed by nine benzene rings, whereas the aromatic core of asphaltenes is around 20, and asphaltenes and resins from unstable crudes have more aromatic rings in their cores compared with relatively more stable crudes. Recently, Ghloum and Oskui (2004) have reported larger structures for Kuwaiti crude with a total number of carbons atoms of 220, where at least half of them are in aromatic rings. The number of aromatic rings was estimated to be 42 and naphthenic rings 114.

Asphaltenes can be defined not only in terms of their solubility properties, but also in terms of chemical basis. Buenrostro-Gonzalez et al. (2001a) have stated that the balance between the propensity of asphaltenes for stacking by van der Waals interactions through π-bond and the steric disruption due to the peripheral alkyl substitution determines which molecules will fall into the asphaltenes solubility classification. Without alkane steric disruption, the solubility of large aromatic ring systems drops very low. These authors have measured the molecular size of asphaltenes of different sources by depolarization fluorescence, e.g., asphaltenes from Mexican, Venezuelan, and Arabian crudes, and the results indicate a molecular size of about 17.5 Å. Fluorescence depolarization data were analyzed by using the equation (Groenzin and Mullins, 1999; 2000):

$$\tau_c = \frac{V\eta}{kT} \tag{1.1}$$

where τ_c is the rotational correlation time, V the molecular volume, η the solvent viscosity, k the Boltzman constant, and T the temperature. Comparison with known molecular sizes for various model compounds show that the mean molecular weights are ~750 g/mol with small differences depending on the corresponding crude oil. Very low concentrations of asphaltenes in toluene (~6 mg/L) are used, preventing aggregation effects.

1.2 VARIABLES AFFECTING THE ASPHALTENES PRECIPITATION

Content and composition of asphaltenes depend mainly on the nature of crude, contact time, temperature, pressure, type of solvent, and solvent/crude ratio used (Speight et al., 1984; Andersen and Birdi, 1990). Various methodologies have been used for asphaltenes separation, which use different precipitation conditions. Some of them have been standardized, i.e., ASTM (American Society for Testing and Materials)-D-3279, ASTM-D-4124, IP 143, etc.

Asphaltenes precipitation is influenced by the nature of the medium in which they are contained. Asphaltenes are polar, but they have acidic and basic compounds that interact with polar resin adsorbed on asphaltenes. Resins help to stabilize the micelles; however, when a paraffinic solvent is added, the asphaltene–resin equilibrium is disrupted and an increase of asphaltene monomers in the bulk phase occurs. At some extent, concentration of asphaltenes will reach the onset concentration and precipitation will occur (Al-Sahhaf et al., 2002). Precipitation of asphaltenes from Western Canadian bitumen was modeled by Alboudwarej et al. (2003) using the mole fraction, molar volume, and solubility parameters for different components. This model predicted very well the onset and amount of asphaltenes precipitation with different *n*-alkanes.

1.2.1 Nature of the Crude

The geographic region and the depth of well where the crude is extracted are important parameters and influence the API gravity and quality of the crude. These two parameters also have impact on the content and properties of asphaltenes separated from crudes (Speight and Moschopedis, 1981). Figure 1.6 shows how API gravity influences the asphaltenes content of different origins of crude oil.

The composition of crude oils can change with time due to local variations where the crude is extracted. Figure 1.7 shows how the metals content varies with the asphaltene content, which suggests a possible correlation. The higher the asphaltene concentration, the higher the vanadium content (Reynolds, 1990).

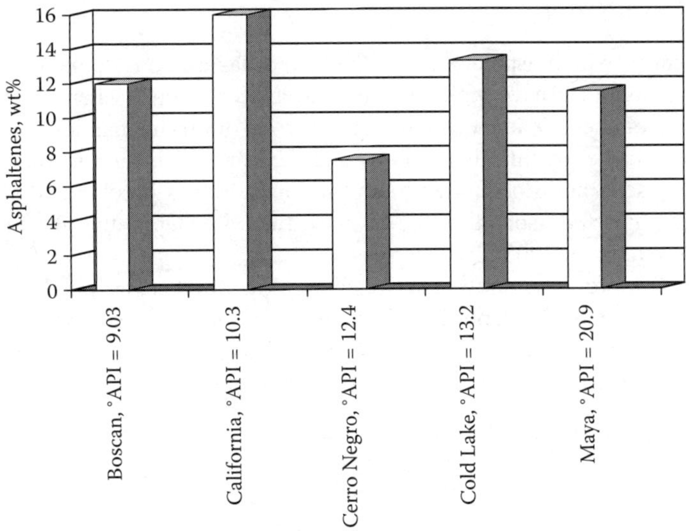

FIGURE 1.6 Asphaltenes content from different crude oils. (*Data taken from several sources.*)

Vanadium, an important well-known metal responsible for problems during processing of petroleum, which increases with decreasing API gravity, indicating that more contaminated crudes will be available in the future and more emphasis needs to be put in removing these contaminants for the production of cleaner fuels.

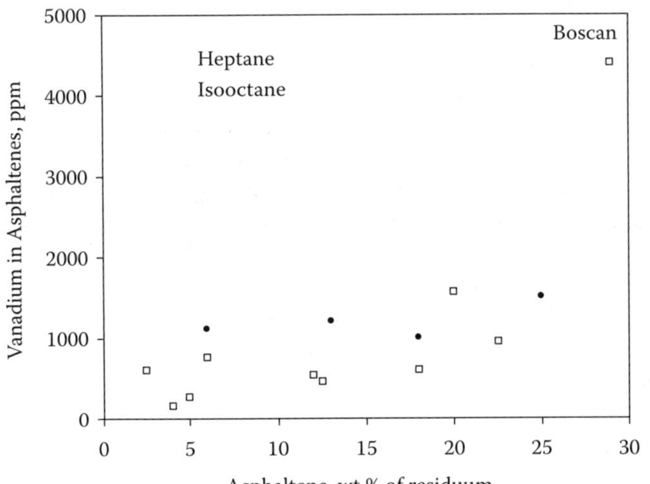

FIGURE 1.7 Contaminants present in different crude oils.

1.2.2 Solvent Type

The solvent in which asphaltenes are dissolved has a good aromatic power, but because of this, they have high molecular weight especially when solvent favors aggregation of asphaltene molecules. The entropy of mixing cannot keep them in solution when the medium is not favorable, i.e., more aliphatic medium. In this case, phase separation occurs and solid-like material is precipitated. The solid state of precipitated asphaltenes is because they are below the glass-transition temperature (Sirota, 2005).

Solvents more commonly used include low molecular weight paraffins. Changes in asphaltenes properties can occur as a consequence of the solvent used. Mitchell and Speight (1973) have established that the substances which precipitate asphaltenes in higher amount are in the following order:

Terminal olefin < *n*-paraffins < *iso*-paraffins

The best solvents for dissolving asphaltenes are

Cyclic paraffins < Aromatics

Good solvents for asphaltenes also include pyridine, dichloromethane, chloroform, and carbon tetrachloride.

Corbett and Petrossi (1978) have found that the amount of asphaltenes precipitated is more or less constant from *n*-heptane to higher carbon number precipitants, hydrocarbon stability; and repeatability of experiments are also quite high when using *n*-heptane. For these reasons, *n*-heptane is the most preferred solvent.

As the carbon number of the paraffin is increased the amount of asphaltenes precipitated is reduced (Fuhr et al., 1991). The solvent also influences the properties of asphaltenes. With the use of *n*-heptane, asphaltenes are heavier and more polar. On the other hand, when using *n*-pentane, asphaltenes are less polar and the molecular weight is lower; however, the amount of asphaltene precipitated is higher (Andersen, 1990).

Hu and Guo (2001) have used Caoquiao Chinese crude and tested different kinds of precipitants with various precipitant-to-oil ratios and confirmed that the amount of asphaltenes precipitated decreased as the molecular weight of *n*-alkanes increased. Figure 1.8 shows the results reported by these authors. It can be observed that the amount of asphaltenes precipitated with n-C_5 is about twice that precipitated with n-C_{12}.

Similar solvent-to-oil ratios with n-C_5, n-C_6, n-C_7, n-C_8, and n-C_{10} at 26°C were employed by Rassamdana et al. (1999) with API gravity crude of ~30°. The difference between the amount of precipitated asphaltenes with n-C_5 and n-C_{10} was around 35%.

1.2.3 Contact Time

Contact time during asphaltenes precipitation varies depending on the particular method used. For example, the ASTM-D-3279 method uses contact times of 15 to 20 min under reflux conditions with *n*-heptane. The IP-143 standard method

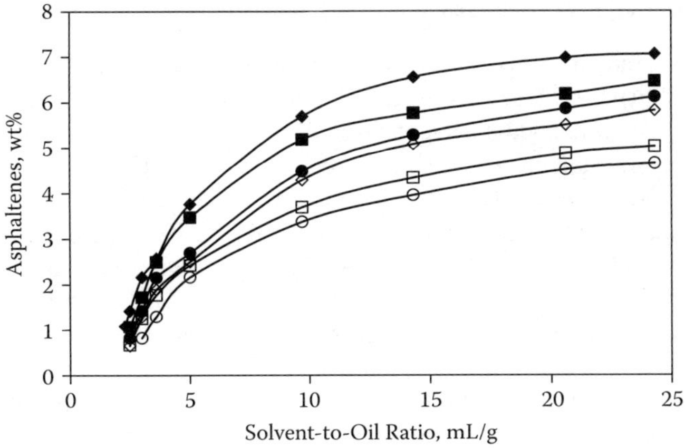

FIGURE 1.8 Amount of asphaltenes precipitated at 20°C using different *n*-alkanes precipitants: (♦) *n*-C_5; (■) *n*-C_6; (●) *n*-C_7; (◇) *n*-C_8; (□) *n*-C_9; (○) *n*-C_{12}. (Adapted from Hu and Guo, 2001. *Fluid Phase Equilib.* 192 (1-2): 13–25.)

requires 1 h of reflux with *n*-heptane. Speight (1999) has recommended between 8 and 10 h for contact time.

Centeno et al. (2004) have used the ASTM-D-3279 standard method modifying the contact time for precipitation and corroborated that at least 8 h are required for having a constant amount of asphaltenes by using this method (Figure 1.9). When increasing contact time, more resins are removed with *n*-heptane, leaving the asphaltenes free to precipitate because the medium becomes unfavorable for them.

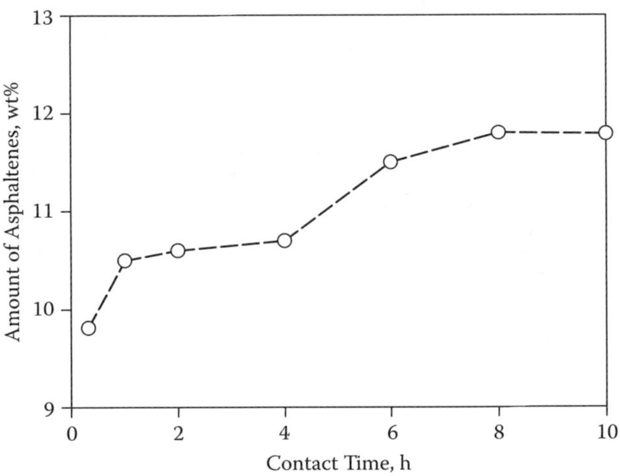

FIGURE 1.9 Amount of asphaltenes precipitated with ASTM-D-3279 standard method as function of contact time. (From Centeno et al., 2004. *J. Mex. Chem. Soc.* 48 (3): 186–195. With permission.)

1.2.4 Solvent-to-Oil Ratio

Solvent-to-oil ratio also depends on the method used, i.e., ASTM-D-3279 standard method establishes that the solvent-to-oil ratio is 100:1. Speight and Moschopedis (1981) have recommended a solvent-to-oil ratio of 40:1 for asphaltenes precipitation when using light hydrocarbons. Ancheyta et al. (2002) have reported that a solvent-to-oil ratio of 60:1 (v/wt) is appropriate for precipitating asphaltenes; *n*-pentane and a refinery solvent (*iso*-pentane: 30.5 mol%, *n*-pentane: 40.9 mol%, C_3–C_4: 3.4 mol%, and C_6–C_7: 25.9 mol%) were employed. The behavior of asphaltenes precipitation was reported to be different with both solvents. More asphaltenes were precipitated with the refinery solvent due to the presence of lighter hydrocarbons and the high content of *iso*-paraffins, which precipitates more asphaltenes. Figure 1.10 shows the amount of precipitated asphaltenes as a function of the solvent-to-oil ratio for both solvents. It is seen from the figure that a constant amount of asphaltenes is obtained with a solvent-to-oil ratio of 60:1. The trend obtained with solvent refinery is similar to *n*-pentane.

1.2.5 Temperature

Some results reported in the literature show opposite trends in the amount of precipitated asphaltenes with increasing temperature. Some authors have stated that at higher temperatures less amounts of asphaltenes are precipitated (see Figure 1.11) as a function of solvent-to-oil ratio (Fuhr et al., 1991; Hu and Guo, 2001).

Andersen (1994) observed that the amount of precipitated asphaltenes with *n*-heptane increased with a reduction of the temperature in Boscan and Kuwaiti

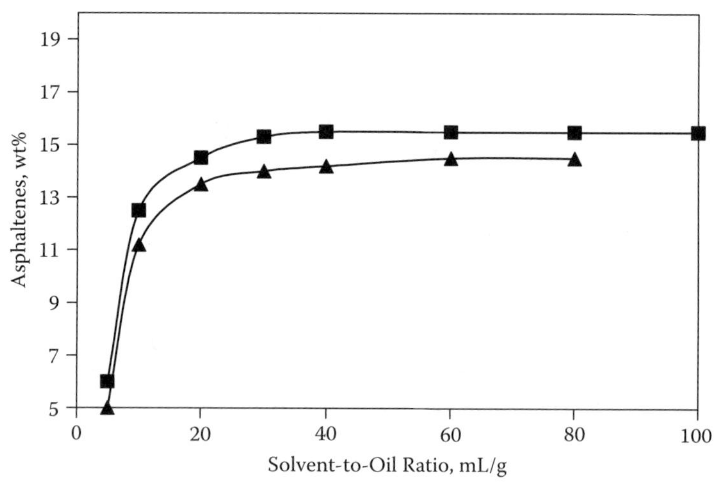

FIGURE 1.10 Effect of solvent-to-oil ratio on asphaltenes precipitation: (■) Refinery solvent; (▲) *n*-C_5. (From Ancheyta et al., 2002. *Energy Fuels* 16 (5): 1121–1127. With permission.)

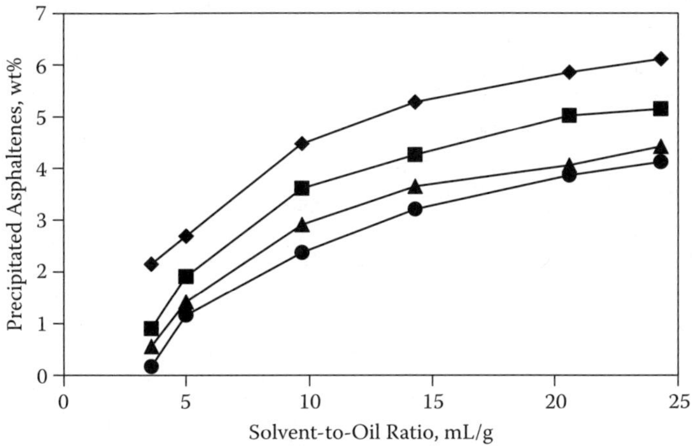

FIGURE 1.11 Variation of the amount of precipitated asphaltenes with increasing temperature using n-C_7: (♦) 25°C; (■) 35°C; (▲) 50°C; (●) 65°C. (Adapted from Hu and Guo, 2001. *Fluid Phase Equilib.* 192 (1-2): 13–25.)

crude oils. Akbarzadeh et al. (2005) also obtained the same conclusion for different crude oils and bitumens using *n*-pentane and *n*-heptane as solvents. González et al. (2006) reported the precipitation of lower amounts of asphaltenes from Brazilian crudes when increasing the temperature using *n*-decane and *n*-dodecane. The amount of asphaltenes separated at 20°C was higher than at 50°C for both solvents. The same behavior was found by Ghloum and Oskui (2004) who reported that the tendency of asphaltene precipitation decreases as the temperature increases. This fact could be attributed to a change in asphaltenes solubility because, as the temperature increases, their solubility in crude also increases, thus delaying their precipitation. Fuhr et al. (1991) stated that the aggregate molecular weight of asphaltenes measured by vapor pressure osmometry (VPO) with pyridine increases as the temperature of precipitation also increases.

On the other hand, some other authors have found the opposite trend in precipitation of asphaltenes with increasing temperature, that is, the higher the temperature, the higher amount of precipitated asphaltenes. This behavior has been attributed to a disruption of the polar interactions between asphaltenes. When disruption begins, these polar groups in asphaltenes are exposed to the nonpolar solvent (*n*-alkane medium); then the phase separation occurs, which precipitates more asphaltenes. Leontaritis and Mansoori (1988) and Kabir and Jamaluddin (1999) have also reported an enhanced asphaltenes precipitation by temperature. Ali and Al-Ghannam (1981) observed a decrease in the amount of precipitated asphaltenes with increasing temperature in the case of light crude (Kirkuk crude, 36° API), and an increase in the amount of precipitated asphaltenes in respect to that obtained at room temperature for heavy crude (Qaiyarah crude, 16° API) at reflux temperature (Figure 1.12). The amount of asphaltenes at reflux temperature is almost similar to that obtained at 0°C. This is an indication that the amount of

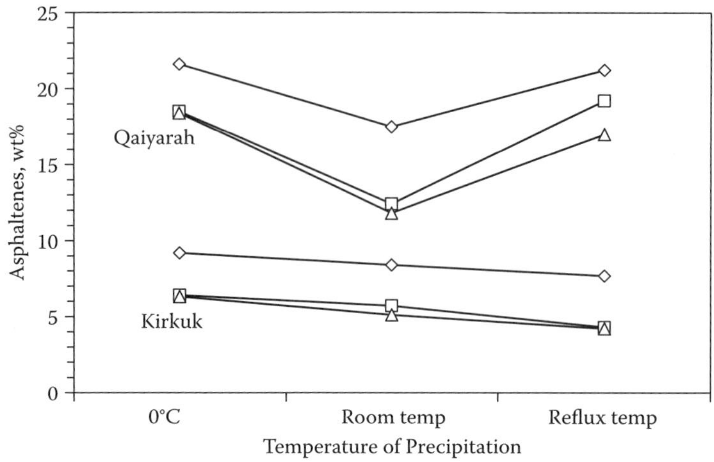

FIGURE 1.12 Precipitation conditions with different solvents and temperatures. (◊) n-C_5; (□) n-C_6; (Δ) n-C_7. (Adapted from Ali and Al-Ghannam, 1981. *Fuel* 60 (1): 1043–1046.)

precipitated asphaltenes varies depending on the nature of the crude. This is especially true with thermally unstable crude oils in which temperature can induce some changes when precipitating the asphaltenes at reflux conditions.

The content of metals in asphaltenes of three crude oils (CO-1: 30.3° API, CO-2: 21.4° API, and CO-3: 12.2° API) precipitated at different temperature was reported by Pineda et al. (2007) and the results are shown in Figure 1.13, which presents the relationship between asphaltenes content and metals (Ni + V) in asphaltenes. Dashed lines indicate the range in which asphaltenes from each crude oil are found from 40° to 100°C. In spite of having a higher amount of asphaltenes, the heaviest crude (CO-3) possesses similar amounts of metals compared with CO-2 crude oil, which is due to the nature of the crude. The higher the precipitation temperature, the higher the amount of metals (Ni + V) in precipitated asphaltenes. This could mean that more porphyrinic compounds in which Ni and V are inserted are precipitated along with asphaltenes as the temperature increases.

1.2.6 Pressure

Studies related to variation of pressure in asphaltene precipitation are scarce. Pasadakis et al. (2001) found that the amount of dissolved asphaltenes in the crude diminishes as pressure goes from the initial reservoir pressure to the bubble point of the crude and further increases as pressure diminishes. On the other hand, Hirschberg et al. (1984) stated that the asphaltene solubility increased with increasing pressure up to the bubble point. Above this point, asphaltene solubility diminished. Browarzik et al. (2002) studied the influence of pressure on the onset of flocculation of asphaltenes in the region from 1 to 300 bar.

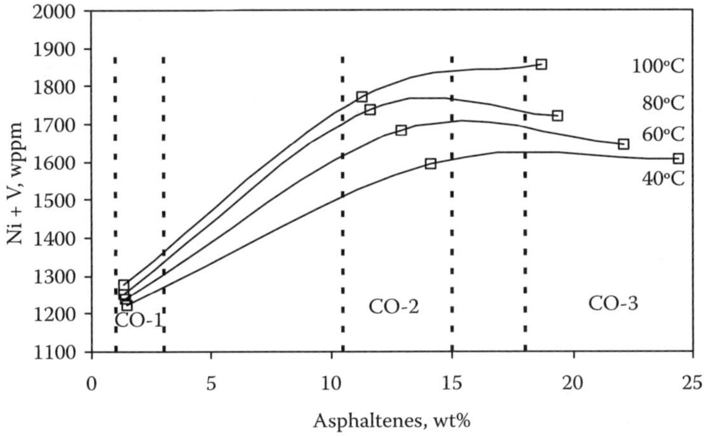

FIGURE 1.13 Relationship between asphaltenes in crude oils and metals in asphaltenes at different precipitation temperatures. (Pineda et al., 2007. *Petrol. Sci. Tech.* 25 (1-2): 105–119. With permission.)

Centeno et al. (2004) evaluated the precipitation of asphaltenes in a pressurized system adapted with stirring and heating. Two pressures were studied at four temperatures keeping constant the solvent-to-oil ratio, stirring rate, and contact time. Conditions for precipitating asphaltenes were: pressure of 15 and 25 kg/cm^2; temperature of 40, 60, 80, and 100°C; stirring rate of 750 rpm; contact time of 30 min; and solvent-to-oil ratio of 5:1 (mL/g).

A comparison of elemental composition between asphaltenes precipitated with the ASTM-D-3279 standard method and with the pressurized system is presented in Table 1.1. It can be observed that almost the same asphaltenes properties are obtained with the standard method (atmospheric pressure and 10 h contact time) and the pressurized system at the following precipitation conditions: pressure of 25 kg/cm^2, temperature of 60°C, stirring rate of 750 rpm, solvent-to-oil ratio of 5:1, and contact time of 30 min. H/C (hydrogen to carbon) atomic ratio is very close in both systems (1.116 with pressurized system versus 1.117 with the standard method). Sulfur and nitrogen contents are practically the same. Not only pressure, but also temperature influences the amount and composition of precipitated asphaltenes. The advantage of using the pressurized system is the amount of precipitated asphaltenes that can be obtained. The precipitated asphaltenes are recovered as a solid by filtration under vacuum.

Pineda et al. (2007) have used the same pressurized system to study asphaltenes precipitation of three crude oils (CO-1, CO-2, and CO-3). The main variables of asphaltenes precipitation were correlated with some of their properties. The studied range of variables comprised two pressures (15 and 25 kg/cm^2), four temperatures (20, 40, 60, and 100°C), four solvent-to-oil ratios (2:1, 3:1, 4:1, and 5:1), and contact times between 0.5 to 6 h. To show how asphaltenes amount changes, different tests were carried out with CO-2 crude varying the contact time at a constant

TABLE 1.1
Chemical Composition of Asphaltenes Precipitated with ASTM-D-3279 Method and with a Pressurized System

Variable	T,°C	C, wt%	H, wt%	N, wt%	S, wt%	H/C Atomic Ratio
Pressure, kg/cm^2			Pressurized System			
25	40	82.31	7.78	1.60	7.51	1.134
	60	82.25	7.65	1.65	7.86	1.116
	80	82.17	7.48	1.60	8.32	1.092
	100	81.71	7.42	1.57	8.05	1.090
15	40	82.41	7.72	1.66	8.13	1.124
	60	82.33	7.55	1.73	7.85	1.100
	80	82.26	7.45	1.63	8.18	1.087
	100	82.00	7.41	1.65	8.15	1.084
Contact Time, h			ASTM-D 3279 Method at Atmospheric Pressure			
10	n-C_7 reflux	82.26	7.66	1.66	7.87	1.117

Source: Centeno, G. et al., 2004. *J. Mex. Chem. Soc.* 48 (3): 186–195. With permission.

solvent-to-oil ratio of 5:1. In another test, the contact time was kept constant and solvent-to-oil ratio was varied. The results are shown in Table 1.2. It is seen that asphaltenes content is almost constant. It is likely that pressure does precipitate completely all asphaltenes even at 0.5 h. It is also observed that keeping constant the contact time at 0.5 h and varying the solvent-to-oil ratio, asphaltenes amount increases and it remains almost constant from a ratio of 3:1 to 5:1. For assuring the correct precipitation of asphaltenes at least a ratio of 3:1 is required; however, the solvent-to-oil ratio of 5:1 was preferred to precipitate all asphaltenes with n-C_7. The three crudes studied in this work have different asphaltene contents, which can be classified as high asphaltenes content (CO-3), middle asphaltenes content (CO-2), and low asphaltenes content (CO-1). Figure 1.14 shows how asphaltenes content diminishes with increasing temperature. Reduction of carbon and hydrogen content was found to be smooth up to 80°C, but above this temperature the change is more noticeable. Nitrogen and sulfur contents are more or less constant, particularly in the range of 60° to 100°C. It can be thought that asphaltenes from CO-1 are less complex than asphaltenes from the other crudes and probably possess a fewer number of carbons in the alkyl side chains.

The combined effect of pressure and temperature at constant pressure caused the asphaltene amounts to decrease as temperature increases. This is in agreement with previous observations (Fuhr et al., 1991; Andersen, 1994; Hu and Guo, 2001). Pineda et al. (2007) developed various correlations in which dependent

TABLE 1.2
Effect of Contact Time and Solvent-to-Oil Ratio on Asphaltenes from CO-2

Time, h	Pressure, kg/cm²	Temperature, °C	Solvent-to-oil Ratio, mL/g	Asphaltenes, wt%
0.5				11.40
1.0				11.90
2.0				11.20
2.5	25	60	5:1	11.42
3.0				11.48
4.0				11.80
5.0				11.30
6.0				11.45
0.5			2:1	9.67
	25	60	3:1	11.41
			4:1	11.51
			5:1	11.40

Source: Pineda, L.A. et al., 2007. *Petrol. Sci. Tech.* 25 (1-2): 105–119. With permission.

variables were the amount of precipitated asphaltenes, metals content, and elemental composition of asphaltenes, and independent variables were temperature and pressure. The following correlation was reported that minimizes the error between experimental and predicted values:

$$Y = A + BT + CT^2 + DT^3 + EP \tag{1.2}$$

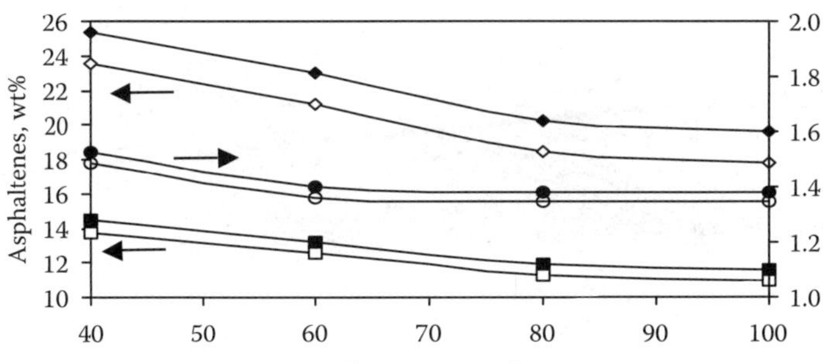

FIGURE 1.14 Effect of pressure and temperature on asphaltenes precipitations: CO-1: rhombus; CO-2: frames; CO-3: triangles; (○, □ , ◇) 15 kg/cm²; (●, ■, ◆) 25 kg/cm². (Pineda et al., 2007. *Petrol. Sci. Tech.* 25 (1-2): 105–119. With permission.)

TABLE 1.3
Constants of Equation (1.2) for Predicting the Yield of Asphaltenes of Different Crudes

Constant	Crude CO-1	Crude CO-2	Crude CO-3
A	2.4342	1.1592×10^{1}	1.5749×10^{1}
B	-4.0591×10^{-2}	1.6892×10^{-1}	4.9033×10^{-1}
C	4.9144×10^{-4}	-3.9158×10^{-3}	-1.0099×10^{-2}
D	-1.9635×10^{-6}	2.1294×10^{-5}	5.3102×10^{-5}
E	1.2267×10^{-3}	2.2318×10^{-2}	6.0433×10^{-2}

Source: Pineda, L.A. et al., 2007. *Petrol. Sci. Tech.* 25 (1-2): 105–119.

where Y is the desired property (wt% of precipitated asphaltenes; wt% of C, H, N, and S; and wppm of Ni and V), T is the temperature (°C), P is the precipitation pressure (kg/cm^2), and A, B, C, D, and E are correlation constants. According to this equation, temperature is the variable that influences in deeper sense all properties of asphaltenes, which agrees with experimental information. Constants obtained for this equation are shown in Table 1.3 for the yield of asphaltenes. A comparison of experimental values of asphaltenes content in CO-1 crude oil with those obtained with the correlation given by Equation (1.2) is presented in Figure 1.15, in which both pressure and

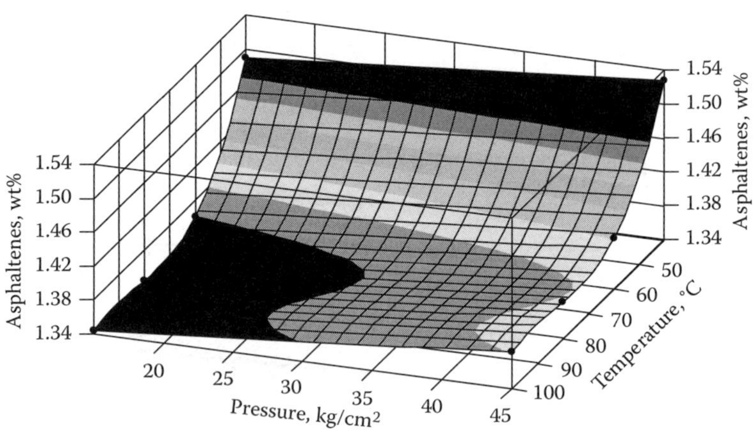

FIGURE 1.15 Comparison of experimental and predicted asphaltene contents in CO-1 crude oil: (●) experimental, (lines) predicted. (Pineda et al., 2007. *Petrol. Sci. Tech.* 25 (1-2): 105–119. With permission.)

temperature effects are plotted. It is seen that predicted values agree quite well with experimental ones.

1.3 REVERSIBILITY OF ASPHALTENES PRECIPITATION

Various studies have reported that asphaltenes precipitation can be a reversible process, in some cases, being influenced by pressure or temperature. Studies developed by Pfeiffer and Saal (1940) several decades ago revealed that irreversibility occurred in the asphaltenes precipitation by means of observations of colloidal behavior of asphaltenes. Partially reversible precipitation of asphaltenes was observed at room temperature tests by Rassamdana et al. (1996) who stated that precipitated asphaltenes were redissolved and kept in solution. Ramos et al. (1997) have also carried out experiments at room temperature and verified that asphaltenes precipitation/dissolution during liquid titration is a reversible process; however, at room temperature it takes more time. Andersen and Stenby (1996) concluded that asphaltenes partially redissolved with increasing temperature. They observed that the higher the temperature, the higher the redissolution of asphaltenes in the range of 24 to 80°C.

Hirschberg et al. (1984) stated that precipitation of asphaltenes is reversible at 0.37 MPa. Hammami et al. (2000) carried out experiments at a pressure and temperature typical of a field production system indicating that precipitation is a reversible process by alternating depressurization and repressurization steps. Repressurization experiments showed, at least partially, the reversibility of asphaltene precipitation. Ghloum and Oskui (2004) studied the precipitation and redissolution of asphaltenes upon the addition and removal of CO_2 and light alkanes at 20.7 MPa and room temperature observing that asphaltene precipitation is reversible. Pina et al. (2006) conducted studies of decompression and recompression in crude and found variation in the content of asphaltenes dissolved in a sample as function of pressure. Hysteresis occurred and the initial amount of precipitated asphaltenes at 65 MPa was not the same after decompression and recompression, which indicated that redissolution of asphaltenes was likely present. However, the same authors performed another experiment with different crude in which decompression and recompression took place starting at 70 MPa. In this case, hysteresis was not observed and the same amount of precipitated asphaltenes was obtained after decompressing and recompressing the crude. According to the authors, asphaltenes precipitation can be considered as a reversible process, at least partially. Differences in results are addressed to the degree of destabilization of the asphaltenes. The greater the degree of destabilization, the less reversible is precipitation of asphaltenes.

Hysteresis was also observed in asphaltene precipitation and redissolution. Complete reversibility can be obtained when asphaltenes are destabilized as little as possible. When destabilization occurs, asphaltenes can combine and form large aggregates; however, these big macromolecules cannot be redissolved completely unless heating is applied to return the asphaltenes to their original size. Beck et al. (2005) proposed that hysteresis could be due to an energy barrier.

1.4 ELEMENTAL COMPOSITION OF ASPHALTENES

Elemental composition of asphaltenes from different origins varies in a narrow range, i.e., carbon content around 82 ± 3 wt% and hydrogen content around 8.1 ± 0.7 wt% (Speight and Moschopedis, 1981). León et al. (1999) found that asphaltenes from unstable crude oils exhibit a deficit of hydrogen in their structure having greater aromaticity with highly condensed aromatic rings.

Ancheyta et al. (2002) have carried out characterization of asphaltenes precipitated from Maya, Isthmus, and Olmeca Mexican crude oils by elemental analysis. These results together with those from other crudes are reported in Table 1.4. Variations in asphaltenes composition with changing the solvent were compared. Higher H/C atomic ratios were obtained with *n*-pentane because lighter species are precipitated, which are more paraffinic in nature. Oxygen, nitrogen, and sulfur contents are higher with *n*-heptane. Metals content is more or less similar in asphaltenes from Isthmus and Olmeca. The smallest difference in composition using both solvents is observed in Olmeca asphaltenes. This is likely due to the small size of asphaltenes of this light crude, which makes its composition almost constant in spite of using different solvents.

Mass balances performed with heteroatoms content in asphaltenes and in the whole crude with data for asphaltenes precipitated with n-C_5 allowed for determining the content of asphaltenic heteroatoms, which are seen in Figure 1.16. It is seen that a large part of the heteroelements is contained in asphaltenes. Olmeca's asphaltenes have less sulfur, nitrogen, and oxygen content in their structures. This can be attributed to its very low asphaltenes content in which metals surely concentrate. For the three crudes, metals are contained in at least 50% of the asphaltene structure. A plot of heteroatom content against asphaltenes content in the three crude oils is shown in Figure 1.17. A nice linear behavior is observed in all cases, which indicated heteroatoms content is proportional to asphaltenes content in crude oils.

It has been reported that the concentration of Ni and V in asphaltenes increases linearly with increasing API gravity of crudes (Siddiqui, 2003). Other authors have concluded that with removal of asphaltenes, a decrease of the Conradson carbon residue (CCR) and pour point of crude oils are observed (Sharma et al., 2007).

Organic nitrogen compounds are a subject of interest to petroleum refiners due to the detrimental role in catalyst deactivation and product stability. Usually, basic nitrogen compounds (i.e. quinoline, acridine) have been considered stronger inhibitors for the HDT reactions than the nonbasic moieties (i.e. carbazole, indole), thus identification and quantification of nitrogen species have great importance particularly for catalytic processes. When applying gas chromatography–mass spectrometry (GC-MS), the derivative compounds of quinoline show signals at *m/z* (mass-to-charge ratio) 129 for quinoline, *m/z* 143 for methylquinolines, *m/z* 157 for C_2-alkylquinolines, *m/z* 179 for benzoquinolines, and *m/z* 193 for methylbenzoquinolines. Carbazoles are also present in heavy cuts. When carbazoles are analyzed, different signals are obtained, i.e., the ion abundances of methylcarbazoles appear

TABLE 1.4
Elemental Analysis and Metal Content of Asphaltenes

Property	Maya (Mexico)		Isthmus (Mexico)		Olmeca (Mexico)		Canada (Athabasca)		Iran (Iranian Heavy)		Iraq (Kirkuk)		Kuwait (Khafji)	
	n-C_5	n-C_7	n-C_5	n-C_7	n-C_5	n-C_7	n-C_5	n-C_7	n-C_5	n-C_7	n-C_5	n-C_7	n-C_5	n-C_7
Crude oil														
API Gravity	20.2		32.9		38.9		8.3		31.0		36.1		28.5	
Asphaltenes yield (wt%)	14.10	11.32	3.63	3.34	1.05	0.75		15.0		6.0		1.3		4.0
Asphaltenes														
Elemental analysis (wt%)														
Carbon	81.23	81.62	83.90	83.99	86.94	87.16	79.5	78.4	83.8	84.2	81.7	80.7	82.4	82.0
Hydrogen	8.11	7.26	8.00	7.30	7.91	7.38	8.0	7.6	7.5	7.0	7.9	7.1	7.9	7.3
Oxygen	0.97	1.02	0.71	0.79	0.62	0.64	3.8	4.6	2.3	1.4	1.1	1.5	1.4	1.9
Nitrogen	1.32	1.46	1.33	1.35	1.33	1.34	1.2	1.4	1.4	1.6	0.8	0.9	0.9	1.0
Sulfur	8.25	8.46	6.06	6.48	3.20	3.48	7.5	8.0	5.0	5.8	8.5	9.8	7.4	7.8
Atomic ratios														
H/C	1.198	1.067	1.144	1.043	1.092	1.016	1.21	1.16	1.07	1.00	1.16	1.06	1.14	1.07
O/C	0.009	0.009	0.006	0.007	0.005	0.006	0.036	0.044	0.021	0.012	0.010	0.014	0.014	0.017
N/C	0.014	0.015	0.013	0.014	0.013	0.013	0.013	0.015	0.014	0.016	0.008	0.010	0.009	0.010
S/C	0.038	0.039	0.027	0.029	0.014	0.015	0.035	0.038	0.022	0.026	0.039	0.046	0.034	0.036
Metals (wppm)														
Nickel	269	320	155	180	82	158								
Vanadium	1217	1509	710	747	501	704								

Source: Data taken from multiple references.

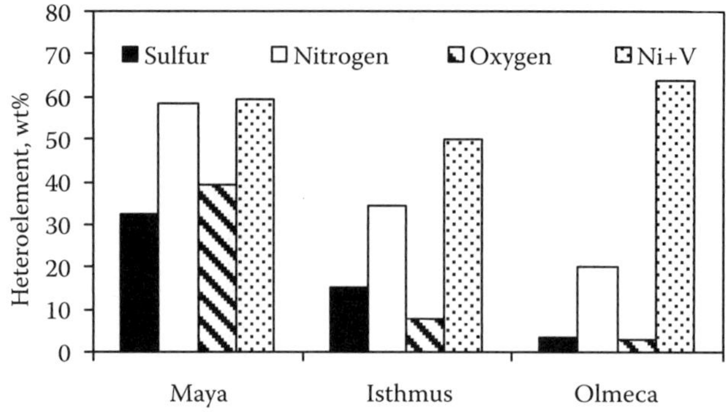

FIGURE 1.16 Percentages of asphaltenic heteroatoms in different crude oils with *n*-pentane as solvent. (Ancheyta et al., 2002. *Energy Fuels* 16 (5): 1121–1127. With permission.)

at *m/z* 181, C_2-carbazoles at *m/z* 195, and benzocarbazoles at *m/z* 217 (Bennett and Love, 2000). N and S containing molecules commonly remain in the nonvolatile residue during the GC-MS analysis, which is due to their location in aromatic rings. When hydroprocessing residue or heavy oils, nitrogen species remain unchanged because they are very resistant to being processed and tend to concentrate. Nitrogen in multiple aromatic rings is also present. Most of the nitrogen in asphaltenes is found to be present in aromatic forms, with very small amounts as saturated amine. The pyrrole form of nitrogen is more abundant than the pyridine form in asphaltenes as reported by Mitra-Kirtley et al. (1993).

Schmitter et al. (1984) and Dorbon et al. (1984) have identified carbazole, methyl, dimethyl, and benzocarbazoles by means of gas chromatography. After

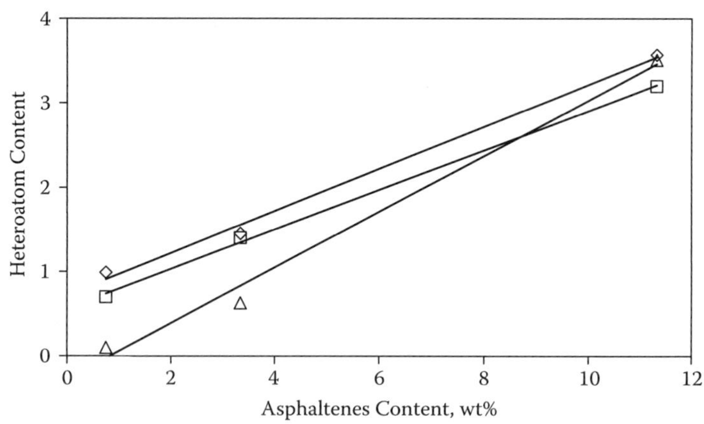

FIGURE 1.17 Asphaltenes in *n*-heptane against heteroatoms content in different crude oils: (Δ) Ni+V in wppm × 10^2; (□) N in wt% × 10^{-1}; (◇) S in wt%. (Ancheyta et al., 2002. *Energy Fuels* 16 (5): 1121–1127. With permission.)

catalytic hydrotreating, it was found that benzocarbazoles are easier to remove than carbazoles in the presence of high hydrogen pressure at 10 MPa and at 360°C, whereas 1-methyl carbazol is removed at higher temperatures, such as 380°C and 5 MPa of hydrogen pressure. Alkyl anilines are the most abundant bases in hydrotreated samples.

Nitrogen in lower molecular mass asphaltenes is mainly present in alkylated benzocarbazoles and in smaller amount as substituted carbazoles ($C_nH_{2n-21}N$, $C_nH_{2n-15}N$, $C_nH_{2n-27}N$, and $C_nH_{2n-23}N$). Nitrogen atoms are in basic pyrrolic form that may lead to the formation of strong intermolecular hydrogen bonds in asphaltenes (Siddiqui, 2003). Nitrogen is also present in vanadyl porphyrins containing 22–52 carbons and small amounts of vanadyl prophyrin tetracarboxylic acids ($C_nH_{2n-24}N_4O_8$) (Strausz et al., 2002). The nature of Athabasca asphaltenes has been analyzed by ion exchange chromatographic fractionation and the values found were: acids (39.8 wt%), basics (23.3 wt%), neutrals (20.2 wt%), and amphoterics (16.9 wt%). This indicates a prevalent acid nature of asphaltenes from Athabasca bitumen. Acids contained more O, and bases more N and S. Thus, the highest O content occurred in the strongest acid subfraction and the highest N content in the strongest basic one.

Speight (1986) correlated the microcarbon residue (MCR) formation as a function of the nature of asphaltenes and found that MCR values decrease in the order of amphoteric, basic, whole asphaltenes, acid, and neutral constituents. Speight (1999) has reported that the molecular weight measured by VPO in pyridine at different temperatures and extrapolated to room temperature for Athabasca asphaltenes were:

Amphoteric (2350 g/mol) > Basic (2250 g/mol) > Acid (1910 g/mol) > Neutrals (1420 g/mol)

Dutta and Holland (1984) and Jada and Salou (2002) have reported that the strong acidity is related to compounds, such as carboxylic and benzoic acids; weak acidity to the presence of phenol, indole, and carbazole; strong basicity is due to pyridine and weak basicity is due to the presence of pyrazine, dimethylformamide, and dimethylsulfoxide. Hypothetical structures of basic and acid asphaltenes are observed in Figure 1.18. Yen (1979) has suggested some structures of asphaltenes present in fractions with different polarities.

Sulfur is present in petroleum fractions as benzo and dibenzothiophenes, and naphthene benzothiophenes, as reported by Drushel (1978). Rose and Francisco (1988) have also identified benzo and dibenzothiophenes in petroleum fractions, and Mullins (1995) stated that all sulfur of asphaltenes is the organic thiophenic type. Kelemen et al. (1990) also reported that most of the sulfur in heaviest fractions of petroleum is thiophene type. According to Hunt et al. (1997), sulfur species present in Maya residue are benzo and dibenzothiophenes. The content of S in asphaltenes is in the range of 6 to 8 wt%, which is higher than in maltenes (3 to 5 wt%). Other heteroatoms, such as Ni, V, O, and N, are present in higher content in asphaltenes than in maltenes (Wiehe, 1992). Some structures of asphaltenes

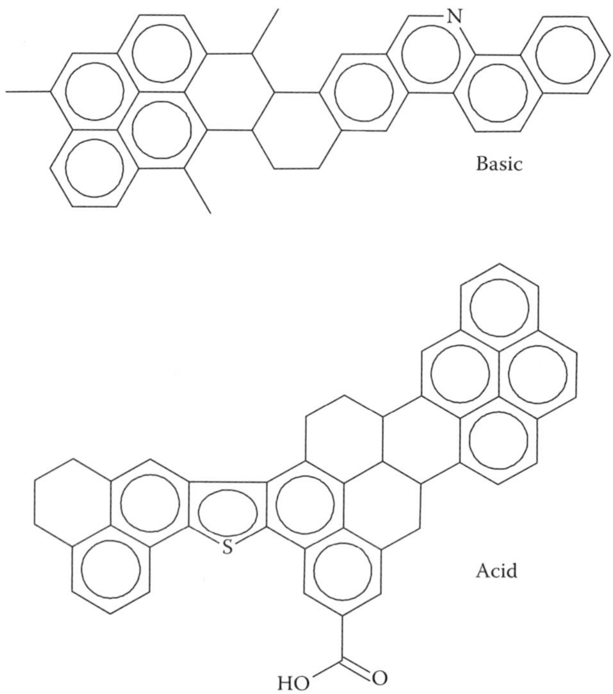

FIGURE 1.18 Hypothetical structures of asphaltenes from acid and basic fractions.

containing sulfur and nitrogen species are depicted in Figure 1.19, where radicals defined as R_1 to R_4 correspond to alkyl or aromatic bondings.

Earlier studies carried out by Moschopedis and Speight (1976a, 1976b), George et al. (1979), and Rose and Francisco (1987) have identified oxygen in carboxylic, phenolic, and ketonic structures. The oxygen in asphaltenes may be present as acidic hydroxyl groups. This was confirmed by the poor interaction of phenol solution with asphaltenes.

Asphaltenes can be separated into low- and high-molar weight asphaltenes if acetone extraction in Soxhlet is used. Different constituents are present in both fractions. For example, high-molar weight (acetone-insolubles) Athabasca asphaltenes have acyclic sulfur linkages, which bond core segments. Low-molar mass asphaltenes (acetone-solubles, pentane-insolubles) do not have this kind of sulfur linkages. Both asphaltene fractions contain low-molecular weight, sulfide-bound appendages comprising saturates, aromatics, and resins, as reported by Peng et al. (1997). Asphaltenes contain ester and ether bonds in which alcohols and carboxylic acids are bonded to the core by C–O and also C–C bonds.

Characterization of low-molecular weight asphaltenes (acetone-soluble fraction) from Athabasca bitumen indicates that specific oxygen species are identified including: fluorenones and substituted fluorenones; tri- and penta-cyclic terpenoid carboxylic

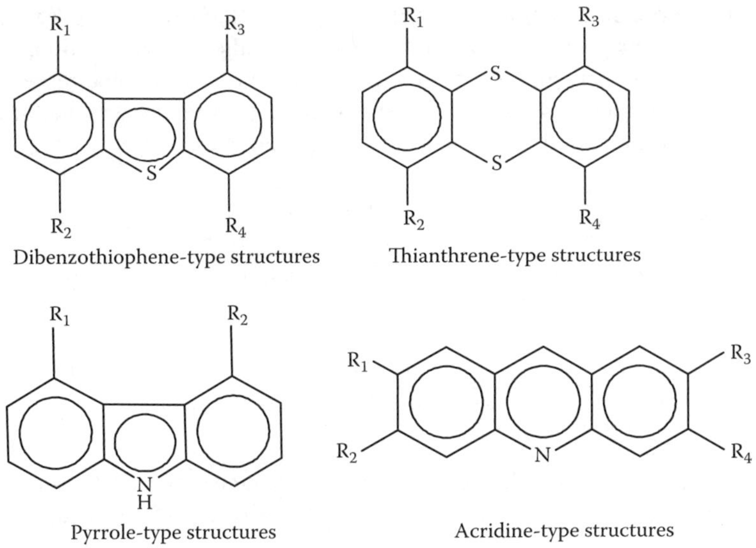

FIGURE 1.19 Sulfur and nitrogen species supposed present in residue.

acids; carboxylic acids of dibenzothiophene, *n*-alkanes, anthracene, and dibenzofuran; and bi-, tri-, tetra-, and hexa-cyclic terpenoid sulfoxides (Frakman et al., 1990). The most abundant oxygen species were the terpenoid sulfoxides and carboxylic acids.

All of these species correspond to ~1.3% of the asphaltenes amount. However, characterization of resins from Athabasca bitumen has shown similar oxygen compounds by which some structural similarities between low-molecular weight asphaltenes and resins from Athabasca bitumen are present.

1.5 ASPHALTENES DENSITY

Asphaltene density can be determined indirectly from measurements of asphaltene–toluene solution densities. Yarranton and Masliyah (1996) measured the concentration of asphaltenes in toluene in the range of 0 to 1.14 wt%. Experiments were carried out on the whole asphaltene fraction and on the insoluble fraction precipitated from hexane–toluene solutions of 20, 25, and 33 vol% toluene. The behavior of regular solution can be invoked and density can be determined indirectly by plotting the reciprocal mixture density and asphaltene mass fraction according to the equation:

$$\frac{1}{\rho_M} = \frac{1}{\rho_T} + \left(\frac{1}{\rho_A} - \frac{1}{\rho_T}\right) x_A \tag{1.3}$$

where

$$\rho_A = \frac{1}{S + I} \tag{1.4}$$

The terms ρ_M, ρ_T, and ρ_A are the average densities (kg/m^3) of the mixture, toluene and asphaltenes, respectively, and x_A is the asphaltene mass fraction. S and I are the slope and interception of the reciprocal mixture density plot, respectively. The authors assumed that asphaltene density increases linearly with molar mass. Density can be expressed in terms of molecular weight as:

$$\rho_i = 0.017\text{MW}_i + 1080 \tag{1.5}$$

Additionally, molar volume (ν) is calculated as follows in terms of molecular weight (MW):

$$\nu_\text{i} = \frac{1{,}000\text{MW}_i}{0.017\text{MW}_i + 1080} \tag{1.6}$$

For Athabasca bitumen-derived asphaltenes, the density is 1,162 kg/m^3. Other attempts to calculate the asphaltenes density is by using equations of state such as the Peng-Robinson equation. Values obtained with this equation gave an asphaltene density of 1,158 kg/cm^3, which is very close to the previous value reported by Mehrotra et al. (1985).

When density cannot be determined experimentally, an alternative method to estimate the specific gravity (SG) can be used. The molecular weight (MW) of the species along with the carbon number (C_n) and the aromaticity of the fraction (f_a) are correlated and expressed in the equation:

$$\text{SG} = 0.846\ \exp(0.6807 f_{ar}) + 0.0003 C_n - \frac{1}{0.72 C_n\ \exp(-3.26 f_{ar}) + 1.2} \tag{1.7}$$

The aforementioned correlation was used by Peramanu et al. (1999) based on the previous work of Kokal et al. (1993) and Yarborough (1979). Carbon number can be obtained from the approximation established by Whitson (1983):

$$C_n = \frac{1}{14}[\text{MW} + 6] \tag{1.8}$$

f_{ar} is assumed to be constant in the fraction; however, this is only true in lighter fractions and not in asphaltenes. f_{ar} and C_n are obtained experimentally by ^{13}C NMR, but when aromaticity is not constant through the whole fraction it is necessary to establish a way to calculate it. In this respect, Peramanu et al. (1999) have correlated the carbon number with the aromaticity of the fraction for asphaltenes from Athabasca and Cold Lake bitumens given the following correlations:

For Athabasca bitumen:

$$f_{ar} = -0.0717\ \ln(C_n) + 0.73 \tag{1.9}$$

For Cold Lake bitumen:

$$f_{ar} = -0.0647\ \ln(C_n) + 0.73 \tag{1.10}$$

Special care must be taken when aromaticity of asphaltenes is considered as constant. Aromaticity of asphaltenes is a particular property of each sample and it is related with the geographic origin of the crude where asphaltenes come from.

More recently, Luo and Gu (2007) stated that, by measuring the density of the oil and the maltenes (oil without asphaltenes), it is possible to obtain the asphaltenes density by knowing also the weight fraction of asphaltenes (w_{asph}), by means of the formula:

$$\rho_{\text{asph}} = \frac{w_{\text{asph}}}{\dfrac{1}{\rho_{\text{oil}}} - \dfrac{1-w_{\text{asph}}}{\rho_{\text{maltene}}}} \tag{1.11}$$

This approach was applied to heavy oil from the Lloydminster area in Canada and it was determined that maltene and oil densities are $\rho_{\text{maltene}} = 962$ kg/m^3 and $\rho_{\text{oil}} =$ 988 kg/m^3, respectively, at ambient conditions. The weight fraction of asphaltenes isolated with *n*-pentane is $w_{\text{asph}} = 0.145$. From Equation (1.11), it was found that $\rho_{\text{asph}} = 1{,}175$ kg/m^3. This value is very similar to that reported by Akbarzadeh et al. (2004) who found the density of 1,181 kg/m^3 for Lloydminster asphaltenes based on indirect measurements from the densities of mixtures of asphaltenes in toluene for the same crude (Yarranton and Masliyah, 1996). Indirect measurements of asphaltenes density were also applied by Fahim et al. (2001).

Akbarzadeh et al. (2004) measured densities of asphaltenes from different sources, which are shown in Figure 1.20. It can be seen that densities do not vary in a wide range. Only 5% difference is found in the values in spite of high

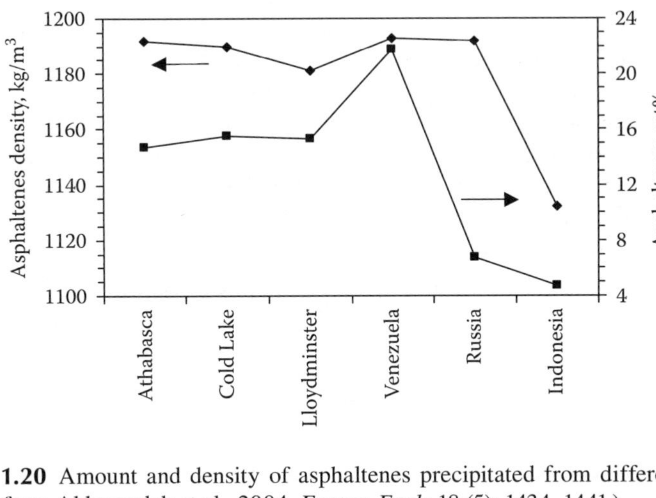

FIGURE 1.20 Amount and density of asphaltenes precipitated from different sources. (Adapted from Akbarzadeh et al., 2004. *Energy Fuels* 18 (5): 1434–1441.)

differences in the amount of precipitated asphaltenes, e.g., asphaltenes content of Venezuelan, Russian, and Cold Lake crude oils are quite different; however, the densities are practically the same.

Volume fraction of asphaltenes (φ) was also reported to be obtained if densities of asphaltenes and samples are known along with the mass fraction of asphaltenes according to the equation:

$$\varphi = \frac{\rho_{\text{sample}}}{\rho_{\text{asph}}} w_{\text{asph}} \tag{1.12}$$

Asphaltene density is a parameter almost unreported in the literature and its value, together with other characterizations, increases the knowledge of this important macromolecule.

1.6 CHARACTERIZATION OF ASPHALTENES

Due to the complex nature of asphaltenes, it is mandatory to carry out different characterization analyzes to understand their structure. Characterization techniques are vast and involve simple to complex analyses using sophisticated equipment. The purpose of this section is to describe some relevant analyses applied to asphaltenes in order to have a better idea of their molecular structure.

1.6.1 Infrared Spectroscopy

Usually, crude oil is fractionated in a column where different fractions are obtained. Aromatic distribution in vacuum gas oil presents structures from one to three polyaromatic, whereas vacuum residues contain from five to six polyaromatic cycles (Merdrignac and Espinat, 2007). Thus, the higher the boiling point of the cut, the higher the aromatic content. Multiring cycles can also be analyzed by GC-MS; however, the main issue is the amount of material that remains as nonvolatile residue. When asphaltenes are characterized by GC-MS, not all the sample is volatile as pointed out by Pillon (2001), who found that only 74 wt% of the sample was analyzed. Several authors have conducted analyses of asphaltenes by infrared spectroscopy (IS) and diffuse reflectance infrared (DR-IR) (Buenrostro-Gonzalez et al., 2001b, 2002; Aske et al., 2001; Miura et al., 2001; Seidl et al., 2004; Elsharkawy et al., 2005). Rodrigues Coelho et al. (2007) have demonstrated the existence of a linear correlation in the infrared (IR) intensities of the symmetric and asymmetric aromatic hydrogens in methyl-substituted arenes, in the 2,900 to 3,100 cm^{-1} region, and of the out-of-plane deformation in the 700 to 900 cm^{-1} region. A summary of their observations is shown in Table 1.5. FT-IR analysis of asphaltenes shows the presence of different groups such as $-OH$, $-CH_3$, $-CH_2CH_3$. Hydrogen bonded to phenols is present in the range of 3,100 to 3,300 cm^{-1}. If bands are not present at around 2,500 cm^{-1} but the sample possesses sulfur, it might be present as thioether, thiophene, or any other $-C-S-C-$ type of structure (Elsharkawy et al., 2005).

TABLE 1.5
Functional Groups Present in Asphaltenes by Infrared Spectroscopy

Functional Group	Absorption Band (cm^{-1})
–OH, –NH stretch	3600–3300
OH contributing to different hydrogen bonds[a]	
OH–π hydrogen bond	3530
Self-associated *n*-mers (n>3)	3400
OH–ether O hydrogen bonds	3280
Tightly bound cyclic OH tetramers	3150
OH–N (acid/base structures)	2940
COOH dimmers	2640
Aromatic hydrogen	3050
Aliphatic hydrogen	2993, 2920
–CH, $-CH_2$, $-CH_3$ stretching regions	3000–2800
–SH stretching regions	~2500
C=O	1800–1600
Keton (C=O stretching)	1735–1705
Aldehide (C=O stretching)	1740–1730
Conjugated C=C	1650, 1600
Aromatic C=C	1602
–CH, $-CH_2$, $-CH_3$ bending regions	1450–1375
Methyl bending vibrations	1377
Ether or ester group	1306
Ester linkage	1032
Sulfoxide groups	1030
C–S, C–O, C–N stretching regions	~1000
Aromatic C–H bending	900–700
Two adjacent H	810
1 adjacent H	900–860
2 adjacent H	860–800
3 adjacent H	810–750
4 adjacent H	770–735
5 adjacent H	710–690 (or 770–730)
Alkyl chain longer than four methylene groups	725–720

[a] Obtained by diffuse reflectance infrared.
Source: Data taken from multiple references.

Infrared spectra of asphaltenes show characteristic bands for hydrocarbons. Symmetric and asymmetric stretching of C–H aromatics bounds νCH_{AR} are found at 3,057 to 3,000 cm^{-1}, symmetric and asymmetric stretching of C–H aliphatic bounds δCH_3–CH_2 at 2,922 to 2,852 cm^{-1}, deformation bands of methyl δCH_3 at 1,375 to 1,365 cm^{-1}, and methylene γCH_3–CH_2 at 1,460 to 1,440 cm^{-1}, aromatic bending of mono-substituted γCH_{AR1} at 870 to 860 cm^{-1}, di- and tri-substituted $\gamma CH_{AR2,3}$ at 810 to 800 cm^{-1}, and tetra-substituted δCH_{AR4} at 760 to 740 cm^{-1}, and stretching of four methyl groups $\gamma CH_{2,n}$ at 727 to 722 cm^{-1}. These values have been reported previously by Buenrostro-Gonzalez et al. (2002), Langhoff et al. (1998), Christy et al. (1989), Conley (1972), and Dyer (1965).

Signals at 3,585 cm^{-1} correspond to oxygenated groups and they are detected in samples less than 0.01 wt% of concentration. Signals at 1,609 to 1,580 cm^{-1} are assigned to aromatic C=C stretching. Signals at 1,600 cm^{-1} have been reported by Borrego et al. (1996) and they have been assigned to the same stretching.

FT-IR has not only been used to determine different groups in asphaltenes, but also it has been employed to quantify the amount of asphaltenes in crude oils, as reported by Wilt et al. (1998) who used a partial least squares model to predict the amount of asphaltenes from 42 different crude oils. The model had an r^2 value of 0.95 and a standard error of 0.92 wt%. This technique provides a faster analysis method without using solvents compares favorably with the current laboratory procedures.

1.6.2 Molecular Weight of Asphaltenes

The molecular weight of asphaltenes has been immersed in a great dilemma by different research groups. The main discrepancies are due to the experimental technique employed for its measurement and, in some cases, the conditions for the analysis, such as solvent or temperature. Vellut et al. (1998) have done a review of the different techniques for determining the molecular weight of macromolecules. They have distinguished between several techniques dividing them into thermodynamic colligative methods, separation methods, and spectroscopic methods. In this section, reported results of molecular weight of asphaltenes were analyzed by different techniques, which are summarized as: vapor pressure osmometry (VPO), size exclusion chromatography, and mass spectroscopy (MS).

1.6.2.1 Vapor Pressure Osmometry (VPO)

Molecular weight of asphaltenes has generated several controversies because of the broad range of values that have been obtained with different techniques having variations even for a factor of 10. One of the most used techniques is vapor pressure osmometry (VPO), which gives high values of molecular weight. This is due to aggregation of asphaltenes in poor solvents that tend to increase the molecular weight. VPO allows for determining an absolute value for the number-average molecular weight. Values obtained for low molecular weight species and nonpolar compounds are reliable by using this technique. Even the values obtained by VPO are similar to those obtained by mass spectroscopy when analyzing resins and crude oils (Yang and Eser, 1999).

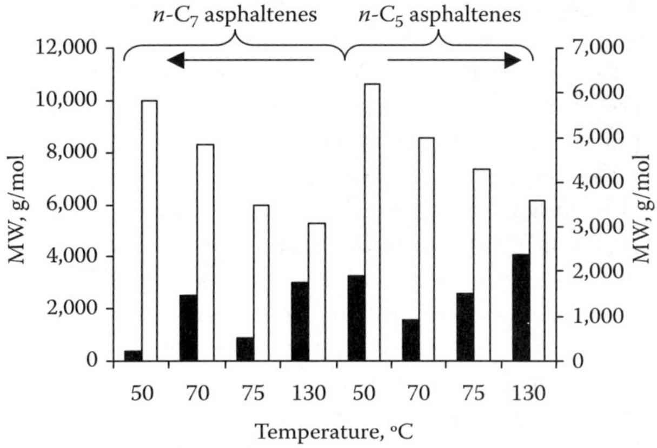

FIGURE 1.21 Estimated molecular weight of monomer and aggregate asphaltenes in two different solvents: (■) monomer (□) aggregate. (Adapted from Yarranton et al., 2000. *Ind. Eng. Chem. Res.* 39 (8): 2916–2924.)

When analyzing asphaltenes, different values have been obtained depending on the solvent used. For example, Acevedo et al. (1992a) have reported values from 1,500 to 12,300 g/mol measured by VPO using pyridine as a solvent. Values of asphaltene molecular weight from Athabasca and Cold Lake bitumens ranged from 4,000 to 10,000 g/mol (Yarranton et al., 2000). It was also reported that molecular weight of a monomer of asphaltene measured by VPO is around 1,800 g/mol. This value was calculated from the interception of a plot of measured molecular weight versus asphaltenes concentration and extrapolated at zero concentration. Extrapolation can be achieved only when low concentration of asphaltenes is present, i.e., less than 3 g/L. The molecular weight of asphaltenes extracted with *n*-pentane and *n*-heptane in their "monomeric" and aggregate way was also compared. Figure 1.21 shows the values obtained by VPO on Athabasca asphaltenes and toluene and 1,2-dichlorobenzene as solvents. Asphaltenes are present in aggregate going from two to six monomers depending on the composition and temperature at which analysis is carried out.

Acevedo et al. (2005) have carried out analysis of asphaltenes by VPO with nitrobenzene at 100°C varying the concentration of solutions in the range of 0 to 6 g/L. Good correlation between VPO and mass spectrometry was found. According to Speight (1987), solvent type and its polarity, temperature, and concentration of asphaltenes play an important role when dealing with VPO. For obtaining reliable molecular weights of asphaltenes by this technique, it is recommended that samples be analyzed at each of three different concentrations and temperatures. In a plot of molecular weight versus concentration, the data for each temperature are extrapolated at zero concentration, and the zero concentration data at each temperature also extrapolated at room temperature (Speight, 1994).

Moschopedis et al. (1976) have also proposed the use of nitrobenzene and high temperatures ranging from 100 to 150°C for determining the molecular weight of asphaltenes. Low molecular weight was obtained and the results agree with those obtained by structural determinations by ^{1}H NMR. However, Wiehe (1992) observed by optical microscopy that nitrobenzene could not dissolve completely the asphaltenes and, therefore, the molecular weight is not utterly true. Instead of nitrobenzene, this author has proposed the use of *o*-dichlorobenzene. By using this solvent, consistent results were obtained. Molecular weight of asphaltenes decreases linearly with decreasing the concentration. The difference in molecular weight at 70 and 130°C is minimal (3,380 versus 3,400 g/mol, respectively) when both values are extrapolated to zero concentration. The independence of the measurement with *o*-dichlorobenzene at two different temperatures indicates that this solvent is better than toluene. The use of toluene as solvent gave a molecular weight of 4,900 g/mol. High temperature and high dissociating power of the solvent are necessary for obtaining good results with this technique. Figure 1.22 shows the linear dependence of molecular weight as a function of asphaltenes concentration.

Wiehe (2007) has argued that molecular weight cannot be as low as established by other techniques, i.e., fluorescence depolarization by which it is possible to obtain molecular weights lower than 1,000 g/mol because, if this would be possible, it should be expected that a significant fraction of the asphaltenes can be evaporated during high temperature vacuum distillation and during thermal processing. According to the author, it is more important to have consistency in results and, by means of VPO, those consistent measurements of asphaltene molecular weight can be obtained. The author has established that the use of

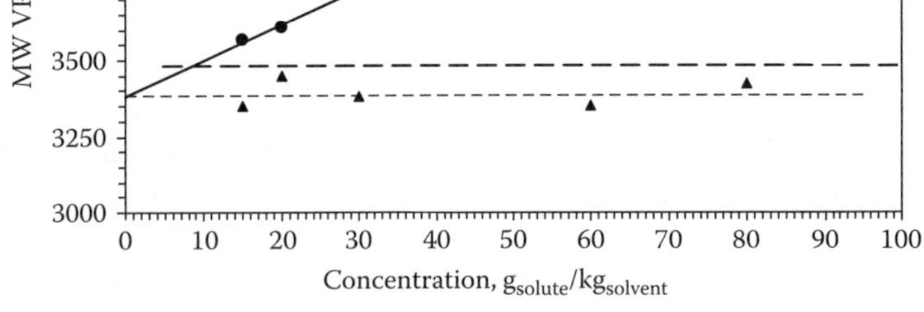

FIGURE 1.22 Asphaltene association broken by increasing the dilution and temperature for Arab heavy asphaltenes: (●) 70°C; (▲) 130°C. (From Wiehe, I.A., 1992. *Ind. Eng. Chem. Res.* 31 (2): 530–536. With permission.)

o-dichlorobenzene at 130°C provides the best conditions for measuring asphaltene molecular weight by VPO. He also conducted a series of experiments in which the aim was to separate different components of the residue and each of them was submitted to thermal conversion. Each fraction yielded other fractions as shown in Figure 1.23. Resins form asphaltenes by free radicals while original asphaltenes also form coke quickly. It can be seen that the aromaticity (reduction of H/C atomic ratio) diminishes as products become heavier. Thus, coke has the lower H/C atomic ratio.

Other authors have also carried out analysis by VPO. Sharma et al. (2007) have determined VPO molecular weight of soluble and insoluble fractions of vacuum residues. Molecular weight of soluble fraction (maltenes) is around 700 to 1,600 g/ mol depending on the sample, whereas molecular weight of insoluble fraction (asphaltenes) is between 1,800 and 2,900 g/mol. The source of asphaltenes and maltenes was Heera and Jodhpur Indian vacuum residues. Based on molecular weights of the soluble and insoluble fractions, the following equation for determining the molecular weight of the whole residue in terms of its fractions was proposed:

$$MW_r = \frac{1}{\dfrac{w_i}{MW_i} + \dfrac{w_s}{MW_s}} \tag{1.13}$$

where MW_r, MW_i, and MW_s are the number average molecular weight of the residue, insolubles, and solubles, respectively. w_i and w_s are the weight fraction of insoluble and soluble fractions in vacuum residues, respectively. Experimental and calculated values from VPO measurements deviate ±10%, which can be considered as acceptable.

Peramanu et al. (1999) have separated the Athabasca and Cold Lake bitumens into saturates, aromatics, resins, and asphaltenes, and obtained molecular weights of each fraction by VPO. The results are presented in Figure 1.24 in which the content of each fraction and its molecular weight are observed. It can be seen that asphaltene is the smallest fraction in weight but it has the highest molecular weight. The higher the complexity of the fraction, the higher the molecular weight. Saturates and aromatics have almost the same molecular weight. Although the molecular weight of resins is not too different in both crudes, molecular weight of asphaltenes makes a difference in spite of having similar asphaltenes content (17.28 wt% in Athabasca bitumen versus 15.25 wt% in Cold Lake bitumen). The complexity of asphaltenes in each case is responsible for differences in molecular weight.

Ancheyta et al. (2002) have determined asphaltene molecular weight by VPO. The values were reported as aggregate molecular weight because the solvent used for measurements was toluene, which is unable to separate aggregates of asphaltenes in solution. The observed trend indicated that aggregate molecular weight of asphaltenes is higher if n-C_7 is used compared with that obtained with n-C_5. The aggregate molecular weight of asphaltenes from three different crude oils is shown in Figure 1.25.

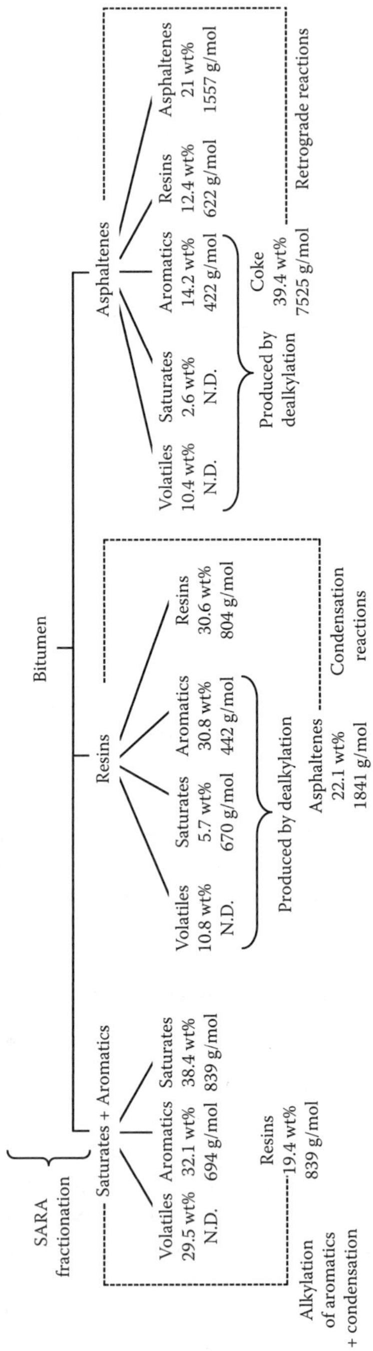

FIGURE 1.23 Molecular weight and yield of fractions obtained by thermolysis (Adapted from Wiehe, I.A. 2007. Ind. *Eng. Chem. Res.* 31 (2): 530–536.)

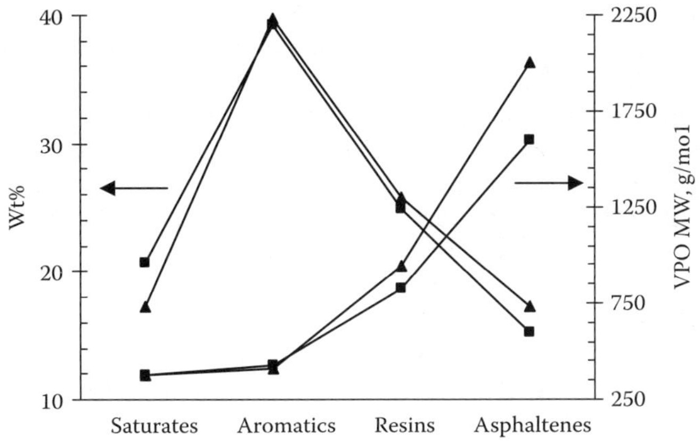

FIGURE 1.24 SARA (saturates, aromatics, resins, and asphaltenics) fractionation and molecular average weight of petroleum fractions by VPO: (▲) Athabasca, (■) Cold Lake. (Adapted from Peramanu et al., 1999. *Ind. Eng. Chem. Res.* 38 (8): 3121–3130.)

Comparison of VPO measurements made with toluene and 1,2-dichlorobenzene as solvents indicated that molecular weights obtained with 1,2-dichlorobenzene are 2.2 times lower than values obtained with toluene (Yarranton and Masliyah, 1996). Different ratios have been reported in the literature, i.e., Acevedo et al. (1992b)

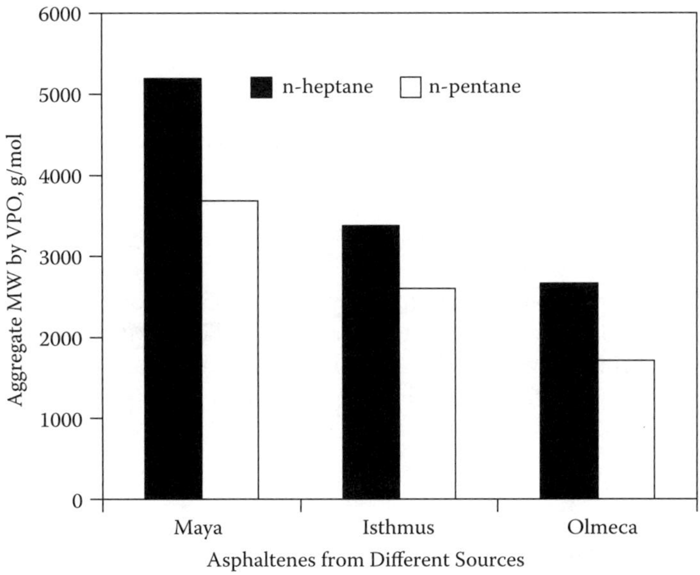

FIGURE 1.25 Aggregate molecular weight of asphaltenes by VPO using toluene as solvent. (From Ancheyta et al., 2002. *Energy Fuels* 16 (5): 1121–1127. With permission.)

found a ratio of 3.6, Moschopedis et al. (1976) a ratio of 1.6, and Schabron and Speight (1997) a ratio of 2. Molecular weights determined by VPO and mass spectrometry (MS) have given different values. For Ratawi asphaltenes, the molecular weight obtained by VPO was 2,360 and 814 g/mol by MS; for Alaskan North Slope asphaltenes the values obtained were 3,248 by VPO and 1,270 g/mol by MS (Hunt et al., 1997). The high molecular weight obtained by VPO has been attributed to different degrees of molecular stacking. Takanohashi et al. (1998) applied molecular dynamics in coal and found that associates are destroyed when pyridine is used. Methanol and benzene did not break the aggregates. Methanol is able to break the hydrogen bonds in the aggregates, but this is not enough to dissociate them. Brandt et al. (1995) have studied the size stacking of asphaltenes in different solvents. In 1-methylnaphthalene at temperatures between 90 and 130°C, the agglomerate size is between 1 and 3 unit sheets, whereas in solvents, such as toluene and dichloromethane, the values are in between 7 and 13 unit sheets. In the solid state, the size of stacking is between 5 and 8 unit sheets. The creation of agglomerates is a consequence of the temperature and solvent aromaticity on asphaltene alkyl chains.

Molecular weight of asphaltenes and residue obtained by VPO in toluene along with microcarbon residue (MCR) served to establish a correlation in which whole asphaltenes, residue, aromatic and polar fractions, and chromatographic fractions from various sources were considered as fractional contributions. Values that do not fall near the straight line were excluded from the correlation, so that with 44 points, the following trend line was obtained (Schabron and Speight, 1997):

$$\text{MCR (wt\%)} = 0.0198\,(\text{VPO MW in toluene}) - 0.42; \quad r = 0.943 \tag{1.14}$$

The elemental composition of asphaltenes, residue, and different fractions were also correlated with MCR content to give a linear dependence as function of the H/C atomic ratio. Thirty-three points were used and the following correlation was determined:

$$\text{MCR (wt\%)} = 173\,(\text{asphaltene H/C atomic ratio}) - 105; \quad r = 0.975 \tag{1.15}$$

Figure 1.26 shows the relationship between MCR, molecular weight, and H/C atomic ratio for asphaltenes. It is seen that increasing MCR content, MW of asphaltenes increases, whereas H/C atomic ratio diminishes indicating more aromatic residual asphaltenes. As summary, advantages and disadvantages of VPO include:

Advantages:

- VPO gives true values of molecular weight for lighter fractions of petroleum, even resins.
- At high temperatures with a highly dissociating solvent for asphaltenes, the molecular weight will be reliable and will not correspond to an aggregate molecular weight.

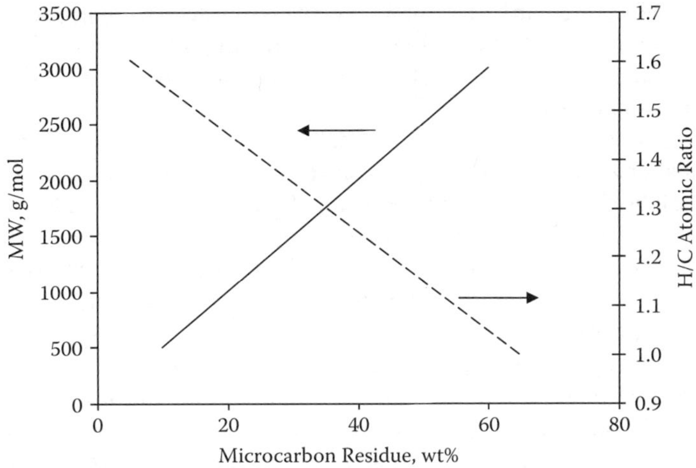

FIGURE 1.26 Molecular weight and H/C atomic ratio versus microcarbon residue in asphaltenes and residue. (Adapted from Schabron and Speight, 1997. *Am. Chem. Soc., Div. Fuel Chem.* 42 (2): 386–389.)

Disadvantages:

- If the solvent is not adequate, asphaltenes molecular weight will be an aggregate molecular weight.
- The increase of temperature in the equipment is limited by the solvent used.

1.6.2.2 Size Exclusion Chromatography (SEC)

Size exclusion chromatography (SEC) is one of the most frequently used techniques for obtaining bitumen and asphaltene molecular weight distributions. For the determination of molecular weight distributions by means of SEC, it is necessary to have a calibration curve that relates molecular weight to elution volume (Reerink and Lijzenga, 1975). Problems related to SEC have been reported elsewhere (Bartle et al., 1984; Champagne et al., 1985; Mulligan et al., 1987). This technique allows for separating components based on their molecular size, not on their mass (Domin et al., 1999). A standard for calibration is required that imitates the shape of the sample, but due to asphaltene polydispersity, there isn't available one polymer that mimics the entire shape of all molecules present in asphaltenes. The most used polymer for this purpose is polystyrene, (Nali and Manclossi, 1995). However, it has been reported recently that there are strong interaction forces between polystyrene and asphaltenes as detected by the atomic force microscopy (AFM) that showed that asphaltenes can interact and most likely do interact with the SEC column-packing beads (Behrouzi and Luckham, 2008).

SEC can be used for monitoring the evolution of the removal of contaminants, such as sulfur and metals (mainly Ni and V), from hydroprocessing of heavy petroleum, which are harder to remove, i.e., more refractory compared with light fractions (Koinuma et al., 1997). In this sense, the importance of SEC in analyzing changes that asphaltenes and other heavy fractions of petroleum suffer during hydroprocessing is evident.

Under hydrotreating conditions, asphaltenes size is changed. During reaction, asphaltenes may suffer rearrangements and changes in their structures and dissociation into smaller aggregates before conversion could be feasible. Merdrignac et al. (2006) have reported the evolution of the molecular weight of asphaltenes and compared the changes that occur in fixed-bed reactor and ebullated-bed reactor. A continuous evolution between the feedstock and the products of the fixed and ebullated-bed reactors was observed. The conversion of residue influences widely the molecular weight of asphaltenes contained in residue, i.e., molecular weight of asphaltenes from the fixed-bed process decreased as conversion of residue was gradually increased; however, large aggregates could be present in the remaining asphaltenes. Dissociation of large aggregates of asphaltenes at lower conversion like that obtained in fixed-bed is less pronounced than in ebullated-bed. Chromatograms of asphaltenes from the feedstock (Buzurgan vacuum residue) with tetrahydrofuran (THF) as solvent showed a bimodal profile, which can be due to two types of aggregates: (1) highly associated asphaltenes with higher molecular weight and (2) less associated asphaltenes with lower molecular weight. Hence, after hydroprocessing asphaltenes, the bimodal profile tends to be unimodal. Trejo et al. (2007) also observed this behavior when using THF as a mobile phase. It can be observed from the chromatogram that the elution time for Maya crude asphaltenes starts at 7.5 min, but a shoulder exists at 8.5 min (Figure 1.27), corresponding to higher molecular weight asphaltenes. Molecular weight distribution continues up to 12 min.

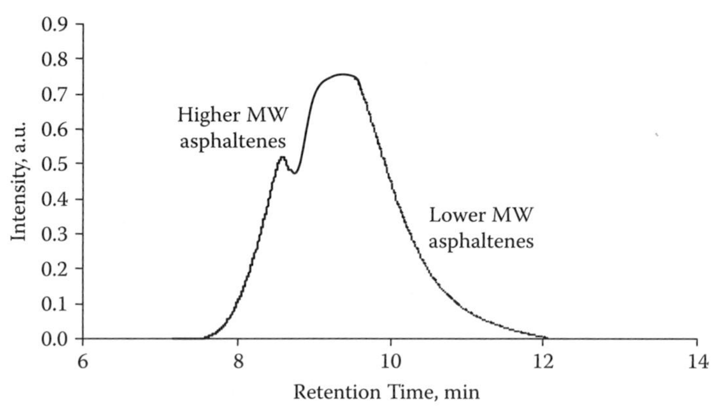

FIGURE 1.27 SEC chromatogram of Maya crude asphaltenes. (Trejo et al., 2007. *Energy Fuels* 1 (4): 2121–2128. With permission.)

An explanation about the bimodal SEC profiles has been given by Dabir et al. (1996) who postulated that if the asphalt- or asphaltene-containing solution has not been aged enough, a bimodal MW distribution is obtained, which is due to the existence of two different types of aggregates with distinct structures and mechanisms of formation. However, after some time, the MW distribution is unimodal.

The inconvenience for using toluene as solvent in SEC analysis is that nonsize effects may dominate if the solute is adsorbed on the stationary phase, which is observed when chromatograms tailing to lower molecular weight. Johnson et al. (1997) solved the problem of the nonsize effect by employing *N*-methyl-2-pyrrolidinone (NMP) as a mobile phase, which allowed them to operate the SEC column at higher temperatures. Bartholdy et al. (2001) analyzed asphaltenes by SEC with NMP as the solvent and they did not observe material that eluted after the permeation limit such as toluene does when used as a mobile phase. NMP has been proven to be a good solvent for coal or products derived from thermal processes, according to Guillén et al. (1991). However, this solvent can only dissolve partially petroleum asphaltenes and the dissolution process may be very slow. The SEC chromatograms with NMP as solvent show two characteristic peaks as reported by Ascanius et al. (2004) and Karaca et al. (2004). In addition, when comparing THF and NMP as eluents, it is seen that an early eluting peak observed with NMP is not present with THF (Morgan et al., 2005). The second peak corresponds to smaller molecular weights. Depending on the nature of the sample and the particular type of column, this peak corresponds to molecular weights from 3,000 to 9,000 g/mol. Molecular weight up to 3,000 g/mol is within a factor of 2 to 2.5 when comparing with the maximum ion abundance obtained by mass spectrometry. There is a valley of approximately zero intensity between the excluded peak and the retained peak observed for almost all samples and fractions when using NMP as a mobile phase and a possible explanation is a transition from one type of molecular conformation to another (Herod et al., 2007; Morgan et al., 2005; Li et al., 2004; Karaca et al., 2004). Figure 1.28 shows SEC chromatograms of Maya asphaltene and a heptane-soluble fraction (maltenes) with NMP. The heptane-soluble fraction contained smaller proportions of excluded material than whole asphaltenes and its retained peak shifted to longer elution times than for asphaltenes indicating the lesser complexity of maltene molecules, which have lower molecular weight (Millan et al., 2005). However, when using NMP, aliphatics cannot be dissolved. Instead of using NMP, heptane as eluent is preferred for dissolving aliphatics. For this reason, Al-Muhareb et al. (2006) have developed an alternative system using heptane with a evaporative light-scattering detector, which allows for determining alkanes up to C_{50} without needing to isolate aliphatics from aromatics.

The use of SEC and element specific detection by inductively coupled plasma emission spectrometry (SEC-ICP) has suggested that V and S correlate indicating that large V compounds may have S in their structures. Both heteroelements also correlate well with asphaltenes content. At low metals removal during thermal hydroprocessing these compounds are eliminated only partially; however, at high demetallization levels, complete elimination of these contaminants occurs.

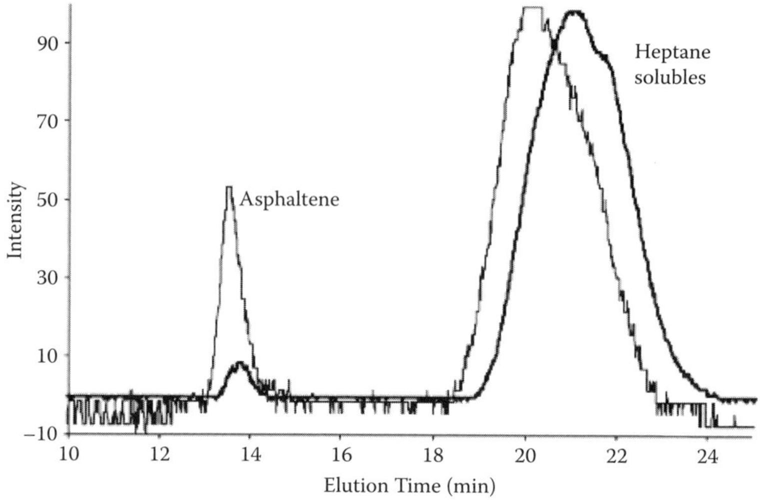

FIGURE 1.28 SEC chromatograms of Maya asphaltenes and maltenes (heptane soluble fraction). (From Millan et al., 2005. *Cata. Today* 109 (1-4): 154–161. With permission.)

Figure 1.29 shows some suggested structures for vanadium compounds present in residue, which can be substituted with sulfur bondings. R_1 and R_2 indicate the positions that sulfur can occupy inside the tetra-coordinated porphyrin structure. If only R_1 is substituted with a sulfur atom instead of nitrogen atom, like in a common metallo-porphyrin whereas R_2 is nitrogen, then the resulting structure is a four-coordinated pyramidal vanadyl complex, such as vanadyl dithiocarbamates and vanadyl thiolates (Reynolds and Biggs, 1987). If both R_1 and R_2 are sulfur atoms, then the porphyrin is a disubstituted structure that could link to other porphyrins for obtaining a dimer structure. Sulfur occupying the positions R_1 and R_2 is thiophenic type. A metallo-porphyrin can interact with sulfur of alkyl or thiophene of other estructures, i.e., S is directly bound in the axial coordination position or not, but in this last case, there is an association between sulfur and vanadium. Figure 1.29 also shows the possible interactions with sulfur where R_3 and R_4 are alkyl or aromatic carbons.

The main advantages and disadvantages of SEC are:

Advantages

- SEC gives a full distribution of molecular weight of heavy samples, such as bitumen or asphaltenes.
- Reliable distributions of molecular weight are obtained if a proper solvent is used and good qualitative results can be obtained.

Disadvantages

- Nonsize effects and adsorption of the solute in the column packing can be present.

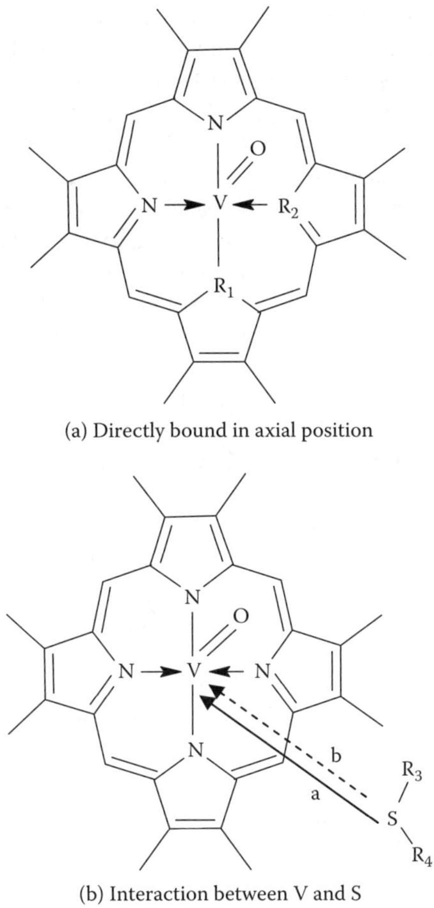

FIGURE 1.29 Different representations of vanadyl porphyrins with probable S-bonds. R_1 and R_2 are possible S substituents, R_3 and R_4 are alkyl or aromatic carbons.

- There is not a good standard for calibration due to the polydispersity and heterogeneity of heavy samples.
- Distribution of molecular weight of heavy fractions is dependent on the solvent used.

1.6.2.3 Mass Spectroscopy

A useful technique that has emerged in recent years with applications to heavy fractions of petroleum is mass spectroscopy (MS). This technique gives distribution of molecular weight for asphaltenes and other fractions. It is desirable that nonfragmenting ionization techniques be used in mass spectroscopy. Domin et al. (1999) stated that in order to obtain reliable distributions of molecular weight by this method, the following points are important to cover: (1) to vaporize all molecules and (2) to ionize all molecules without fragmentation; however, it is problematic when analyzing asphaltenes because of their complexity and polydispersity.

Laser desorption mass spectroscopy (LDMS) tends to underestimate values of molecular weight for complex samples. Another technique based on mass spectroscopy is matrix assisted laser desorption/ionization (MALDI), which uses a compound as matrix. This method was first introduced by Karas and Hillenkamp (1988) and has been used for biomacromolecules, polymers, and even fractions of petroleum. The use of a matrix turns MALDI into a soft technique because the matrix disperses the heat created by the laser beam and it is distributed avoiding the fragmentation of the sample. A good correspondence has been reported between MALDI and SEC by Suelves et al. (2003) who found agreement from 200 to 3,000 g/mol. The material excluded from SEC, which corresponds to 1 to 2 wt% of the total sample, is undetermined and would correspond according to MALDI to a sample from 3,000 to 10,000 g/mol. These species of high molecular weight may act as a nucleation center for the aggregation of the smaller molecules in poor solvents.

Herod et al. (2000) listed the most important issues to be taken into account when using MALDI to analyze molecular weight, which can be summarized as:

1. Increasing the ion-extraction voltage provides a higher ionization, which helps to detect species with larger molecular weight.
2. Bigger molecular weights can be detected if reflector is removed, which improves sensitivity to detection of higher molecular weights.
3. The signal-to-noise ratio in spectra is higher as the number of co-added scans is increased.
4. Composition of matrix, sample, and mixing of both of them could influence the molecular weight distribution.

Merdrignac et al. (2004) concluded that the molecular weight distribution of asphaltenes fraction obtained by MALDI exhibited molecular weights from 100 to 1,000 g/mol. Low molecular weight is not only due to molecular fragmentation, but also polydispersity plays an important role. Seki and Kumata (2000) reported different spectra of asphaltenes obtained from demetallized Kuwait atmospheric residue at various temperatures. LDMS of asphaltenes from the feedstock and the product treated at 370 and 390°C were broad and polydisperse indicating that in spite of having reaction the conditions are not severe enough to promote a significant change in asphaltenes below 400°C. Asphaltenes under these conditions presented a major peak centered on *m/z* 1,100 and at *m/z* 600 g/mol indicating that compounds with small molecular weight were detected. However, above 400°C, changes were more evident. At 410°C, the peak was displaced to lower molecular weight below *m/z* 1,000 g/mol. At 430°C, two peaks were observed (Figure 1.30) and two possible explanations were suggested: (1) the peaks derive from a new fraction of asphaltenes, which are formed during hydrodemetallization (HDM) reactions, and (2) one peak is originated by asphaltenes having one aromatic skeleton and the other peak is due to asphaltenes having two aromatic skeletons.

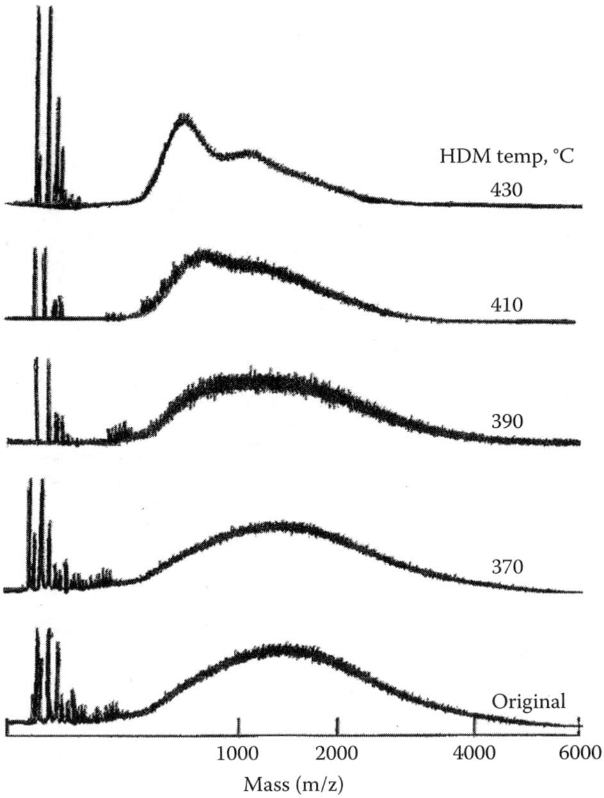

FIGURE 1.30 LD-MS spectra of nonhydrotreated and hydrotreated asphaltenes. (From Seki and Kumata, 2000. *Energy Fuels* 14 (5): 980–985. With permission.)

Trejo et al. (2007) have also reported LDMS and MALDI spectra of Maya crude asphaltenes, which are shown in Figure 1.31. The maximum ion abundance was observed from *m/z* 1,500 to *m/z* 2,000 g/mol by LDMS and *m/z* 2,000 g/mol by MALDI when using sinapinic acid as a matrix. The part of the MALDI spectrum, which is missing at lower molecular weights, could be due to evaporation of the smaller and lighter asphaltene molecules.

As one can see, different characterization techniques are available to obtain molecular weight of asphaltenes, which can be reliable if selection of operating conditions and solvent treatment is done properly. A summary of molecular weight values of asphaltenes reported by several authors is presented in Table 1.6 (Vellut et al., 1998). Different methods, such as tonometry, viscosimetry, and ultrafiltration have been also reported by Briant and Hotier (1983) to measure the molecular weight and size of asphaltenes clusters in different diluents mixtures. It has been concluded that solute concentration and composition of the diluents mixture directly influence molecular weight.

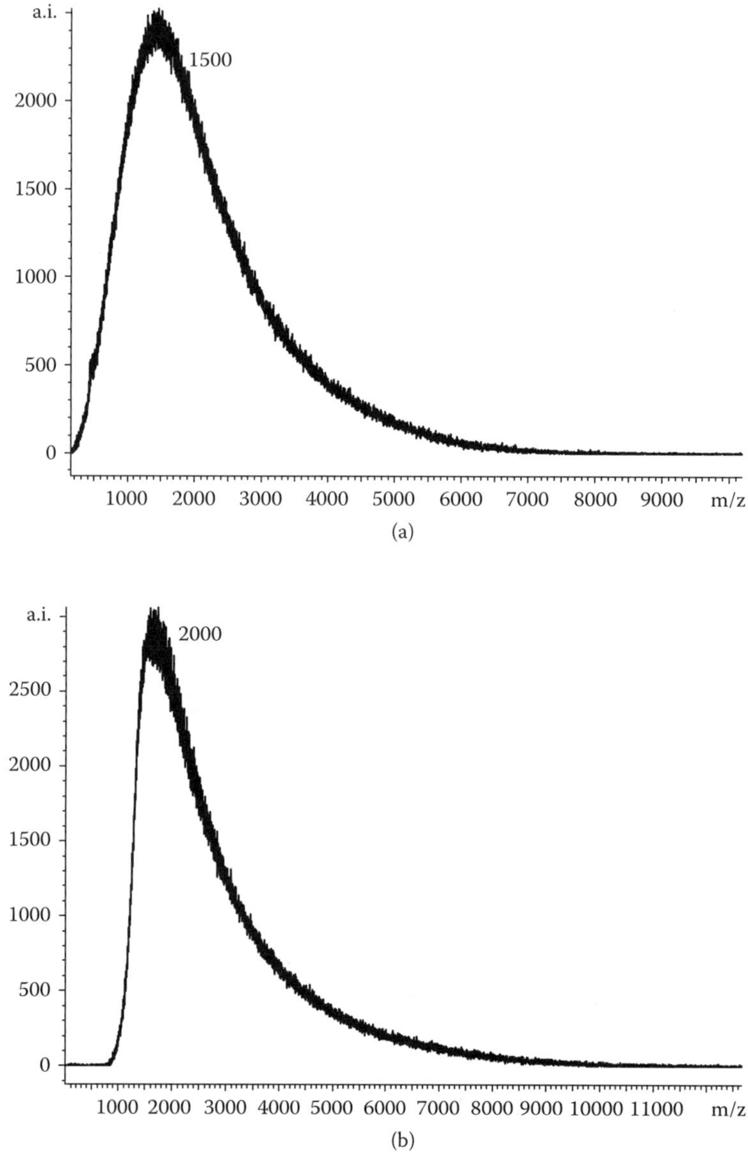

FIGURE 1.31 LDMS (A) and MALDI (B) spectra of Maya crude asphaltenes. (From Trejo et al., 2007. *Energy Fuels* 1 (4): 2121–2128. With permission.)

1.6.3 Ultraviolet (UV) Fluorescence

Combination of FT-IR, UV-vis spectrophotometry, and synchronous fluorescence spectrometry has been applied by Guan and Zhu (2007) to study the distribution of aromatic rings in asphaltenes and resins, where strong interactions between the

TABLE 1.6
Comparison of Different Techniques for Obtaining Molecular Weight of Asphaltenes

References	MW Range g/mol	Concentration Range	MW Type	Experimental Uncertainty	Macromolecules Analyzed
		Thermodynamic Colligative Methods			
		Ebulliometry			
Vellut et al. (1996)	150–1000	1–5% (wt/wt)	Mn	<2%	Maltenes, atmospheric residue, resins
		Cryometry			
Moschopedis et al. (1976)	600–10,000	1–15% (wt/wt)	Mn	—	Asphaltenes
		Vapor Pressure Osmometry			
Moschopedis et al. (1976)	1000–5000	2–7%(wt/wt)	Mn	—	Asphaltenes
Briant and Hotier (1983)	1000–10,000	5–55 g/L (C_6H_6+C_7)	Mn	5–20%	Asphaltenes
Blondel-Teoluk et al. (1995)	150–1000	0.005–0.05 (mol/mol)	Mn	2%	Heavy crude (in benzene)
		Osmometry (Osmotic Pressure or Membrane Osmometry)			
Moschopedis et al. (1976)	80,000	—	Mn	—	Asphaltenes
		Viscosimetry			
Reerink (1973)	5000–50,000	—	Mw	—	Asphaltenes
Briant and Hotier (1983)	7000	4–25 g/L (benzene)	Mn	>5%	Asphaltenes
		Separation Methods			
		Size Exclusion Chromatography			
Reerink and Lijzenga (1975)	10,000–30,000	—	Mn	—	Asphaltenes
Dabir et al. (1996)	800–11,000	—	Mw, Mn	—	Asphaltenes, asphalts

(Continued)

TABLE 1.6 (CONTINUED)
Comparison of Different Techniques for Obtaining Molecular Weight of Asphaltenes

References	MW Range g/mol	Concentration Range	MW Type	Experimental Uncertainty	Macromolecules Analyzed
		Ultracentrifugation			
Lemerle et al. (1984)	2500	0.002–0.009% (wt/wt)	Mw	—	Asphaltenes
		Ultrafiltratrion			
Speight et al. (1985)	80,000–140,000	–	Mw	—	Asphaltenes
		Spectroscopic Methods			
		Light Scattering			
Gottis and Lelanne (1989)	4000–134,000	1% (THF)	–	–	Asphaltenes
Espinat et al. (1984)	4000–134,000	6–10% (wt/wt)	Mn	10%	Asphaltenes
Espinat (1991)	16,000–89,000	0.7–1.2% (pyridine, benzene)	Mw	–	Asphaltenes
		Mass spectroscopy			
Bouquet and Brument (1990)	200–450	–	Mw	3–7.5%	Heavy hydrocarbons

Note: Mn = Number average molecular weight, Mw = Weight average molecular weight.
Source: Adapted from Vellut, D. et al., 1998. *Oil Gas Sci. Tech. – Rev. IFP* 53 (6): 839–855.

two are carried out by hydrogen bonds. UV absorption and synchronous fluorescence spectrometry of asphaltenes and resins were used to find their structures and it was concluded that the main difference between asphaltenes and resins is the amount of conjugated aromatic rings. Resins have generally less than five aromatic rings in a conjugated aromatic unit and asphaltenes have more than five aromatic rings, both of them having a peri-condensed structure.

Ascanius et al. (2004) determined that the insoluble fraction of asphaltenes in NMP did not emit much fluorescence over the range of wavelengths studied. They concluded that all methods based on UV-visible and fluorescence spectroscopy do not analyze completely the entire asphaltenes. The lack of fluorescence can be attributed to the structural constitution of the sample or strong molecular interactions that lead to significant red shifts and quenching of fluorescence intensity. More aliphatic species have been found in the insoluble fraction of asphaltenes, whereas more aromatic structures are contained in the soluble fraction in NMP. However, there is an increase of aromatic chromophores when increasing insolubility in NMP, as reported by Al-Muhareb et al. (2007). Paul-Dauphin et al. (2007) also concluded that the largest chromophores present in asphaltenes are insoluble in NMP, but soluble in chloroform, and they can be observed by UV-fluorescence spectroscopy, and insoluble fraction in NMP is probably attached to very large aliphatic groups. It was reported that heptane-insoluble materials (i.e., asphaltenes) do not fluoresce strongly when UV-fluorescence is applied and this method is unable to detect high mass materials, as stated by Millan et al., 2005. Trejo et al. (2007) reported UV fluorescence results using NMP as a mobile phase and asphaltenes from virgin Maya crude. Greater differences were observed between UV-A (absorption) and UV-F (fluorescence) chromatograms at 350, as can be seen in Figure 1.32. nm. When using a UV-A detector, the higher solvent power of

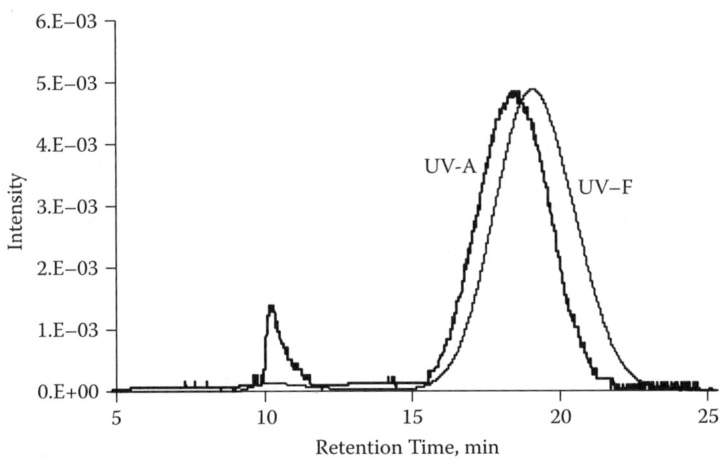

FIGURE 1.32 SEC profiles with UV-F and UV-A in NMP at 350 nm of asphaltenes from Maya crude oil. (From Trejo et al., 2007. *Energy Fuels* 1 (4): 2121–2128. With permission.)

NMP allowed detection of large-sized material eluting at the exclusion limit of the chromatographic column, which was observable from 6.5 min. The UV-A signal shows narrower distribution than the UV-F chromatogram, probably because the smallest aromatics have the strongest fluorescence. The lack of fluorescence can be ascribed to structural reasons or due to strong molecular interactions leading to quenching of fluorescence intensity and significant red shifts, as reported before (Ascanius et al., 2004). There is a consensus between different researchers in which it has been stated that UV fluorescence is unable to detect molecular weights greater than 3,000 g/mol, as highlighted by Herod et al. (2007). In addition, a progressive shift of UV fluorescence spectral maximum intensity and molecular weight with decreasing mobility of fractions on the planar chromatographic plate has been reported, which suggests a progressively larger polynuclear aromatic system (Herod and Kandiyoti, 1995).

Strausz et al. (2002) stated that fluorescence properties vary depending on the molecular weight of the sample. In the case of highly dilute fractions of Athabasca asphaltenes, the lowest molecular weight fraction was the most fluorescent and the highest molecular weight fraction scarcely exhibited fluorescence. In addition, the rotational motion of two or more aromatic chromophores connected by varying lengths of polymethylene and/or sulfide and/or C–C bridges cannot be represented by rigid geometrical objects. Some issues have been summarized as:

1. Not all species in asphaltenes absorb in the spectral range of 300 to 700 nm and principal absorbers are aromatic compounds present in asphaltenes.
2. Not all species that absorb necessarily fluoresce because fluorescence is a competitive process of an intramolecular level.
3. Quenchers could include moieties, such as $-CO_2H$, heavy atoms of sulfur, metal salts and complexes, clay organics, intramolecular H-bonded complexes, etc., as reported by other authors (Cowan and Drisko, 1976; Brauman, 2000).

On the other hand, Suelves et al. (2003) have indicated according to their observations that the shift of maximum intensity of fluorescence toward longer wavelength with increasing molecular size along with the molecular weight obtained by SEC and MALDI could indicate that molecules that fluoresce contain larger aromatic systems instead of being aggregates of small molecules.

1.6.4 X-Ray Diffraction

X-ray diffraction (XRD), developed by Warren (1941), Franklin (1950), and Cartz et al. (1956), has been used by many researchers to study the structure of coal (Ergun and Tiensuu, 1959), carbon black structures (Alexander and Sommer, 1957), small aromatic systems in noncrystalline polymers (Ruland, 1967), and pitch fractions (Shiraishi et al., 1972). Other earlier studies were carried out by Yen et al. (1961) who discovered a practical use of this technique for petroleum asphaltenes, which is helpful in determining the spacing between aromatic layers corresponding to the graphene band (or 002 band), which appears around 26° and the spacing between aliphatic layers that appears around 20°.

XRD can provide valuable information concerning the internal structure of asphaltenes. This technique is valuable for extracting crystallite parameters of the molecules associated in the aggregates (Shirokoff et al., 1997). Aromatic cores of asphaltenes can be stacked and form a pile of aromatic sheets. This region is available to XRD analysis because the stacking makes it possible for crystallites to form. However, other crystallite parameters in the polar aromatics, naphthene aromatics, and saturate fractions also can be analyzed by XRD, as reported by Siddiqui et al. (2002), who measured these properties in Arabian asphalts.

From XRD, a hypothetical crystallite can be drawn to represent the asphaltene structure, as proposed by Schwager et al. (1983). It is asserted that condensed aromatic sheets are stacked on top of each other with the sheets parallel and with aliphatic chains or naphthenic rings protruding from the edges. The following crystalline parameters can be obtained: L_a = diameter of aromatic sheet plus α-carbons of alkyl chains (Å); L_c = average height of the stack of aromatic sheets perpendicular to the plane of the sheet (Å); d_m = interaromatic layer distance (Å); d_γ = interchain or internaphthene layer distance (Å); M = average number of aromatic sheets associated in a stacked cluster.

Determination of crystalline parameters for the interlayer distance between aromatic sheets (d_m) is estimated from the maximum of the 002 band using the Bragg relation (Siddiqui et al., 2002):

$$d_m = \frac{\lambda}{2\sin\theta} \tag{1.16}$$

where λ is the wave length (Å) and θ is the angle in which the peak is centered (degrees). The average height of the stack of aromatic sheets perpendicular to the plane of the sheet (L_c) requires measuring the full-width half maximum (FWHM):

$$L_c = \frac{\lambda}{\omega\cos\theta} \tag{1.17}$$

where ω is the FWHM. The number of aromatic sheets in a stacked cluster (M) is given by the equation:

$$M = \frac{L_c}{d_m} + 1 \tag{1.18}$$

In many cases, results indicate that the average interlayer distance (d_m) ranges from 3.5 to 3.7 Å, and d_γ is between 4.4 and 5.4 Å. A schematic view of a probable stacking of asphaltenes is shown in Figure 1.33. XRD gives quantitative intensity curves and structural parameters can be obtained from the shape as well as the position of peaks, the determination of crystallographic parameters for asphaltenes is based on numerous assumptions and also depends upon the approach in the

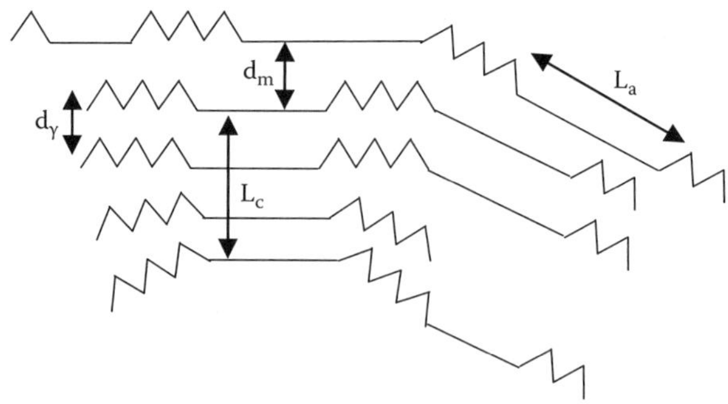

FIGURE 1.33 Cross-sectional view of asphaltene model; zigzag structures represent the configuration of alkyl chains or naphthenic rings and the straight lines represent the edge of flat sheets of condensed aromatic rings. (From Schwager et al., 1983. *Anal. Chem.* 55 (1): 42–45. With permission.)

data treatment (Andersen et al., 2005). In the interpretation of the data, special care must be taken in oversimplifying the very complex asphaltenic system. The stacking is only one part of the asphaltene aggregates, and more complex association schemes must be considered. The 002 peak for most asphaltenes is located around 26° indicating an interlayer spacing of approximately 3.55 Å. Different interlayer distances have been reported by several authors (Kumar and Gupta, 1995; Adams et al., 1998; Bouhadda et al., 2000; Hishiyama and Nakamura, 1995; Shirokoff et al., 1997; Buenrostro-Gonzalez et al., 2002; Suresh Babu and Seehra, 1996; and Alvarez et al., 1999) and compiled by Andersen et al. (2005) and shown in Table 1.7.

Tanaka et al. (2004) studied the changes that Maya, Khafji, and Iranian Light asphaltenes and their vacuum residues suffered by heating directly the sample on the x-ray diffractometer specifically designed to incorporate a rotating anode and a high-power x-ray generator. It was found, in general, that with increasing temperature from 30° to 300°C, the value of d_m increased from ~3.5 Å to ~3.6 Å and L_c decreased from ~26 Å to 16 Å and, hence, M decreased from 8 to 5 aromatic layers. Figure 1.34 shows a diffractogram of Maya asphaltenes with varying temperature of analysis. It can be noticed that as temperature increases, the graphene band tends to disappear also as consequence of the changes suffered during heating and disruption of stacked aromatic layers. The size of core aggregates deduced from XRD was consistent with the disk-shaped particles, as reported by Lin et al. (1997). Asphaltenes from Maya crude were also analyzed by Trejo et al. (2007) having a d_m value of 3.53 Å; however, the average number of stacked aromatic layers was significantly higher ($M = 20$) measured at room temperature. In this case, only asphaltenes from virgin Maya crude were analyzed, whereas Tanaka et al. (2004) studied asphaltenes from vacuum residue of Maya crude, which indicates the importance of the feedstock and its influence on crystalline parameters of asphaltenes.

TABLE 1.7
Reported Magnitudes of XRD Derived Parameters of Different Carbon and Hydrocarbon Material

Material	d_{002} Band, Å	L_c, Å	L_a, Å
Graphitized Assam coking coal	3.338	220	110
Single-crystal graphite	3.354		
High-purity graphite	3.357	458	324
Thermal graphite	3.364	710	990
Turbostratic film	3.437	16.4	40.4
Semicoke (H/C = 0.75)	3.52	35	
Hassi-Messaoud asphaltenes	3.56	16.5	9.5
Baxterville asphaltenes	3.57	19	10
Arab Berri (H/C = 1.02)	3.60	22.7	13
RT-asphaltene	3.70	14.3	9.9

Source: Andersen, S.I. et al., 2005. *Energy Fuels* 19 (6): 2371–2377. With permission.

1.6.5 Nuclear Magnetic Resonance

Nuclear magnetic resonance (NMR) is a very powerful technique because it allows for different components to be identified. When analyzing complex hydrocarbons, it is really useful to know how many carbons are present both in an are present in

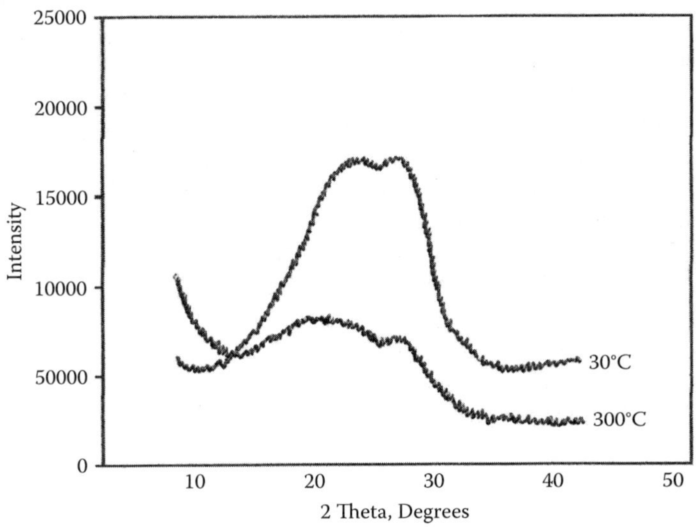

FIGURE 1.34 X-ray diffraction (XRD) patterns of Maya asphaltenes at different temperatures. (From Tanaka et al., 2004. *Energy Fuels* 18 (4): 1118–1125. With permission.)

both aromatic or aliphatic way. One technique used by Patt and Shoolery (1982) and Jakobsen et al. (1982), called the Attached Proton Test used in NMR, consists of the scalar coupling between carbon and a proton that allows differentiation between carbon that is close to an even proton number (quaternary carbons and methylene groups) or an odd proton number (methyne and methyl groups). Classical analysis by means of NMR is commonly used and it is called the "direct ^{13}C method." With this method, identification of aromatic and aliphatic carbons is possible. Aromatic carbons (quaternary aromatic carbon) are found by integration of the chemical shift area between 160 and 100 ppm, whereas aliphatic carbons (CH, CH_2, CH_3, quaternary aliphatic carbon) are found at 70 to 0 ppm.

Ibrahim et al. (2003) have used NMR to analyze asphaltenes from Kuwait crudes and concluded that they consist of a number of polycondensed aromatic units, normally between 5 and 9, which are linked by alkyl chains having a length of 4 to 6 carbons with or without heteroatoms. In some studies of hydroprocessing, the aromaticity factor of asphaltenes is almost constant up to 400°C, after that f_a increased smoothly in asphaltenes from reacted product at 430°C (Seki and Kumata, 2000). A notorious change occurred when the temperature during hydrotreating is higher than 400°C indicating that the shortening of alkyl chains attached to aromatic core is taking place. This is in accordance with Merdrignac et al. (2006), who stated that remaining carbon in asphaltenes during conversion becomes more aromatic, increasing the aromaticity. The increase in the aromatic carbon content is explained in terms of faster conversion of aliphatic chains.

Asphaltenes from Maya vacuum residue were analyzed by Zajac et al. (1994) with NMR. Aromatic carbon was divided into three types: (1) peripheral aromatic carbons attached to protons, (2) peripheral aromatic carbons attached to aliphatic carbon, and (3) internal or bridgehead carbons. All of these carbons are necessary to suggest the size of an average aromatic cluster. The condensed ring sections of asphaltene sample gave an average dimension of 11.1 Å ± 1.4 Å. A possible schematic representation of asphaltenes corresponds to a condensed aromatic core having approximately nine aromatic rings with attached naphthenic rings where sulfur and nitrogen structures are forming part of the core. The main structure for asphaltenes from Maya vacuum residue is continental type.

Structural data can be obtained from NMR spectra by applying the following equations, as reported by Calemma et al. (1995):

$$f_a = \frac{C_{\text{aromatic}}}{C_{\text{aromatic}} + C_{\text{aliphatic}}} \tag{1.19}$$

where f_a is the aromaticity factor, $C_{aromatic}$ and $C_{aliphatic}$ are the amount of aromatic and aliphatic carbon, respectively,

$$n = \frac{C_{\text{aliphatic}}}{C_{\text{substituted aromatic carbon}}} \tag{1.20}$$

where n is the average length of alkyl chains and $C_{\text{substituted aromatic carbon}}$ is all aromatic carbon which has an alkyl substituent attached to an aromatic core,

$$A_s = \frac{\text{Percent of substituted aromatic carbon}}{\text{Percent of nonbridge aromatic carbon}} \times 100 \tag{1.21}$$

where A_s is the percentage of substitution of aromatic rings,

$$R_a = \frac{C_{\text{aromatic}} - (C_{\text{unsubstituted aromatic carbon}} + C_{\text{substituted aromatic carbon}})}{2} - 1 \tag{1.22}$$

where R_a is the number of aromatic rings, $C_{\text{unsubstituted aromatic carbon}}$ is all aromatic carbon without any alkyl substitution. Other equations to calculate the substitution and condensation index are (Merdrignac et al., 2006)

$$SI = \frac{C_{\text{quat-sub}}}{C_{\text{aro-total}} - C_{\text{quat-cond}}} \tag{1.23}$$

where SI is the substitution index, $C_{quat\text{-}sub}$ is a quaternary carbon substituted by an alkyl chain, $C_{\text{aro-total}}$ is the total aromatic carbon content, and $C_{\text{quat-cond}}$ is the total amount of quaternary carbons linked to other quaternary carbons,

$$CI = \frac{C_{\text{aro-total}} - C_{\text{quat-cond}}}{C_{\text{aro-total}}} \tag{1.24}$$

where CI is the condensation index. Figure 1.35 is an example of different types of carbon.

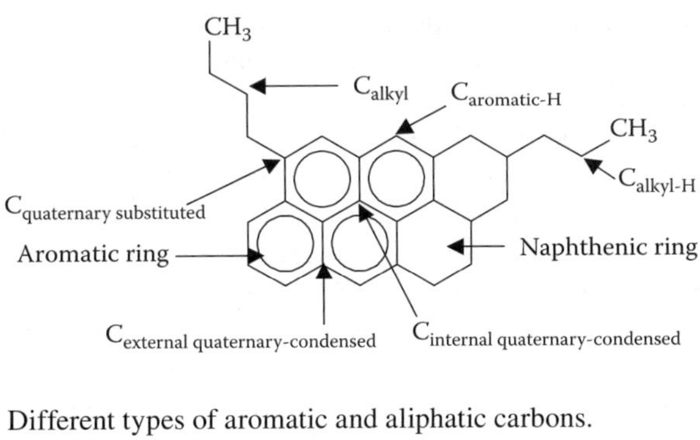

FIGURE 1.35 Different types of aromatic and aliphatic carbons.

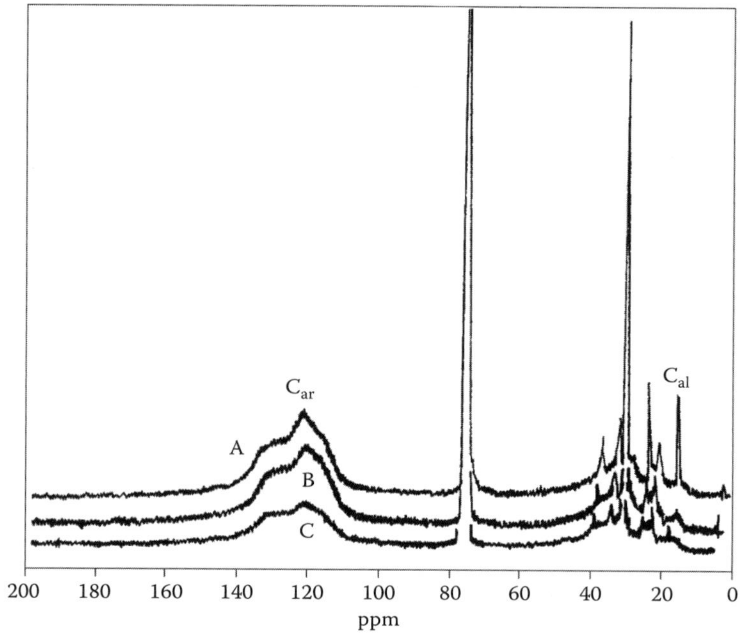

FIGURE 1.36 NMR spectra of asphaltenes using *n*-pentane as solvent: (A) Olmeca, (B) Isthmus, and (C) Maya. (From Ancheyta et al., 2002. *Energy Fuels* 16 (5): 1121–1127. With permission.)

In spite of having different geographical origins, many asphaltenes present similar *CI* and *SI* with around seven condensed aromatic cycles (Calemma et al., 1995).

Regarding different geographical origin, asphaltenes from Maya, Isthmus, and Olmeca crude have been analyzed by NMR (Ancheyta et al., 2002), as shown in Figure 1.36. The use of different solvents to precipitate asphaltenes has influences on chemical properties as well. When using heptane, more aromatic asphaltenes with larger alkyl chains are obtained as measured by NMR. In addition, aggregate molecular weight by VPO is also higher when using heptane compared with pentane (Table 1.8).

Zhang et al. (2007) have analyzed resins and asphaltenes by NMR to quantify the average number of aromatic rings and obtained values of 5.6 and 8.2, respectively. The aromatic core in asphaltenes is mainly built by benzene-polycarboxylic acids giving a peri-condensed structure. Bansal et al. (2007) conducted analyses of Residue Fluid Catalytic Cracking (RFCC) samples and found that alkyl chain length of aromatic and saturate fractions is between the normal and isoparaffinic hydrocarbons (normal paraffinic/isoparaffinic ~1.2-1.4), whereas the substitution index of aromatic rings reaches 47% and the condensation index is around 25%.

In conclusion, by combining structural information obtained by NMR, functional group analyses, and that obtained by x-ray diffraction, it is possible to

TABLE 1.8
Structural Parameters for Average Molecule of Asphaltenes

	Maya		Isthmus		Olmeca	
Property	$n\text{-}C_5$	$n\text{-}C_7$	$n\text{-}C_5$	$n\text{-}C_7$	$n\text{-}C_5$	$n\text{-}C_7$
Aggregate MW (VPO)	3680	5190	2603	3375	1707	2663
f_a	0.47	0.52	0.59	0.57	0.61	0.62
n	7.4	6.8	4.8	5.0	4.1	5.5
R_a	35	62	34	45	24	40
A_s	35.6	38.9	37.9	41.0	39.1	32.9

Source: Ancheyta, J. et al., 2002. *Energy Fuels* 16 (5): 1121–1127.

derive hypothetical average structures for asphaltene molecules by which NMR is one of the most powerful tools to understand asphaltene structure at the molecular level.

1.6.6 Small Angle Scattering

Asphaltenes are present in crude oils as a colloidal structure and an appropriate technique for characterizing these complex systems is by the use of "small angle neutron and x-ray scattering." Analysis can be performed in different solvents capable of dissolving asphaltenes from heavy crude or its vacuum residue from different geographic origins at high temperatures (Espinat et al., 1993). It is very important to completely understand the colloidal nature of heavy crudes and their products especially where problems created during production, transport, and refining of the crude occur. There is a lack of knowledge related with the molecular weight of heavy fractions, composition, and size of aggregates and the evolution and changes that heavy fractions, including asphaltenes, suffer with increasing temperature or pressure during the processing of the crude in refineries. A good tool for analyzing complex fraction of heavy crudes is small angle scattering, which can be by means of x-ray or neutron scattering. Small angle x-ray scattering (SAXS) is an efficient technique that determines in aromatic rings, whereas small angle neutron scattering (SANS) provides information related to the volume of the particles (Bardon et al., 1996). SANS can be used to determine the shape and size of asphaltenes in different solvents at the expense of the nature of solvent, concentration, and temperature where measurements are carried out.

Asphaltenes in solution can be represented as polymer-like solutions in which asphaltenes are solvated and resins surround them (Laux et al., 1997; Hirschberg et al., 1984). It has been also proposed that asphaltenes are suspended in maltenes as a solvent; however, the system was considered to be a molecular solution rather than a suspension of small particles (Altgelt and Harle, 1975).

To study the behavior of the aggregated asphaltenes different kinds of analyses have been applied. For example, Fenistein and Barré (2001) analyzed (using SAXS) asphaltenes separated by ultracentrifugation, finding that the radii of gyration and molar mass of the separated asphaltenes range over a wide scale. Savvidis et al. (2001) investigated the internal structure of the asphaltenes powder concluding that the aggregates are clearly a compact organization of asphaltenic material that is subject to sedimentation, which induces the visible macroscopic phase separation. Xu et al. (1995) studied Athabasca asphaltenes with concentrations of 5 and 15 wt% dissolved in toluene at room temperature and found that sample was polydispersed spheres with a radius of 33 Å. Pierre et al. (2004) investigated the influence of asphaltene concentration on the structure determined by this technique and found that asphaltenes have an identical structure no matter the concentration. Other studies with SAXS have been carried out by Giavarini et al. (2000), Barré et al. (1997), Bardon et al. (1996), Espinat (1991), Espinat et al. (1984), and Herzog et al. (1988).

SANS has been applied by Gawrys and Kilpatrick (2005), who found that the shape of asphaltenic aggregates obtained from Hondo, Canadon Seco, and Arab Heavy asphaltenes is better approached by an oblate cylinder model. Spiecker et al. (2003) carried out tests in which fractionation with *n*-heptane and toluene was firstly applied to Hondo, B6, Canadon Seco, and Arab Heavy asphaltenes and then SANS of the aggregates was run. It was found that the less soluble fractions formed aggregates that were considerably larger than the more soluble fractions. Asphaltene aggregate size increased with decreasing solvent aromaticity up to the solubility limit and, after that, the aggregate size decreased with adding *n*-heptane. The solubility mechanism for Canadon Seco asphaltenes was reported to have the largest aggregates, which could be influenced by aromatic π-bonding interactions due to the low H/C ratio and low nitrogen content. B6 and Hondo asphaltenes formed aggregates with similar size. In this case, the aggregation phenomenon is related to polar interactions due to the high H/C atomic ratio and nitrogen content. It was found that Arab Heavy asphaltenes had the smallest aggregates in the mixture of *n*-heptane and toluene. In this case, the combination of some other tools, such as elemental analysis and VPO for molecular weight, helped to better understand the mechanism by which asphaltenes aggregate in presence of *n*-heptane and toluene.

Fenistein et al. (1998) also performed fractionation with *n*-heptane and toluene along with intrinsic viscosities. When increasing the volume percentage of heptane in the mixture of heptane and toluene, it has been observed that intrinsic viscosities of asphaltene aggregates diminished with low heptane percentage (10 to 20 vol%) and then it increased as the flocculation point was reached. By using SANS, it was revealed that asphaltene aggregates possess an opened structure in toluene that becomes slightly denser after adding a small amount of *n*-heptane. When *n*-heptane is greater than 15 vol% in the mixture, there is a growth in the molecular weight and size of the aggregates, which continues up to the precipitation threshold.

Ravey et al. (1988) carried out an interesting comparison on how asphaltenes modify their size and morphology after catalytic hydrotreating. They found that hydrotreating Boscan crude and studying the size of its asphaltenes before and after reaction that a dramatic change occurred. The concentration of asphaltenes in studied solution of tetrahydrofuran was 1.1 wt% for hydrotreated and nonhydrotreated asphaltenes and the average diameter of disc-type aggregates diminished down to 0.9 nm taking into account that the mean diameter size of asphaltenes dispersed in a good solvent is in the range of 6 to 20 nm. These observations are important for designing a proper catalyst for hydrotreating. Thus, deep characterization of asphaltenes can be used for a better design of hydrotreating catalysts to process heavy feedstocks because the size and shape of asphaltenes play an important role in the catalyst textural properties. Asphaltenes can be polydisperse in nature and, for this reason, a good catalyst for hydroprocessing would include macropores to allow for the diffusion of these particles with big sizes to go inside the pores (Plumail et al., 1983). Liu et al. (1995) conducted experiments with asphaltenes from Ratawi crude in concentrations up to 80 wt% in toluene. This study analyzed asphaltenes in solution at high concentration. Several other SANS analyses have been carried out by Roux et al. (2001), Sheu et al. (1992), Storm and Sheu (1995), Storm et al. (1993), and Overfield et al. (1988; 1989).

SAXS and SANS are commonly applied with asphaltenes dissolved in different substances. In this regard, Tanaka et al. (2003) studied the changes in the structures of petroleum asphaltene aggregates *in situ* with SANS isolating asphaltenes from Maya, Khafji, and Iranian Light crudes that were dissolved in different solvents (decalin, 1-methylnaphthalene, or quinoline). SANS analyses were measured at temperatures from 25 to 350°C and showed different topological features, which were different depending on the asphaltenes' origin or solvent used. Asphaltenes aggregate in all solvents as a prolate ellipsoid at 25°C and become smaller with increasing temperature. For example, a compact sphere with a radius of 25 Å at 350°C was observed. In this case, size and morphology of asphaltenes were observed to be dependent on the temperature at which analysis was performed. In the case of Maya asphaltenes, a fractal network was created using decalin as the solvent. This structure remained even at 350°C. The fractal network may be related to the high coking tendency of Maya asphaltenes. According to Thiyagarajan et al. (1995), when changing the solvent to 1-methylnaphthalene, the aggregates appear to be rodlike-shaped having dimensions of 2×15 nm. When using toluene as a solvent, aggregates of asphaltenes are spherical having a radius near 5 nm (Sheu et al., 1992).

Maya asphaltenes in 1-methylnaphthalene showed that the colloidal particles were rod-shaped at lower temperatures. The rod-shaped particles possess uniform radius, but they are polydispersed in their length. When analyzing solutions of asphaltenes dissolved in that solvent at room temperature, asphaltenes aggregate as rod-like particles having a radius of 18 Å, but with variable length going to 500 Å. As temperature of analysis is increased to 100°C, the length of the aggregates decreases significantly, but the radius remains almost the same. When temperature reaches 150°C, two kinds of particles are present: one is spherical having a radius of 12 Å and the other is ellipsoidal with semiaxes of 33 and 12 Å. However, with increasing the temperature to 320°C, the ellipsoidal

FIGURE 1.37 Shape of asphaltenes (5 wt%) in 1-methylnaphthalene by SANS at different temperatures. (From Hunt and Winans, 1999. Paper presented at 218th American Chemical Society National Meeting, New Orleans, Aug. 22–26. With permission.)

particles are diminished in concentration. From 340 to 400°C, only spherical particles with a radius of 12 Å are present. When cooling the sample to room temperature, no signal was observed indicating irreversibility during the process (Thiyagarajan et al., 1995). Measurements of viscosity of asphaltenes dissolved in benzene have also taken into account ellipsoidal particles considering that asphaltenes are solvated by one molecular layer of benzene, as reported by Takeshige (2001). Changes at high temperature could imply that covalent bonds formation is taking place. A schematic view of the size and shape of asphaltene particles as function of temperature of analysis is given in Figure 1.37 (Hunt and Winans, 1999).

Results obtained with asphaltenes dissolved with different solvents give us an idea about their size and shape and the morphological changes they suffer when analyzed at high temperature. For example, molecular weight decreases as temperature of analysis increases using organic solvents (Thiyagarajan et al., 1995; Roux et al., 2001). However, these changes could not be the same when treating with real crude or its residue, even with differences in their morphologies with respect to those reported for asphaltenes dissolved in a number of solvents. To bring this issue to light, some studies with asphaltenes dissolved in resins and with vacuum residue were completed by Espinat et al. (1993) and Bardon et al. (1996). SAXS studies have shown that asphaltenes are present, similar to disk-type systems within resins suspensions. According to Espinat et al. (1993), asphaltenes could be the result of the association of various lamellar particles containing aromatic and paraffinic sections. The aggregation is due to the dipole–dipole interactions, according to Maruska and Rao (1987) who measured the dielectric constants of asphaltenes to be ranging from 5 to 7 and having more than one dipole per asphaltene molecule. The diameter of the dipole is comprised from 3 to 6 Å. Bardon et al. (1996) have confirmed the disc-shaped morphology of aggregates of asphaltenes by viscosimetry. SAXS analysis determined that the Safaniya vacuum residue diluted with toluene (10 and 34 wt% w/w) behaves like pure solutions of asphaltenes and resins in this solvent. However, when analyzing pure Safaniya vacuum residue, fluctuations in electronic density are present due to the different kinds of molecules, including aromatics and paraffins. It was also compared to the behavior of a vacuum residue artificially done that contains around 14 wt% of asphaltenes and 86 wt% of maltenes with pure Safaniya vacuum residue. It found good correspondence at low Q-values, but a big discrepancy was obtained at high Q-values indicating that the artificial system did not behave identically to the pure vacuum residue. The scattering measured at 200°C is very close to that obtained at room temperature. Density fluctuations are present even at high temperature. At room temperature, fluctuations can be due to aggregation of asphaltenes and resins or paraffins crystallization. At 300°C, there is an increase of scattered intensity as a consequence of probable thermal cracking in some regions, which shows less-dense regions being similar to liquid density. A network is created most likely by a three-dimensional arrangement created by associations; however, at higher temperatures, there is a Newtonian behavior of the vacuum residue because some links are broken in the regions with high density.

From data derived from SANS, SAXS, and XRD, Tanaka et al. (2004) have proposed an hypothetical asphaltene aggregate hierarchy based on: (1) core aggregates formed by π–π interactions giving stacked asphaltenes with an average size of 20 Å; (2) medium aggregates, which are secondary aggregates of core aggregates as result of interactions with the medium, maltenes, oils, or solvents that have a size of 50 to 500 Å; and (3) fractal aggregates, which are secondary aggregates of core aggregates that result from diffusion-limited cluster aggregation (DLCA) or reaction-limited cluster aggregation (RLCA), independent of any media having sizes bigger than >1000 Å. A model was proposed based on these assumptions as shown in Figure 1.38.

1.6.7 Microscopic Analysis

Microscopic analysis is a very useful characterization technique for understanding the structure and changes that asphaltenes suffer at the molecular level. Microscopic characterization has been used earlier to identify the structure of bitumen or asphaltenes, as reported by Dickie et al. (1969) and Donnet et al. (1977). High resolution transmission electron microscopy (HRTEM), scanning electron microscopy (SEM), scanning tunneling microscopy (STM), and confocal laser-scanning microscopy can be used to obtain a detailed picture of asphaltenes.

Analyses by HRTEM with model compounds have been performed by Sharma et al. (2002). These model compounds have similarities with asphaltenes because they have aromatic structures and in some cases an attached alkyl chain is present. It was concluded that alkyl chains are responsible for the rupture of the stacking. Similar observations can be applied to asphaltenes that have alkyl chains with different length. Camacho-Bragado et al. (2002) have studied the asphaltene structure of a purified sample by removing resins, and using scanning transmission electron microscopy (STEM), determined that the main heteroatoms that constitute asphaltene structure were S, V, and Si. With the use of HRTEM, it is possible to measure the interlayer distance between aromatic sheets. These aromatic sheets are the same that are detected by XRD in the graphene band corresponding to the stacked aromatic cores. The authors found that the distance between aromatic layers by HRTEM was 0.39 nm corresponding to amorphous structures. The image viewed was similar to a cauliflower structure. However, when a same point over the asphaltene structure

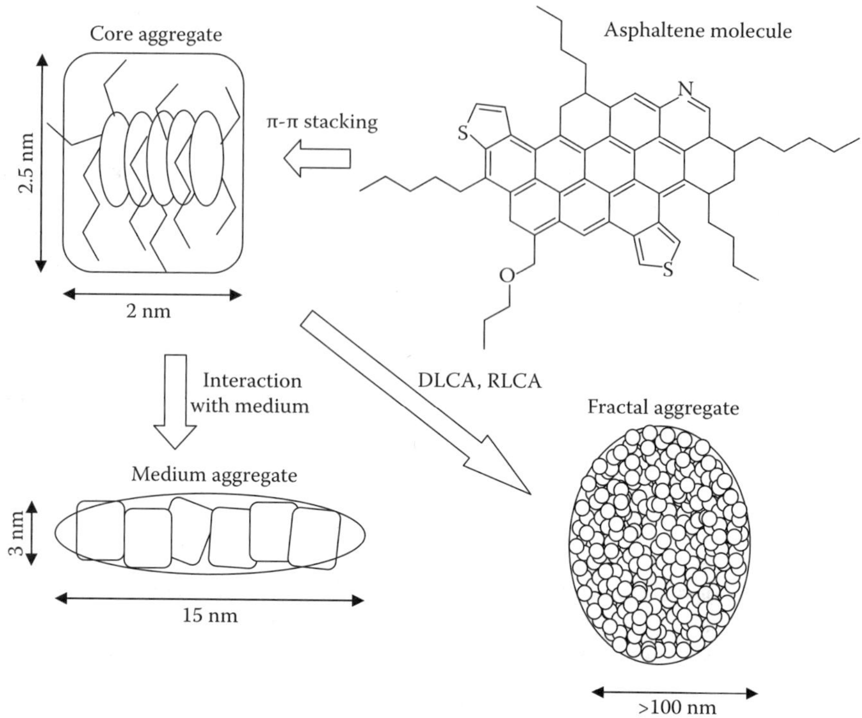

FIGURE 1.38 Hypothetical representation of the hierarchy in asphaltene aggregates based on XRD, SAXS, and SANS data. (From Tanaka et al., 2004. *Energy Fuels* 18 (4): 1118–1125. With permission.)

is electron irradiated, a modification of the structure was found as the irradiation time increases. The initial structure was a cauliflower-type fullerene, but the same structure, after 1 min, changed periodically to become an onion fullerene. Finally, this structure disintegrated. The changes in asphaltene structure are due to destruction of aliphatic chains caused by electron irradiation that allows the aromatic core to stack and form another type of structure. In other regions of the sample, a two-shell structure also has been identified having a size of 1.5 nm and it has been assigned to a $C_{60}-C_{240}$ molecule. This has been supported on observations made by Mordkovich et al. (2000) who reported the existence of a two-shell fullerene with a size of 1.4 nm, which is very similar to that reported by Camacho-Bragado et al. (2002).

Trejo et al. (2009) have observed rounded edges morphology with concentric structures having radius of approximately 10 nm in asphaltenes from Maya crude, which are not very well defined in the outer layers. This morphology is due to the presence of alkyl chains that impede the stacking of aromatic layers for a long extent, giving a poorly ordered arrangement, as shown in Figure 1.39. They also identified changes in the asphaltene morphology after electron irradiation applied to virgin Maya asphaltenes, as shown in Figure 1.40 and reported by

FIGURE 1.39 TEM image of Maya crude asphaltenes. (From Trejo et al., 2009. Energy *Fuels*. 23(1): 429–439. With permission.)

Camacho-Bragado et al. (2002). In this case, asphaltenes were washed with heptane for 15 h in Soxhlet equipment for removing resins. Figure 1.40a shows the original sample with a tangled structure with edges similar to a cauliflower. After 5 min (Figure 1.40b) under electron irradiation, it is observed that the sample is now less tangled at the edge and the initial structure suffered substantial changes that resulted finally (Figure 1.40c) in another type of rearrangement with very defined and sharp edges without tangled structures at the edge. Asphaltenes were also purified with a mixture of solvents (toluene and heptane) under Soxhlet reflux to separate asphaltenes by solubility and to study their morphology. Two different mixtures of solvents were used (67/33 and 33/67 toluene/heptane in volume basis). The first mixture of solvents separates the most insoluble asphaltenes and the second mixture, the most soluble ones. However, it is convenient to analyze the changes that asphaltenes exhibit as a consequence of the use of different solvents for precipitation. Figure 1.41 shows that the most insoluble asphaltenes present are tangled structures similar to a cauliflower. Structures are not well defined and poor stacking is observed only at the edges because asphaltenes still preserve their alkyl chains that are responsible for disrupting the stacking. The interlayer distance is 0.353 nm, which agrees with the XRD for the d_{002} graphene band in the aromatic section (Andersen et al., 2005). Figure 1.42 reveals the poor stacking exhibited by the more soluble asphaltenes. Only some layers are piled up with the characteristic dimensions of amorphous asphaltenes (~0.353 nm). It was not possible to distinguish any well-ordered structure in this asphaltene sample and it can be stated that purification with solvents separated more tangled structures in the case of most insoluble asphaltenes compared with the most soluble ones. Cauliflower-like structures were not present in this sample, which means that

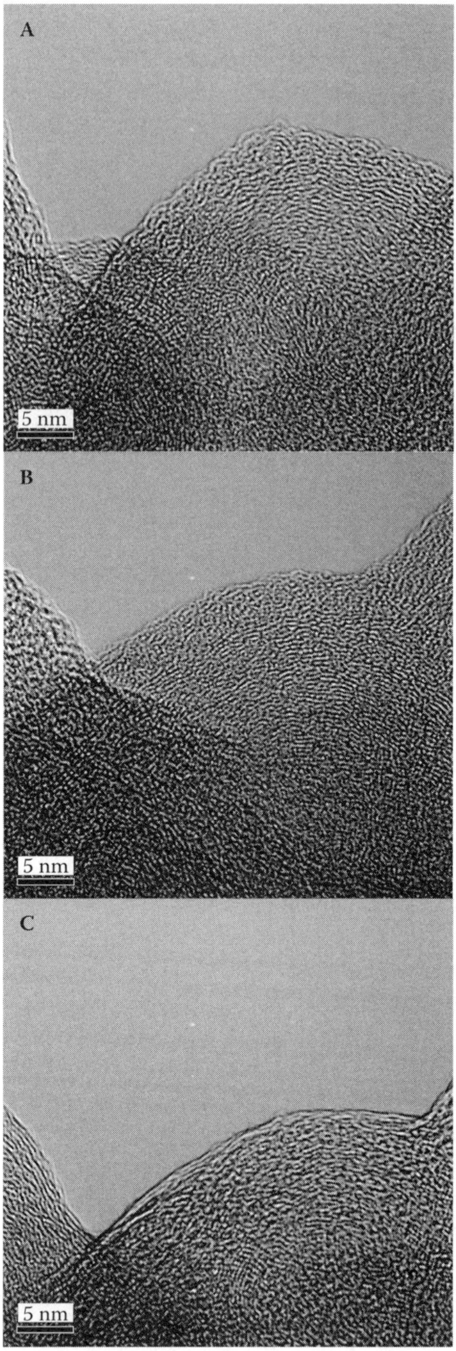

FIGURE 1.40 Modification of the structure of asphaltenes after some minutes under electron irradiation. (From Trejo et al., 2009. *Energy Fuels.* 23(1): 429–439. With permission.)

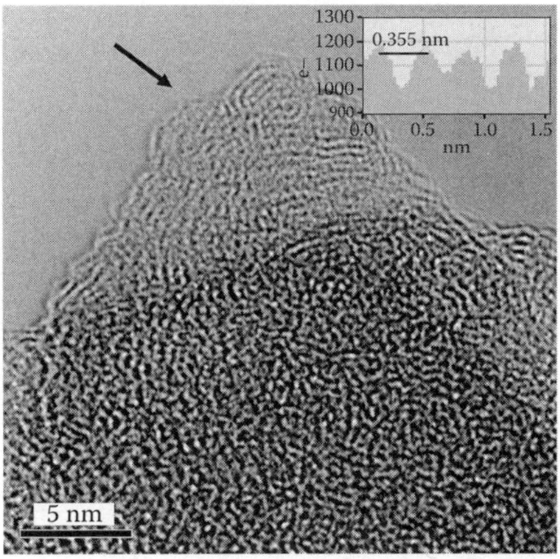

FIGURE 1.41 TEM image of asphaltenes from Maya crude washed with a mixture of toluene/heptane (67/33 vol%). (From Trejo et al., 2009. *Energy Fuels*. 23(1): 429–439. With permission.)

these structures are more commonly separated in the heaviest fraction (separated with a mixture of 67/33 toluene/heptane).

Some attempts have been carried out to compare structural parameters obtained from XRD and HRTEM, but only a qualitative comparison can be made (Sharma et al., 2000). For comparison purposes, only fringes in the linear portion obtained at high resolution by HRTEM must be considered because the layer diameter is larger when measured with HRTEM than with XRD. For this reason, HRTEM images need to be taken in a thin section that partly transmits the electron beam.

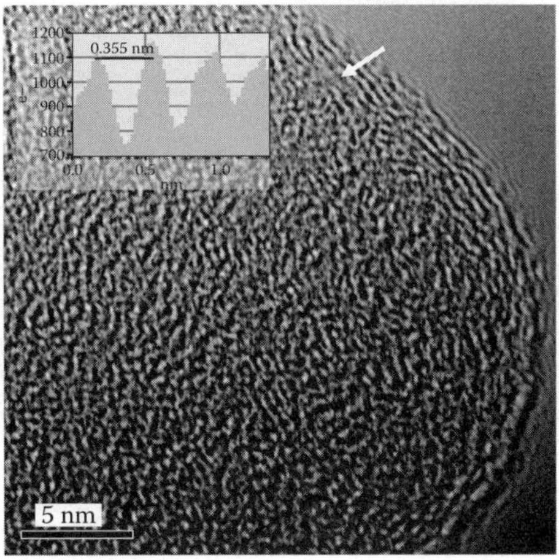

FIGURE 1.42 TEM image of asphaltenes from Maya crude washed with a mixture of toluene/heptane (33/67 vol%). (From Trejo et al., 2009. Energy *Fuels*. 23(1) 429–439. With permission.)

The thickness of the sample can originate overlapping, especially when the sample is not fully ordered (Oberlin, 1989).

The use of low vacuum scanning electron microscopy on asphaltenes has been reported by Pérez-Hernández et al. (2003) who observed one type of asphaltene being highly porous and another type to be a structure with a smooth surface. Along with TEM analysis, the authors stated that micelles have a size between ~350 and ~550 nm. Sánchez-Berna et al. (2006) also observed two morphologies of asphaltenes, i.e., one is a compact structure and the second one is a porous structure. The elemental composition obtained by energy dispersive spectroscopy (EDX) identified metallic particles principally like Na, Ca, Fe, Al, Cr, K, V, Si, Ti, Ni, Mg, Cu, and P. Camacho-Bragado et al. (2001) detected particles rich in Si and S in the range from 10 to 30 μm and Mg, Cl, K, V, and Fe in minor quantities. The use of different solvents for precipitating asphaltenes is the main variable that controls their morphology. The size of particles ranged from 230 × 130 μm to 730 × 240 μm. These particles presented both smooth and rough surfaces.

Trejo et al. (2009) have identified different morphologies of asphaltenes from Maya crude by SEM, as shown in Figure 1.43. Case *a* corresponds to asphaltenes having irregular particles on its surface. Dimensions of irregular particles are ~8 × 5 μm and of cylindrical particles ~6.1 μm in length and ~1.5 μm in diameter. Case *b* corresponds to irregular-shaped agglomerates of asphaltenes that may occur due

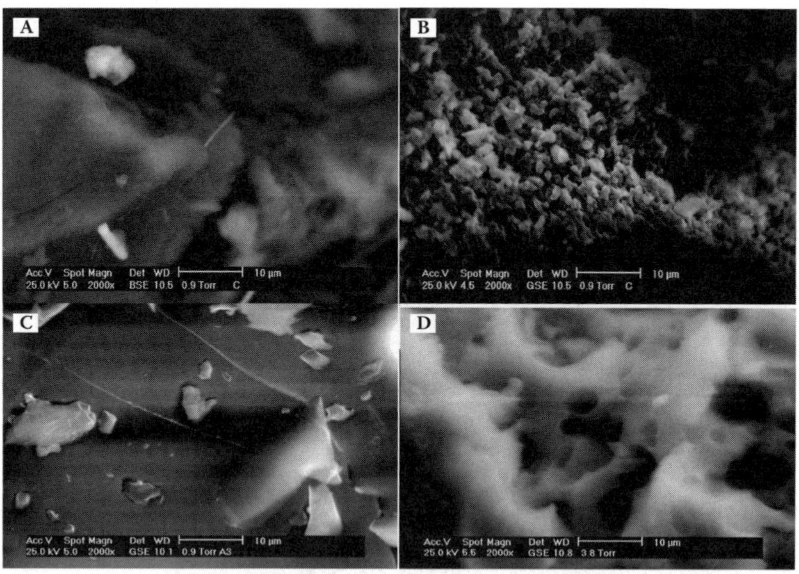

FIGURE 1.43 SEM images of asphaltenes: (A) Asphaltenes having irregular particles on its surface, (B) aggregates of asphaltenes (both washed with n-C_7 under reflux for removing resins), (C) asphaltenes washed with a mixture of toluene/heptane (67/33 vol%), and (D) asphaltenes washed with a mixture of toluene/heptane (33/67 vol%). (From Trejo et al., 2009. *Energy Fuels.* 23(1): 429–439. With permission.)

to the presence of resins. In cases *a* and *b*, asphaltenes were washed with n-C_7 under reflux during 15 h for removing resins. Case *c* shows asphaltenes washed with a mixture of toluene/heptane (67/33 vol%) and having a smooth surface with particles of irregular shape deposited on its surface of three different sizes: one corresponds to smaller particles with an average length of ~2.1 μm, intermediate size particles of ~3.4 μm, and bigger particles of ~10.9 μm. Case *d* presents asphaltenes washed with a mixture of toluene/heptane (33/67 vol%). It is possible to see in this case a continuum phase with small cavities. Asphaltenes obtained in case *c* are the most insoluble in the mixture of solvents and are heavier structures. Asphaltenes in case *d* possess cavities as consequence of the removal of smaller molecules, such as resins that leave void spaces. Small cavities have dimensions of ~3.2 μm and bigger ones of ~10.2 μm on average. Particles observed on the surface in case *c* could originally be inside the cavities, but due to their insolubility, they were separated from there.

Elemental analysis by EDX gives the composition of asphaltenes and elemental mapping. Figure 1.44 shows the presence of different elements (C, O, S, V, Ni, Si, P, and Fe) that are evenly distributed. Carbon, sulfur, and oxygen are the most abundant elements, whereas the remaining elements are present in low concentration, especially Ni, V, and Fe, which are in trace amounts. Phosphorous is also present in asphaltenes, but scarce studies have reported the presence of this element. Sánchez-Berna et al. (2006) have reported the presence of P in asphaltenes from Maya crude oil, but it is not clear how it is bound inside the molecule. The nature of phosphorous seems to be inorganic along with calcium and both of them are probably linked together to form calcium phosphates.

Porous structures with spherical particles forming a network have been also observed by SEM, which is due to removal of oil. Measurements of the smallest particles had a diameter around 100 nm (Loeber et al., 1996). Other authors (Sánchez-Berna et al., 2006) have shown SEM images of asphaltenes where the influence of solvent and precipitation temperature is very evident, especially at boiling point of n-heptane, where precipitated asphaltenes form flakes linked among them, as shown in Figure 1.45. Void spaces are due to elimination of oil, which could be occupying those spaces.

Another useful microscopic technique is STM, which is a direct real space probe. Zajac et al. (1997) have performed a detailed microscopy analysis of Maya asphaltenes with a very diluted concentration (0.001 to 0.003 wt%), and, along with ^{13}C NMR, have obtained an approach of asphaltenes structure. They measured separated entities using their method and obtained an average full-width dimension of these entities to be 10.4 Å ± 1.9 Å. Agreement between STM and ^{13}C NMR is achieved because the condensed ring sections of the asphaltene sample obtained with NMR gave an average dimension of 11.1 Å ± 1.4 Å (Zajac et al., 1994). The STM is also an ideal tool for determining the configuration of adsorbates on solid surfaces at the molecular level (Chiang, 1997). Watson and Barteau (1994) observed a kind of periodicity in images of Ratawi asphaltenes, which was mainly attributed to aggregation phenomenon reported with other techniques, such as SANS (Sheu et al., 1991).

Confocal laser-scanning microscopy, along with fluorescence microscopy, allows for observing directly asphaltenes in their natural medium, such as bitumen.

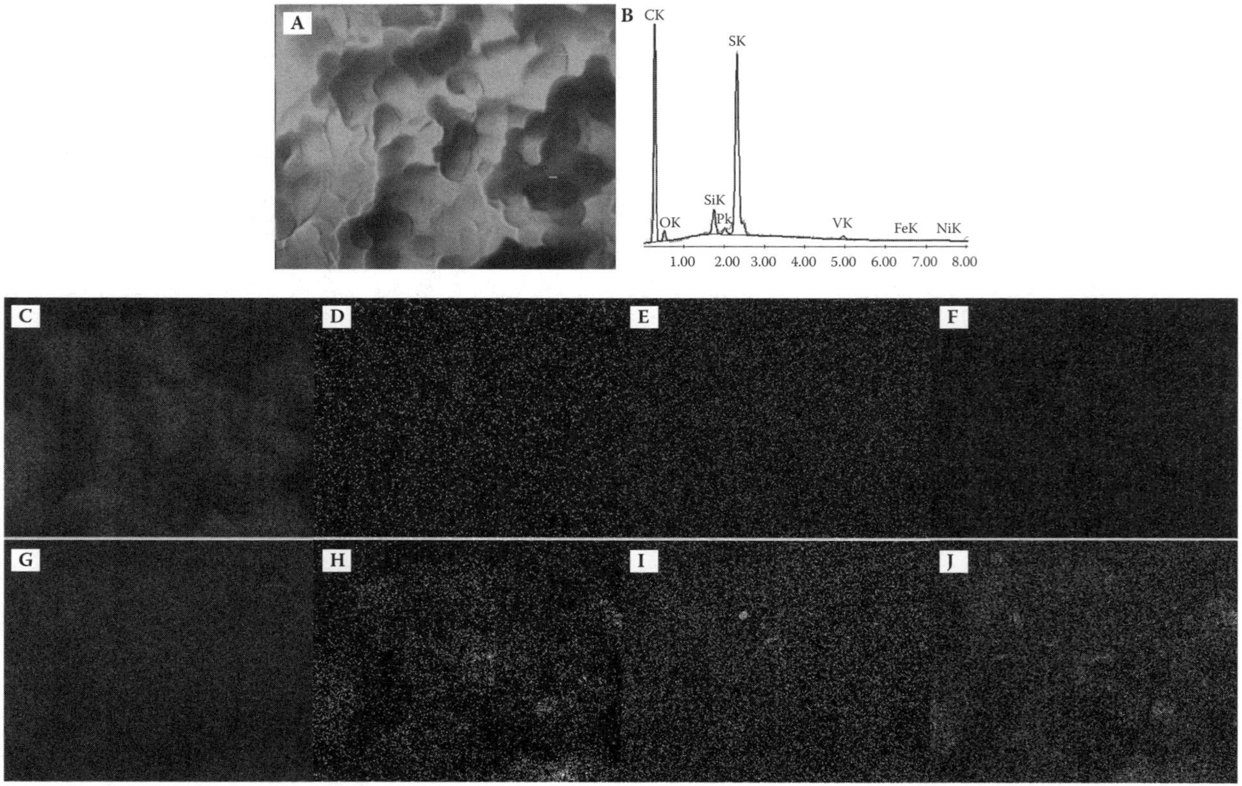

FIGURE 1.44 SEM-EDX elemental mapping of different elements. Virgin asphaltenes: (A) original picture, (B) elemental analysis, (C) C*k* mapping, (D) Ni*k* mapping, (E) V*k* mapping, (F) Fe*k* mapping, (G) S*k* mapping, (H) O*k* mapping, (I) P*k* mapping, and (J) Si*k* mapping. (From Trejo et al., 2009. *Energy Fuels*. *23*(1): 429–439. With permission.)

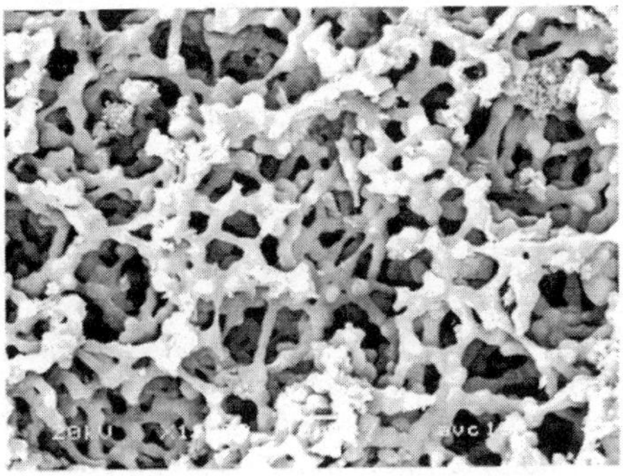

FIGURE 1.45 Asphaltenes showing flake chains. Void spaces are due to de-oiling process. (From Sánchez-Berna et al., 2006. *Petrol. Sci. Tech.* 24 (9): 1055–1066. With permission.)

Bearsley et al. (2004) reported that fluorescence properties of asphaltenes can be used to directly observe their level of dispersion and structure in bitumen. The size of fluorescing particles detected ranged from 2 to 7 μm. Atomic force microscopy (AFM) is another technique that can provide atomic and molecular resolution (Binnig et al., 1986; Mou et al., 1996). Bitumen was first observed by Loeber et al. (1996) with AFM. The importance of this technique is that an analyzed sample can conserve its solid-state morphology. When applying this analysis to bitumen, a dispersed phase with a bee-like appearance is observed (Figure 1.46), which is attributed to asphaltenes (Masson et al., 2006). This observation was also made by Pauli et al. (2001).

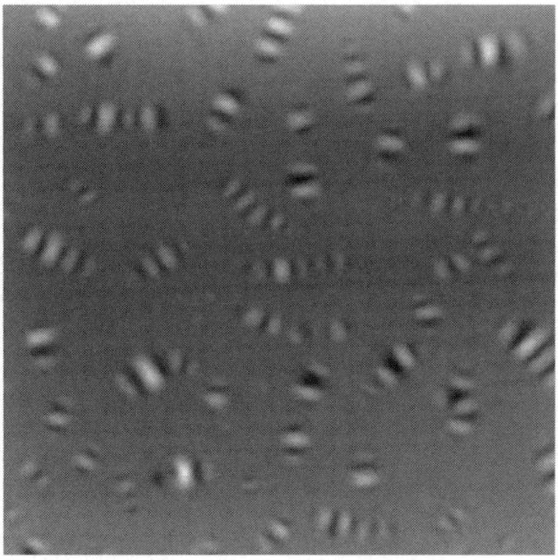

FIGURE 1.46 Topographic AFM image of bitumen (15 × 15 μm). The height variation is ~80 nm. (From Masson et al., 2006. *J. Microscopy* 221 (1): 17–29. With permission.)

In summary, microscopic analysis allows us to study asphaltene structures at the molecular level. Technical advances from optical microscopy have brought new tools for understanding the complex structure of asphaltenes, even in their natural medium and solid states.

1.7 ROLE OF RESINS IN STABILIZING ASPHALTENES

Micelles (aggregates of asphaltene molecules) are surrounded by resins, which act as peptizing agents keeping asphaltenes under colloidal dispersion within crude oil. Resins and asphaltenes exert attraction by means of hydrogen bonds through heteroatoms and dipole–dipole interactions between polar groups in asphaltenes and resins. The most polar part of resins is oriented to the asphaltenes' core and the aliphatic part is in contact with oils. The paraffinic components of the resins make the transition to the relatively nonpolar bulk of the crude where individual molecules also exist in true solution (Hunt, 1996). There is general agreement that more important than asphaltenes amount in the crude is the resins/asphaltenes ratio, which undoubtedly is related to stability of the crude (Hammami et al., 1998).

Resins are components present in deasphalted oil that are strongly adsorbed on surface-active materials, such as Fuller's earth, alumina, silica (Goual and Firoozabadi, 2002), or modified silica (Aske et al., 2001; Bollet et al., 1981) being desorbed with the use of pyridine, chloroform, or toluene-methanol. Resins are considered as the fraction of crude which is soluble in alkanes, but insoluble in liquid propane. They are a type of compounds that cannot be desorbed from a solid support by using straight alkanes (*n*-pentane or *n*-heptane) despite the fact that they are soluble in these solvents. Resins contribute to the stability of petroleum and prevent separation of asphaltenes in the crude. This polar fraction, as in the case of asphaltenes, can generate up to 35 wt% of coke on the catalyst surface leading to deactivation. Resins are necessary for the micelle formation, which contributes to keep the colloidal stability of petroleum (Andersen and Speight, 2001).

Murgich and Strausz (2001) have stated that crude oil cannot be described just as a sol formed by solid asphaltene particles dispersed by resins or as a simple micellar system of asphaltene and resin molecules. Different physical shapes of molecular aggregates could be present, i.e., solid particles formed by asphaltenes and resins and micelles weakly bound, which have short lifetimes. These different aggregates may coexist within the crude oil and exchange components with others.

Resins need to be isolated from crude for analysis; in this regard, standard methods are available for separating crude into different fractions, such as ASTM-D-4124. Resins are obtained as the last effluent obtained from a chromatographic column in which solvents, such as methanol/toluene and trichloroethylene, are used. Once resins have been isolated, analytical tests are carried out in order to understand their composition. Chromatographic fractionation permits a more complete analysis of structural features of resins, such as pointed out by Islas-Flores et al. (2006), who stated that the complexity of resins is due to the fact they are a polydisperse mixture of aromatic, aliphatic, and polar hydrocarbons whose chemical structures change gradually following a continuous distribution.

Different groups are present in resins as in the case of asphaltenes. Infrared spectroscopy has detected species such as acid functions carbonyl, sulfoxide, ether, and ester groups. Pyrroles and indoles are also contained in resins. The multiplicity of functional groups allows resins for experiencing different reactions, for example, oxidation of resins in the presence of benzene yields compounds which are similar to those obtained from asphaltenes (Speight, 1999). Resins can be sulfonated giving oil-or water-soluble products. Reactions with sulfur and nitric acid are possible yielding complex mixtures of products. A summary of reactions that resins suffer has been discussed by Andersen and Speight (2001).

Speight (1999) has reported that the composition of resins does not vary in a long range. The carbon content is around 85 ± 3 wt%, hydrogen content is 10.5 ± 1 wt%, oxygen content is 1.0 ± 0.2 wt%, and nitrogen content is 0.5 ± 0.15wt%. Sulfur content is more variable going from 0.4 to around 5 wt%. Less aromatic carbon is also present in resins and it is inferred either that napthenic rings are present in larger amounts in resins than asphaltenes or alkyl chains are attached to aromatic cores of resins. However, this is not a rule of thumb because long alkyl chains are not always present in resins (Buenrostro-Gonzalez et al., 2001b).

The asphaltenes-to-resin ratio can be used to determine the stability and instability of crude. Resins may be present with asphaltenes because resins adsorb onto asphaltenes' surface from the supernatant liquid or by entrapment during the flocculation process. Successive precipitations are able to remove adsorbed resins. As reported by Speight et al. (1984), the precipitation sequence requires the addition of benzene or toluene to asphaltenes for dissolution (10 mL/g asphaltenes) and then the addition of precipitant to the solution (50 mL per each mL toluene or benzene). This sequence must be applied three times for obtaining asphaltenes free of resins.

If resins are adsorbed onto asphaltenes during the precipitation process, then washing the precipitate must free asphaltenes of resins. In this regard, no standardized methods are available in literature for purifying asphaltenes. Alboudwarej et al. (2002) have compared different washing procedures for removing resins from asphaltenes. Methods such as ASTM-D-4124, IP 143, Speight's methodology for precipitation and purification of asphaltenes (Speight, 1999), and a new washing method in Soxhlet proposed by the authors are compared. They found that with increased washing, asphaltene yield decreased and asphaltene density slightly increased, also the molar mass increased and the solubility of asphaltenes precipitated decreased. An increase in density is a consequence of removing lighter compounds; however, the change in density is less than 1 wt%. Increase in molar mass of asphaltenes is due to elimination of resins making asphaltenes self-associate to a greater extent and the average molar mass increases in consequence of this. The longer the period of washing the higher the molar mass of asphaltenes. Decrease of asphaltenes yield is also observed when washing in Soxhlet because more adsorbed resins are removed. Less soluble asphaltenes were obtained by increasing the washing. This property is likely due to the elimination of resins that originate self-association of asphaltenes, which turn them into heavier compounds. As appointed by Yarranton and Masliyah (1996), asphaltenes are composed by different chemical species. Some precipitated asphaltenes, which are lighter in nature, can be redissolved in the precipitant medium and be soluble in it. These asphaltenes probably constitute the more resinous fraction. When asphaltenes are fractionated, the properties of the most soluble fraction approach resin properties. A comparison of the different methods is shown in Table 1.9.

Speight's method provides the lightest asphaltenes, which is probably due to more resinous asphaltenes as a consequence of less contact time. Longer contact time yields more asphaltenes precipitated, as in the case of Soxhlet washing. The sensitivity of analytical methods could be influenced by the presence of resins on asphaltenes when resins are not completely removed from asphaltenes.

Asphaltenes are polar in nature and, for this reason, when colloidal asphaltene particles are dispersed in nonpolar liquid-like petroleum without using stabilizing agent, particles will flocculate leading to precipitation of asphaltenes. Resins not only stabilize asphaltene dispersion, but they also can cause instability. Resins-to-asphaltenes ratio has been claimed to be an indicator of stability when its value is higher than unity. However, Buenrostro-Gonzalez et al. (2001b) stated that this is not completely true because the stability of the crude also depends on the content of saturates. If saturates are present in high quantity, the medium will be adverse for asphaltenes and precipitation will occur in spite of having high resins-to-asphaltenes ratio. Polarity of resins is intermediate between asphaltenes and crude oil. Resins absorb onto the surface of the asphaltene colloidal particles and act as steric dispersants. For this reason, alkyl-benzene compounds inhibit asphaltene fouling. Figure 1.47a/b explains the stability/instability of asphaltene particles. In case A, resins disperse very well asphaltenes avoiding their aggregation, but in case B, low amount of resins is not enough for preventing asphaltenes association, which leads to asphaltenes precipitation later.

Pereira et al. (2007) have carried out tests in which a dual effect is observed. They concluded that if resins have a weak tendency to adsorb, such as in Hamaca and Guafita crudes, asphaltenes will be stabilized by resins. On the other hand, when resins have a strong tendency to adsorb (as in the case of Furrial crude), asphaltenes tend to become unstable.

In this respect, different compounds and fractions of petroleum that prevent the asphaltenes precipitation are: dodecyl resorcinol > dodecyl benzene sulfonic acid > nonyl phenol > resins > toluene > deasphaltened oil. The poorest inhibitors of asphaltenes precipitation are toluene and deasphaltened oil (Al-Sahhaf et al., 2002). When resins and asphaltenes are present together, resin–asphaltene interactions could be preferred over dimer interactions, i.e., asphaltene–asphaltene or resin–resin (Speight, 2004). Carnahan et al. (1999) confirmed that resins from Boscan crude oil have a stabilizing effect on asphaltenes from this crude and even in Hamaca crude oil.

Molecular mechanics and dynamics calculations can be used for estimating the stabilization energy for asphaltene and resin associates. Association is mainly due to van der Waals forces between molecules. According to Rogel (2000), it is possible to suppose that in the interaction of asphaltenes and resins, very

TABLE 1.9
Comparison of Different Methods for Extracting Asphaltenes and Properties Obtained for Asphaltenes from Athabasca Bitumen

Method	Solvent Addition Step	Equilibrium Step	Washing Step	Yield[a] (wt%)	Solids[b] (wt%)	Density[c] (kg/m^3)	Molar Mass[d] (g/mol)
ASTM D4124	*n*-heptane S/B = 100 cm^3/g 1 h boiling (98°C)	24 h at 23°C	Soxhlet washing at 30°C for 24 h	9.3	5.7	1215	9200
IP 143	*n*-heptane S/B = 30 cm^3/g 1 h boiling	None	Soxhlet washing at 30°C for 1–2 h	8.7	5.6	1203	8300
Speight	*n*-heptane at 23°C S/B = 30 cm^3/g	8-10 h at 23°C	Disperse in toluene and reprecipitate with heptane three times at 23°C	9.2	5.6	1190	6300
Soxhlet	(1) *n*-heptane at 23°C S/B = 40 cm^3/g (2) *n*-heptane at 23°C S/B = 4 cm^3/g	24 h at 23°C for each step	Soxhlet washing at 30°C for 72 h	9.8	5.3	1192	9100

[a] Mass percent of bitumen containing solids.
[b] Mass percent of asphaltenes.
[c] Solid-free asphaltenes.
[d] Molar mass at 10 kg/m^3 of asphaltenes in toluene by VPD.
Solvent-to-bitumen ratio.

Source: From Alboudwarej, H. et al. 2002. *Energy Fuels* 16 (2): 462–469. With permission.

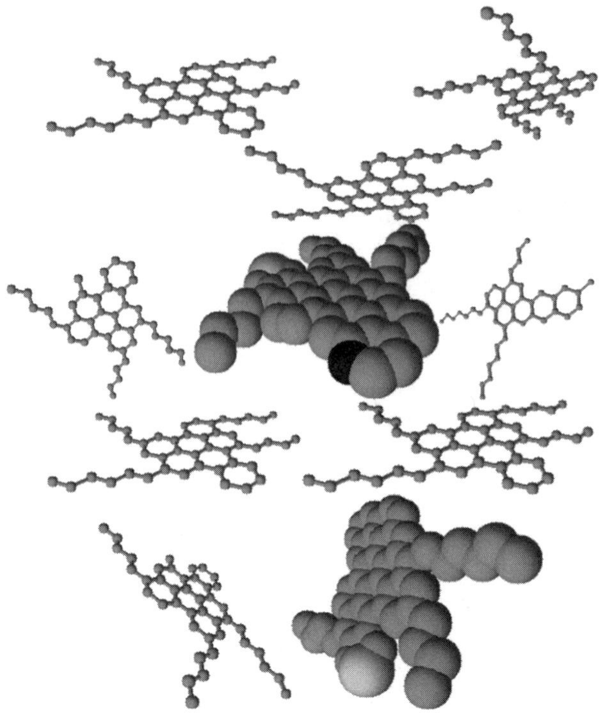

FIGURE 1.47A Effect of resins on asphaltenes, which provide steric repulsion and protect them against aggregation and further precipitation.

few structural changes take place during aggregation. The associations can be established as:

Asphaltene-asphaltene dimer + resins → asphaltene-resin dimer + asphaltene

A quasilinear relationship between the interaction energies and the molecular weight indicating that the van der Waals forces are predominant in the interaction of asphaltenes and resins and with solvents capable to establish hydrogen bonds.

If the molecular weight of resins and lighter compounds in fractions of petroleum, such as saturates and aromatics, is known along with the amount of saturates, aromatics, resins, and asphaltenes, the following equation to determine molecular weight of maltenes from its components (saturates, aromatics, resins) can be established (Vellut et al., 1998):

$$MW_{\text{maltenes}} = \frac{\%malt}{\dfrac{\%res}{MW_{\text{res}}} + \dfrac{\%aro}{MW_{\text{aro}}} + \dfrac{\%sat}{MW_{\text{sat}}}} \tag{1.25}$$

where MW_{maltenes} is the molecular weight of maltenes; *%malt, %res, %aro*, are the weight percent of maltenes, resins, and aromatics, respectively; and MW_{res}, MW_{aro},

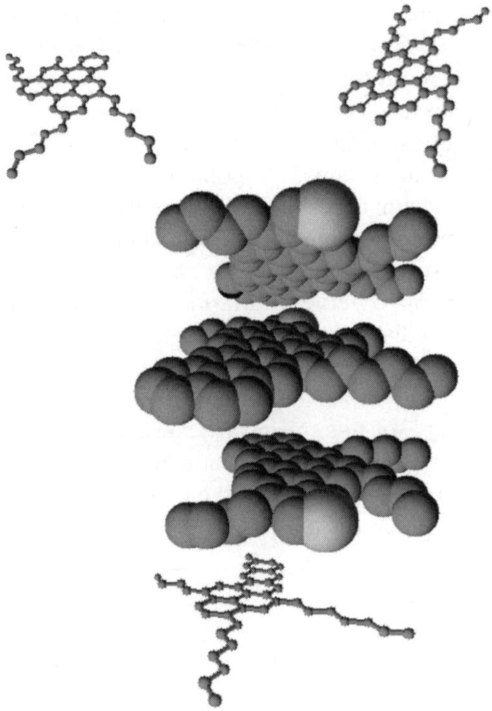

FIGURE 1.47B Representation of unprotected asphaltenes that tend to aggregate due to low amount of resins.

MW_{sat}, are the molecular weight of resins, aromatics, and saturates, respectively. Differences between experimental values of molecular weight of maltenes and those calculated by Equation (1.25) were less than 12%. Molecular weight was obtained by ebulliometry; however, the formula could be applied to other analyses, such as VPO, since molecular weight obtained by this technique is true when dealing with lighter components of petroleum. Special care must be taken when fractionating by applying the SARA (saturates, aromatics, resins, and asphaltenics) technique in order to obtain the correct amount of each fraction.

1.8 CONCLUDING REMARKS

The importance of asphaltenes in crude oils and residua has been discussed in this chapter. Different analytical tools are available to characterize asphaltenes, which include traditional analysis, such as elemental (ultimate) analysis or metals content, and more sophisticated techniques, which involve NMR, XRD, SANS/SAXS, and/or microscopic analysis SEM/TEM that give more information at the molecular level. The understanding of asphaltene structure has changed over time. Nowadays, the central idea about asphaltenes is not the same as two decades ago.

Characterization of asphaltenes provides a powerful tool to comprehend probable changes that asphaltenes will suffer during upgrading of petroleum fractions. Analytical data can be used for improving catalysts or designing new processes compatible with heavier feedstocks processed in refineries.

Molecular weight of asphaltenes is a property which has caused controversy for many years. New findings appear to point to molecular weights smaller than 1,000 g/mol. It is not the aim of this book to state which technique gives the most accurate molecular weight of asphaltenes; instead of that, we decided to discuss briefly some of the most used techniques for obtaining this important parameter. However, as the reader can observe, similarities in molecular weight values have been obtained even with distinct techniques.

It is also very important to remember the role that resins play in stabilizing asphaltenes. Without resins, asphaltenes would be easily separated from the crude and precipitated. Because of this, resins are considered as natural stabilizers of asphaltenes. However, stability of crude is not only dependent on resins, but also on the remaining natural environment of the crude, i.e., saturates and aromatics content and the type of processes used for improving the quality of crudes.

REFERENCES

Acevedo, S., Escobar, G., Gutierrez, L.B., and D'Aquino, J. 1992a. Synthesis and isolation of octylated asphaltene standards for calibration of GPC columns and determination of asphaltene molecular weights. *Fuel* 71 (9): 1077–1079.

Acevedo, S., Escobar, G., Gutierrez, L., and Rivas, H. 1992b. Isolation and characterization of natural surfactants from extra heavy crude oils, asphaltenes and maltenes. Interpretation of their interfacial tension-pH behavior in terms of ion pair formation. *Fuel* 71 (6): 619–623.

Acevedo, S., Gutierrez, L.B., Negrin, G., Pereira, J.C., Mendez, B., Delolme, F., Dessalces, G., and Broseta, D. 2005. Molecular weight of petroleum asphaltenes: A comparison between mass spectrometry and vapor pressure osmometry. *Energy Fuels* 19 (4): 5481560.

Adams, P.M., Katzman, H.A., Rellick, G.S., and Stupian, G.W. 1998. Characterization of high thermal conductivity carbon fibers and a self-reinforced graphite panel. *Carbon* 36 (3): 233–245.

Aguilera-Mercado, B., Herdes, C., Murgich, J., and Müller, E.A. 2006. Mesoscopic simulation of aggregation of asphaltenes and resin molecules in crude oils. *Energy Fuels* 20 (1): 327–338.

Akbarzadeh, K., Alboudwarej, H., Svrcek, W.Y., and Yarranton, H.W. 2005. A generalized regular solution model for asphaltene precipitation from *n*-alkane diluted heavy oils and bitumens. *Fluid Phase Equilibr.* 232 (1-2): 159–170.

Akbarzadeh, K., Dhillon, A., Svrcek, W.Y., and Yarranton, H.W. 2004. Methodology for the characterization and modeling of asphaltene precipitation from heavy oils diluted with *n*-alkanes. *Energy Fuels* 18 (5)1434–1441.

Alboudwarej, H., Akbarzadeh, K., Beck, J., Svrcek, W.Y., and Yarranton, H.W. 2003. Regular solution model for asphaltene precipitation from bitumens and solvents. *AIChE J.* 49 (11): 2948–2956.

Alboudwarej, H., Beck, J., Svrcek, W.Y., Yarranton, H.W., and Akbarzadeh, K. 2002. Sensitivy of asphaltenes properties to separation techniques. *Energy Fuels* 16 (2): 462–469.

Alexander, L.E. and Sommer, E.C. 1957. Systematic analysis of carbon black structures. *J. Phys. Chem.* 60 (12): 1646–1649.

Ali, L.H. and Al-Ghannam, K.A. 1981. Investigations into asphaltenes in heavy crude oils. I. Effect of temperature on precipitation by alkane solvents. *Fuel* 60 (1): 1043–1046.

Al-Muhareb, E.M., Karaca, F., Morgan, T.J., Herod, A.A., Bull, I.D., and Kandiyoti, R. 2006. Size exclusion chromatography for the unambiguous detection of aliphatics in fractions from petroleum vacuum residues, coal liquids, and standard materials, in the presence of aromatics. *Energy Fuels* 20 (3): 1165–1174.

Al-Muhareb, E., Morgan, T.J., Herod, A.A., and Kandiyoti, R. 2007. Characterization of petroleum asphaltenes by size exclusion chromatography, UV-fluorescence and mass spectrometry. *Petrol. Sci. Tech.* 25 (1-2): 81–91.

Al-Sahhaf, T.A., Fahim, M.A., and Elkilani, A.S. 2002. Retardation of asphaltene precipitation by addition of toluene, resins, deasphalted oil and surfactants. *Fluid Phase Equilib.* 194197: 1045–1057.

Altgelt, K.H. and Harle, O.L. 1975. The effect of asphaltenes on asphalt viscosity. *Ind. Eng. Chem. Prod. Res. Dev.* 14 (4): 240–246.

Alvarez, A.G., Martínez-Escandell, M., Molina-Sabio, M., and Rodríguez-Reinoso, F. 1999. Pyrolysis of petroleum residues: Analysis of semicokes by x-ray diffraction. *Carbon* 37 (10): 1627–1632.

Ancheyta, J., Centeno, G., Trejo, F., Marroquín, G., García, J.A., Tenorio, E., and Torres, A. 2002. Extraction and characterization of asphaltenes from different crude oils and solvents. *Energy Fuels* 16 (5): 1121–1127.

Andersen, S.I. 1990. Association of petroleum asphaltenes and related molecular—study on interactions and phase equilibria. PhD thesis, Technical University of Denmark.

Andersen, S.I. 1994. Dissolution of solid Boscan asphaltenes in mixed solvents. *Fuel Sci. Tech. Int.* 12 (11–12): 1551–1577.

Andersen, S.I. and Birdi, K.S. 1990. Influence of temperature and solvent on the precipitation of asphaltenes. *Fuel Sci. Tech. Int.* 8 (6): 593–615.

Andersen, S.I. and Stenby, E.H. 1996. Thermodynamics of asphaltene precipitation and dissolution investigation of temperature and solvent effects. *Fuel Sci. Tech. Int.* 14 (1–2): 231–287.

Andersen, S.I. and Speight, J.G. 2001. Petroleum resins: Separation, character, and role in petroleum. *Petrol. Sci. Tech.* 19 (1–2): 1–34.

Andersen, S.I., Jensen, J.O., and Speight, J.G. 2005. X-ray diffraction of subfractions of petroleum asphaltenes. *Energy Fuels* 19 (6): 2371–2377.

Ascanius, B.E., Merino-Garcia, D., and Andersen, S.I. 2004. Analysis of asphaltenes subfractionated by *N*-methyl-2-pyrrolidone. *Energy Fuels* 18 (6): 1827–1831.

Aske, N., Kallevik, H., and Sjöblom, J. 2001. Determination of saturate, aromatic, resin, and asphaltenic (SARA) components in crude oils by means of infrared and near-infrared spectroscopy. *Energy Fuels* 15 (5): 1304–1312.

ASTM D 3279. Standard test method for n-heptane insolubles.

ASTM D 4124. Standard test methods for separation of asphalt into four fractions.

Bansal, V., Krishna, G.J., Chopra, A., and Sarpal, A.S. 2007. Detailed hydrocarbon characterization of RFCC feed stocks by NMR spectroscopic techniques. *Energy Fuels* 21 (2): 1024–1029.

Bardon, C., Barré, L., Espinat, D., Guille, V., Li, M.H., Lambard, J., Ravey, J.C., Rosenberg, E., and Zemb, T. 1996. The colloidal structure of crude oils and suspensions of asphaltenes and resins. *Fuel Sci. Tech. Int.* 14 (1–2): 203–242.

Barré, L., Espinat, D., Rosenberg, E., and Scarsella, M. 1997. Colloidal structure of heavy crudes and asphaltenes. *Oil Gas Sci. Tech.–Rev. IFP* 52 (2): 161–175.

Bartholdy, J., Lauridsen, R., Mejlholm, M., and Andersen, S.I. 2001. Effect of hydrotreatment on product sludge stability. *Energy Fuels* 15 (5): 1059–1062.

Bartle, K.D., Mulligan, M.J., Taylor, N., Martin, T.G., and Snape, C.E. 1984. Molecular mass calibration in size-exclusion chromatography of coal derivatives. *Fuel* 63 (11): 1556–1560.

Bearsley, S., Forbes, A., and Haverkamp, R.G. 2004. Direct observation of the asphaltene structure in paving-grade bitumen using confocal laser-scanning microscopy. *J. Microscopy* 215 (2): 149–155.

Beck, J., Svrcek, W., and Yarranton, H. 2005. Hysteresis in asphaltene precipitation and redissolution. *Energy Fuels* 19 (3): 944–947.

Behrouzi, M. and Luckham, P.F. 2008. Limitations of size-exclusion chromatography in analyzing petroleum asphaltenes: A proof by atomic force microscopy. *Energy Fuels* 22 (3): 1792–1798.

Bennett, B. and Love, G.D. 2000. Release of organic nitrogen compounds from kerogen via catalytic hydropyrolysis. *Geochem. Trans.* 1: 61–67.

Binnig, C., Quate, F., and Gerber, C. 1986. Atomic force microscope. *Phys. Rev. Lett.* 56 (9): 930–933.

Bollet, C., Escalier, J-C., Souteyrand, C., Caude, M., and Rosset, R. 1981. Rapid separation of heavy petroleum products by high-performance liquid chromatography. *J. Chromatography A* 206 (2): 289–300.

Borrego, A., Blanco, C., Prado, J., Díaz, C., and Guillén, M.H. 1996. ^{1}H NMR and FTIR spectroscopic studies of bitumen and shale oil from selected Spanish oil shales. *Energy Fuels* 10 (1): 77–84.

Bouhadda, V., Bendedouch, D., Sheu, E., and Krallafa, A. 2000. Some preliminary results on a physico-chemical characterization of a Hassi Messaoud petroleum asphaltene. *Energy Fuels* 14 (4): 845-853.

Brandt, H.C.A., Hendriks, E.M., Michels, M.A.J., and Visser, F. 1995. Thermodynamic modeling of asphaltene stacking. *J. Phys. Chem.* 99 (26): 10430–10432.

Brauman, J.I. 2000. Twist and fluoresce. *Science* 290 (5490): 286–287.

Briant, J. and Hotier, G. 1983. Research on the state of asphaltenes in hydrocarbon mixtures: Size of molecular clusters (in French). *Oil Gas Sci. Tech.–Rev. IFP* 38 (1): 83–100.

Browarzik, D., Kabatek, R., Kahl, H., and Laux, H. 2002. Flocculation of asphaltenes at high pressure. II. Calculation of the onset of flocculation. *Petrol. Sci. Tech.* 20 (3-4): 233–249.

Buenrostro-Gonzalez, E., Andersen, S.I., Garcia-Martinez, J.A., and Lira-Galeana, C. 2002. Solubility/molecular structure relationships of asphaltenes in polar and non-polar media. *Energy Fuels* 16 (3): 732–741.

Buenrostro-Gonzalez, E., Espinosa-Peña, M., Andersen, S.I., and Lira-Galeana, C. 2001a. Characterization of asphaltenes and resins from problematic Mexican crude oils. *Petrol. Sci. Tech.* 19 (3–4): 299–316.

Buenrostro-Gonzalez, E., Groenzin, H., Lira-Galeana, C., and Mullins, O.C. 2001b. The overriding chemical principles that define asphaltenes. *Energy Fuels* 15 (4): 972–978.

Calemma, V., Iwanski, P., Nali, M., Scotti, R., and Montanari, L. 1995. Structural characterization of asphaltenes of different origins. *Energy Fuels* 9 (2): 225–230.

Camacho-Bragado, G.A., Romero-Guzmán, E.T., and José-Yacamán, M. 2001. Preliminary studies of asphaltene aggregates by low vacuum scanning electron microscopy. *Petrol. Sci. Tech.* 19 (1–2): 45–53.

Camacho-Bragado, G.A., Santiago, P., Marín-Almazo, M., Espinosa, M., Romero, E.T., Murgich, J., Rodríguez-Lugo, V., Lozada-Cassou, M., and José-Yacamán, M. 2002. Fullerenic structure derived from oil asphaltenes. *Carbon* 40 (15): 2761–2766.

Carbognani, L., Orea, M., and Fonseca, M. 1999. Complex nature of separated solid phases from crude oils. *Energy Fuels* 13 (2): 351–358.

Carnahan, N.F., Salager, J.-L., Anton, R., and Dávila, A. 1999. Properties of resins extracted from Boscan crude oil and their effect on the stability of asphaltenes in Boscan and Hamaca crude oils. *Energy Fuels* 13 (2): 309–314.

Cartz, L., Diamond, R., and Hirsch, P.B. 1956. New x-ray data on coals. *Nature* 177 (4507): 500–502.

Centeno, G., Trejo, F., Ancheyta, J., and Carlos, A. 2004. Precipitation of asphaltenes from Maya crude in a pressurized system (In Spanish). *J. Mex. Chem. Soc.* 48 (3): 186–195.

Champagne, P.J., Manolakis, E., and Ternan, M. 1985. Molecular weight distribution of Athabasca bitumen. *Fuel* 64 : 423–425.

Chiang, S. 1997. Scanning tunneling microscopy imaging of small adsorbed molecules on metal surfaces in an ultrahigh vacuum environment. *Chem. Rev.* 97: 1083–1096.

Christy, A.A., Dahl, B., and Kvalheim, O.M. 1989. Structural features of resins, asphaltenes and kerogen studied by diffuse reflectance infrared spectroscopy. *Fuel* 68 (4): 430–435.

Conley, R. *Infrared Spectroscopy*. 1972, 2nd ed. Ally and Bacon Editors, Boston, 92–210.

Corbett, L.W. and Petrossi, U. 1978. Differences in distillation and solvent separated asphalt residua. *Ind. Eng. Chem. Proc. Des. Dev.* 17 (4): 342–346.

Cowan, D.O. and Drisko, R.L. 1976. *Elements of Organic Photochemistry*, Plenum Press, New York, 2.

Dabir, B., Nematy, M., Mehrabi, A.R., Rassamdana, H., and Sahimi, M. 1996. Asphalt flocculation and deposition. III. The molecular weight distribution. *Fuel* 75 (14): 1633–1645.

Demirbaş, A. 2002. Physical and chemical characterizations of asphaltenes from different sources. *Petrol. Sci. Tech.* 20 (5–6): 485–495.

Dickie, J.P., Haller, M.N., and Yen, T.F. 1969. Electron microscopic investigations on the nature of petroleum asphaltics. *J. Colloid Interface Sci.* 29 (3): 475–484.

Domin, M., Herod, A.A., Kandiyoti, R., Larsen, J.W., Lazaro, M-J., Li, S., and Rahimi, P. 1999. A comparative study of bitumen molecular-weight distributions. *Energy Fuels* 13 (3): 552–557.

Donnet, J.-B., Ducret, J., Kennel, M., and Papirer, E. 1977. Electron microscopic observations of the morphology of bitumens. *Fuel* 56 (1): 97–100.

Dorbon, M., Ignatiadis, I., Schmitter, J-M., Arpino, P., Guiochon, G., Toulhoat, H., and Huc, A. 1984. Identification of carbazoles and benzocarbazoles in a coker gas oil and influence of catalytic hydrotreatmen on their distribution. *Fuel* 63 (4): 565–570.

Drushel, H.V. 1978. Trace sulfur determination in petroleum fractions. *Anal. Chem.* 50 (1): 76–81.

Dutta, P.K. and Holland, R.J. 1984. Acid-base characteristics of petroleum asphaltenes as studied by non-aqueous potentiometric titrations. *Fuel* 63 (2): 197–201.

Dyer, J.R. 1965. *Applications of Absorption Spectroscopy of Organic Compounds*. Prentice-Hall, Upper Saddle River, NJA, 23–52.

Elsharkawy, A.M., Al-Sahhaf, T.A., Fahim, M.A., and Yarranton, H.W. 2005. Characterization of asphaltenes and resins separated from water-in-crude oil emulsions formed in Kuwaiti oil fields. Paper presented at the 2005 AIChE Annual Meeting. Cincinnati, OH, October 30 to November 4.

Ergun, S. and Tiensuu, V.H. 1959. Alicyclic structures in coals. *Nature* 183 (4676): 1668–1670.

Espinat, D. 1991. Application of light, x-ray and neutron diffusion techniques to the study of colloidal systems. Part three: microemulsions, solid materials, liquids and miscellaneous systems with industrial uses (in French). *Oil Gas Sci. Tech.–Rev. IFP* 46 (6): 759–803.

Espinat, D., Ravey, J.C., Guille, V., Lambard, J., Zemb, T., and Cotton, J.P. 1993. Colloidal macrostruture of crude oil studied by neutron and x-ray small angle scattering techniques. *J. Physique IV* 3 (8): 181–184.

Espinat, D., Tchoubar, D., Boulet, R., and Freund, E. 1984. Study of heavy petroleum products by small angle X-ray scattering, in *Characterization of Heavy Crude Oils and Petroleum Residues*. Editions Technip, Paris, 147–152.

Fahim, M.A., Al-Sahhaf, T.A., and Elkilani, A.S. 2001. Prediction of asphaltene precipitation for Kuwaiti crude using thermodynamic micellization model. *Ind. Eng. Chem. Res.* 40 (12): 2748–2756.

Fenistein, D. and Barré, L. 2001. Experimental measurement of the mass distribution of petroleum asphaltene aggregates using ultracentrifugation and small-angle x-ray scattering. *Fuel* 80 (2): 283–287.

Fenistein, D., Barré, L., Broseta, D., Espinat, D., Livet, A., Roux, J.N., and Scarsella, M. 1998. Viscosimetric and neutron scattering study of asphaltene aggregates in mixed toluene/heptane solvents. *Langmuir* 14 (5): 1013–1020.

Frakman, Z., Ignasiak, T.M., Lown, E.M., and Strausz, O.P. 1990. Oxygen compounds in Athabasca asphaltene. *Energy Fuels* 4 (3): 263–270.

Franklin, R.E. 1950. Influence of the bonding electrons on the scattering of x-rays by carbon. *Nature* 165 (4185): 71–72.

Fuhr, B.J., Cathrea, C., Coates, L., Kalra, H., and Majeed, A.I. 1991. Properties of asphaltenes from a waxy crude. *Fuel* 70 (11): 1293–1297.

Gawrys, K.L. and Kilpatrick, P.K. 2005. Asphaltenic aggregated are polydisperse oblate cylinders. *J. Colloidal Interface Sci.* 288 (2): 325–334.

Gawrys, K.L., Spiecker, P.M., and Kilpatrick, P.K. 2002. The role of asphaltene solubility and chemistry on asphaltene aggregation: Conversion chemistry of petroleum residua. *Am. Chem. Soc., Div. Petrol. Chem.–Prepr.* 47 (4): 332–335.

George, R., Ritchie, S., Roche, R.S., and Steedman, W. 1979. Pyrolysis of Athabasca tar sands: Analysis of the condensable products from asphaltene. *Fuel* 58 (7): 523–530.

Ghloum, E.F. and Oskui, G.P. 2004. Investigation of asphaltene precipitation process for Kuwaiti reservoir. *Petrol. Sci. Tech.* 22 (78): 1097–1117.

Giavarini, C., Mastrofini, D., Scarsella, M., Barré, L., and Espinat, D. 2000. Macrostructure and rheological properties of chemically modified residues and bitumens. *Energy Fuels* 14 (2): 495–502.

González, G., Sousa, M.A., and Lucas, E.F. 2006. Asphaltenes precipitation from crude oil and hydrocarbon media. *Energy Fuels* 20 (6): 2544–2551.

Goual, L. and Firoozabadi, A. 2002. Measurements of asphaltenes and resins and dipole moment in petroleum fluids. *AIChE J.* 48 (11): 2646–2663.

Groenzin, H. and Mullins, O.C. 1999. Asphaltene molecular size and structure. *J. Phys. Chem. A* 103 (50): 11237–11245.

Groenzin, H. and Mullins, O.C. 2000. Molecular size and structure of asphaltene from various sources. *Energy Fuels* 14 (3): 677–684.

Guan, R.-L. and Zhu, H. 2007. Study on components in Shengli viscous crude oil by FTIR and UV-Vis spectroscopy. *Guang Pu Xue Yu Guang Pu Fen Xi/Spectroscopy and Spectral Analysis* 27 (11): 2270–2274.

Guillén, M.D., Blanco, J., Canga, J.S., and Blanco, C.G. 1991. Study of the effectiveness of 27 organic solvents in the extraction of coal tar pitches. *Energy Fuels* 5 (1): 188–192.

Hammami, A., Ferworn, K.A., Nighswander, J.A., Over, S., and Stange, E. 1998. Asphaltene crude oil characterization: An experimental investigation of the effect of resins on the stability of asphaltenes. *Petrol. Sci. Tech.* 16 (3–4): 227–249.

Hammami, A., Phelps, C., Monger-McClure, T., and Little, T.M. 2000. Asphaltene precipitation from live oils: An experimental investigation of onset conditions and reversibility. *Energy Fuels* 14: 14–18.

Herod, A.A. and Kandiyoti, R. 1995. Fractionation by planar chromatography of a coal tar pitch for characterisation by size-exclusion chromatography, UV fluorescence and direct-probe mass spectrometry. *J. Chromatog. A* 708 (1): 143–160.

Herod, A.A., Bartle, K.D., and Kandiyoti, R. 2007. Characterization of heavy hydrocarbons by chromatographic and mass spectrometric methods: An overview. *Energy Fuels* 21 (4): 2176–2203.

Herod, A.A., Lazaro, M.-J., Domin, M., Islas, C.A., and Kandiyoti, R. 2000. Molecular mass distributions and structural characterization of coal derived liquids. *Fuel* 79 (3–4): 323–337.

Herzog, P., Tchoubar, D., and Espinat, D. 1988. Macrostructure of asphaltene dispersions by small-angle x-ray scattering. *Fuel* 67 (2): 245–250.

Hirschberg, A., de Jong, L.N.J., Schipper, B.A., and Meijer, J.G. 1984. Influence of temperature and pressure on asphaltene flocculation. *SPE J.* 24 (3): 283–293.

Hishiyama, Y. and Nakamura, M. 1995. X-ray diffraction in oriented carbon films with turbostratic structure. *Carbon* 33 (10): 1399–1403.

Hu, Y.-F. and Guo, T.-M. 2001. Effect of temperature and molecular weight of *n*-alkane precipitants on asphaltene precipitation. *Fluid Phase Equilib.* 192 (1–2): 13–25.

Hunt, A. 1996. Uncertainties remain in predicting paraffin deposition. *Oil Gas J.–Rev. IFP* 94 (31): 96–103.

Hunt, J.E. and Winans, R.E. 1999. An overview of resid characterization by mass spectrometry and small angle scattering techniques. Paper presented at 218th American Chemical Society National Meeting, New Orleans, LA, August 22 to 26.

Hunt, J.E., Winans, R.E., and Miller, J.T. 1997. Characterization of asphaltenes from processed resides. *Am. Chem. Soc., Div. Fuel Chem.* 42 (2): 427–430.

Ibrahim, Y.A., Abdelhameed, M.A., Al-Sahhaf, T.A., and Fahim, M.A. 2003. Structural characterization of different asphaltenes of Kuwaiti origin. *Petrol. Sci. Tech.* 21 (5–6): 825–837.

IP 143 1985. Asphaltenes (*n*-heptane insolubles) in petroleum products. Standards for petroleum and its products. Institute of Petroleum. London.

Islas-Flores, C.A., Buenrostro-Gonzalez, E., and Lira-Galeana, C. 2006. Fractionation of petroleum resins by normal and reverse phase liquid chromatography. *Fuel* 85 (12–13): 1842–1850.

Jada, A. and Salou, M. 2002. Effects of the asphaltene and resin contents of the bitumens on the water-bitumen interface properties. *J. Petrol. Sci. Eng.* 33 (1): 185–193.

Jakobsen, H.J., Sørensen, O.W., Brey, W.S., and Kanyha, P. 1982. The "magic angle" for the differentiation between CH_3 and CH multiplicities in ^{13}C spin-echo *J*-modulation experiments. *J. Magnet. Resonance* 48 (2): 328–335.

Johnson, B.R., Bartle, K.D., Herod, A.A., and Kandiyoti, R. 1997. N-methyl-2-pyrrolidinone as a mobile phase in the size-exclusion chromatography of coal derivatives. *J. Chromatog. A* 758 (1): 65–74.

Kabir, C.S. and Jamaluddin, A.K.M. 1999. Asphaltene characterization and mitigating in south Kuwait's Marrat reservoir. Society of Petroleum Engineers (SPE). Middle East Oil Conference 11, February 20–23, Bahrein, Bahrein, 177–185.

Karaca, F., Islas, C.A., Millan, M., Behrouzi, M., Morgan, T.J., Herod, A.A., and Kandiyoti, R. 2004. The calibration of size exclusion chromatography columns: Molecular mass distributions of heavy hydrocarbon liquids. *Energy Fuels* 18 (3): 778–788.

Karas, M. and Hillenkamp, F. 1988. Laser desorption ionization of proteins with molecular masses exceeding 10 000 daltons. *Anal. Chem.* 60 (20): 2299–2301.

Kelemen, S.R., George, G.N., and Gorbaty, M.L. 1990. Direct determination and quantification of sulphur forms in heavy petroleum and coals: 1. The x-ray photoelectron spectroscopy (XPS) approach. *Fuel* 69 (8): 939–944.

Koinuma, Y., Kushiyama, S., Aizawa, R., Kobayashi, S., Uemasu, I., Mizuno, K., and Shimizu, Y. 1997. Distribution of heteroatoms in asphaltenes separated from Khafji residue before and after hydrotreatment as studied by GPC fractionation. *Am. Chem. Soc., Div. Petrol. Chem.–Prep.* 42 (2): 331–335.

Kokal, S.L., Sayeg, S.G., and George, A.E. 1993. Phase equilibria of crude oils using the continuous thermodynamic approach. *Can. J. Chem. Eng.* 71: 130–140.

Kumar, M. and Gupta, R.C. 1995. Graphitization study of Indian Assam coking coal. *Fuel Proc. Tech.* 43 (2): 169–176.

Langhoff, S., Bauschlicher, C., Hudgins, D., Sandford, S., and Allamandola, L. 1998. Infrared spectra of substituted polycyclic aromatic hydrocarbons. *J. Phys. Chem. A* 102 (9): 1632–1646.

Laux, H., Rahimian, I., and Butz, T. 1997. Thermodynamics and mechanism of stabilization and precipitation of petroleum colloids. *Fuel Proc. Tech.* 53 (1–2): 69–79.

León, O., Rogel, E., Espidel, J., and Torres, G. 1999. Structural characterization and self association of asphaltenes of different origins. Paper presented at AIChE Spring National Meeting. Houston, March 14–18, 37–43.

Leontaritis, K.J. and Mansoori, A. 1988. Asphaltene deposition: A survey of field experiences and research approaches. *J. Petrol. Sci. Eng.* 1: 229–239.

Li, W., Morgan, T.J., Herod, A.A., and Kandiyoti, R. 2004. Thin-layer chromatography of pitch and a petroleum vacuum residue. Relation between mobility and molecular size shown by size-exclusion chromatography. *J. Chromatog. A* 1024 (1–2): 227–243.

Lin, M.Y., Sirota, E.B., and Gang, H. 1997. Neutron scattering characterization of asphaltene particles. *Am. Chem. Soc., Div. Fuel Chem.* 42 (2): 412–415.

Liu, Y.C., Sheu, E.Y., Chen, S.H., and Storm, D.A. 1995. Fractal structure of asphaltenes in toluene. *Fuel* 74 (9): 1352–1356.

Loeber, L., Sutton, O., Morel, J-M., and Muller, V.G. 1996. New direct observations of asphalts and asphalt binders by scanning electron microscopy and atomic force microscopy. *J. Microscopy* 182 (1): 32–39.

Luo, P. and Gu, Y. 2007. Effects of asphaltene content on the heavy oil viscosity at different temperatures. *Fuel* 86 (7–8): 1069–1078.

Martínez, M.T., Benito, A.M., and Callejas, M.A. 1997. Thermal cracking of coal residues: Kinetics of asphaltene decomposition. *Fuel* 76 (9): 871–877.

Maruska, H.P. and Rao, B.M.L. 1987. The role of polar species in the aggregation of asphaltenes. *Fuel Sci. Tech. Int.* 5 (2): 119–168.

Masson, J.-F., Leblond, V., and Margeson, J. 2006. Bitumen morphologies by phase-detection atomic force microscopy. *J. Microscopy* 221 (1): 17–29.

Mehrotra, A.K., Sarkar, M., and Svrcek, W.Y. 1985. Bitumen density and gas solubility predictions using the Peng-Robinson equation of state. *AOSTRA J. Res.* 1 (4): 215–229.

Merdrignac, I. and Espinat, D. 2007. Physicochemical characterization of petroleum fractions: The state of the art. *Oil Gas Sci. Tech. – Rev. IFP* 62 (1): 7–32.

Merdrignac, I., Desmazières, B., Laprevote, O., and Terrier, P. 2004. Secondary effects in SEC analysis of oil asphaltenes evidenced by cross SEC separations and SEC-MS coupling. *Preprints of AIChE 2004 Spring National Meeting*, April, New Orleans, LA, 1922–1930.

Merdrignac, I., Quoineaud, A.-A., and Gauthier, T. 2006. Evolution of asphaltene structure during hydroconversion conditions. *Energy Fuels* 20 (5): 2028–2036.

Millan, M., Behrouzi, M., Karaca, F., Morgan, T.J., Herod, A.A., and Kandiyoti, R. 2005. Characterising high mass materials in heavy oil fractions by size exclusion chromatography and MALDI-mass spectrometry. *Catal. Today* 109 (1–4): 154–161.

Mitchell, D.L. and Speight, J.G. 1973. Solubility of asphaltenes in hydrocarbon solvents. *Fuel* 52 (2): 431–434.

Mitra-Kirtley, S., Mullins, O.C., Elp, J.V., George, S.J., Chen, J., and Cramer, S.P. 1993. Determination of the nitrogen chemical structures in petroleum asphaltenes using XANES spectroscopy. *J. Am. Chem. Soc.* 115 (1): 252–258.

Miura, K., Mae, K., Li, W., Kusakawa, T., Morozumi, F., and Kumano, A. 2001. Estimation of hydrogen bond distribution in coal through the analysis of OH stretching bands in diffuse reflectance infrared spectrum measured by in-situ technique. *Energy Fuels* 15 (3): 599–610.

Mordkovich, V.Z., Umnov, A.G., and Inoshita, T. 2000. Nanostructure of laser pyrolysis carbon blacks: Observation of multiwall fullerenes. *Int. J. Inorg. Mat.* 2 (4): 347–353.

Morgan, T.J., Millan, M., Behrouzi, M., Herod, A.A., and Kandiyoti, R. 2005. On the limitations of UV-fluorescence spectroscopy in the detection of high-mass hydrocarbon molecules. *Energy Fuels* 19 (1): 164–169.

Moschopedis, S.E. and Speight, J.G. 1976a. Investigation of hydrogen bonding by oxygen functions in Athabasca bitumen. *Fuel* 55 (3): 187–192.

Moschopedis, S.E. and Speight, J.G. 1976b. Oxygen functions in asphaltenes. *Fuel* 55: 334–336.

Moschopedis, S.E., Fryer, J.F., and Speight, J.G. 1976. Investigation of asphaltene molecular weights. *Fuel* 55 (3): 227–232.

Mou, J., Czajkowsky, D.M., Sheng, S., Ho, R., and Shao, Z. 1996. High resolution surface structures of E. coli GroES oligomer by atomic force microscopy. *FEBS Lett.* 381 (1–2): 161–164.

Mulligan, M.J., Thomas, K.M., and Tytko, A.P. 1987. Functional group fractionation and characterization of tars and pitches: Use of size exclusion chromatography and DMF as the mobile phase. *Fuel* 66 (11): 1472–1480.

Mullins, O.C. 1995. Sulfur and nitrogen molecular structures in asphaltenes and related materials quantified by XANES spectroscopy, in *Asphaltenes: Fundamentals and Applications,* Sheu, E.Y. and Mullins, O.C., Eds., Plenum Press, New York, Chap. 2.

Murgich, J. and Strauz, O. 2001. Molecular mechanics of aggregates of asphaltenes and resins of the Athabasca oil. *Petrol. Sci. Tech.* 19 (1–2): 231–243.

Murgich, J., Rodríguez, J., and Aray, Y. 1996. Molecular recognition and molecular mechanics of micelles of some model asphaltenes and resins. *Energy Fuels* 10 (1): 68–76.

Nali, M. and Manclossi, A. 1995. Size exclusion chromatography and vapor pressure osmometry in the determination of asphaltene molecular weight. *Petrol. Sci. Tech.* 13 (10): 1251–1264.

Oberlin, A. 1989. High-resolution TEM studies of carbonization and graphitization, in *Chemistry and Physics of Carbon*, Vol. 22, Thrower, P.A., Ed., Marcel Dekker, New York, 1–135.

Overfield, R.E., Sheu, E.Y., Sinha, S.K., and Liang, K.S. 1988. SANS study of asphaltenes aggregation. *Am. Chem. Soc., Div. Petrol. Chem. – Prepr.* 33 (2): 308–313.

Overfield, R.E., Sheu, E.Y., Sinha, S.K., and Liang, K.S. 1989. SANS study of asphaltene aggregation. *Fuel Sci. Tech. Int.* 7 (5–6): 611–624.

Pasadakis, N., Varotsis, N., and Kallithrakas, N. 2001. The influence of pressure on the asphaltenes content and composition in oils. *Petrol. Sci. Tech.* 19 (9–10): 1219–1227.

Patt, S.L., Shoolery, J.N. 1982. Attached proton test for carbon-13 NMR. *J. Magnet. Resonance* 46 (3): 535–539.

Paul-Dauphin, S., Karaca, F., Morgan, T.J., Millan-Agorio, M., Herod, A.A., and Kandiyoti, R. 2007. Probing size exclusion mechanisms of complex hydrocarbon mixtures: The effect of altering eluent compositions. *Energy Fuels* 21 (6): 3484–3489.

Pauli, A.T., Branthaver, J.F., Robertson, R.E., Grimes, W., and Eggleston, C.M. 2001. Atomic force microscopy investigation of SHRP asphalts. *Am. Chem. Soc., Div. Petrol. Chem.–Prepr.* 46 (2): 104–110.

Peng, P., Morales-Izquierdo, A., Hogg, A., and Strausz, O.P. 1997. Molecular structure of Athabasca asphaltene: Sulfide, ether, and ester linkages. *Energy Fuels* 11 (6): 1171–1187.

Peramanu, S., Pruden, B.B., and Rahimi, P. 1999. Molecular weight and specific gravity distributions for Athabasca and Cold Lake bitumens and their saturate, aromatic, resin, and asphaltene fraction. *Ind. Eng. Chem. Res.* 38 (8): 3121–3130.

Pereira, J.C., López, I., Salas, R., Silva, F., Fernández, C., Urbina, C., and López, J.C. 2007. Resins: The molecules responsible for the stability/instability phenomena of asphaltenes. *Energy Fuels* 21 (3): 1317–1321.

Pérez-Hernández, R., Mendoza-Anaya, D., Mondragón-Galicia, G., Espinosa, M.E., Rodríguez-Lugo, V., Lozada, M., and Arenas-Alatorre, J. 2003. Microstructural study of asphaltene precipitated with methylene chloride and *n*-hexane. *Fuel* 82 (8): 977–982.

Pfeiffer, J.P. and Saal, R.N.J. 1940. Asphaltic bitumen as colloid system. *J. Phys. Chem.* 44 (2): 139–149.

Pierre, C., Barré, L., Pina, A., and Moan, M. 2004. Composition and heavy oil rheology. *Oil Gas Sci. Tech.–Rev. IFP* 59 (5): 489–501.

Pillon, L.Z. 2001. Effect of experimental conditions and solvents on the precipitation and composition of asphaltenes. *Petrol. Sci. Tech.* 19 (5–6): 673–683.

Pina, A., Mougin, P., and Béhar, E. 2006. Characterisation of asphaltenes and modeling of flocculation – State of the art. *Oil Gas Sci. Tech.–Rev. IFP* 61 (3): 319–343.

Pineda, L.A., Trejo, F., and Ancheyta, J. 2007. Correlation between properties of asphaltenes and precipitation conditions. *Petrol. Sci. Tech.* 25 (1–2): 105–119.

Plumail, J.C., Jacquin, Y., Martino, G., and Toulhoat, H. 1983. Effect of the pore size distribution on the activities of alumina supported Co-Mo catalysts in the hydrotreatment of Boscan crude. *Am. Chem. Soc., Div. Petrol. Chem.–Prepr.* 28 (3): 562–575.

Ralston, C.Y., Mitra-Kirtley, S., and Mullins, O.C. 1996. Small population of one to three fused-aromatic ring moieties in asphaltenes. *Energy Fuels* 10 (3): 623–630.

Ramos, A.C.S., Delgados, C.C., Mohamed, R.S., Almeida, V.R., and Loh, W. 1997. Reversibility and inhibition of asphaltene precipitation in Brazilian crude oils. SPE (Society of Petroleum Engineers) paper 38967 presented at Rio de Janeiro, August 30–September 3.

Rassamdana, H., Dabir, B., Nematy, M., Farhani, M., and Sahimi, M. 1996. Asphalt flocculation and deposition: The onset of precipitation. *AIChE J.* 42 (1): 10–22.

Rassamdana, H., Farhani, M., Dabir, B., Mozaffarian, M., and Sahimi, M. 1999. Asphalt flocculation and deposition. V. Phase behavior in miscible and immiscible injections. *Energy Fuels* 13 (1): 176–187.

Ravey, J.C., Ducouret, G., and Espinat, D. 1988. Asphaltene macrostructure by small angle neutron scattering. *Fuel* 67 (11): 1560–1567.

Reerink, H. and Lijzenga, J. 1975. Gel-permeation chromatography calibration curve for asphaltenes and bituminous resins. *Anal. Chem.* 47 (13): 2160–2167.

Reynolds, J.G. 1990. Trace metals in heavy crude oil and tar and bitumens. Paper presented at Eastern Oil Shale Symposium, November 6–8, Lexington, KY.

Reynolds, J.G. and Biggs, W.R. 1987. Analysis of residuum desulfurization by size exclusion chromatography with element specific detection. *Am. Chem. Soc., Div. Petrol. Chem.–Prepr.* 32 (2): 398–405.

Rodrigues Coelho, R., Hovell, I., Lopez-Moreno, E., Lopes de Souza, A., and Rajagopal, K. 2007. Characterization of functional groups of asphaltenes in vacuum residues using molecular modelling and FTIR techniques. *Petrol. Sci. Tech.* 25 (1–2): 41–54.

Rogel, E. 2000. Simulation of interactions in asphaltenes aggregates. *Energy Fuels* 14 (3): 566–574.

Rogel, E. and Carbognani, L. 2003. Density estimation of asphaltenes using molecular dynamics simulations. *Energy Fuels* 17 (2): 378–386.

Rogel, E., León, O., Espidel, J., and González, J. 1999. SPE paper 53998 presented at SPE Latin American and Caribbean Petroleum Engineering Conference, SPE Caracas, Venezuela.

Rose, K.D. and Francisco, M.A. 1987. Characterization of acidic heteroatoms in heavy petroleum fractions by phase-transfer methylation and NMR spectroscopy. *Energy Fuels* 1 (3): 233–239.

Rose, K.D. and Francisco, M.A. 1988. A two-step chemistry for highlighting heteroatom species in petroleum materials using ^{13}C NMR spectroscopy. *J. Am. Chem. Soc.* 110 (2): 637–638.

Roux, J.N., Broseta, D., and Demé, B. 2001. SANS study of asphaltene aggregation: Concentration and solvent quality effects. *Langmuir* 17 (16): 5085–5092.

Ruland, W. 1967. X-ray studies on preferred orientation in carbon fibers. *J. Appl. Phys.* 38 (9): 3585–3589.

Sánchez-Berna, A.C., Camacho-Morán, V., Romero-Guzmán, E.T., and José-Yacamán, M. 2006. Asphaltene aggregation from vacuum residue and its content of inorganic particles. *Petrol. Sci. Tech.* 24 (9): 1055–1066.

Savvidis, T.G., Fenistein, D., Barré, L., and Behar, E. 2001. Aggregated structure of flocculated asphaltenes. *AIChE J.* 47 (1): 206–211.

Schabron, J.F. and Speight, J.G. 1997. Correlation between carbon residue and molecular weight. *Am. Chem. Soc., Div. Fuel Chem.* 42 (2): 386–389.

Schmitter, J.-M., Ignatiadis, I., Dorbon, M., Arpino, M., Guiochon, G., Toulhoat, H., and Hue, A. 1984. Identification of nitrogen bases in a coker gas oil and influence of catalytic hydrotreatment on their composition. *Fuel* 63 (4): 557–564.

Schwager, I., Farmanian, P.A., Kwan, J.T., Weinberg, V.A., and Yen, T.F. 1983. Characterization of the microstructure and macrostructure of coal-derived asphaltenes by nuclear magnetic resonance spectrometry and x-ray diffraction. *Anal. Chem.* 55 (1): 42–45.

Seidl, P.R., Chrisman, E.C.A.N., Silva, R.C., de Menezes, S.M.C., and Teixeira, M.A.G. 2004. Critical variables for the extraction of asphaltenes extracted from vacuum residues. *Petrol. Sci. Tech.* 22 (7–8): 961–971.

Seki, H. and Kumata, F. 2000. Structural change of petroleum asphaltenes and resins by hydrodemetallization. *Energy Fuels* 14 (5): 980–985.

Sharma, A., Groenzin, H., Tomita, A., and Mullins, O.C. 2002. Probing order in asphaltenes and aromatic ring systems by HRTEM. *Energy Fuels* 16 (2): 490–496.

Sharma, A., Kyotani, T., and Tomita, A. 2000. Comparison of structural parameters of PF carbon from XRD and HRTEM techniques. *Carbon* 38 (14): 1977–1984.

Sharma, B.K., Sharma, C.D., Bhagat, S.D., and Erhan, S.Z. 2007. Maltenes and asphaltenes of petroleum vacuum residues: physico-chemical characterization. *Petrol. Sci. Tech.* 25 (1–2): 93–104.

Sheremata, J.M., Gray, M.R., Dettman, H.D., and McCaffrey, W.C. 2004. Quantitative molecular representation and sequential optimization of Athabasca asphaltenes. *Energy Fuels* 18 (5): 1377–1384.

Sheu, E.Y., Liang, K.S., Sinha, S.K., and Overfield, R.E. 1992. Polydispersity analysis of asphaltene solutions in toluene. *J. Colloid Interface Sci.* 153 (2): 399–410.

Sheu, E.Y., Storm, D.A., and De Tar, M.M. 1991. Asphaltenes in polar solvents. *J. Non-Crystal. Solids.* 131–133 (1): 341–347.

Shiraishi, M., Kobayashi, K., and Toyoda, S. 1972. Thermal expansion of interlayer spacing and thermal vibrational displacement of carbon atoms in petroleum coke. *J. Mat. Sci.* 7 (11): 1229–1232.

Shirokoff, J.W., Siddiqui, M.N., and Ali, M.F. 1997. Characterization of the structure of Saudi crude asphaltenes by x-ray diffraction. *Energy Fuels* 11 (3): 561–565.

Siddiqui, M.N. 2003. Infrared study of hydrogen bond types in asphaltenes. *Petrol. Sci. Tech.* 21 (9–10): 1601–1615.

Siddiqui, M.N., Ali, M.F., and Shirokoff, J. 2002. Use of x-ray diffraction in assessing the aging pattern of asphalt fractions. *Fuel* 81 (1): 51–58.

Sirota, E.B. 2005. Physical structure of asphaltenes. *Energy Fuels* 19 (4): 1290–1296.

Speight, J.G. 1986. Polynuclear aromatic systems in petroleum. *Am. Chem. Soc., Div. Petrol. Chem. –Prepr.* 31 (3): 818–825.

Speight, J.G. 1987. Initial reactions in the coking of residua. *Am. Chem. Soc., Div. Petrol. Chem –Prepr.* 32 (2): 413–418.

Speight, J.G. 1994. Chemical and physical studies of petroleum asphaltenes, in *Asphaltenes and Asphalts. 1*, Yen, T.F. and Chilingarian, G.V., Eds., Elsevier, The Netherlands, 35–37.

Speight, J.G. 1999. *The Chemistry and Technology of Petroleum,* 3rd ed., Marcel Dekker, New York, 412–467.

Speight, J.G. 2004. Petroleum Asphaltenes. Part 1. Asphaltenes, resins and the structure of petroleum. *Oil Gas Sci. Tech.–Rev. IFP* 59 (5): 467–477.

Speight, J.G. and Moschopedis, S.E. 1981. On the molecular structure of petroleum asphaltenes, in *Chemistry of Asphaltenes*, Bunger, J.W. and Li, N.C., Eds., Advances in Chemistry Series No. 195, American Chemical Society, Washington, D.C., 195, 1–15.

Speight, J.G., Long, R.B., and Trowbridge, D. 1984. Factors influencing the separation of asphaltenes from heavy petroleum feedstocks. *Fuel* 63 (5): 616–620.

Spiecker, P.M., Gawrys, K.L., and Kilpatrick, P.K. 2003. Aggregation and solubility behavior of asphaltenes and their subfractions. *J. Colloid Interface Sci.* 267 (1): 178–193.

Storm, D.A. and Sheu, E.Y. 1995. Characterization of colloidal asphaltenic particles in heavy oil. *Fuel* 74 (8): 1140–1145.

Storm, D.A., Sheu, E.Y., and DeTar, M.M. 1993. Macrostructure of asphaltenes in vacuum residue by small-angle x-ray scattering. *Fuel* 72 (7): 977–981.

Strausz, O.P., Mojelsky, T.W., Faraji, F., Lown, E.M., and Peng, P. 1999a. Additional structural details on Athabasca asphaltene and their ramifications. *Energy Fuels* 13 (2): 207–227.

Strausz, O.P., Mojelsky, T.W., Lown, E.M., Kowalewski, I., and Behar, F. 1999b. Structural features of Boscan and Duri asphaltenes. *Energy Fuels* 13 (2): 228–247.

Strausz, O.P., Peng, P., and Murgich, J. 2002. About the colloidal nature of asphaltenes and the MW of covalent monomeric units. *Energy Fuels* 16 (4): 809–822.

Suelves, I., Islas, C.A., Millan, M., Galmes, C., Carter, J.F., Herod, A.A., and Kandiyoti, R. 2003. Chromatographic separations enabling the structural characterisation of heavy petroleum residues. *Fuel* 82 (1): 1–14.

Suresh Babu, V. and Seehra, M.S. 1996. Modeling of disorder and x-ray diffraction in coal-based graphitic carbons. *Carbon* 34 (10): 1259–1265.

Takanohashi, T., Iino, M., and Nakamura, K. 1998. Simulation of interaction of coal associates with solvents using the molecular dynamics calculations. *Energy Fuels* 12 (6): 1168–1173.

Takeshige, W. 2001. Hydrodynamic shape and size of Khafji asphaltene in benzene. *J. Colloid Interface Sci.* 234 (2): 261–268.

Tanaka, R., Hunt, J.E., Winans, R.E., Thiyagarajan, P., Sato, S., and Takanohashi, T. 2003. Aggregates structure analysis of petroleum asphaltenes with small-angle neutron scattering. *Energy Fuels* 17 (1): 127–134.

Tanaka, R., Sato, E., Hunt, J.E., Winans, R.E., Sato, S., and Takanohashi, T. 2004. Characterization of asphaltene aggregates using x-ray diffraction and small-angle x-ray scattering. *Energy Fuels* 18 (4): 1118–1125.

Thiyagarajan, P., Hunt, J.E., Winans, R.E., Anderson, K.B., and Miller, J.T. 1995. Temperature-dependent structural changes of asphaltenes in 1-methylnapthalene. *Energy Fuels* 9 (5): 829–833.

Trejo, F., Ancheyta, J., Morgan, T.J., Herod, A.A., and Kandiyoti, R. 2007. Characterization of asphaltenes from hydrotreated products by SEC, LDMS, NMR, and XRD. *Energy Fuels* 21 (4): 2121–2128.

Trejo, F., Ancheyta, J., and Rana, M.S. 2009. Structural characterization of asphaltenes obtained from hydroprocessed crude oils by SEM and TEM, *Energy Fuels*, 23 (1): 429–439.

Vellut, D., Jose, J., Béhar, E., and Barreau, A. 1998. Comparative ebulliometry: A simple, reliable technique for accurate measurement of the number average molecular weight of macromolecules. Preliminary studies on heavy crude fractions. *Oil Gas Sci. Tech.–Rev. IFP* 53 (6): 839–855.

Warren, B.E. 1941. X-ray diffraction in random layer lattices. *Phys. Rev.* 59 (9): 693–698.

Watson, B.A. and Barteau, M.A. 1994. Imaging of petroleum asphaltenes using scanning tunneling microscopy. *Ind. Eng. Chem. Res.* 33 (10): 2358–2363.

Whitson, C.H. 1983. Characterizing hydrocarbon plus fractions. *SPE J.* 23: 683–694.

Wiehe, I.A. 1992. A solvent-resid phase diagram for tracking resid conversión. *Ind. Eng. Chem. Res.* 31 (2): 530–536.

Wiehe, I.A. 2007. In defense of vapor pressure osmometry for measuring molecular weight. *J. Dispers. Sci. Tech.* 28 (3): 431–435.

Wilt, B.K., Welch, W.T., and Rankin, J.G. 1998. Determination of asphaltenes in petroleum crude oils by Fourier transform infrared spectroscopy. *Energy Fuels* 12 (5): 1008–1012.

Xu, Y.N., Koga, Y., and Strausz, O.P. 1995. Characterization of Athabasca asphaltenes by small-angle x-ray scattering. *Fuel* 74 (7): 960–964.

Yang, M-G., and Eser, S. 1999. Fractionation and molecular analysis of a vacuum residue asphaltenes. *Am. Chem. Soc., Div. Fuel Chem.* 44 (4): 768–771.

Yarborough, L. 1979. Application of generalized equation of state to petroleum reservoir fluids, in *Equations of State in Engineering*, Chao, K.C. and Robinson, R.L., Eds., Advances in Chemistry Series, Vol. 182. American Chemical Society, Washington, D.C., 385–439.

Yarranton, H.W., Alboudwarej, H., and Jakher, R. 2000. Investigation of asphaltene association with vapor pressure osmometry and interfacial tension measurements. *Ind. Eng. Chem. Res.* 39 (8): 2916–2924.

Yarranton, H.W. and Masliyah, J.H. 1996. Molar mass distribution and solubility modeling of asphaltenes. *AIChE J.* 42 (12): 3533–3543.

Yen, T.F. 1979. Structural difference between petroleum and coal-derived asphaltenes. *Am. Chem. Soc., Div. Petrol. Chem.–Prepr.* 24 (4): 901–909.

Yen, T.F., Erdman, J.G., and Pollack, S.S. 1961. Investigation of the structure of petroleum asphaltenes by x-ray diffraction. *Anal. Chem.* 33 (11): 1587–1594.

Zajac, G.W., Sethi, N.K., and Joseph, J.T. 1994. Molecular imaging of petroleum asphaltenes by scanning tunneling microscopy: Verification of structure from ^{13}C and proton nuclear magnetic resonance data. *Scann. Microscopy* 8 (3): 463–470.

Zajac, G.W., Sethi, N.K., and Joseph, J.T. 1997. Maya petroleum asphaltene imaging by scanning tunneling microscopy: Verification of structure from ^{13}C and proton nuclear magnetic resonance. *Am. Chem. Soc., Div. Fuel Chem.* 42 (2): 423–426.

Zhang, H., Yan, Y., Sun, W., and Wang, J. 2007. Structural description of polyaromatic nucleus in residue. *China Petrol. Proc. Petrol. Tech.* 1 (3): 35–42.

Zhao, S., Kotlyar, L.S., Woods, J.R., Sparks, B.D., Hardacre, K., and Chung, K.H. 2001. Molecular transformation of Athabasca bitumen end-cuts during coking and hydrocracking. *Fuel* 80 (8): 1155–1163.

2 Hydroprocessing of Heavy Oils

2.1 INTRODUCTION

The petroleum refining industry is entering into a significant era due to the depletion of light petroleum and an increase in the heavy or extra heavy crude oil exploration and production. Most of the light crude oil sources are almost extinguished, and exploration is turning to the heavy or extra heavy oil. This is the case particularly in Mexico, where the commercial production of light crude oils is nearly ended, and exploration and production of heavy and extra heavy crude oils is increasing year by year. To process these crudes, the development and use of technologies for hydroprocessing of heavy oils has been indolent for the past few years mainly due to the availability of light, sweet crudes all over the world. As the crude oil quality decreases, refineries have been under pressure to meet the market demand for high quality fuels because heavy crudes have such a low yield of light fractions while, at the same time, the yield of bottom of the barrel (residue) is higher. Therefore, it is foreseen that some refineries in the near future will increasingly replace light crudes with heavy or extra heavy crude oils, particularly in countries like Canada, Venezuela, and Mexico.

Apart from the quality of different crude oils, the price also varies substantially (Figure 2.1). The figure focuses on the price variation with API (American Petroleum Institute) gravity of crude oil in the past few decades. The price of crude oil in international markets is unprecedented, where the trade index has exceeded continuously since 2000 (Figure 2.1 inset). The increasing global demand for energy sustained the high price of crude oil, when, at the same time, enduring tropical storms, e.g., Wilma, Katrina, etc., political uncertainty, and wars in many oil-producing countries, which significantly affected the oil prices. An increase in petroleum prices also affects other industries, which raises inflation by several factors. Thus, the oil market comes under the influence of a number of issues and it can rapidly go from a situation of abundance to a situation of shortage and vice versa. The crude oil prices are mostly affected by the reflection of the United States because, worldwide, it is the third largest oil producer, first in consumer demand (42% of the top 10 consumers) as well as an aggressive importer (36% of the top 10 importers) (EIA, 2007). Japan is the second largest importer (17% of the top 10 countries) and third in consumption of crude oil in the world. Thus, Japan could also play a role in the price of oil as well. The top 10 big business countries (exporters–importers) are shown along with their place and capacity in Figure 2.2. The United States has huge facilities to process heavy crude, which makes it more important than any other country, particularly throughout

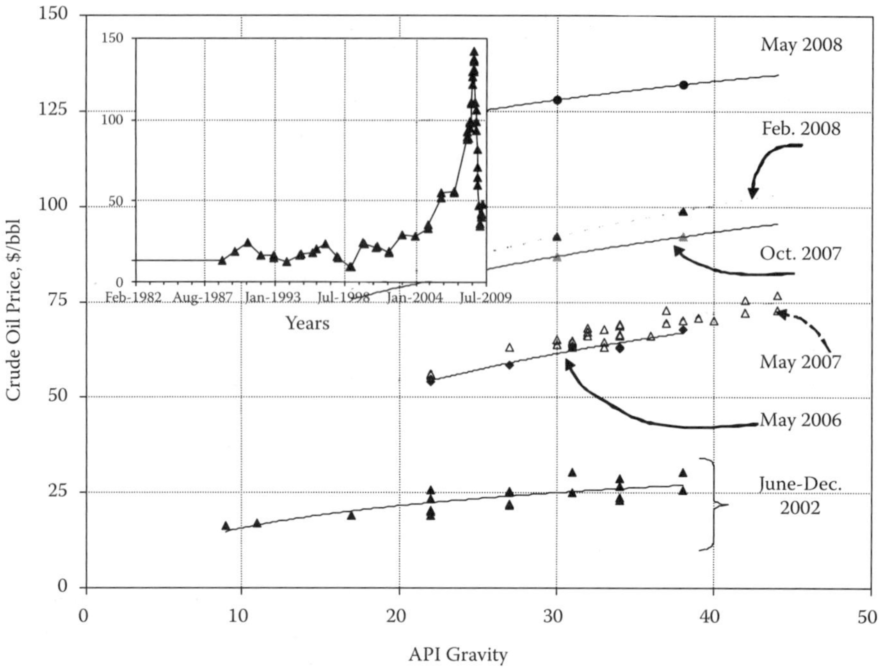

FIGURE 2.1 Fluctuation in crude oil prices.

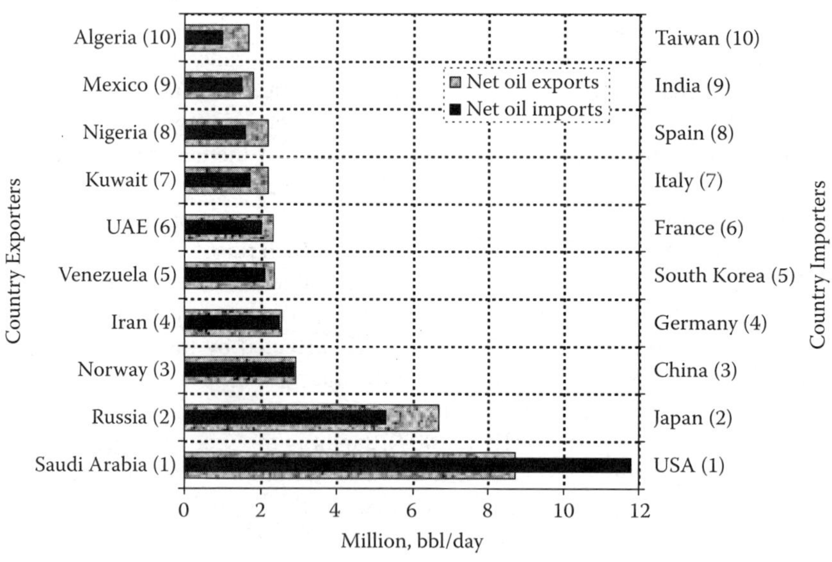

FIGURE 2.2 Crude oil market and top 10 countries with net exports and imports. (From EIA, June 14, 2007.)

North America, that send crude oil to United States for upgrading or refining and, then, import back the refined product.

On the other hand, some analysts have begun to say that if the price average is at around 70 US$/barrel, it wouldn't be logical to put money into new processing plants. If oil prices are to stay at 85 US$/barrel or above, it wouldn't be realistic to explore for these petroleum sources. Apart from another emerging source of competition for existing petrochemicals are materials coming from renewable resources (e.g., vegetable oils), whose long-term prospects have been considerably brightened by huge hikes in oil prices. It is also noted that animal fats are a much less expensive food for biodiesel fuel than vegetable oils, e.g., they are about half the price of soybean oil. Thus, biomass sources are becoming competitive for chemical production when oil is at or above 80 US$/barrel. The recent and rapid increase in crude oil price (2000 to 2008) is facing a new situation, in that the market price of crude oil (1.8 to 2.0 US$/kg) is higher than that of biomass-derived hydrocarbon molecules. Moreover, it is expected that the cost of these molecules derived from carbohydrates or vegetable oils is fairly stable compared with that of petroleum fuels and even tends to decrease steadily with time (Huber et al., 2006; Gallezot, 2007). Apart from these benefits, the commercial applicability of biomass is still in the primary stages and requires a decade or so to become fully commercial applicable. Nevertheless, the fossil fuels are still the principal source in meeting the present energy demand and the source of energy that created the Industrial Revolution in the nineteenth century and improved the living standard of all humans.

In general, the direct processing of heavy crude oil will not produce commercial fuel, rather it is a primary process to reduce excess metal and asphaltene content and, subsequently, it will be effectively used as FCC (fluid catalytic cracking) and hydrocracking feedstocks or as synthetic crude oil (Furimsky, 2007; Speight, 1999; Billon and Bigeard, 2001; van Veen, 2002; Rana et al., 2007e; Ancheyta et al., 2005a). The composition of these crudes and residua is more complex than that of light oils and their precise chemical and physical constitution is not well understood. This is due to the lack of analytical procedures that are capable of dealing with high concentrations of metals, sulfur, nitrogen, and oxygen compounds, mostly in the form of asphaltenes and resins (Table 2.1). A simple parameter, such as molecular H/C (hydrogen to carbon) atomic ratio, exhibits a decreasing trend from light crudes toward heavy crudes. Also, the heavier distillate fractions, such as asphaltene and resins, are the principal carriers of metals, nitrogen, and sulfur, as seen in Table 2.2.

A comparison of properties and composition of three crudes is given in Table 2.3 and Table 2.4, respectively. It is clearly seen that the quality of crude oil declines and the gasoline and diesel yield decreases when decreasing the API gravity of crude oil. Historically, the API gravity of crude oil has been reduced almost by an average of 0.15° per year, but the decline was accelerated after 1995 to around 0.22° per year. The variation of API gravity widely affects the chemical composition of crude oil, e.g., a decrease in API gravity of 0.4° increases sulfur

TABLE 2.1
Variation of Complex Hydrocarbon Composition with Origin of Crude Oil and Their API Gravity

Country, Crude	°API	Resin, wt%	Asphaltene, wt%
Canada, Athabasca	8.3	14.0	15.0
Venezuela, Boscan	10.2	29.4	17.2
Canada, Cold Lake	10.2	25.0	13.0
Mexico, Panuco	11.7	26.0	12.5
Iraq, Qayarah	15.0	36.1	20.4
Canada, Lloyminster	15	38.4	12.9
USA, MS, Baxterville	16.0	8.9	17.2
Russia, Balachany	31.7	6.0	0.5
Russia, Bibi-Eibat	32.1	9.0	0.3
USA, TX, Mexia	36.0	5.0	1.3
Iraq, Kirkuk	36.1	15.5	1.3
Mexico, Isthmus	37.8	8.1	1.3
USA, OK, Tonkawa	40.8	2.5	0.2
France, Lagrave	43.0	7.5	4.0
Algeria, Hassi Messaoud	45.0	3.3	0.15

Source: Adapted from several literature-based data.

content to about 900 wppm, as shown in Figure 2.3. Thus, crude oil sources are contaminating more and more with time even coming from the same source of crude, the quality and composition varying with time (Figure 2.4). This is, however, the case of a particular source, but due to the huge consumption of middle distillate, the decline in light-to-heavier crude is understandable throughout the world. In this regard, some analysts reported that a petroleum scenario has already begun to unfold and "world production is flat at the moment," which seems to be a controversial opinion because others predict it to bottom out about 2010, while other analysts see no change until 2035 (Francis, 2004). Of course, various "experts" have been predicting the end of the oil age for more than 100 years. And even now, no one really knows how much oil is left in the ground. Some reports forecast that there are certainly declines, but oil will remain as the principal source of fuel. Worldwide oil production (shown in Figure 2.5) is continuously increasing. The forecast of oil; production may vary, but one thing is indisputable: the nature of crude oil will vary from heavy to extra heavy oil; thus, for using such a crude oil; the upgrading becomes a necessary step in order to obtain valuable fuels.

There are two distinct sectors in the petroleum industry, *upstream* and *downstream*, as seen in Table 2.5, while the *midstream* sector manages crude oil stores, markets, and transports the commodities. The focus of the downstream

TABLE 2.2
Distribution of Sulfur (S), Nitrogen (N), Nickel (Ni), and Vanadium (V) in Hydrocarbon (HC) Fraction of Crude Oil and Atmospheric Residue (AR)

			Distribution, wt% of the total					
			Whole Crude				AR, 345°C+	
		HC fraction	S	N	V	Ni	Ni	V
Maya	Saturates	20.7	0.9	3.3	–	–	–	–
	Aromatics	26.5	24.6	8.2	0.4	3.3	2.7	2.7
	Resins	29.9	39.0	39.6	17.9	17.7	13.0	13.1
	Asphaltenes	20.6	36.3	48.9	81.7	79.0	84.3	85.6
Kern River	Saturates	21.8	<1	2.7	–	–	–	–
	Aromatics	28.7	30.7	4.2	7.5	4.5	1.8	2.7
	Resins	37.6	60.3	77.2	52.8	63.0	22.8	16.7
	Asphaltenes	5.5	8.8	15.8	39.8	32.5	75.4	80.6
Arbian Heavy	Aromatics	31.0	29.6	8.4	3.4	10.4	5.2	1.6
	Saturates	20.1	<1	6.7	–	–	–	–
	Resins	31.2	46.3	43.8	25.2	28.2	14.2	11.8
	Asphaltenes	12.2	23.9	41.1	71.4	61.8	80.6	86.6

Note: HC: hydrocarbon, AR: atmospheric residue.
Source: Quann et al., 1988. *Adv. Chem. Eng.* 14: 95–259. With permission.

and upstream sectors for each oil-producing country may vary depending on the quality of crude oil. Significant advances have been made in these sectors of petroleum over the past few decades. The conversion of heavy and extra crude oil is carried out by using different processes based on carbon rejection and hydrogen addition technologies. These processes (noncatalytic and catalytic) either upgrade

TABLE 2.3
Physical Properties of Three Different Crudes Oils

Properties	Heavy (Maya)	Light (Isthmus)	Super Light (Olmeca)
API, gravity	21.3	33.3	39.4
S, wt%	3.52	1.80	0.93
N, wt%	0.32	0.14	0.06
Ni + V, wppm%	322.5	99.7	10.4
Asphaltene, wt%	12.7	3.06	0.41
Rams carbon, wt%	10.8	4.13	2.22

TABLE 2.4
Composition of Isthmus and Maya Crude Oils (vol%)

Petroleum Fraction	Temperature Range (TBP), °C	API = 33.3 Isthmus	API = 21.3 Maya
Gases	<40	0.7	0.5
Gasoline	40–170	23.0	15.0
Jet fuel	170–230	11.7	8.4
Kerosene	230–290	10.5	8.0
Light straight run gas oil	290–360	11.4	9.1
Heavy straight run gas oil	360–538	25.3	21.4
Residue	+538	17.4	37.6

the crude oil or upgrade as well as refine simultaneously. A summary of these processes is given in Table 2.6, along with their key points. The major concern during the processing is to obtain desired product yields, as well as quality, for the downstream business. These products are gasoline, diesel, jet fuel, home heating oil, lubricating oils, and petrochemical streams, which are used to synthesize many bits and pieces that are used in daily life.

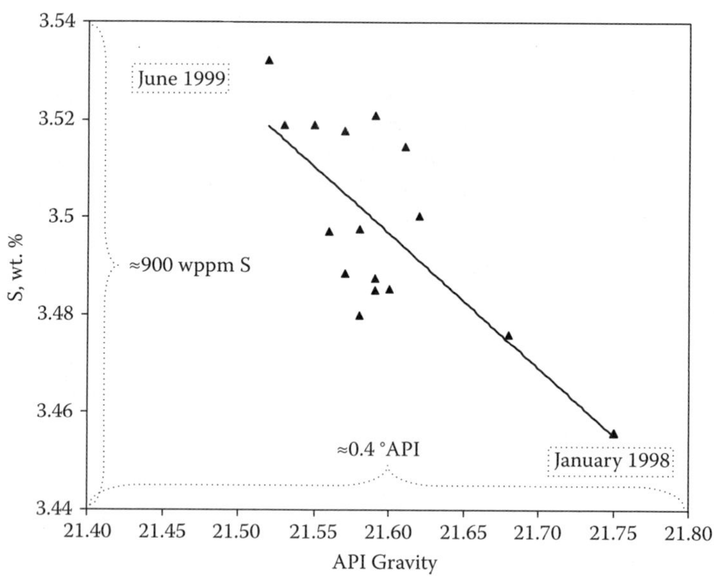

FIGURE 2.3 Effect of variation in Maya crude oil properties (API gravity and sulfur content) during 1½ year's time.

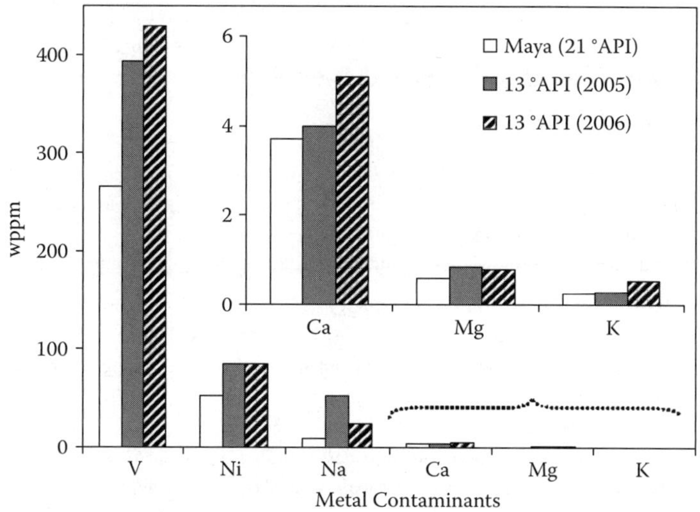

FIGURE 2.4 Variation of metal contents of various crude oils and with time at the well source.

In general, the crude oil upgrading technologies are based on *catalytic* and *noncatalytic* technologies (Figure 2.6). The primary processes are usually applicable to obtain synthetic crude oils from heavy oils containing relatively large amounts of contaminants (N, O, S, metals, etc.). The remaining amount of impurities must be removed in secondary processes, which are able to produce transport

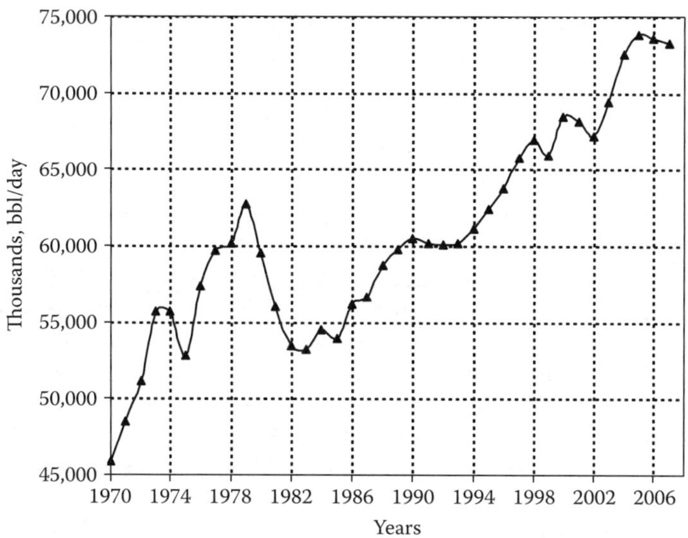

FIGURE 2.5 World crude oil production. (From EIA, June 14, 2007.)

TABLE 2.5
Principle Petroleum Oil Sectors and Their Role

Downstream Sector (Refining/Upgrading)

This sector covers oil refineries, product quality, and its distribution

It handles consumers through thousands of products, such as gasoline, diesel, jet fuel, heating oil, asphalt, lubricants, petrochemicals, and pharmaceuticals

Petroleum product sales focus on wholesale, industrial, and commercial customers

Upstream Sector (Exploration and Production)

This sector includes exploration and development involved in the search of petroleum

The development includes underground or underwater oil and gas fields, and drilling of exploratory wells

Principal hydrocarbon analysis:

- API gravity, viscosity, recovery efficiency, quality, and transport
- Flow assurance problems in well bores and topsides including asphaltene or wax precipitation, contaminants, corrosion, slugging, hydrates foaming, and emulsion breaking
- Early identification of oil quality issues allows the downstream to develop targeted solutions and prepares refineries for upcoming feedstock changes

fuels. It is also possible that, in between primary and secondary processing, nitrogen elimination can be used; the preremoval of basic and/or heterocyclic nitrogen compounds from synthetic crude oil greatly enhances the hydroprocessing rates and improves quality of hydrotreated products. The process technologies are principally different on the basis of the feedstock to be processed and operating conditions (Furimsky, 1988; Gosselink, 1998; Ancheyta, 2007). The catalysts used by the different licensees are listed in Table 2.6. In a first step, the conversion of contaminated crude oils can be carried out by using one of the following approaches:

- Thermal processes, such as visbreaking, delayed coking, fluid- and flexi-coking (carbon rejection)
- Removal of carbon (deasphalting)
- Hydrogen addition (catalytic hydrotreating and hydrocracking)

A brief comparison of various technologies based on these methods is shown in Figure 2.7a to Figure 2.7c, which present the product quality obtained in different processes, operating cost of three different crudes along with product revenues, and properties of middle-distillate feedstocks obtained from Athabasca synthetic crudes. The use of thermal processes for heavy feeds is a good option, but due to the low liquid yield and low quality of products, these processes have lost support among refiners, while hydroprocessing has gained more attention. Recently, the hydroprocessing application range was extended to cover feeds with higher content of contaminants and lower API gravity. This was possible by the use

TABLE 2.6
Various Catalytic, Noncatalytic, and Physical Processes and Their Comparison

Licensor	Process Technology	Process Applicability	Key Points
		Catalytic	
Chevron (Lummus)	RDS/VRDS HDT	Burn in mild HDC mode	
Haldor Topsøe	Mild HDC/VGO HDT	Pretreat cracker feed or work as mild hydrocracker	
UOP LLC	RDS	AR and VR derived feedstock	High process price High yield of liquid Better quality products Coke formation that deactivates catalysts High H_2 pressure
Shell Global Solution	HYCON	AR and VR derived feedstock	
Axens (inel IFP)	T-Star H-Oil LC-fining	Best for complex feeds such as coker VGO and DAO	
Axens (inel IFP)	Hyval fixed-bed Swing reactor concept	Accomplish with 30–50% conversion of VR	
		Noncatalytic	
ABB Lummus/ UOP/ Foster Wheeler	Delayed coking		
ABB Lummus	Visbreaking	Treat high metal, asphaltenes, and high Conradson carbon feedstocks (CCR)	Low quality products Low process price High yield of coke High yield of gas Liquid yield low High temperature Low pressure
ExxonMobil	Fluid Coking		
Conoco-Phillips/ Halliburton KBR	Flexicoking		
Chevron Texaco	Gasification		
		Physical Processes	
	Distillation	H/C separation process	
Foster Wheeler	Solvent extraction	Precipitate asphaltene	
UOP	Propane Deasphalting	Solvent/oil ratio is 4:1	Bottom-of-the-barrel upgrading at low cost Supercritical solvent recovery Asphalt and pitch can be used as a gasified to make power,
Halliburton KBR	Solvent Deasphalting	Recovery of solvent	
	Solvent Dewaxing	Eliminate most dirty part of crude oil	
		SDA is a separation process, not a conversion process	

(Continued)

TABLE 2.6 (CONTINUED)
Various Catalytic, Noncatalytic, and Physical Processes and Their Comparison

Licensor	Process Technology	Process Applicability	Key points
Exxon	Blending (Syncrude)		
Petro-Canada		Blending is sensitive to the two different compositions due to compatibility	Improve API gravity Provides substantial benefits to downstream Improve the price of crude oil
Conoco Phillips		It only improves flow properties	
Nexen		It is a process that has to mix good in bad	
Mocal Energy & Murphy Oil			

Source: Adapted from several literature-based data.

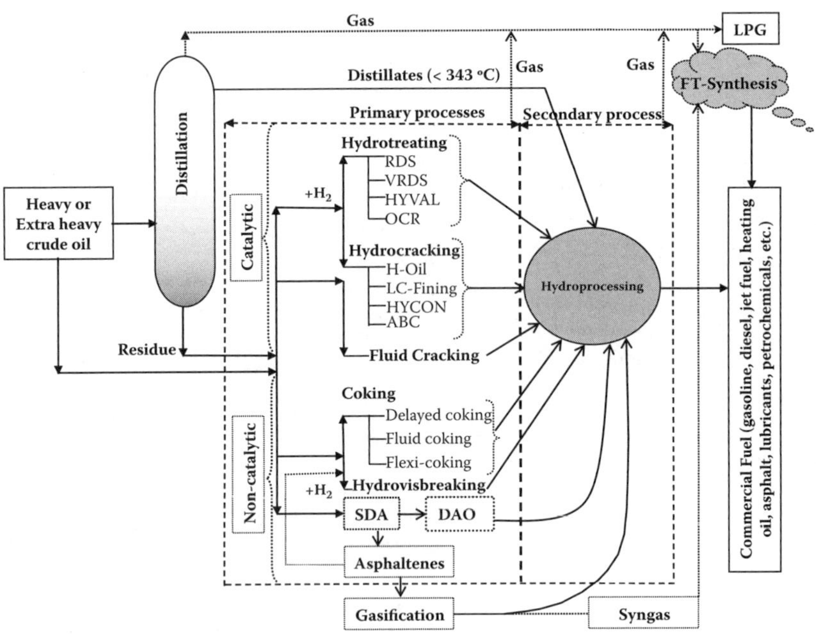

FIGURE 2.6 Processes and new developments for heavy, extra heavy crude oil, and residue upgrading.

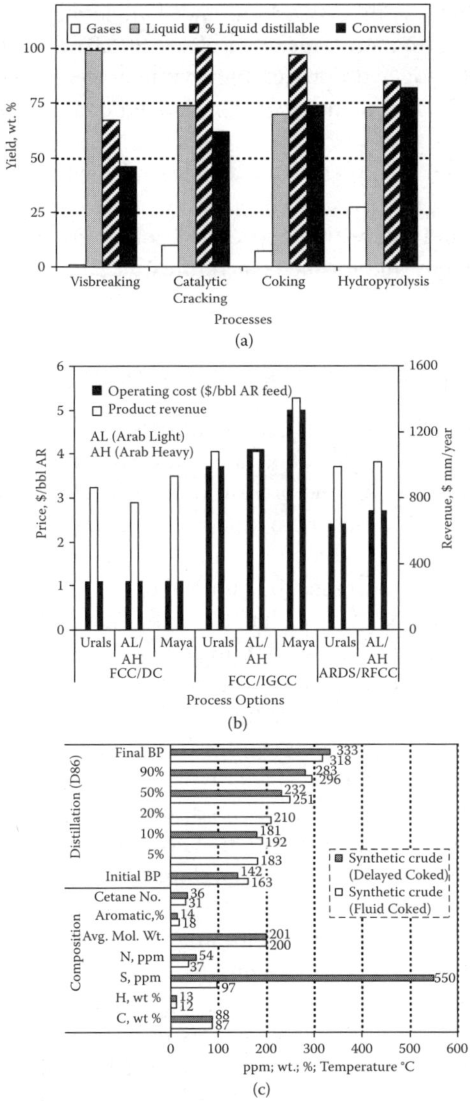

FIGURE 2.7 (a) Effect of different processes on product yield and total conversion. (From Penner et al. 1982. *Energy* 7 (7): 567–602. With permission.) (b) Estimated capital and operating costs for different process options along with product revenue for three atmospheric residue (AR) upgrading options. (From Phillips and Liu, 2003. *Hydrocarb. Eng.* 8 (9): 63–68. With permission.) (c) Properties of middle-distillate feedstocks from Athabasca synthetic crudes. (From Wilson et al. 1986. *Ind. Eng. Chem. Prod. Res. Dev.* 25: 505–511. With permission.)

of catalysts with improved stability working under moderate conversion regimes to minimize sediment formation (Ancheyta et al., 2003). Therefore, within the petroleum industry, catalytic hydroprocessing continues to be the core method for upgrading of bottom of barrel feedstocks (Kressmann et al., 1998; Marafi et al., 2008; Rana et al., 2007g).

The aim of this chapter is to present and discuss relevant aspects of hydroprocessing for the upgrading of heavy or extra heavy crude oils, while considering the role of asphaltenes and catalyst properties and composition, in order to obtain high-quality fuels for today and for the near future.

2.2 COMPOSITION OF HEAVY PETROLEUM FEEDS

One theory of petroleum formation is that it consists of plants and animals that remained buried for eons under thick layers of rock, which is largely responsible for the relatively high organic carbon present in the oil. Thus, due to the subterranean origin of petroleum, it must be extracted by means of wells. Usually the petroleum from a new well will come up to the surface under its own pressure; then, as time passes, the crude oil must be pumped out or forced to the surface by injecting hot water, air, natural gas, etc., into the well. The crude oil is usually sent from the well to a refinery by pipelines or tanker ships.

Generally, the physical properties and exact chemical composition of crude oil vary from one source to another. The hydrocarbon components are separated from each other by heating or fractionated distillation of petroleum according to boiling points. The lighter fractions, especially gasoline and diesel, are in greatest demand and are used to produce commercial fuels. However, bottom of barrel (heavy residua) obtained by direct distillation of crude oils needs further processing. Hydrocracking processes have been developed to convert heavy fractions into lighter and more valuable products, in which heat, pressure, and certain catalysts are used to break up the large molecules of heavy hydrocarbons into small molecules of light hydrocarbons. To have an idea about chemical composition of heavy fractions present in crude oils, Table 2.2 lists the results of saturates, aromatics, resins, and asphaltenes (SARA) contents in different samples. The most complex fraction of petroleum is asphaltene, which is a brownish-black hydrocarbon natural mixture used commonly in road paving, roofing, etc. It varies from a solid to a semisolid, has great tenacity, melts when heated, and, when ignited, it will leave very little or no ash.

Rising oil prices are not just affecting how far people can drive, prices are also affecting where and what people are driving on. In this regard, "what" corresponds to the quality of petroleum, while "where" to the quality of roads, which further depends on the quality of asphalt used during the construction. Asphalt is a complex ingredient in crude oil, which *like* crude oil, its price has also increased about 40 to 50% in the past few years. An increase in asphalt price mainly affects nonpetroleum-producing countries because the development of a country depends

TABLE 2.7
Properties and Amount of Asphaltenes, Resins, and Oils in Light, Heavy Crudes, and Residue

		Hydrocarbons			Contaminants		
Classification	API Gravity	Asphaltene, wt%	Resins, wt%	Oils, wt%	S, wt%	N, wt%	Metals (Ni + V), ppm
Extra light	>50	0 –<2	0.05– 3	–	0.02–0.2	0.0–0.01	<10
Light crude	22–32	<0.1–12	3–22	67–97	0.05–4.0	0.02–0.5	10–200
Heavy crude	10–22	11–25	14–39	24–64	0.1.–5.0	0.2–0.8	50–500
Extra heavy	<10	15– 40	–		0.8– 6.0	0.11.3	200–600
Residue	–	15–30	25–40	<49	–	–	100–>1000

on the number, as well as quality, of travels or national highways. The higher asphalt costs are forcing planners to scale down the number of road construction work or to postpone it indefinitely.

In general, crude oil enters into the refinery through the distillation column, which separates different fractions by their volatility, either in an approximately 345°C+ atmospheric or in 565°C+ vacuum cut points, yielding nondistillable residues called *atmospheric residua* (AR) or *vacuum residua* (VR), respectively. These nondistilled cuts further concentrate with contaminants. For most of the heavy crude oils, the content of AR or VR is relatively very high; therefore, some of the refiners are considering direct processing of whole crude (Ancheyta and Rana, 2008). Most of the properties of these heavy fractions are well characterized by using different methods, such as those reported in detail by Riazi (1989).

Typical classification of crude oil is defined based on API gravity and the amount of different contaminants and hydrocarbons composition (mainly asphaltenes and resins), as shown in Table 2.7. The fraction and composition of these crudes are quantitatively determined by using true boiling point (TBP) (Figure 2.8a), and by other methods (Figure 2.8b). Moreover, the compositions extensively vary with the fractionation (Figure 2.8c). The molecular composition of heavy fractions is difficult to analyze, nevertheless, it is easier for lighter fractions (< 210°C). For instance, characterization of Maya crude fractions was carried out by gas chromatography (flame ionization detector, FID), while its sulfur components analysis was performed by flame photometric detector (FPD), as shown in Figure 2.9 and Figure 2.10, respectively. The composition of distillates presents different groups of sulfur, such as mercaptane, thiophene, benzothiophene, dibenzothiophene (DBT), and alkylated DBT, particularly 4, 6 dimethyldibenzothiophene (4,6, DMDBT). Alkyl-substituted DBTs are difficult to convert into H_2S due to the sterically hindered adsorption of these compounds on the catalyst surface. It appears from Figure 2.8c and Figure 2.10 that more than 50%

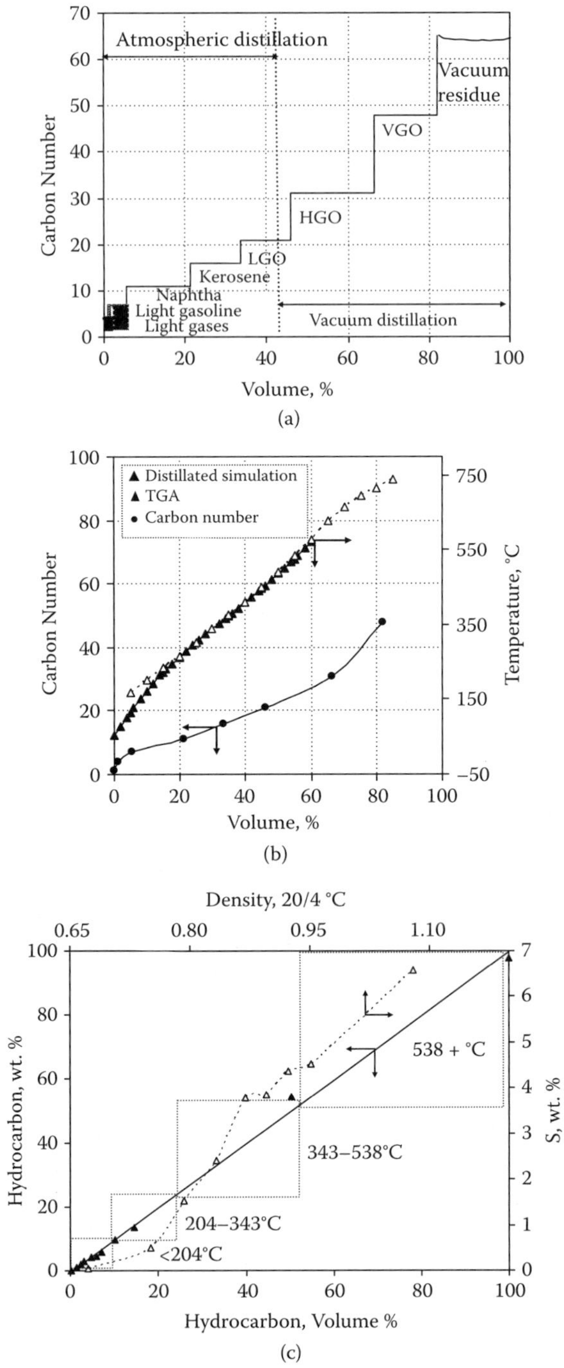

FIGURE 2.8 Composition of (a) Alaska, (b) Maya, and (c) Mexican 13° API crude oils.

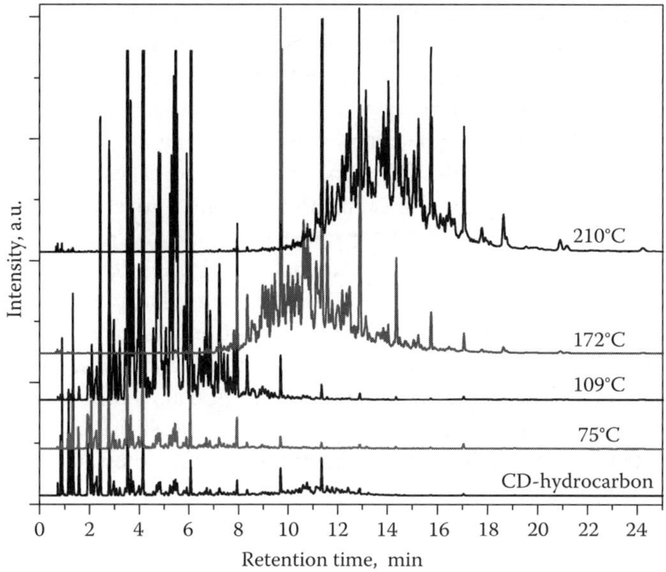

FIGURE 2.9 Effect of distillation temperature on hydrocarbon composition.

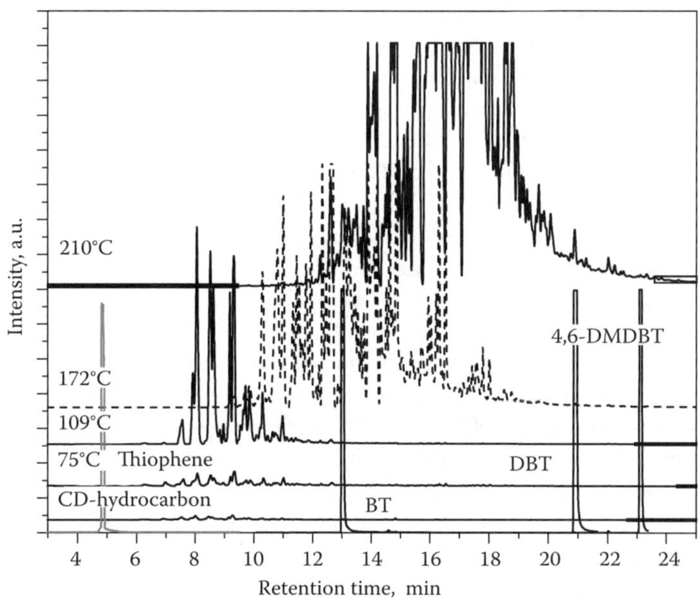

FIGURE 2.10 FPD gas-chromatographic traces showing distribution of the different organo-sulfur compounds at various distillation temperatures (Maya crude oil). The lower trace represents the feed, and the traces of reference compounds represent family of organo-sulfur compounds, such as thiophene, benzothiophene, dibenzothiophene, and 4,6-dimethyldibenzothiophene.

of sulfur species are associated with the asphaltene and resin molecules, which is more likely in the form of alkyl-substituted DBTs. Therefore, as noted by several authors (Shih et al., 1992; Gates and Topsøe, 1997; Kabe et al., 1997; Landau et al., 1996), the challenge is to find more active catalysts for hydrodesulfurization (HDS), and this challenge can be stated with the finding of more active catalysts for HDS of substituted 4,6-dimethyldibenzothiophene. These sulfur molecules are saturated in the heavy fractions, thus, the suitable catalyst requires large pore diameter.

Therefore, designing a catalyst is a complex task in order to have good dispersion of active metal as well as large pore diameter. Developing a catalyst for heavy oil is even more complex, since it requires tremendous industrial effort as well as fundamental research, which has been mostly focused on understanding the reactivity in attempts to design a more efficient catalyst for hydroprocessing of crude oil. The complex organic sulfur molecules present in heavy petroleum cannot be identified easily by gas chromatography due to their nature and limitations of the equipment.

2.2.1 Role of Asphaltene

The major parts of heavy feeds contain aggregates of resins and asphaltenes dissolved in oil fraction held together by weak physical interactions. With resins being less polar than asphaltenes, but more polar than oil, equilibrium between the micelles and the surrounding oil leads to homogeneity and stability of the colloidal system. If the amount of resins decreases, the asphaltenes will coagulate and form sediments. Asphaltenes are complex polar structures with a polyaromatic character containing metals (mostly Ni and V) that cannot be properly defined according to their chemical properties, while they are usually defined according to their solubility properties. Asphaltenes are hydrocarbon compounds that precipitate from petroleum by addition of light paraffins in crude oil or residue. Asphaltenes precipitated with *n*-heptane have a lower H/C ratio than those precipitated with *n*-pentane, while asphaltenes obtained with *n*-heptane are more polar, have a greater molecular weight, and display higher N/C (nitrogen/carbon), O/C (oxygen/carbon), and S/C (sulfur/carbon) ratios than those obtained with *n*-heptane (Sharma et al., 2007a).

Asphaltenes are constituted by condensed aromatic nuclei carrying alkyl groups, alicyclic systems, and heteroelements (Dickie and Yen, 1967; Tynan and Yen, 1970; Merdrignac and Espinat, 2007; Sharma et al., 1999, 2000, 2007b). Asphaltene molecules are grouped together in systems of up to five or six sheets, which are surrounded by the so-called maltenes (all those structures different from asphaltenes that are soluble in *n*-heptane). The exact structure of asphaltenes is difficult to obtain and several structures have been proposed for the asphaltenes present in different crudes (Beaton and Bertolacini, 1991). The length of the alkyl chains in asphaltenes has been the subject of different studies. Mojelsky et al.

(1992) found chains of 3 to 4 carbon atoms, while Speight (1999) found alkylic chains of up to 30 carbon atoms. Other studies on the structure of asphaltenes have been performed (Miller et al., 1998; Mullins and Groenzin, 1999; Matsushita et al., 2004). An asphaltene molecule may be 4 to 5 nm in diameter, which is too large to pass through micropores or even some mesopores in a catalyst. Metals in asphaltene aggregates are believed to be present as organometallic compounds associated to the asphaltene sheets, making the asphaltene molecule heavier than its original structure. The characteristics of heavy or extra heavy oils are quite different, which are mainly due to the higher concentration and nature of asphaltene present in each crude oil.

2.3 UPGRADING OF HEAVY OILS

As reported in Section 2.1, catalyst technologies have been progressing (Kamiya, 1991), but more work has to be done with respect to the catalyst activity, selectivity, and stability as well as in process and reactor technologies. In the case of hydroprocessing for heavy crude oil upgrading, its profitability is a function of product value, i.e., price differential between feed (heavy crude oil) and product (upgraded crude oil). With respect to the selection of desired products from residue or heavy crude oil, hydroprocessing technologies are preferred over other kind of processes, e.g., thermal processes. Upgrading of crude oils using fixed-bed reactor technologies has also some limitations in handling heavy petroleum at high conversion level due to sediments formation and asphaltenes precipitation. Therefore, for the research of crude oil upgrading, catalyst properties are forced toward large pore catalysts, which can retain an optimum quantity of metals and a good level of sulfur and nitrogen removals. It is also demonstrated that the steady improvements in catalysts have led to longer run lengths and lower operating temperatures. However, the opinions vary about the catalyst use, e.g., nature of catalyst; definitely, existing catalysts may not work as effectively as required. In this respect, some researchers thought that ring-opening (acidic) catalyst technologies were already well advanced, while others thought that better selectivity for hydrogen addition and hydrocracking (e.g., less gas production) is needed by using the acid-base nature of catalysts (Leyva et al., 2007).

The principal objective of crude oil upgrading is to obtain a feedstock that is relatively low in nitrogen, sulfur, metals, asphaltenes, and Conradson carbon contents compared with conventional crudes. Upgraded crude produces potentially good hydrocracking or FCC feedstocks, which are susceptible to this contamination (Table 2.8). However, it should be noted that these impurities remaining in synthetic crude oil (SCO) are the most difficult and will be removed through secondary hydroprocessing (Figure 2.6).

Hydroprocessing of heavy oil is principally based on the different components of the catalyst and its preparation method. Catalyst properties, such as catalyst nature (acidic or basic), textural properties of support, nature of catalytic sites,

TABLE 2.8
Effect of Feed Contamination on Hydrocracking (HCR) and Fluid Catalytic Cracking (FCC)

Contaminants	Effect on HCR or FCC Catalyst	Process Remedies
Metals	Structural damage of zeolite (US-Y)	HDM
	Vanadium passivation	
	Metal deactivation on catalytic sites	
Nitrogen	Deactivation on acid sites	HDN
	Competitive adsorption on catalytic sites	
Sulfur	SOx formation (flue gas)	HDS
Carbon (low H/C ratio)	Coke formation	HYD
	Carbon deposition on catalytic sites	(high H_2 pressure)
	Pore plugging (diffusion limitations)	

Source: Ramirez et al. 2007. *Hydroprocessing of Heavy Oil and Residua*, Taylor & Francis, New York. With permission.

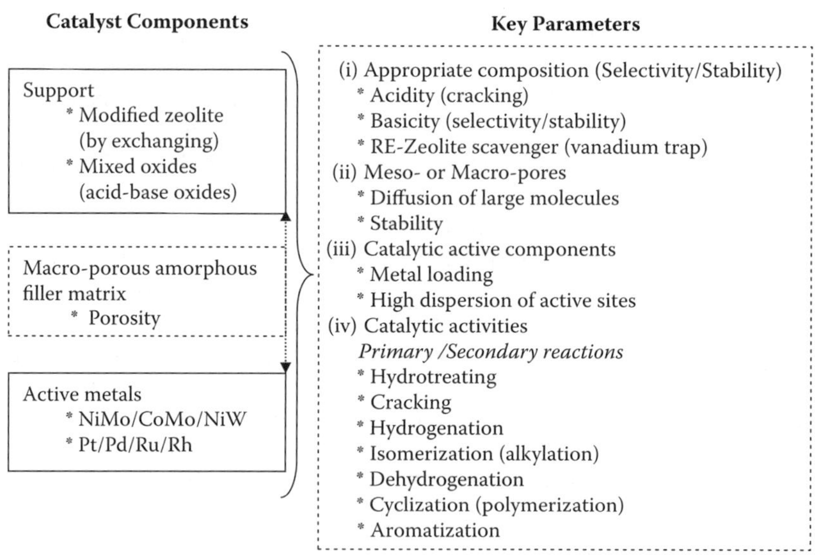

FIGURE 2.11 Concept of acid-base supported catalyst for hydroprocessing of heavy oil. (RE denotes rare-earth exchanged zeolites of US-Y, KL, ZSM-5, NaX, mordenite, etc.) (From Leyva et al. 2007. *Ind. Eng. Chem. Res.* 46 (23): 7448–7466. With permission.)

and their interaction with support, have important effects on the catalyst development (Figure 2.11) (Leyva et al., 2007). It should be highlighted that there are very few studies that have been reported in the literature for heavy oil hydroprocessing with variation of catalyst composition and its preparation.

Hydrogen addition technologies are mature enough to upgrade crude oils and residua (LePage, 1992; Gray, 1994; Scherzer and Gruia, 1996; Schuetze and Hofmann, 1997; Dickenson et al., 1997), which are high investment processes, and differentiate in reactor design, where the catalyst is either used in fixed-beds, ebullated-beds, or allowed for online catalyst addition/withdrawal. For heavy oil hydroprocessing, due to the high level of contaminants, moving-bed reactor technology is more efficient than fixed-bed. In order to examine more detailed advantages and disadvantages, a summary of various reactors used in hydroprocessing is given in Table 2.9. One criterion for selection of reactors is based on the quality and composition of feedstock. Usually dirty feeds can be effectively processed in an ebullated-bed reactor, while the major disadvantages with fixed-bed

TABLE 2.9
Residue/Heavy Oil Hydroprocessing Flow Reactors, Processes, and Their Operating Conditions

Reactor	Licensor	Advantages	Disadvantages
		Fixed-Bed Reactors	
On-stream catalyst Replacement	Chevron	Low catalyst loss No catalyst attrition	High pressure drop Low liquid flow
UFR–up-flow reactor	Shell	Flexibility for conditions Low cost	Mal distribution (hot spot) Diffusion limitations
		Moving-Bed Reactor	
Bunker reactor–HYCON	Shell	High catalyst utilization	Require attrition-resistant
Swing reactor–HYVAHL	IFP	Well mixing of feed and catalyst High tolerance of metals Can combine with FBR	Catalyst replacement in a batch operation
		Ebullated-Bed Reactor	
H-Oil	Axens-IFP	* Very flexible operation	* High catalyst attrition loss
T-Star	Chevron	* Catalyst bed expansion * Excellent heat transfer	* High catalyst consumption rate * Sediment formation
LC-Fining	ABB Lummus		

Source: Adapted from several literature-based data.

reactors are catalyst deactivation with time. In general, hydroprocessing requires hydrogen to hydrogenate oil at high pressures and temperatures in the liquid phase because such oils have very high concentration of carbon. The conversion of heavy oils and residue is still subject of discussion because some researchers have reported that the presence of a catalyst does not affect residuum or bitumen conversion and that residuum conversion occurs primarily via thermal reaction pathways (Miki et al., 1983; Mosby et al., 1986; Beaton and Bertolacini, 1991; Sanford and Chung, 1991; Heck et al., 1992), while others reported that light oil fractions are selectively formed during thermal cracking, even during catalytic hydroprocessing (Miki et al., 1983; Beaton and Bertolacini, 1991; Sanford and Chung, 1991). Thermal reactions have been reported to be more important than catalytic reactions at higher temperatures due to the high activation energies for thermal cracking (Khorasheh et al., 1989). In this respect, asphaltene conversion is more complicated because a wide range of molecular changes occurs with temperature, which has been clearly demonstrated by small-angle neutron scattering characterization of Maya crude asphaltene (Tanaka et al., 2003). Thus, it is a theme for further studies to determine if asphaltene conversion follows catalytic conversion or thermal cracking mechanisms. A comparison between asphaltene from Maya crude oil with other asphaltenes from Iranian light oil and Arabian heavy oil (Khafji) showed that asphaltenes from Maya crude oil have a relatively more refractory structure.

Among all available possibilities for treatment of such heavy feedstocks, hydrogen addition processes lead to high hydrogen consumption, but higher liquid yields. These processes, which provide the feedstock for the subsequent FCC process, require the use of well-designed catalysts capable of dealing with the high concentrations of metals and asphaltenes present in the feedstock. Moreover, the multifunctional catalysts used for hydrocracking processes become poisoned by coke deposition, and the heavy metals present in the feed create a hazardous waste, which has to be disposed of properly and safely (Furimsky, 1996). A high catalyst demetallization function is necessary because vanadium destroys the zeolitic catalyst used in the subsequent FCC process. Additionally, the concentration of nitrogen compounds must be maintained to a minimum to avoid poisoning of the catalyst acid sites in this and the subsequent FCC process. Although in the hydrocracking process, the amount of metals is not as critical as in FCC, the elimination of nitrogen compounds is a determinant to avoid poisoning the catalyst acid sites. Finally, one may well recall that even though a good match between the properties of feed and catalyst has to be achieved to obtain high hydroprocessing conversions, catalyst life, and stability, the final selection of a catalyst for a particular process has to be based on price and performance (van Kessel et al., 1987). The price of the catalyst may also depend on the type of active metal used to prepare it. Active metals usually depend on the type of reactor, nature of feedstock, and type of support used for catalyst preparation (Table 2.10). Typically, these catalysts are prepared by introducing Mo(W) first and then Co or Ni promoter atoms. The nature of catalytic sites and activity can be modified

TABLE 2.10
Catalytic Processes Used to Treat a Residual Type of Crude Oil

Process Conditions	Hydrotreating	Mild Hydrocracking	Hydrocracking
Catalyst	CoMoS, NiMoS, NiWS	Ni, Co, Mo, W, V, Pd, Pt, etc. (single or in combination)	Ni, Co, Mo, W, V, Pd, Pt, etc. (single or in combination)
Support	Al_2O_3, mixed oxides	Al_2O_3, Al-Si, zeolite, mixed oxides	Al-Si, mixed oxides, X or Y type zeolites with amorphous Al and/or Si
Temperature, °C	382415	300–450	390–482
P_{H2}, MPa	4–10	8.5–20.0	12.0–20.5
LHSV, h^{-1}	0.4–1.5	0.5–2.5	0.3–1.5
Catalyst life	2–5 years	2–5 years	1.5–3 years

Source: Adapted from several literature-based data.

by addition of urea, chelating agents, and so on, during the preparation (Rana et al., 2007d; 2007f). Usually, preparation of heavy oil catalysts is relatively easier due to the low loading of active metals, thus, active metals have been considered always in monolayer or well dispersed.

In order to have a beneficial effect of catalytic processes by enhancing quality and quantity of liquid yield from heavy oil and residue conversion, proper reactor, reaction conditions, feedstock composition along with catalyst type and composition have to be selected. The fuel product compositions from secondary hydrotreating of synthetic crude (Athabasca Syncrude) are compared in Figure 2.12a and Figure 2.12b for hydrocracked and coker products. The coker products have much higher sulfur and nitrogen contents relative to the hydrocracking, which decreases fuel quality as well as price. On the other hand, product selectivity also varies in great extent when changing experimental conditions shown in Figure 2.13, which includes the compositional analysis of the feedstock. The various processing conditions show some of the effects of temperature and pressure. It is noted that concentration of aromatics in distillate products are affected by the shift of thermodynamic equilibrium while high temperatures and low pressures favor the formation of aromatics. The major works in this area have been reported for atmospheric or vacuum residues as feedstock using hydroprocessing, which has been characterized by a high degree of metallic contaminants, sulfur, Conradson carbon, and asphaltene contents (Furmisky, 2007). However, the direct heavy crude oil (whole petroleum) hydroprocessing is not common as of yet, but there are few research groups that are working on it. The detailed studies for heavy crude oil hydroprocessing are reported in subsequent sections.

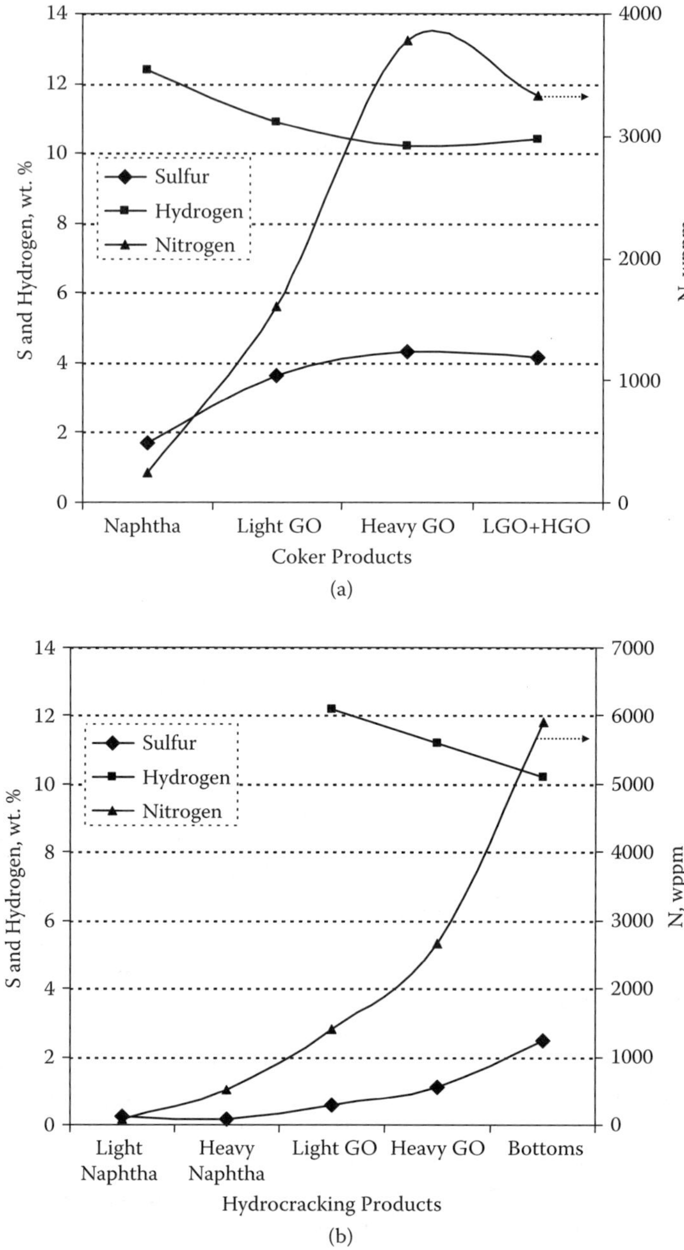

FIGURE 2.12 Typical product composition and its contaminants of upgrading products from (a) thermal-cracking and (b) hydrocracking of bitumen. (From Yui, S. 1999. *Can. Chem. News* July/August 25–27. With permission.)

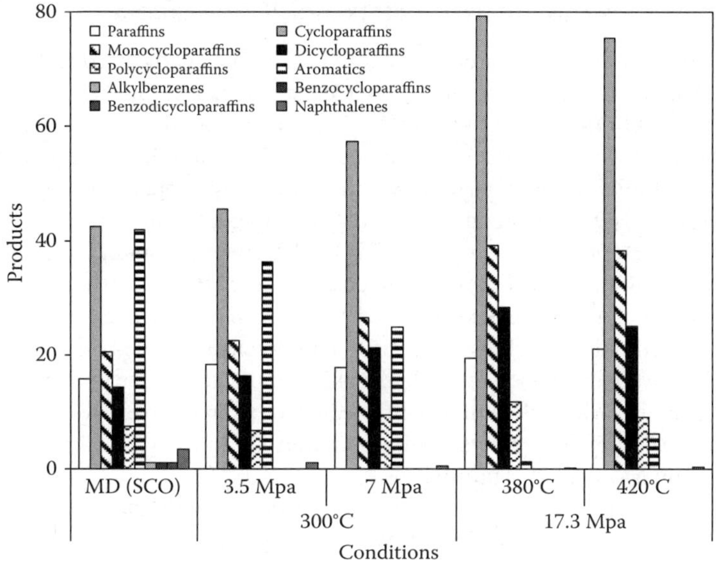

FIGURE 2.13 Fuel product compositions from synthetic crude—selectivity after secondary hydrotreating with variation of reaction conditions. (From Wilson et al. 1986. *Ind. Eng. Chem. Prod. Res. Dev.* 25: 505–511. With permission.)

2.4 COMPONENTS OF HYDROPROCESSING CATALYST

The hydrotreating catalyst consists of a support and active metals in a balanced composition. The support provides a textural nature as well as acid-base properties to the catalyst, while active metals give the desired catalytic activity. Two of the most important properties of catalysts are acidity and textural properties, which are the innate properties of the support. Support plays a particularly important role to the geometry and surface provided for active metal and their interaction with the support (Breysse et al., 2003; Muralidhar et al., 2000, 2003; Okamoto et al., 2003; Luck, 1991; Shimada, 2003; Saih and Segawa, 2003; Topsøe et al., 1996). Apart from these properties, support also provides acid-base nature of the catalyst, which can be selected on the basis of the reaction demanded and the selectivity of the products, e.g., Brønsted acid sites promote cracking of the reacting molecules while basic components like MgO and ZrO_2 in the support enhance the hydrogenolysis selectivity, but the catalyst remains low active toward the hydrogenation (Leyva et al., 2007). The detailed account of such properties on catalyst for heavy oil hydroprocessing are described in the later section of this chapter; however, it is anticipated that such catalyst properties are not exactly the same as reported in most of the literature for model molecules or light petroleum distillates hydrodesulfurization due to the complex nature of contaminants present in the feedstock.

2.4.1 Supports and Catalysts Preparation/Composition

The supports can be either prepared commercially or synthesized in the laboratory. In general, Al_2O_3 and its mixed oxides supports (Al_2O_3-TiO_2, Al_2O_3-SiO_2, Al_2O_3-ZrO_2, Al_2O_3-B_2O_3, Al_2O_3-MgO, etc.) are prepared in the laboratory by using the homogeneous precipitation method. Dry extrudates of commercial alumina boehmite and other additives are commonly prepared using 2 to 10% (v/v) HNO_3 in H_2O for peptization of alumina. Extrudates are subsequently dried and calcined at 550°C for 4 h. Examples of details of laboratory-prepared supports along with their conditions are given elsewhere (Rana, 2007e; Leyva, et al., 2007). The use of mixed oxide supported catalysts in hydroprocessing is not new; initially mixed oxides such as SiO_2-ZrO_2-TiO_2 (Hansford, 1964) and Al_2O_2-TiO_2 (Jaffe, 1968) were developed for hydrocracking applications due to their appropriate acidic nature. Molybdenum was impregnated by an incipient wetness method and Ni or Co promoted catalysts were prepared via the sequential impregnation procedure on Mo-loaded catalysts.

Hydroprocessing catalysts are usually prepared with the amorphous and crystalline supports upon which active metals (principally Mo, W, Ni, Co) are deposited that diminishes to some extent support acidity as well as promotes hydrogenolysis and hydrogenation functions of the catalysts. Molybdenum and tungsten sulfides create catalytic sites by forming very small nanoscale crystallites, as seen in Figure 2.14, which are attached on their flat or edge side to the support surface. The orientation of MoS_2 clusters by using HRTEM (high resolution transmission electron microscopy) has been reported to be standing upright as flakes on Al_2O_3 support with edge bonding (Pratt et al., 1990), while Srinivasan et al. (1992) disagreed and insisted that distinguishing between edge-bonded and basal-bonded MoS_2 clusters is difficult due to the porous structure of support, and concluded that no MoS_2 clusters were edge-bonded to catalyst supports, such as Al_2O_3, TiO_2, ZrO_2, and SiO_2.

The relationship between the morphology and orientation of active sites along with the catalytic activities have been investigated for MoS_2 on various supports (Daage and Chianelli, 1994; Sakashita et al., 2001; Araki et al., 2002), in which MoS_2 clusters with low stacking favored hydrogenation and those with high stacking were hydrogenolysis oriented. As an effect of support, titania containing supported catalysts showed promising results at least for model compounds and middle distillate hydrotreating, which contain relatively higher-layered MoS_2 clusters and possess higher intrinsic activities than single-layered MoS_2 clusters (Shimada, 2003). Literature reports are consistent with the comparison between Co-Mo-S (type I) and Co-Mo-S (type II) (Topsøe et al., 1981); Co-Mo-S has higher activity due to smaller electronic interactions and a larger amount of Mo exposed on the surface of support than single-layered Co-Mo-S. The effect of TiO_2 in the support has been attributed to high dispersion of active metal (Ramirez et al., 1993, Leliveld et al., 1997) or high degree of sulfidation (Okamoto et al., 1989; Yoshinaka and Segawa, 1998). These fundamental aspects of hydrotreating catalysts are either ideal just after sulfidation or catalysts used

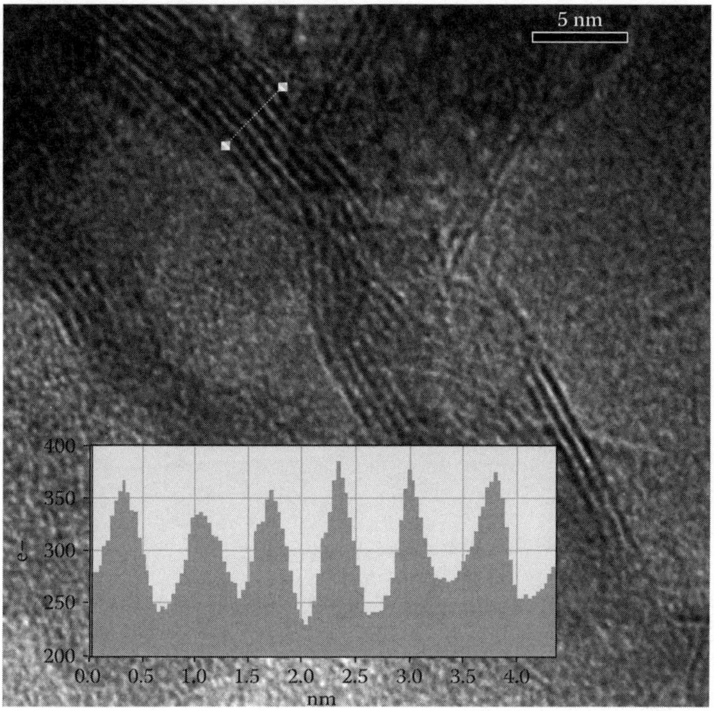

FIGURE 2.14 HRTEM micrographs of MoS_2-supported catalysts over mesoporous silica.

for light feedstock, such as straight-run gas/oil, model molecules, etc., while in the case of heavy oil these sites may be modified during the hydrodemetallization (HDM) due to metal deposition. Hence, one has to consider progressive variation in the catalytic sites during time-on-stream. As a summary of this fundamental discussion about the generation of catalytic sites, it can be pointed out that the support and its acid-base properties exhibit a wide range of variation, which is usually controlled to some extent by the preparation method.

2.4.2 Role of Catalytic Sites in Heavy Oil Reaction Mechanism

The study of reaction mechanisms for hydroprocessing of heavy oil is quite complex. That is why its knowledge is very limited, particularly when feedstock has very high amounts of Ni and V, which are deposited during the reaction and modify the catalytic sites. There are literature reports with scarce experimental data or with insufficient explanations, nonetheless, a few reports have been published with variation of heavy oil reaction conditions, such as temperature, hydrogen pressure, and H_2S partial pressure (Rankel, 1981; Ware and Wie, 1985a, 1985b; Bonné et al., 2001; Rana et al., 2007c). One observation from these studies is that it is primarily difficult to separate the effect of catalytic reaction sites and process conditions. There are few other literature reports on the chemical reaction in which hydroprocessing mechanism has been found to be not only dependent on the catalytic sites (acid/base and metal), but also is affected by the physical processes (diffusion into the pores, adsorption, surface reaction of reactants on the catalytic sites, desorption and diffusion of products), which are controlled by textural properties of the catalyst.

Nevertheless, based on catalytic results, an important HDM reaction mechanism has been proposed (Figure 2.15, scheme I). The role of sulfided catalytic

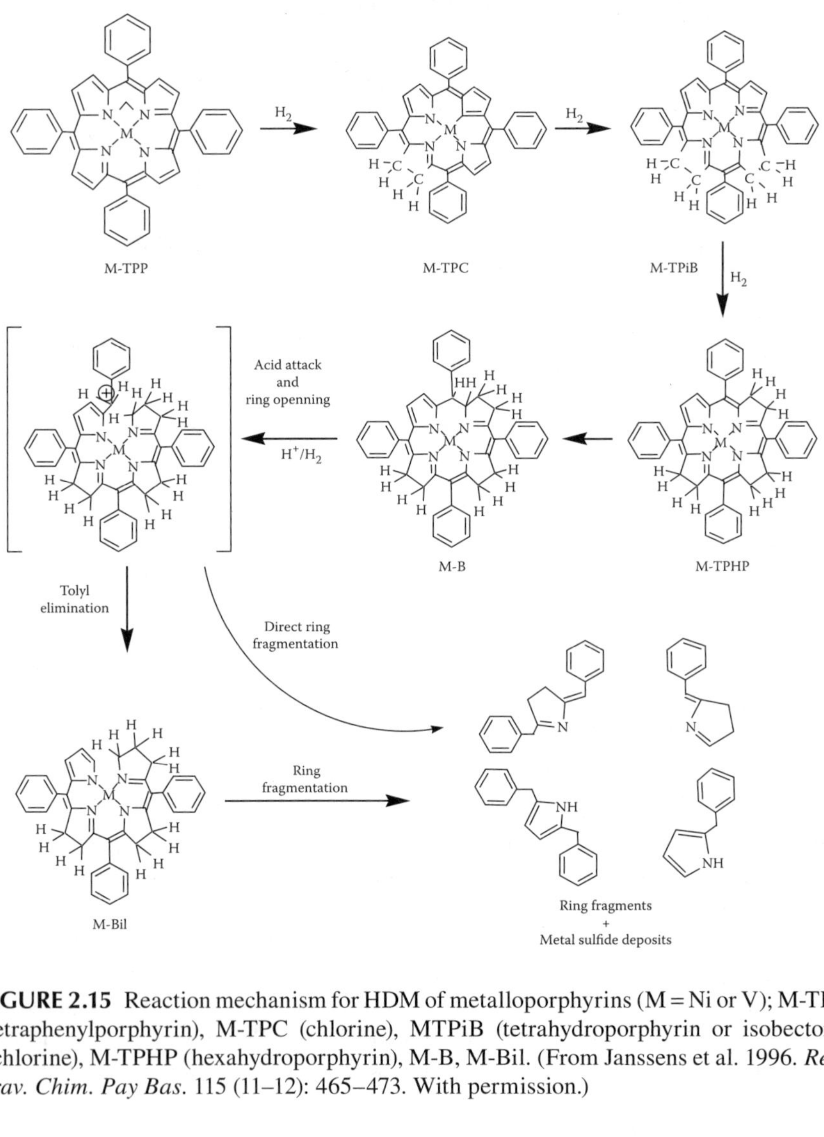

FIGURE 2.15 Reaction mechanism for HDM of metalloporphyrins (M = Ni or V); M-TPP (tetraphenylporphyrin), M-TPC (chlorine), MTPiB (tetrahydroporphyrin or isobectoriochlorine), M-TPHP (hexahydroporphyrin), M-B, M-Bil. (From Janssens et al. 1996. *Rec. Trav. Chim. Pay Bas.* 115 (11–12): 465–473. With permission.)

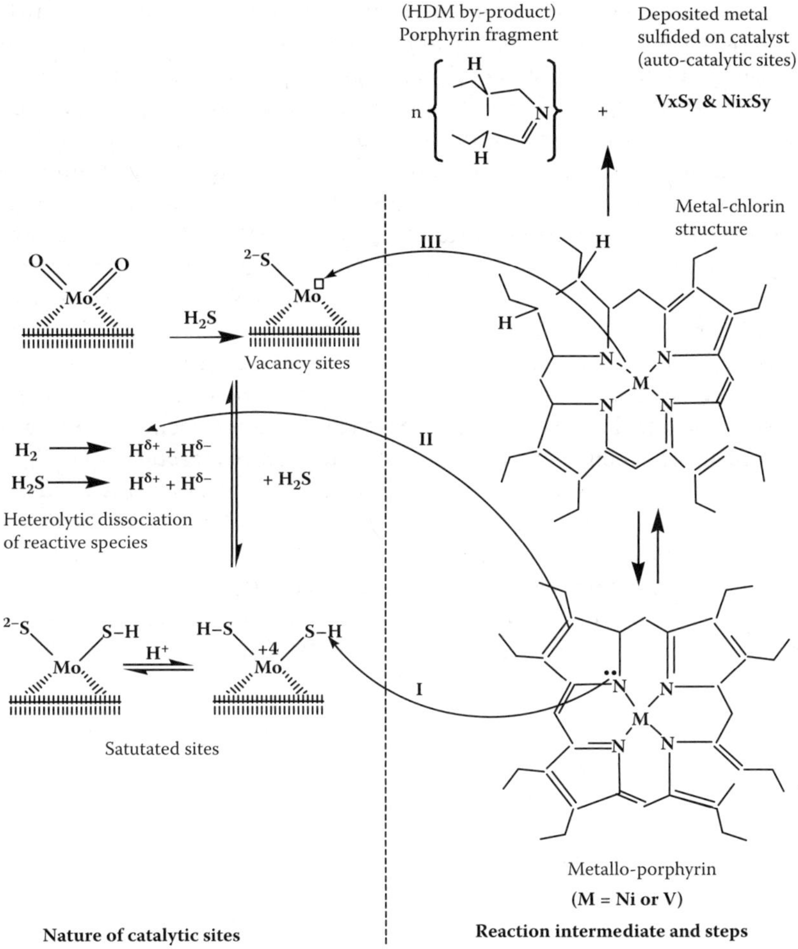

FIGURE 2.16 Proposed reaction mechanism for HDM in presence of high partial pressure of H_2S: (I) adsorption of metal-porphyrin ring to saturated sites (delocalization of electron density), (II) hydrogenation of porphyrine structure, and (III) demetalation of the metalloporphyrine complex molecule. (From Rana et al. 2007c. *Fuel* 86: 1263–1269. With permission.)

sites Coordinated unsaturate site (CVS) was not explained in depth (Janssens et al., 1996). Thus, a further modification in this reaction mechanism was suggested by Furimsky and Massoth (1999) using H_2S during the HDM process. In this regard, recently, an analogous mechanism was developed by Rana et al. (2007c) for the HDM reaction on a sulfided catalyst at an enhanced H_2S partial pressure, as shown in Figure 2.16 (scheme II). The main difference between schemes 1 and 2 was the fragmentation of the transition metal ring in the last step. HDM of the V and Ni mechanisms was proposed to involve the adsorption (step I) and hydrogenation (step II) of a transition-metal complex to form the corresponding chlorin structure (one of the double bonds of pyrrol is hydrogenated) (Rana et al., 2007c). The hydrogenation of the pyrrol ring is easy, and this step can be obtained even at high H_2 (H_2S) partial pressure even without a catalyst (Ancheyta et al., 2005b; Furimsky and Massoth, 1999); however, the presence of a Brønsted acid site (proton) further facilitates the adsorption and hydrogenation, which induces significant flexibility in the metal-ligand and diminishes the stability of the transition–metal complex. The final step (step III) in the HDM of porphyrins includes the cleavage of a metal–N bond, which involves hydrogenolysis and requires the presence of an anion vacancy or CUS of the sulfided catalyst. These results of nickel or vanadium removal were recently demonstrated experimentally at an accelerated H_2S partial pressure (Rana et al., 2007c), which proceeds through Brønsted acid sites that are characteristic of the sulfohydryl (-SH) and CUS. Based on this study, increasing the amount of H_2S had no inhibition effect on HDM and HDA (hydrodeasphaltenization) reactions, whereas HDS (hydrodesulfurization) and HDN (hydrodenitrogenation) reactions were inhibited considerably (Figure 2.17) (inhibition by H_2S). The effect of H_2S for an HDM model molecule was also reported in literature with similar kinds of results (Bonné et al. 2001; Rankel, 1981; Ware and Wei, 1985a; 1985b). According to Bonné et al. (2001), H_2S coordinates to the central transition-metal atom (Ni or V) and, as a result, weakens the metal–N bond, whereas Rana et al. (2007c) proposed that sulfohydryl groups are responsible for the destabilization of the metal–N bond

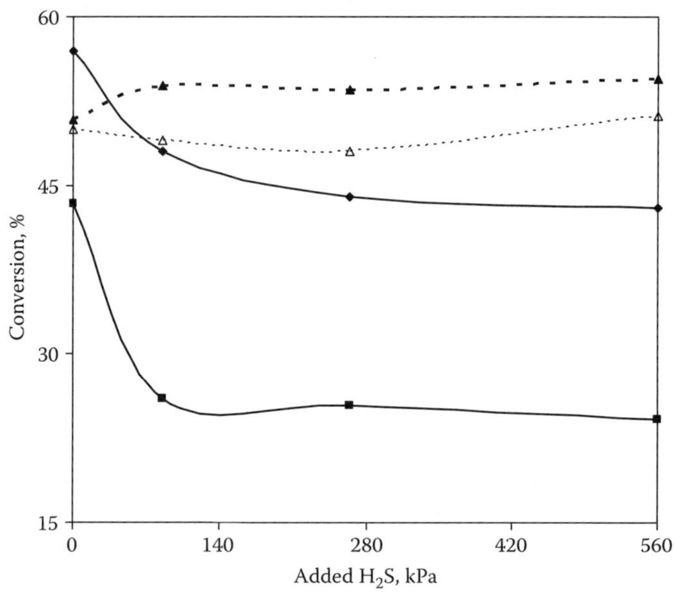

FIGURE 2.17 Effect of added H_2S partial pressure on steady state conversion of Maya crude hydroprocessing: (▲) HDM, (△) HDAs, (◆) HDS, and (■) HDN. (From Rana et al. 2007c. *Fuel* 86: 1263–1269. With permission.)

(Rana et al., 2000, 2004b, 2004c). It is also expected that an increased presence of -SH groups enhances the cracking of the asphaltene molecule or HDAs (Rana et al., 2007c). Apart from hydrotreating, due to the complex nature of feedstock, the reaction mechanism may also compete for several other instantaneous parallel reactions, such as hydrocracking, isomerization, alkylation, dehydrogenation, cyclization, etc., as shown in Figure 2.11.

2.5 EFFECT OF CATALYST PROPERTIES ON HYDROPROCESSING ACTIVITIES

Crude oils with high amounts of residual hydrocarbon fractions can be processed directly by hydroprocessing to produce upgraded synthetic crude oil (SCO), the so-called upgrading of heavy oils, which has been addressed as a feedstock that provides reasonable characteristics for FCC and hydrocracking (Furimsky, 2007; Ancheyta et al., 2005a). Although this alternative would require larger equipment to process the entire crude volume rather than only the residual fraction, some advantages are worth mentioning here: better priced synthetic crude oil would be produced, the problem of secondary processing would diminish, residue transportation would be avoided, among others. On the other hand, good stability of the catalyst along with optimum cracking activity will be required to process and convert this kind of dirty feedstock. The effect of catalyst properties and catalyst stability are both aspects of managing catalyst performance. Hydroprocessing catalysts prior to the reaction is sulfided in order to have an optimum level of catalytic sites (i.e. CUS), which remain active in the presence of H_2S and H_2 pressures.

2.5.1 Effect of Textural Properties of Catalysts

Textural properties are well recognized to be the most important characteristics of heavy oil processing catalysts, which are usually manipulated during support preparation. Therefore, selection or preparation of support is an important step in order to obtain heavy oil hydroprocessing catalysts with suitable textural properties due to large and complex molecule diffusion into the inner surface of catalytic sites. The internal surface area and porosity allow full excess of the surface active sites, which corresponds to the catalytic activity and selectivity of the catalyst; however, the overall performance of the heavy oil catalysts relies on a wide range of pores. The effect of hydrolyzing agents during the preparation, such as ammonia, ammonium bicarbonate, ammonium carbonate, and urea, on textural properties is shown in Figure 2.18, which indicates that alumina prepared by ammonium carbonate has ample pore structure compared with other aluminas, which are prepared with different hydrolyzing agents. The wide range of pore diameter was explained due to the CO_3^{-2} ions, which lead to wide pores during calcination (Rana et al., 2004a). There are several other literature reports that reveal that a desired textural properties support or catalyst can be prepared by varying preparation conditions, such as

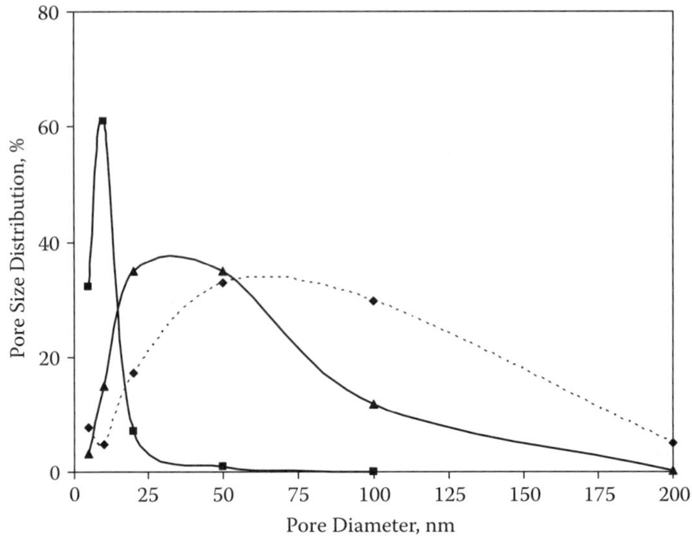

FIGURE 2.18 Effect of support (Al_2O_3) preparation methods on the textural properties of the catalysts: (▲) Al_2O_3-u (urea), (▲) Al_2O_3-acs (ammonia carbonate), and (■) Al_2O_3-am (ammonia). (From Rana et al. 2004a. *Catalysis Today* 98: 151–160. With permission.)

precipitating agents (Huang et al., 1989); by using additives (Trimm and Stanislaus, 1986); acid-base treatment (Absi-Halabi et al., 1993); aging (Johnson and Mooi, 1968; Ono et al., 1983; Chuah et al., 2000); washing, drying, and calcinations (White et al., 1987; Huang et al., 1989). The effect of seeding (Papayannakos et al., 1993), pH swing (Maity et al., 2003, 2004, 2006), and different hydrolyzing agents (Rana et al., 2004a, 2005a) are also reported for such objectives.

In general, alumina supports are precipitated at pH around 8 to obtain pseudo-boehmite with moderate pore volume and unimodal pore diameter, while precipitation at pH > 10 forms barite, having bimodal pore size distribution (Huang et al., 1989). The bimodal type of support can also be prepared by using additive or combustible fibers during the extrusion of boehmite (Tischer, 1981; Tischer et al., 1985; López-Salinas et al., 2005), and the effect of steaming agents on the alumina extrudate also been studied (Absi-Halabi et al., 1993; Walendziewski and Trawczynski 1993, 1994). These supports or different effects were observed at laboratory scale, while they have also been reported for commercial scale preparation of alumina in an integrated minipilot plant (Kaloidas et al., 2000). The impurities trapped in the precursor may influence the support pore structure as well as chemical properties, such as isoelectric point (IEP), which are responsible for active metal dispersion and also the catalytic activity. The textural properties of a catalyst not only varies with the preparation conditions, but also the support composition has an effect (Figure 2.19).

The development of industrial catalysts is an evolutionary process that requires significant pieces of information, which cannot be just textural properties, but

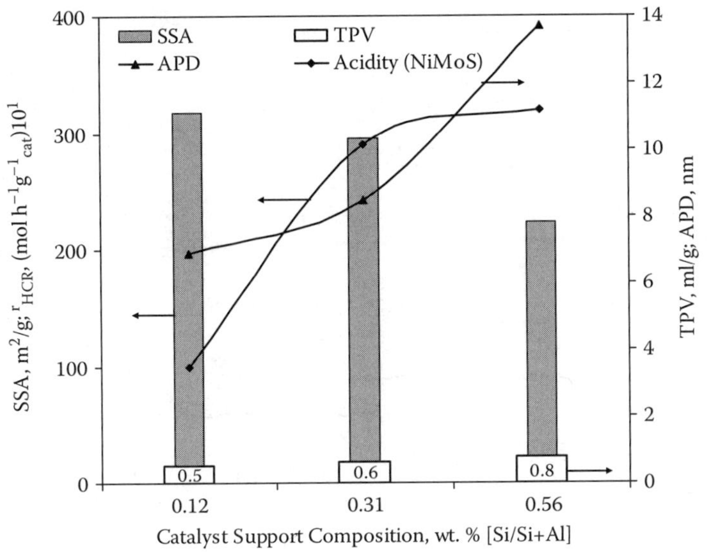

FIGURE 2.19 Effect of support (SiO_2-Al_2O_3) composition on textural properties and acidity of the catalyst.

also the dispersion of catalytic sites and physicochemical properties. The absolute function of the heavy oil processing catalyst is to remove metals and asphaltene. Since HDM and HDAs are believed to be diffusion-limited due to their molecular size, along with other factors, pore diameter of the catalyst is a deciding aspect. Moreover, catalyst surface is characterized by substantial metals penetration during hydroprocessing from the nearest surface to the core of the particle. A large amount of metals penetration and retention on the surface is required in order to define the stability of the catalyst. These large molecules are sensitive to the intrinsic activity, therefore, most of the particle can have a longer effective life and prevent pore mouths plugging on the surface of the catalyst, which will not foul as rapidly due to premature metal deposition (Sie, 1980; Rajagopalan and Lues, 1979; Agrawal and Wei, 1984a, 1984b). Also, larger pores allow easy access for large reactant molecules, whereas the smaller pores contribute more to the higher surface area. A relationship between pore diameter and surface area along with their effect on feedstock nature can be seen in Figure 2.20. In general, the higher the surface area, the greater the active metal dispersion, and the more active is the catalyst, while larger pores provide a better or easy diffusion of reactant molecules. Thus, considering the nature of feedstock (i.e., heavy oil) bimodal or even trimodal pore size (i.e., macro range) distribution may be incorporated into the catalyst. The bimodal pores system has emphasis on the balance between pore diameter and active metal dispersion, which also corresponds to the activity of the large and small molecules that are accessible in the crude oil (as shown in gas chromatography [GC] analysis).

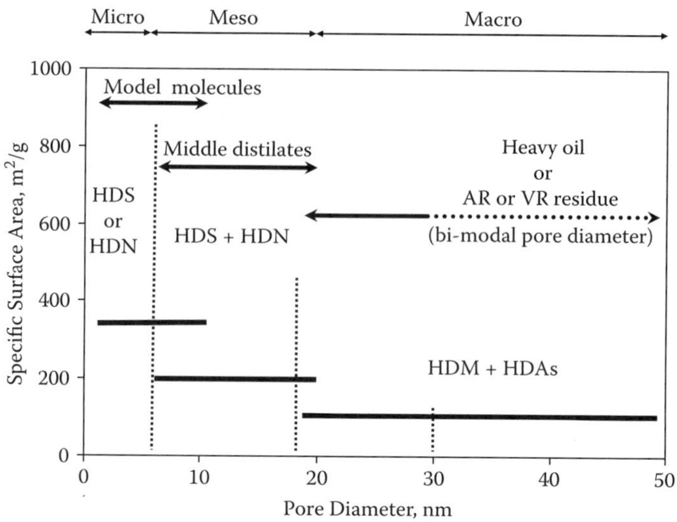

FIGURE 2.20 Relationship between the catalyst textural properties and various feedstocks. (From Ancheyta et al. 2005b. *Catalysis Today* 109: 3–15. With permission.)

In order to observe the effect of catalyst pore diameter on crude oil hydroprocessing, various alumina-supported catalysts were synthesized by using different preparation methods to vary pore diameter. The textural properties of these catalysts varied due to the different preparation method used for support preparation (Rana et al., 2004a). A steady state heavy oil conversion comparison for these catalysts is presented in Figure 2.21. In this figure, the catalyst having average pore diameter at about 22.2 nm showed lower activity, which indicates that pore diameter is not the only factor because activities of this catalyst are comparatively lower (except HDAs) than the other catalysts having smaller pores, probably due to the lack of catalytic sites on the low surface area catalyst. Since the composition of all catalysts is similar, these results reveal that asphaltene conversion is less affected by the metallic function (Co-Mo active sites), rather it is more influenced by the pore diameter of catalyst. The comparison between HDM and HDS shows opposite trends that reveal that an HDM catalyst should be essentially macroporous in nature (Toulhoat et al., 1990; Ancheyta et al., 2005b; Furimsky and Massoth, 1999). Because heavy crude oil contains a significant amount of asphaltenes, which are responsible for catalyst deactivation in hydroprocessing catalysts along with metal deposition, the results also indicated that HDM and HDAs significantly depend on the catalyst pore structure, while the HDS activity may depend on the dispersion of the active metals (Eijsbouts et al., 1993). The catalyst with lower average pore diameter (6.5 nm, pore diameter) shows the lowest activity for HDM and HDAs, while higher activity can be seen for HDS. Thus, HDM and HDAs conversions are limited due to diffusion of complex metalloids and asphaltenes molecules for the catalyst having pore diameter lower than

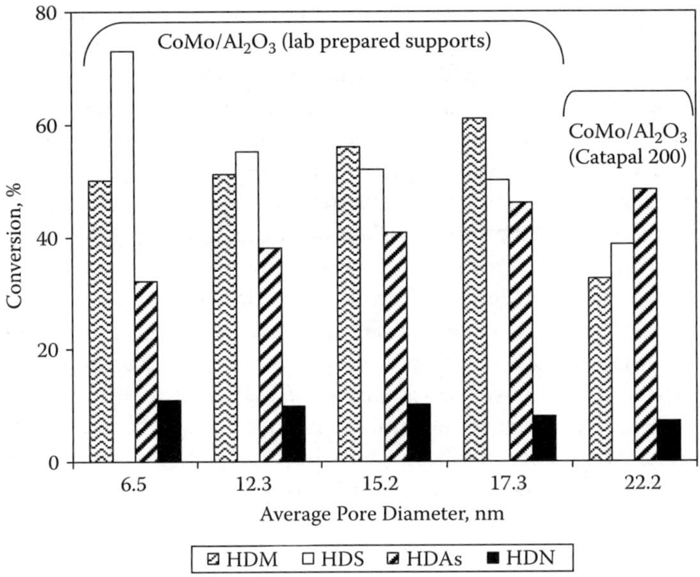

FIGURE 2.21 Effect of average pore diameter (APD) on Maya crude conversion after 60 h TOS (feedstock: Maya + HDS diesel, micro-reactor). (From Rana et al. 2007a. *Petrol. Sci. Technol.* 25 (1–2): 187–200. With permission.)

10 nm. Therefore, the performance of heavy oil HDT process with respect to different functionalities, such as HDM, HDS, and HDAs, is clearly linked to the catalyst porosity and nature of the heavy crude oil. The effect of support preparation on pore size distribution and average pore diameter apparently controls catalytic activities along with metal dispersion of active phases. The effect of pore volume and pore size distribution on catalyst deactivation is discussed later, specifically, related to the catalyst, which includes:

- Metal deposition
- Carbon deposition
- Variation in catalyst textural properties
- Catalyst life cycle
- Balance between catalytic activities (HDS, HDM, and HDAs) and pore diameter
- Effect of surface area, pore volume, and pore diameter on the mechanical strength

Most of these problems occur during hydroprocessing of the heavy oil due to the presence of asphaltenes, which makes treatment more difficult and they have a detrimental effect on the activity and catalyst life. Since asphaltenes are large molecules consisting of highly condensed heterocyclic and aromatic rings that are formed with sulfur, nitrogen, oxygen, vanadium, and nickel, it is anticipated that their diffusion into catalyst pores will be difficult and that generates coke deposition on the catalyst surface particularly near the pore mouth. Additionally, most of the heterometal compounds in the oil are concentrated in asphaltenes, so hydrodemetallization of these compounds leaves deposits of metal sulfides (NixSy and VxSy) in the catalyst pores. These deposited transition metal sulfides cause a decrease in the number of catalytic sites, hinder the transport of reacting molecules to the internal catalyst surface, and eventually cause the complete plugging of the catalyst pore. For these reasons, in the case of heavy oil hydroprocessing, characterization of catalyst textural properties is of great importance given the large molecules to be processed. Thus, the accumulation of these metal sulfide deposits governs the decline of the catalyst activity and determines the catalyst life. Numerous studies have been made for diffusion in fine porous materials, e.g., Renkin (1954) and Anderson and Quinn (1974) proposed theoretical correlations representing the effective diffusion coefficients as functions of pore diameter and molecular diameter. For hindered diffusion of asphaltenes, Baltus and Anderson (1983) showed a very strong pore size effect on the diffusion coefficient in the range 7 to 50 nm pore radius, while Shbnura et al. (1986) reported a model that can quantitatively describe the mechanism of catalyst deactivation into the pore structure due to vanadium sulfides and coke deposits; they have concluded that adequate pore size of a catalyst must be selected for sufficient diffusion of large molecules of asphaltenes and metals. An imaginary asphaltene molecule may be 4 to 5 nm in diameter, which is too large to pass through a micro- and even some mesopores in the catalyst (Larson and Beuther, 1966). Metals and other contaminants, such as S, N, etc., in the asphaltene aggregates are believed to be present as organo-metallic compounds associated to the asphaltene sheets making the asphaltene molecule heavier than its original structure (Figure 2.22). The characterization studies on Ni and V ethioporphyrins indicates that they have diameters of around 1.6 nm (Fleischer, 1963; Lee et al., 1991).

On the other hand, the effect of pore diameter is not only the factor, but also chemical properties (acidity, dispersion of catalytic sites) that have enormous effect on catalyst activity and the stability. In this sense, Al_2O_3-SiO_2 (AS) mixed oxide NiMo (4.6 wt% Mo and 1.7 wt% Ni)-supported catalysts were tested for heavy crude oil hydroprocessing, whose HDS conversions are seen in Figure 2.23. Considerable different behavior was observed with time-on-stream, particularly for deactivation. The catalyst containing lower silica content (31 wt.%) and smaller pore diameter had much better HDS activity and represented normal deactivation with time-on-stream, while large pore diameter and high silica supported catalyst showed high initial activity, but it deactivated with time at a very fast rate. The deactivation behavior can be attributed to the presence of a higher number of cracking sites of the catalysts (Leyva et al., 2009). The role of acidic sites can be also observed in the case of hydrodemetallization, as shown in Figure 2.24. It is also reported that the acid site generated by the support may have some role in the HDM catalytic activity (Rayo et al., 2007). However, these acid sites are likely to deactivate with a faster rate than

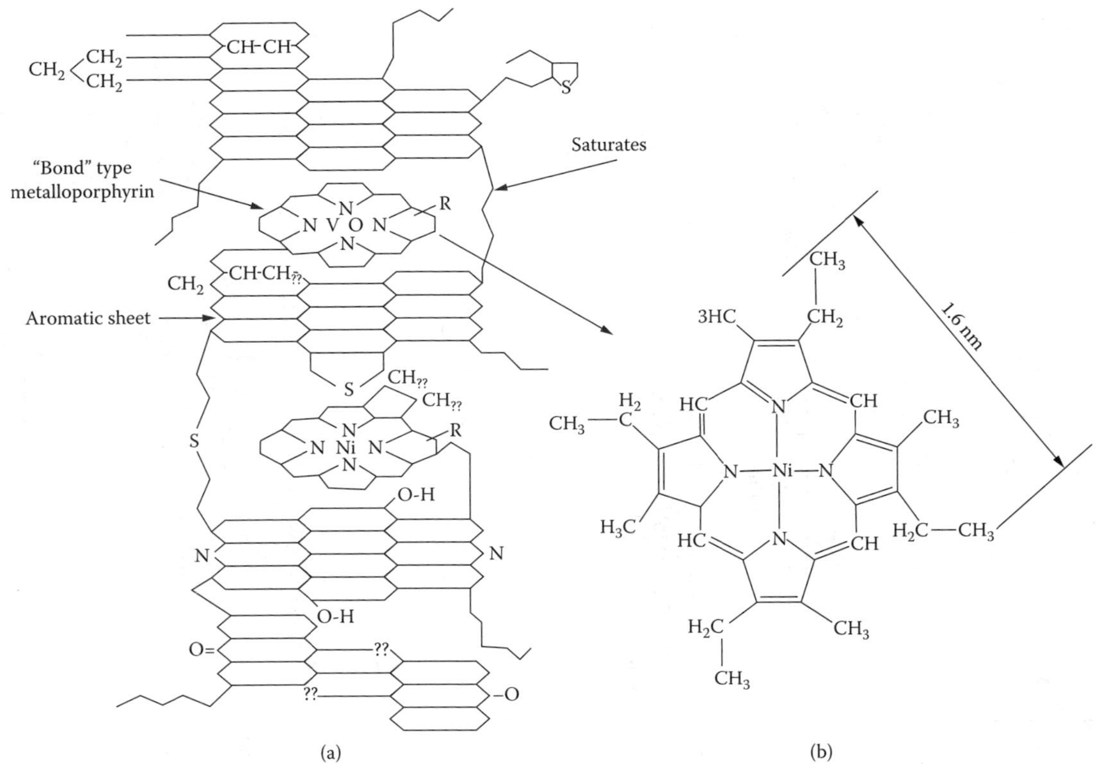

FIGURE 2.22 Hypothetical asphaltene molecule and its interaction with organo-metallic compounds. (a) Adapted from Speight, J.G. 1999. *The Chemistry and Technology of Petroleum*, 3rd ed., Marcel Dekker, New York. With permission.) (b) From Fleischer, E.B. 1963. *J. Am. Chem. Soc.* 85: 146–148. With permission.)

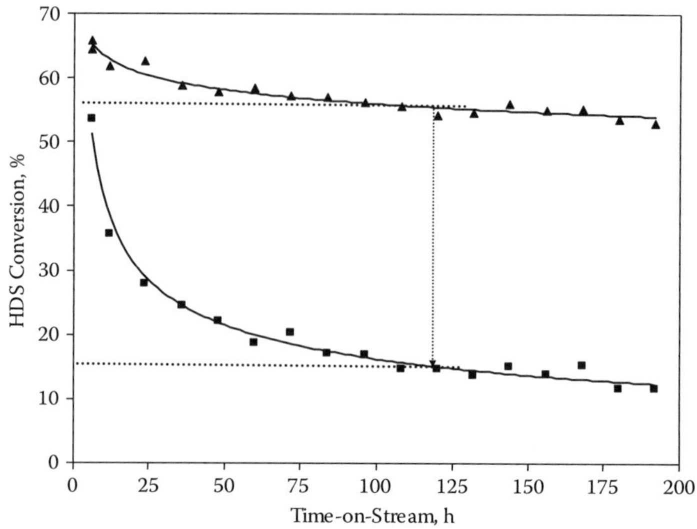

FIGURE 2.23 Effect of support composition on the hydrodesulfurization of Maya heavy crude: (▲) NiMo/AS-31; (■) NiMo/AS-56. (From Leyva et al. 2009. *Catalysis Today* 141(1–2): 168–175. With permission.)

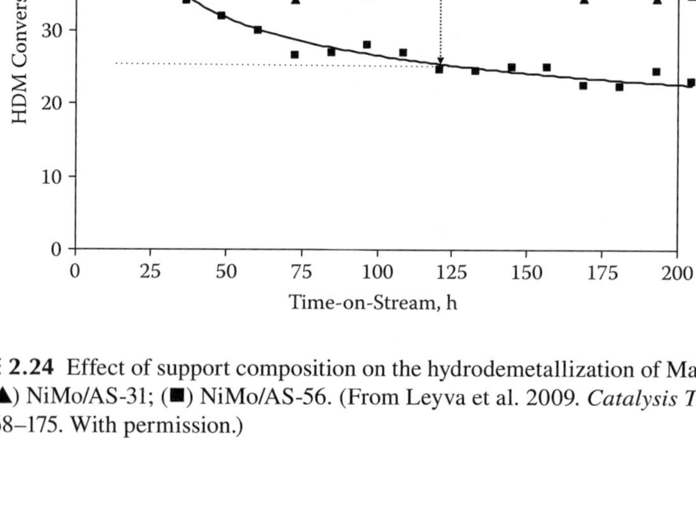

FIGURE 2.24 Effect of support composition on the hydrodemetallization of Maya heavy crude: (▲) NiMo/AS-31; (■) NiMo/AS-56. (From Leyva et al. 2009. *Catalysis Today* 141 (1–2): 168–175. With permission.)

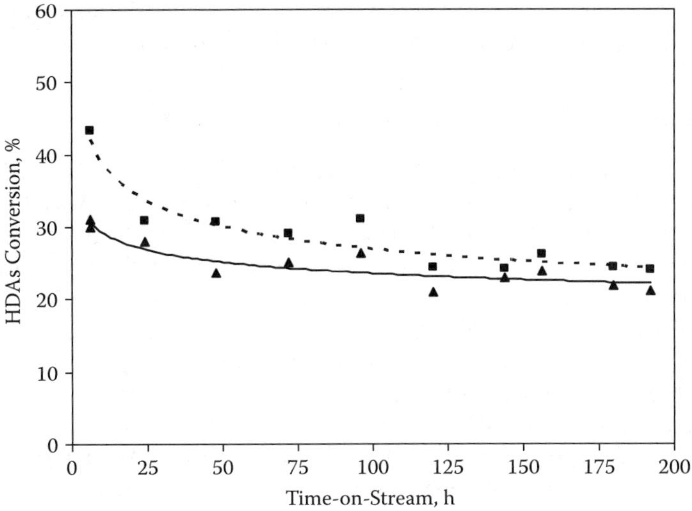

FIGURE 2.25 Effect of support composition on the hydrodeasphaltenization (HDAs) of Maya heavy crude: (▲) NiMo/AS-31; (■) NiMo/AS-56. (From Leyva et al. 2009. *Catalysis Today,* 141(1–2): 168–175. With permission.)

those generated by metallic catalytic sites. The variation in HDM and HDS may be due to the different pore diameter or the dispersion of Mo on these catalysts. It is also reported that the high amount of silica in support has a negative effect on the dispersion of active metals (Rana et al., 2003), which may correspond to the lower activity. Additionally, HDM activity increases with increasing average pore diameter, while HDS activity decreases (Rana et al., 2005b). When pore diameter is large, the surface area becomes smaller, which makes it very difficult to have high dispersion of active metals. In this regard, it is probable that either the catalyst should be bimodal or used for a particular application, such as HDS with high surface area or high dispersion, while HDM can be effectively achieved on the large pore diameter or low surface area catalyst (Eijbsouts et al., 1993).

Contrary to the results of HDS or even HDM, the HDAs activity for these catalysts is relatively different, as shown in Figure 2.25. Although the deactivation with time-on-stream showed similar tendency, the presence of acid sites is clearly evident for asphaltene conversion. The higher acidity-containing catalyst has higher HDAs, but lower HDS and HDM activity. This behavior can be explained in two ways, that it is either due to the pore diameter or due to the acidity. However, the combination of both properties favors the HDAs because the effective diffusion of asphaltene molecules occurs in large pores (macropores) and later enhances the cracking of the molecule. It appears from these results that the catalyst for heavy oil processing does require large pore diameter along with optimum dispersion of active metals, which are responsible for the catalytic reaction. However, the acidity of the catalyst is a secondary parameter for catalyst design and the modification will depend on the selectivity of the desired product.

2.5.2 Effect of Catalyst Composition or Support Effect

Different supports have been prepared in order to find a better catalyst with respect to the textural properties as well as dispersion of active phases. Usually support materials are Al_2O_3, SiO_2, B_2O_3, TiO_2, ZrO_2, MgO, and their mixed oxides, which have been reported previously as support materials for heavy oil hydroprocessing (Rana et al., 2005c, 2005d, 2006, 2007d; Caloch et al., 2004; Ancheyta et al., 2005b; Trejo et al., 2008). Catalytic tests and their supported catalysts (oxide as well as sulfided) characterization clearly show that the support contributes to the activity (Topsøe et al., 1996). In addition, supports also contribute to other important constraints on the catalytic activity (i.e., particles size distribution and aging time support preparation) that have also been studied. With variation of the support preparation method, the catalysts showed variation in surface area, pore volume, and size distribution, and the better textural properties tuned into the relatively good, long-term stability. The oxide supports play a key role toward the dispersion of active metal or their particle size growth. Additionally, the support can participate in the adsorption and exchange of oxygen via defect oxygen sites (i.e., acid sites) in the support lattice.

The benefits of using mixed oxides as supports in heavy oil hydroprocessing catalysts is limited because of relatively higher cost than alumina. On the other hand, mixed oxide supported catalysts have higher acidity, which may enhance the cracking function of catalysts along with the corresponding lower stability. In addition, higher acidity of catalyst has a parallel effect on HDM (Rayo et al., 2004; Rana et al., 2005a, 2005c). It has been shown earlier that the support properties are crucial for defining the desirable porosity of catalysts. The advantage of using mixed oxides supports for hydrotreating is their acid-base properties. This, in turn, improves hydrocracking activity, which is required for achieving desirable conversion of asphaltenes to distillate fractions. At the same time, more acidic supports have an adverse effect on other catalyst functionalities. Therefore, the activity of catalysts for hydroprocessing of heavy feeds must be optimized to achieve a desirable level of hydrocracking with considerable activities of HDM, HDS, and HDN. The effect of acid-base properties of the catalyst on heavy crude oil conversion with time-on-stream is shown in Figure 2.26. The basic supported catalysts (MgO-Al_2O_3 and ZrO_2-Al_2O_3) showed slightly better stability with time-on-stream (TOS) and are more selective for HDM activity than the acidic TiO_2-Al_2O_3-supported catalyst (Caloch et al., 2004; Rana et al., 2005d). At the beginning, the TiO_2-Al_2O_3-supported catalyst was more active due to its higher acidity. The variation in activity was attributed to the MgO, ZrO_2, and TiO_2 incorporated to γ-Al_2O_3. In contrast, the use of catalysts supported on basic supports in hydroprocessing of heavy feeds is limited because their hydrogenation and hydrocracking activities are low, apart from their high stability and good selectivity of HDM (Figure 2.27). Moreover, as an effect of support composition catalytic

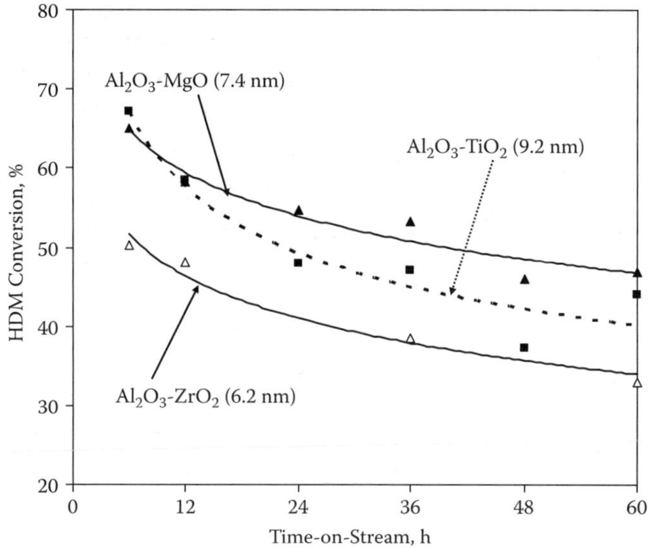

FIGURE 2.26 Effect of the nature of catalyst support on HDM conversion. (From Caloch et al. 2004. *Catalysis Today* 98: 91–98. With permission; and Rana et al. 2006. *Revista Mexicana de Ingenieria Quimica* 5 (3): 227–235. With permission.)

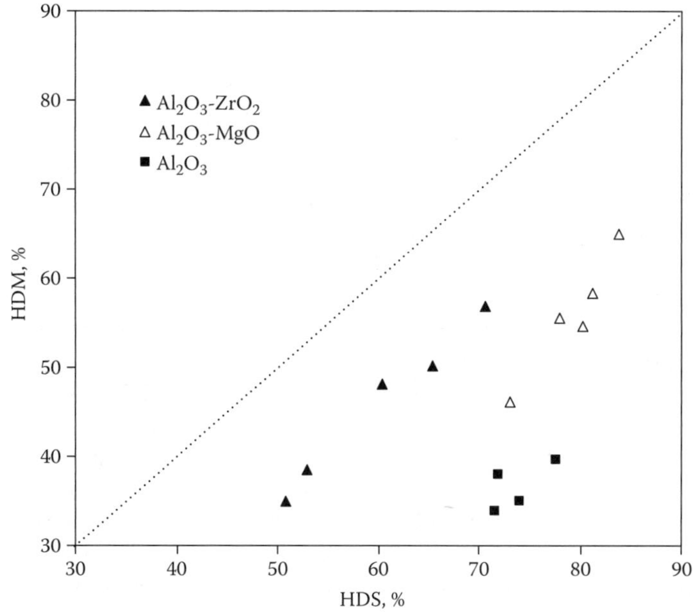

FIGURE 2.27 Hydrogenolysis selectivity for the hydroprocessing of Maya crude oil (380°C and 5.4 MPa) at various support composition. (Rana et al. 2006. *Revista Mexicana de Ingenieria Quimica* 5 (3): 227–235. With permission.)

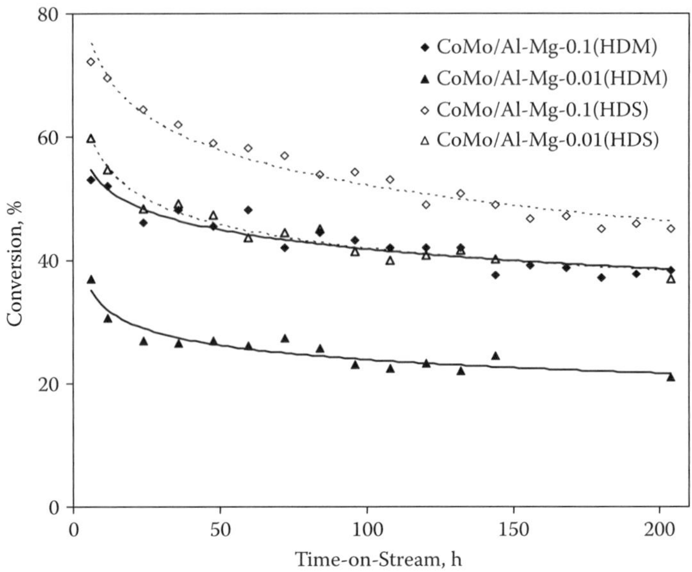

FIGURE 2.28 Effect of MgO content on HDS and HDM activities of heavy oil with time-on-stream at 380°C, 1 LHSV, and 7 MPa. (From Trejo et al. 2008. *Catalysis Today* 130 (2–4): 327–336. With permission.)

activities (HDS and HDM) are affected at wide range, which increase with MgO content (from 1 wt% MgO to 10 wt% MgO) in the catalysts (Figure 2.28). In the case of neutral supports like carbon, CoMo catalysts showed to be less sensitive to poisoning by N and better ability to restrict coke or sediment formation compared with alumina (Fukuyama et al., 2004); however, their hydrocracking activity is also low, which is a subject to the unique modification of carbon.

Among the mixed oxides, TiO_2-Al_2O_3 mixed supported systems are reported as promising support to have acidity as well as favorable geometry to accumulate Mo and Co (Ni), which give a high number of catalytic sites (Topsøe et al., 1996). Apart from that there are some other factors, such as thermal instability, low surface area, and poor mechanical properties, which are not favorable for using TiO_2 for commercial exploitation of heavy oil processing catalysts. Nevertheless, small amounts of TiO_2 into the alumina matrix can offset these unfavorable characteristics. Some properties, like acidity and texture, are reported to be highly dependent on the preparation method of the support material. The activity results in this regard indicate that the catalyst functionalities can be modified by incorporating a small amount of TiO_2 into Al_2O_3, which corresponds to either textural properties or acidity by varying the different Ti incorporation methods (Rana et al., 2005a).

TiO_2-Al_2O_3 supports can be prepared by various ways like co-precipitation, variation of Ti precursor, impregnation, and deposition techniques with similar

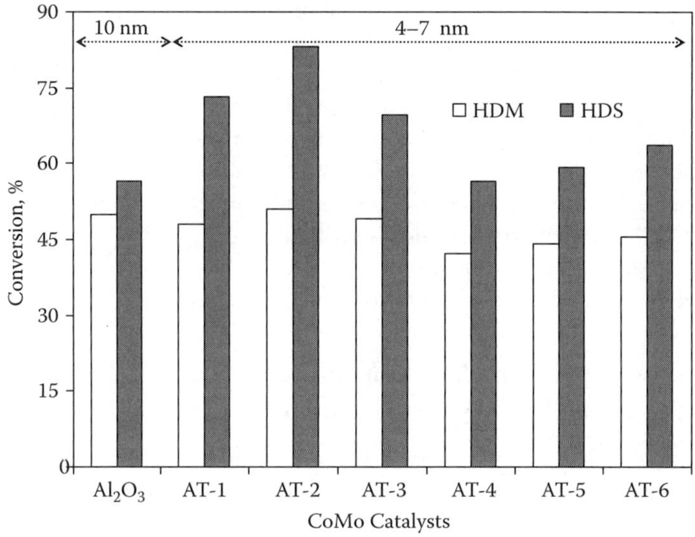

FIGURE 2.29 Comparison between HDS and HDM conversions (120 h TOS) over CoMo/Al_2O_3 and Al_2O_3-TiO_2 (AT) catalysts as function of TiO_2 (10 wt.%) precursor ($TiCl_4$= AT-1, AT-2, and Ti iso-propoxide = AT-3, AT-4, AT-5, and AT-6) and its incorporation method in alumina. (From Rana et al. 2005a. *Catalysis Today* 109: 61–68. With permission.)

concentration of CoMo. A comparison of HDS and HDM conversions (at 120 h TOS) over different TiO_2-Al_2O_3 and alumina-supported CoMo catalysts is shown in Figure 2.29 (Rana et al., 2005a). In all cases, the initial conversion was higher for TiO_2-Al_2O_3-supported catalysts than for the alumina-supported catalyst, but the results of initial conversion are not justifiable for heavy oil processing. The TiO_2-containing catalyst showed enhanced HDS activity (120 h TOS) due to the structural promotional effect of TiO_2, while HDM conversion either remains the same as in alumina or slightly less, which seems to be an effect of average pore diameter (APD) or stability of the catalysts. The TiO_2-Al_2O_3-supported catalysts have an APD in the range of 4 to 7 nm, while the alumina-supported catalyst has 10 nm, which might have a corresponding effect on the stability of the catalyst.

Considering the demand for a cracking catalyst and lighter fraction petroleum yield, the acidic catalyst demand is increasing day by day. Contrarily, bottom of barrel fraction is increasing up to the 50 to 60% in crude oil, which is required to crack into smaller fractions. Therefore, an acidic catalyst along with optimum pore diameter of the catalyst may enhance the possibility of promising catalysts for hydroprocessing of heavy oil. However, the use of an acidic catalyst has also a possibility of severe deactivation on acidic sites. It has been discussed earlier that the properties of supports are crucial for defining the desirable porosity and surface area of the catalysts. Because catalyst performance can be modified by chemical composition of supports, catalysts with attractive hydrocracking activity (bifunctional) and, at the same time, resistant to deactivation and more selective, are desired to enhance valuable product yields. The use of the aforementioned catalyst is reported only for fixed-bed reactors; there is a significant call for the moving bed technology; in this logic, zeolite-based catalysts will be suitable because of their flexibility of easy modification and excess commercial availabilities. However, the application of pure zeolite is limited because of its small pore diameter, particularly for heavy oil hydroprocessing (Rana et al., 2008). On the other hand, the mixing of crystalline zeolite into the amorphous matrix (alumina or silica) is also another difficult task trying to obtain essential mechanical (attrition) properties. The development of a hydroprocessing catalyst with certain hydrocracking activity is dominated by the use of zeolite because of its number and strength of acid sites.

2.5.3 Effect of Catalyst Shape, Size, Mechanical Properties, and Reactor Pressure Drop

The design of a catalyst bed is very much influenced by the structure of the packing bed, which in turn is governed by the catalyst shape, dimension, method of catalyst loading, and its particles. Thus, the reactor applications are required to optimize along with the catalyst properties, such as design of catalyst in terms of shape configuration, internal pores (pore diameter), and available surface area, which has positive effect on catalytic activity and overcomes transport properties of the system. Moreover, at the commercial catalyst design stage, fabrication cost, resistance to crushing and abrasion, as well as dust build-up during the catalyst loading need to be taken into account. Hence, mechanical properties are the underlying factors that have significant influence from a commercial point of view, such as shape, size, and mechanical strength.

Mechanical strength of a catalyst is very important in order to have solid physical strength. In general, commercial reactor bed length is about 90 feet or more. The catalyst is required to drop to this distance during the loading, which necessitates high mechanical strength. The lack of crushing strength generates large amounts of dust during reactor loading and the dust obstructs reactor filters and blocks catalyst pores, which results in loss of catalytic active sites. Conversely, to the large pore diameters or internal pore structure, the solid materials are required intact under demanding conditions, which improves catalyst durability and reduces catalyst replacement costs. Apart from the catalytic activity and if the catalyst does not possess adequate pressure, temperature, and catalyst loading techniques, various methods are used to estimate catalyst crushing strength, which are listed in Table 2.11. The mechanical properties of a catalyst increase with decreasing pore diameter, which lowers the heavy oil conversion. Additionally, the reaction selectivity depends on catalyst diffusion characteristics, and is controlled by the catalyst pore size distribution and particle shape used in the catalyst bed. The breaking of the catalyst particles in a fixed-bed reactor can cause critical problems, such as higher pressure drop, ill-distribution of flow, and shut down of the operation due to collapse of the catalyst bed. The mechanical and textural properties of a catalyst typically provided by the support type, as well as its composition, are mainly aimed during the period of preparation. Different factors like

TABLE 2.11
General Technique Used to Estimate Catalyst Crushing Strength

Test Features	Common Technique	Observation
Pressure shock	Autoclave	% survival of extrudate
Thermal shock	High temperature reactor	% survival of extrudate
Impact Resistant	(1) Individual	% survival of extrudate
	(2) Bulk	
Crushing strength	Flat plate	% survival of extrudate

Source: Mills et al. 2004. *Hydrocar. Eng*. 9, (9): 51–54. With permission.

precipitatc aging, pH, number of pH swing cycles, type of binder, and calcination temperature and time can affect the mechanical properties of the support (Ono et al., 1983; Snel, 1984a, 1984b, 1984c, 1987). The method of incorporating the active components to the support can also affect the mechanical properties of the final catalyst (Trimm and Stanislaus, 1986).

The shape and size of catalysts affect the overall process system, therefore, supports are important parameters in the preparation of commercial hydroprocessing catalysts. To achieve good catalyst performance, it is important to match the size and shape of the catalyst with the properties of the feed and type of reactor. In the case of heavy feeds, special attention has to be put on the shape and size of the catalyst in view of the diffusion problems encountered during the hydroprocessing of the large molecules contained in these feeds. However, optimizing of shape and size of the catalyst or packing material requires a number of considerations, such as extrudates allow maximum quantity of a catalyst in a reactor volume while spheres may offer improved crushing strength and lower pressure drop (ΔP). Currently, rings are the most common (particularly penta-ring and wagon wheels) offering very low pressure drop and high geometric surface area with lower crushing strength. A mixture of these various catalysts shape and size can be effectively used in the reactor. Good flow distribution and low pressure drop are achieved through proper selection of the shape and size of the catalyst as shown in Table 2.12a and Table 2.12b; however, the use also is dependent on whether the feed is light naphtha, gas oil, heavy crude oil, or residue. Thus, the major breakthrough can be obtained using graded bed technologies for the processing of complex or highly contaminated feedstocks. A drawing of a graded reactor is shown in Figure 2.30, in which a large dense stone support is loaded at the top of the reactor, subsequently, guard bed material, process catalyst, small dense stone support, and, at the bottom, a large dense stone support. The graded bed loading overcomes the bed plugging or some extent of the deactivation of the catalyst that corresponds to the longer catalyst life and stability of the catalytic system.

A guard bed (DEMET catalysts) is used for improved protection against fouling and for an improved liquid distribution. Usually the pressure drop that builds

TABLE 2.12A
Catalysts Shape and Their Use in the Type/Place in the Reactor

Catalyst Shape	Method of Preparation	Shape	Size	Use in Reactor
Pellets	High pressure	Cylindrical, rings	2–10 mm	Fixed bed
Extrudates	Squeeze through holes	Irregular lengths, circular, star, or loves cross section	–	Fixed bed, tubular reactor, and ebulating bed
Spheres	Aging liquid drops	Round	1–20 mm	Tubular reactor
Granules	Fusing and crushing particles	–	2–14 mesh	Liquid phase reaction
Flakes	Encapsulated in wax		<100 μm	Fluidized bed
Powder	Spray dried hydrogel	–	–	–

up is caused by formation of deposits at the top-bed of the catalysts. The design of demetallization or guard-bed catalysts is a special case that differs from the design of HDS and HDN catalysts. To reduce the impact of deposition on pressure drop build-up, a gradual change of catalyst activity and sizing is required. To achieve this target, Raschig rings and graded sized catalyst loading is offered. Additionally, sock-loading of the top layers of large sizes of catalysts and support media is recommended, which is specially designed to prevent fouling caused by particulates, gum formation, and metal contaminants in the feed, such as Fe, Ni, V, As, and Si. Usually, these contaminant metals that are removed from the feed by catalytic reactions are deposited onto the catalyst (Thakur and Thomas, 1984; 1985; Furimsky and Massoth, 1999; Ancheyta et al., 2005b; Bartholomew, 1994; Bridge, 1990; Marafi et al., 2007). A large number of pressure drop problems, however, have a similar cause, such as iron deposition, which is originated from corroded upstream equipment and the particles sized ranging from nanometer (nm) to millimeter (mm).

Reactor plugging is a common phenomenon for heavy oil processing due to the deposition of contaminants at the top layer of the catalyst, which generates a difference between the reactor entrance and the outlet of the reactor (i.e., called pressure drop, delta-P, ΔP). It is also characterized by "energy loss," and is an important consideration in the design and operation of fixed-bed systems, which is a subject of great interest, particularly if stipulation feedstock is virgin crude oil. A vast amount of information is reported in the form of empirical and semiempirical correlations, which relate to the pressure drop and hydrodynamic conditions, such as velocity of the fluid and structure of the fixed-bed reactor (Foumeny et al., 1996; Afandizadeh and Foumeny, 2001). Thus, there is a need to develop improved process designs to work in line with graded bed technology and more efficiently utilize high activity and selective catalysts. These designs eliminate the sensitivity to plugging or pressure drop mainly by applying sufficient void fraction at the top of the catalyst bed in a fixed-bed reactor system.

TABLE 2.12B
Various Filling Material or Reactor Inert and Catalyst Support Materials

Inert Solid	Shape	Principal Use of Material
Penta rings media		The low pressure drop of screened materials of random sized and shaped at top catalyst bed
Macro Trap™ guard bed Media	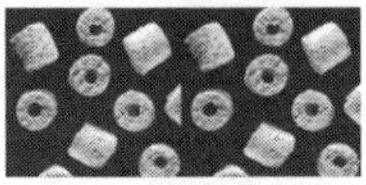	Designed ceramic material with large pores that trap particulate matter from the feedstock and problems of mal-distribution of the feed, pressure drop caused by catalyst bed fouling and extends the life of the catalyst
Denstone® Catalyst Bed	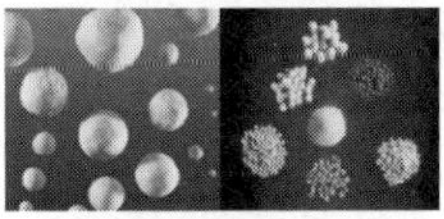	Support Media translates into lower operating costs Maximum resistance to erosion and attrition minimizes the possibility of catalyst contamination
Support Media Grid Blocks		Grid Blocks offer many advantages over conventional secondary support systems utilizing cross partition rings
Raschig Rings		Raschig rings are a form of random packing and are typically used to control the reactor pressure (i.e., ΔP)
Pall Rings	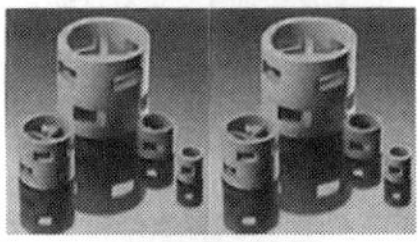	Lower pressure drop, Less total packing volume per reactor, Fouling and plugging resistant, Retrofits increase capacity and lower operational costs.
Rosetter Ring		Ring packing has the advantages of high void ratio, low pressure drop, low mass transfer, high flooding point, uniform gas-liquid contact, small specific gravity, high efficiency of mass transfer
Heilex Ring		

On the other hand, moving or ebullating bed processes are effectively used for heavy oil hydroprocessing in which hydrocarbon feed and H_2 are fed up-flow through a catalyst bed, expanding and back-mixing the bed, minimizing bed plugging, thus reducing ΔP. The ebullated-bed technology utilizes a three-phase system that is gas, liquid, and solid (catalyst). There are two most important ebullated bed processes (H-Oil and LC-Fining), which are similar in concept but different in mechanical details. From the catalyst point of view, moving bed catalysts are

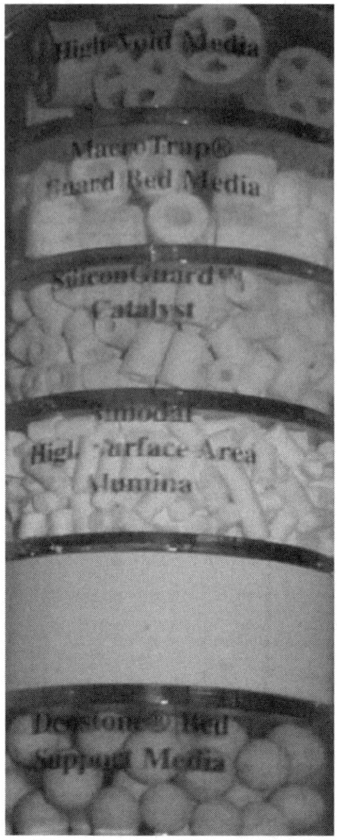

FIGURE 2.30 Example of shapes and sizes of supports and catalysts used in a typical catalytic bed.

chemically quite similar to the fixed-bed catalysts, except that catalyst particles and their mechanical strength and shape should meet the more demanding situation.

It is well reported in the literature that heavy oil hydroprocessing is a diffusion-controlled reaction in which the rate of diffusion is smaller than the rate of reaction. Such a problem can be avoided in most of the cases by modifying pore diameter and shape of the catalyst, which can be achieved by decreasing the particle diameter of the catalyst (LePage et al., 1987). The use of different shapes may promote better diffusional path of reactants from the external surface to the center of the catalyst particle, which can be avoided by the use of different shapes of catalysts. Some of these particle shapes, which may be suitable for hydroprocessing heavy feeds, are shown in Figure 2.31. Total geometric volume and external surface area of lobe-shaped catalytic particles as a function of easy-to-determine geometrical parameters were presented elsewhere (Macias and Ancheyta, 2004). Therefore, refiners may opt to change the shape from cylinder shaped to multilobed catalysts. The preparation details of such shapes are reported in the literature (Richardson, 1989;

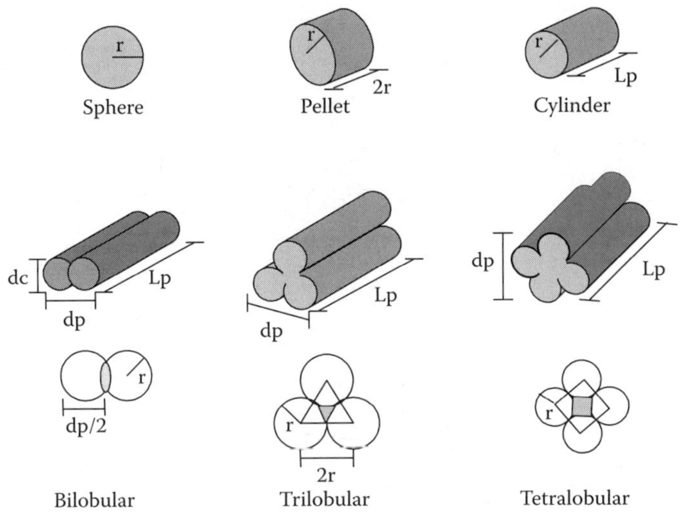

FIGURE 2.31 Particle shapes of industrial hydroprocessing catalysts. (From Macias and Ancheyta 2004. *Catalysis Today* 98: 243–250. With permission.)

LePage et al., 1987; Moulijn et al., 1993). Usually, the multilobed catalyst shapes presenting big pores, small particle diameter, and large external area are preferred (Cooper et al., 1986; Bartholdy and Cooper, 1993). Figure 2.32 shows the effect of catalyst shape (MacroTape Spheres) on reactor pressure drop occurring due to the particulate and metal sulfide (Ni, V, Fe, etc.) deposition (Mills, 2004). It seems that no easy solution is available because small catalyst particles decrease diffusion paths, but cause high-bed pressure drops. On the other hand, catalysts with wide pores are less susceptible to diffusion limitations, but they have lower surface areas. So, to determine what option one must use in a particular hydroprocessing

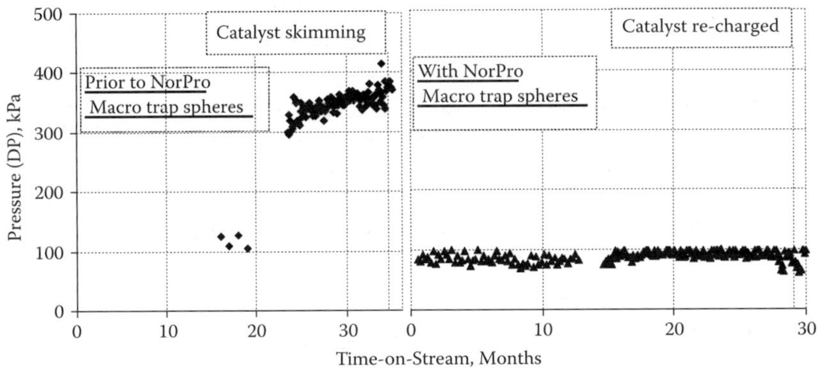

FIGURE 2.32 A layer of MacroTrap media eliminated pressure drop increase. (From Mills, K.J. 2004. *Hydrocarb. Eng.* 9 (9): 51–54. With permission.)

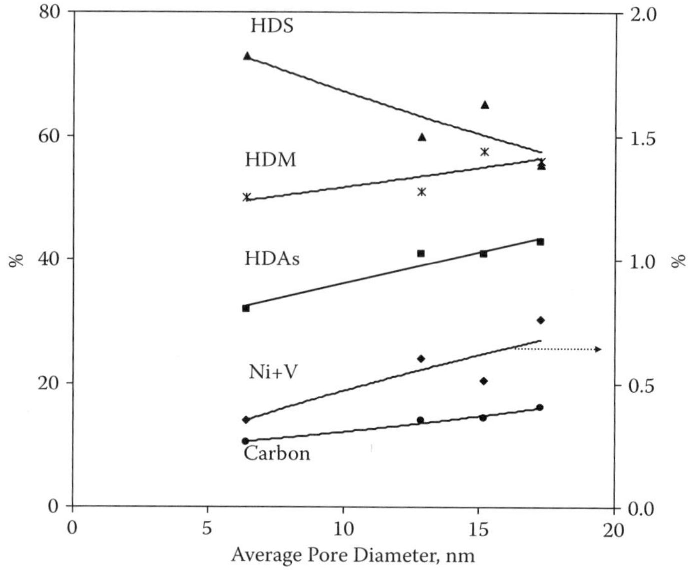

FIGURE 2.33 Effect of catalyst average pore diameter on the catalytic activity (60 h TOS), and carbon and metal deposition. (Rana et al. 2005b. *Catalysis Today* 104: 86–93. With permission.)

operation, an integrated approach that takes into account size, shape, pore size, and catalyst loading must be considered (Cooper et al., 1986). It is important to remember that under severe diffusion limitations, performance can also be improved by an increase in catalyst pore size. To emphasize the importance of pore diameter for some of the reactions taking place during hydroprocessing, Figure 2.33 presents the activity of HDS, asphaltene conversion (HDAs), and metal removal (HDM) for a series of $CoMo/Al_2O_3$ catalysts as function of average pore diameter. The figure shows that for HDS reaction involving small molecules the optimum average pore diameter is about 10 nm. In contrast, for HDAs and HDM reactions the optimum activity lies between 15 and 20 nm. The figure also indicates that small pore volume catalysts are best for model molecules and light feeds while large pore volume and, consequently, low surface area catalysts work better for heavy feeds where HDM and HDAs are much more important. Thus, the diameter of the asphaltene and metal (Ni or V) complex are subject to diffusion limitation because asphaltene micelles have about the same dimensions as the pore of the catalyst, hence, the heavy oil.

2.6 CATALYST LIFE (CATALYST STABILITY)

The catalytic conversion of bottoms of barrel, heavy, and extra heavy crude oil into lighter fuel oils is expected to provide the refining industry with a major technological challenge due to the deactivation nature of catalysts. Moreover, the crude oil nature is shifting toward the heavy and extra heavy, which have very high amounts of asphaltene, metals and other hetero-atoms. During processing of these feedstocks, Ni- and V-porphyrins decompose in catalytic cracking or hydrocracking to yield metals sulfides, which remain deposited either on active site or plug catalyst pores, while the complex hydrocarbon cracking (asphaltenes) leaves carbon deposition (coke content) that also leads to catalyst deactivation (Absi-Halabi et al., 1991; Hauser et al., 2005; Rana et al., 2007b; Marafi and Stanislaus, 1997; Marafi et al., 2007). The deposited contaminants content on the catalyst is not the only factor that influences its deactivation behavior, but also the location (in the micro-, meso-, or macropores at the outer surface of the pore) and the deactivation mode (simple pore filling or pore blockage) of the catalyst must be considered. Different authors have used a simple method to determine the amount of deactivation or eventual pore blockage by coke and metals that prevent access of internal surface of the pore (Newson, 1975; Johnson et al., 1986; Fleicher et al., 1984; Rana, 2005b). The carbon deactivation is initially due to acid sites coverage at the surface while as the coke deposited increases, coke is formed either on the external sites or by growth of coke molecules located on sites near the external surface producing pore blockage. Thus, typically, coke formation is expected to be at or near the pore mouth due to the huge molecular size of asphaltene at reaction conditions inside the pore that decomposes at the pore entrance. Therefore, enhances in pore volume and pore size distribution are important to maximize catalyst utilization when deposition is a problem. Figure 2.34 compares the nitrogen physic-sorption isotherms of fresh and spent catalysts with small (6 nm) and large (17 nm) pores diameter. The variations in the shape of the hysteresis loops were considered as representative of pore mouth deactivation (Rana et al., 2005c). The adsorption/desorption isotherms obtained for the fresh catalyst are similar with a type IV curve with a sharp upturn in the high relative pressure region comparable to a type II curve. This sharp upturn or lack of saturation in the high relative pressure region is indicative of liquid condensation associated with the presence of large pores (macropores) in the catalysts. The hysteresis loop of the adsorption/desorption isotherm of the low pore diameter is relatively different in two catalysts and has a combination of H1 and H3 characters that correspond to cylindrical and slit-shape pore geometries. The adsorption/desorption isotherms of spent catalysts also correspond to a type IV curve, indicating mesoporosity, but in contrast to the fresh catalyst, the hysteresis loops area increases relatively corresponding to a H2-type loop, indicative of "ink-bottle neck"-shaped pores in a dense network (Gregg and Sing, 1982; Webb and Orr, 1997; de Boer, 1958). The buildup of carbon is reversible and catalyst can be regenerated while the metal deposits are irreversible and largely determines the catalyst life. Different aspects of the poisoning by coke and metal in hydroprocessing catalysts have been addressed by Furimsky and Massoth (1999).

Apart from the textural properties, a common spent catalysts characterization is electron microprobe technique (SEM-EDX), which is used to study the deposition of nickel, vanadium, carbon, and other metal compounds onto and into the catalyst during hydroprocessing of bottoms of barrel and heavy crude oils. EDX analysis yields accurate information on surface as well as quantitative results

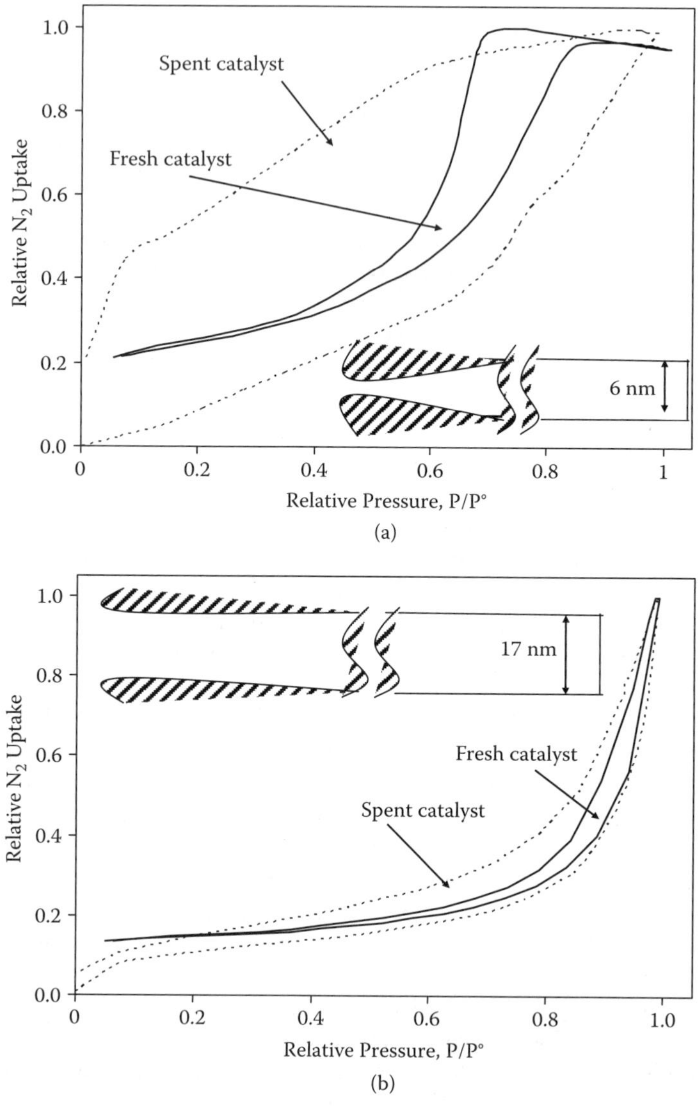

FIGURE 2.34 Fresh and spent catalysts adsorption–desorption isotherms: (a) CoMo/Al_2O_3 (6 nm average pore diameter) and (b) CoMo/Al_2O_3 (17 nm average pore diameter). (Adapted from Rana et al. 2004a. With permission. From Ancheyta et al. 2005b. *Catalysis Today* 109: 3–15. With permission.)

featuring all the elements present in the support catalyst and spent catalyst, specifically, C, O, Al, S, V, Ni, Mo, Co, etc., by their *K*α (in the case of Mo *K*α and *L*) x-ray emission lines (keV), as shown in Figure 2.35. The amounts of deposits on the catalysts are also depending on the nature of specific feedstock and textural characteristics of the catalyst, as seen in Figure 2.36. The metals concentration

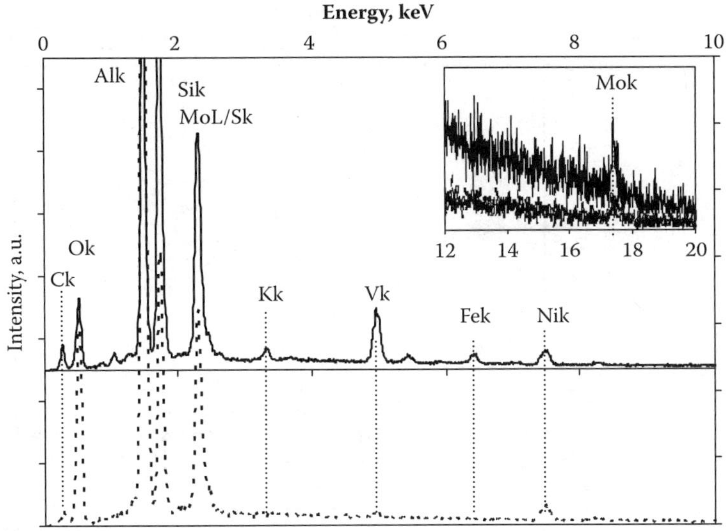

FIGURE 2.35 EDX spectra of deposited elements on spent catalysts [(----) NiMo/AS-31; (—) NiMo/AS-56]. (From Leyva et al. 2009. *Catalysis Today* 141 (1–2): 168–175. With permission.)

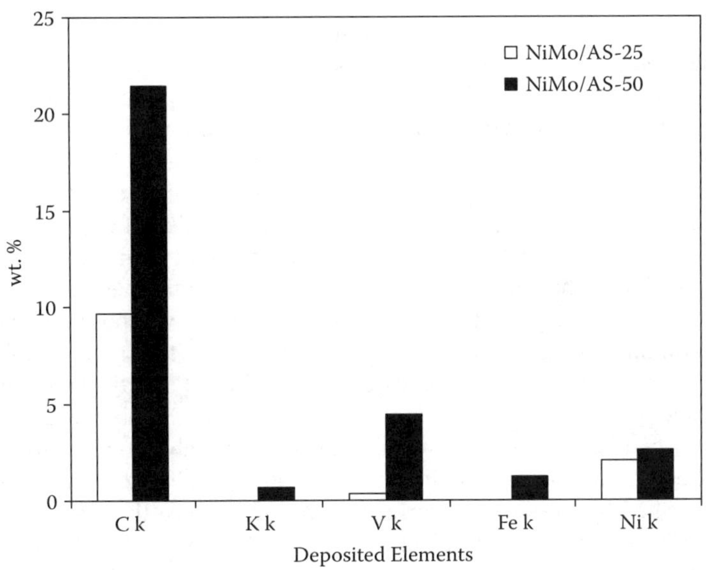

FIGURE 2.36 Quantitative (average) analysis of deposited elements on spent catalysts [(□) NiMo/AS-31; (■) NiMo/AS-56]. (From Leyva et al. 2009. *Catalysis Today* 141 (1–2): 168–175. With permission.)

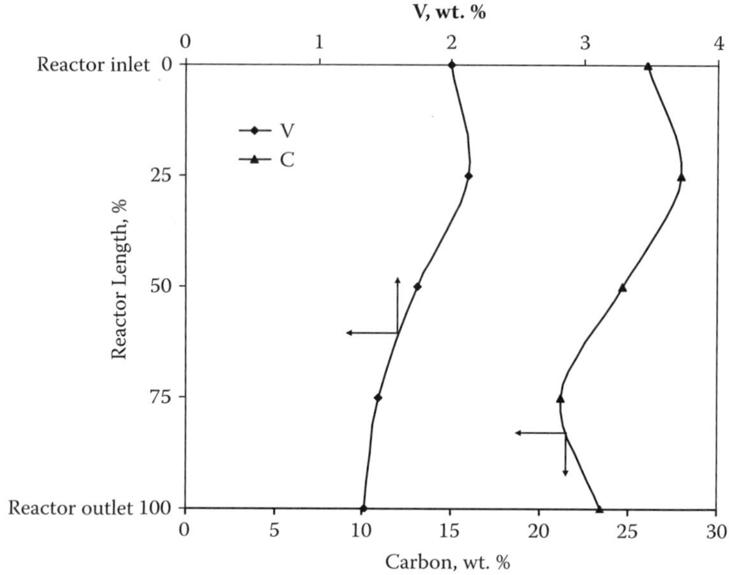

FIGURE 2.37 SEM-EDX analysis of spent catalyst showing carbon and vanadium distribution along with catalyst bed-depth. (From Rana et al. 2008. *Catalysis Today* 130: 411–420. With permission.)

at the point of maximum deposit (catalyst near the reactor inlet) reaches a limiting value before the catalyst bed is fully deactivated predominantly by metals (Ni and V) and carbon deposition during the reaction as presented in Figure 2.37 as a function of reactor length. The figure shows that in the spent catalyst V and carbon are highly concentrated in the first part (top-bed) of the reactor and exhibited maximum values at around 25% of the total reactor bed-depth, which is expected due to the asphaltenes cracking. Since conversion of asphaltene is more controlled by the temperature than the catalytic reaction (Tanaka et al., 2003), it is an indication to the induction period of complex metal molecules breakdown. Similar results were reported by Fleisch et al. (1984) for V deposition, where carbon deposition was increasing with weight percent of catalyst loading. Recently, Takashi et al. (2005) reported that carbon is distributed evenly along the reactor length. The deposition of metal, particularly vanadium, is observed throughout the reactor bed length, while Ni was deposited in traceable concentration only at the top-bed of the catalyst. The distribution profile of sulfided particles of Ni and V is confirmed by EDX on the surface of the catalyst. At the top of the reactor nickel sulfide species were relatively uniformly dispersed while larger particles or higher concentrations of V deposition are present on the surface of extrudate, with smaller amounts at the interior, as shown in Figure 2.38.

The literature has mainly dealt with Ni and V deposition phenomenon, e.g., shell progressive poisoning. In this sense, mapping results for Ni and V are shown in Figure 2.39 and Figure 2.40, respectively. The textural properties (pore diameter)

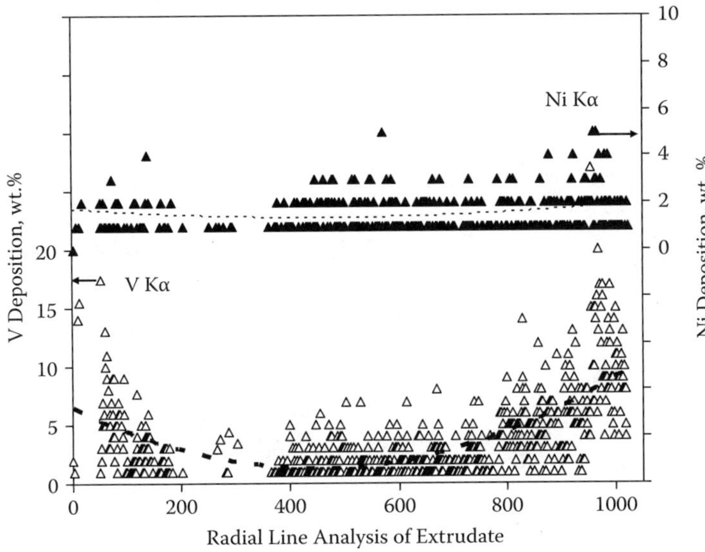

FIGURE 2.38 Radial distribution of vanadium (VKα) and nickel (NiKα) along the extrudate. (From Rana et al. 2008. *Catalysis Today* 130: 411–420. With permission.)

of two spent catalysts demonstrate that the NiMo/AS-56 catalyst, with a larger number of pores that are greater than 25 nm, facilitates the diffusion of Ni and V complex during the hydroprocessing and, as a result, the distribution of Ni and V was homogeneous. The other catalyst (NiMo/AS-31), which has smaller pores (<15 nm), has Ni distribution homogeneous, but the V was deposited more on the surface. These results clearly indicate that diffusion limitation plays an important role to distribute metal deposition, which is further confirmed by the Ni and V metals mapping that are complementary to the deposited profile and the pore diameter of these catalysts. When the poison molecule is large like Ni- or V-porphyrins, its deposition on the catalyst presents site coverage as well as pore blockage. This has been dealt with through "pore-mouth blockage" models, in which pore network structure should be taken into account (Figure 2.41). Profiles of the poison concentration inside the particle will then develop, even in the absence of diffusional limitations, just like in coking (Froment, 2001). In Figure 2.39, V metal deposition is clearly indicating that hydrodevanadization (HDV) is a diffusion limited reaction, while Ni is distributed relatively homogeneous on the catalyst (Leyva et al., 2009; Rana, et al., 2007a). Contrarily to the metal deposition, carbon deposition is relatively distributed homogeneously for both catalysts, as shown in Figure 2.42. The HDM of model compounds of vanadium porphyrin (HDVp) and nickel porphyrin (HDNip) over a sulfided catalyst at several temperatures, hydrogen pressures, and initial porphyrin concentrations indicated that a hydrogenated intermediate leading to deposited metal was found for both reactants. The HDVp is faster than HDNip, which is attributed to a higher intrinsic reactivity of vanadium porphyrin than nickel porphyrin (Chen and Massoth, 1988).

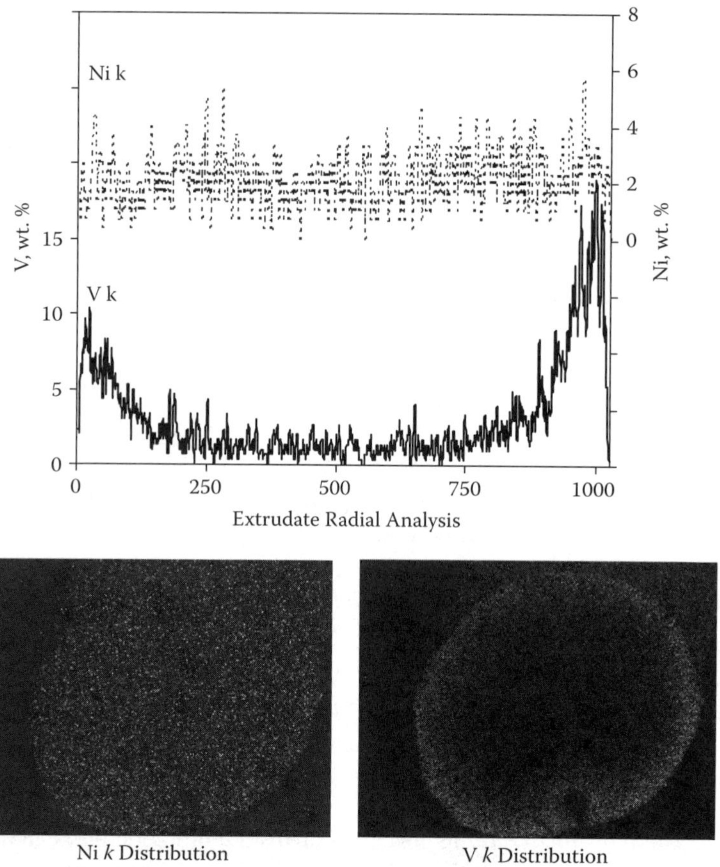

FIGURE 2.39 Ni and V distributions of spent catalyst (NiMo/AS-31-S) by SEM-EDX analysis. (From Leyva et al. 2009. *Catalysis Today* 141 (1–2): 168–175. With permission.)

The deposited metal sulfides species and its qualitative analysis of spent catalysts are confirmed by various spectroscopic techniques, such as x-ray diffractions (XRD) (Figure 2.43) and scanning transition electron microscope (STEM), by using Ni and V mapping images (Figure 2.44). The element image of Ni and V atom distribution is indicating that all of the metals were consistently deposited with time-on-stream. as shown in Figure 2.45. However, the large pore diameter catalyst (CoMo/Mg-Al-0.1) has more Ni and V deposited, which is in agreement to the quantitative results obtained by atomic absorption for these catalysts (Table 2.13). Deposited metal sulfides have been characterized by Smith and Wei (1991a, 1991b, 1991c) using TEM, EDX, STEM, and PS for the HDM of model molecules on a deactivated catalyst, and by Takuchi et al. (1985) by means of XRD, ESR, and SEM, on catalysts used in the HDM of heavy oil, where the V_2S_3 phase was observed in circular or rod-shaped crystallites about 100 nm in length.

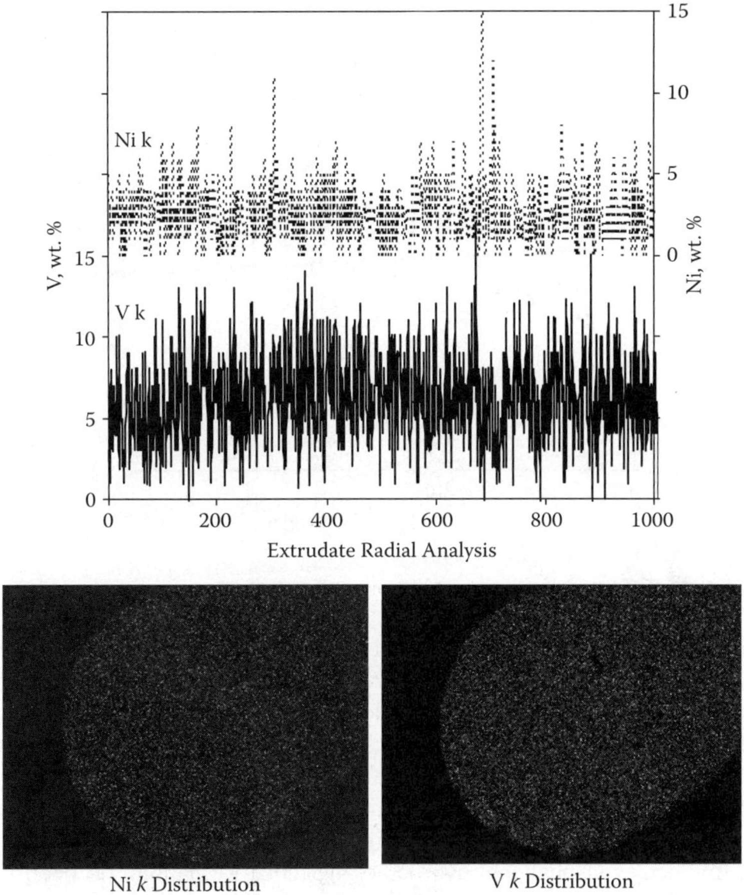

FIGURE 2.40 Ni and V distributions of spent catalyst (NiMo/AS-56-S) by SEM-EDX analysis. (From Leyva et al. 2009. *Catalysis Today* 141 (1–2): 168–175. With permission.)

Toulhoat et al. (1987) reported that nickel is always associated with vanadium [Ni(V_3S_4)] and its crystallite grew perpendicular to the support surface. Apart from these metals Ca, Mg, and Fe sulfides have also been reported to deposit on the exterior surface of the catalyst (Gosselink, 1998), and may play an important role on the plugging of catalyst pores. The typical U-shape of vanadium profile across the catalyst pellet diameter was observed by Tailleur and Caprioli (2005) for an ebullated bed catalyst. However, attempts to predict the effect of metals deposition on catalyst life have included correlation of the experimental data with the help of pore diffusion theory. These efforts have only been partially successful because of the complexity of the system and the unavailability of high-quality experimental data.

It is demonstrated from the spent catalyst characterization that heavy oil catalyst deactivation occurs in step-wise (Trimm and Stanislaus, 1986), such as

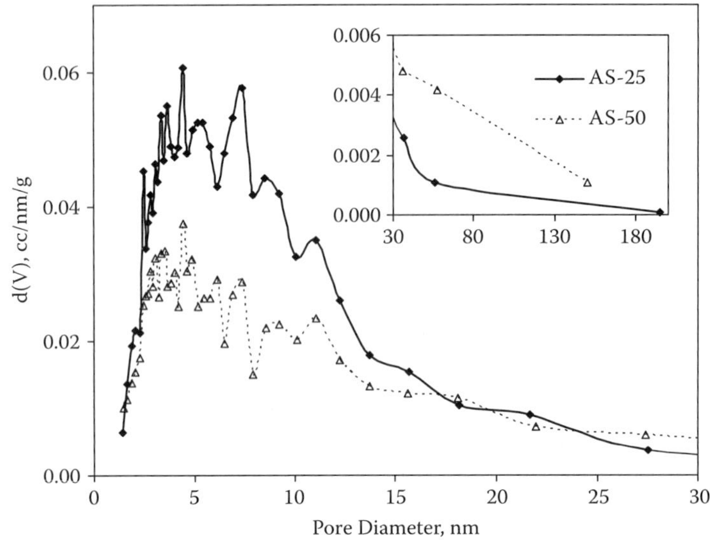

FIGURE 2.41 Effect of support composition on the textural properties.

deactivation of acid site (i.e., rapid deactivation occurring in first hours of the run), secondly, the progressive deactivation (i.e., due to the metal deposition), while the last deactivation period is due to diffusion limitation that normally occurs at the end of the catalyst life. Consequently, it is reported that during the time-on-stream more than 50% of the surface area, pore volume, and pore size distribution of the catalyst is significantly reduced by coke and metal deposition. This step-wise deactivation occurs through the three following modes: (1) at acidic sites due to hydrocarbon molecules (asphaltene) hydrocracking (coke formation), (2) deposition of metal sulfide on the sites of the cavities (or channel intersections), and (3) blockage of the access (pore mouth) of the sites of the pores, which is relatively strapping. It is expected that there is no catalytic (Mo or CoMoS/NiMoS) site poisoning by coke because at the initial stage of reaction coke deposition is expected at the acid site, which is relatively fast while metal deposition is close to the catalytic sites and deactivation takes place relatively at slower rate, as shown in Figure 2.46 as reported by Rana et al. (2008). Thus, to cope up with catalyst deactivation, a suitable processing methodology needs to be employed, such as two-stage heavy oil hydroprocessing using a combination of a hydrotreating catalyst (first stage reactor) and a zeolite or acidic catalyst (second stage reactor) for hydrocracking.

The variations in the shape of the hysteresis loops are considered as representative of pore mouth deactivation (Ancheyta et al., 2005b). An increase in the absolute isotherm (adsorption–desorption) area of two samples may correspond to the relative deactivation of the two catalysts. On the other hand, textural properties of a regenerated catalyst indicated that almost 70 to 75% of the pores can be reproduced during the regeneration, but 25 to 30% of the pores, which can be called metal deactivated pores, remain unaffected. No more carbon is detected

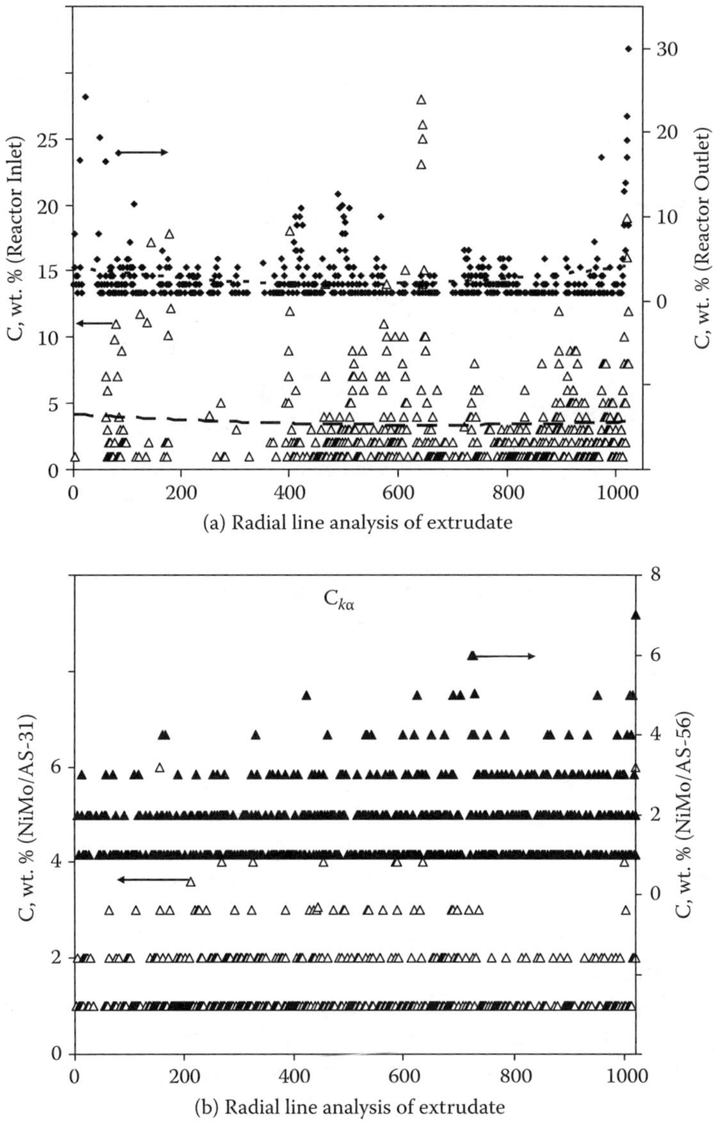

FIGURE 2.42 Deposited carbon concentration profiles on spent catalysts obtained by SEM-EDX (flow reactor; at 380°C and 7.0 MPa; 204 h TOS; Maya heavy crude). (a) Comparison for carbon deposition on top-bed and bottom-bed catalyst. (From Rana et al. 2008. *Catalysis Today* 130: 411–420. With permission.) (b) Effect of catalyst composition on carbon deposition (From Leyva et al. 2009. *Catalysis Today* 141 (1–2): 168–175. With permission.)

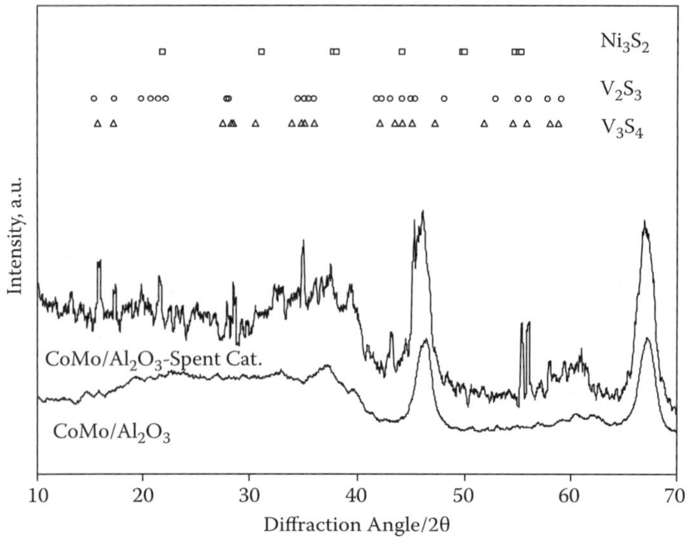

FIGURE 2.43 X-ray diffraction patterns for fresh, spent catalysts and comparison with JCPDS-ASTM data: (Δ) V_3S_4; (O) V_2S_3, and (□) Ni_2S_3. (From Rana et al., 2004a. *Catalysis Today* 98: 151–160. With permission.)

after regeneration (at 550°C for 8 h in the presence of oxygen); therefore, only deposited metals have effect on textural properties (Rana et al., 2005a). However, the position of metals may not be the same as in the unregenerated catalyst (spent catalyst). The recovery of surface area and pore volume is around 70 and 50%, respectively, which is low and that should correspond to the metal deactivation. Therefore, textural properties of the regenerated catalyst cannot be 100% recovered unless deposited metals (Ni + V) could be separated, which appears to be an almost impossible task during the *in situ* regeneration.

Spent catalysts characterization results are enough to establish that metal deposition may change the final state of the catalyst, which is contrary to the definition given in 1895 by Friedrich Wilhelm Ostwald, who recognized catalysis as a phenomenon that was explained in terms of the laws of physical chemistry, and the catalyst was assumed to remain unchanged in the course of the reaction. Metals are deposited and catalyst weight increases with time-on-stream (Figure 2.47), moreover, the weight gain by metal deposition cannot be regenerated and will be part of the catalyst permanently. Therefore, Ostwald's theory may not be followed exactly in the case of heavy oil hydroprocessing catalysis where metal sulfide is deposited during the reaction and the catalyst cannot be regenerated to its initial stage, since the catalyst consumes one of the reaction products (i.e., metal sulfides), which would not be the case of an ideal catalyst (theory).

It is also reported that deposited metal sulfides may create some new sites that may contribute to auto-catalysis (Toulhoat et al., 2005; Rana, 2005b). On the other hand, carbon deposition simultaneously occurs, but that is reversible while

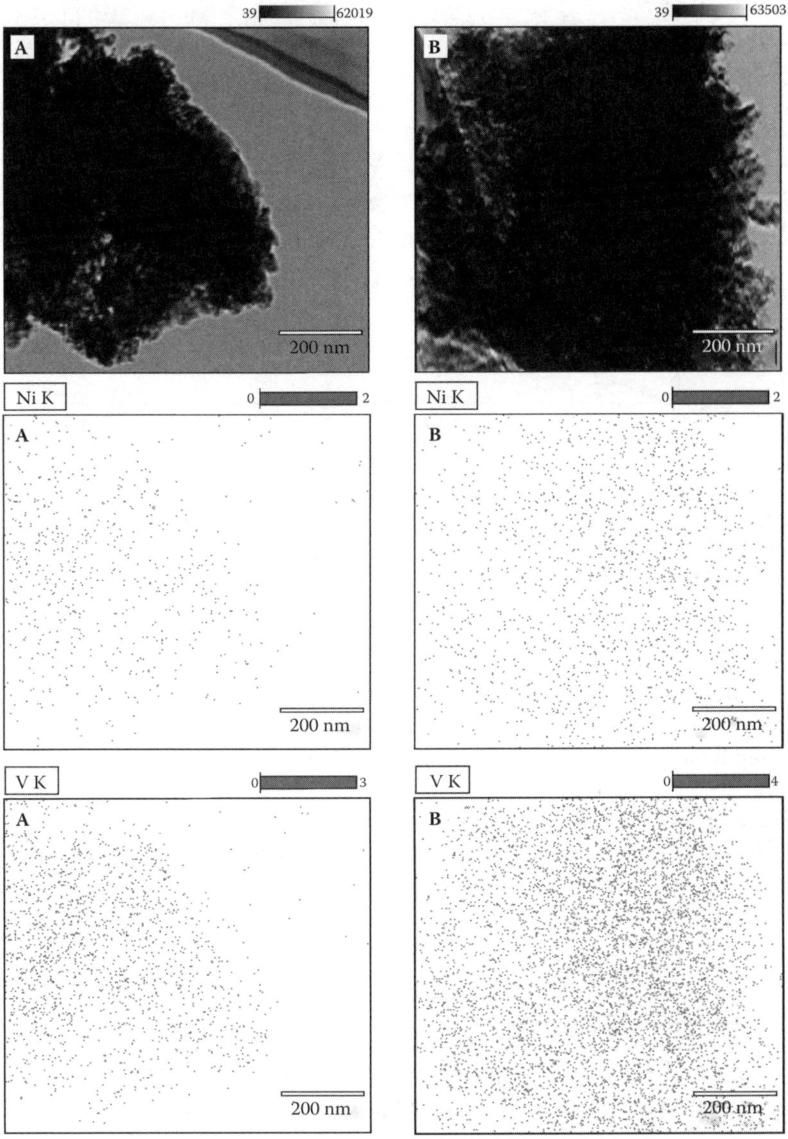

FIGURE 2.44 Scanning transmission electron micrographs (STEM) and qualitative nanomicroanalysis of Ni and V distribution: (A) 1.5 wt% MgO (7.4 nm, APD) and (B) 9.4 wt% MgO (9.5 nm, APD) in support. (From Trejo et al. 2008. *Catalysis Today* 130: 327–336. With permission.)

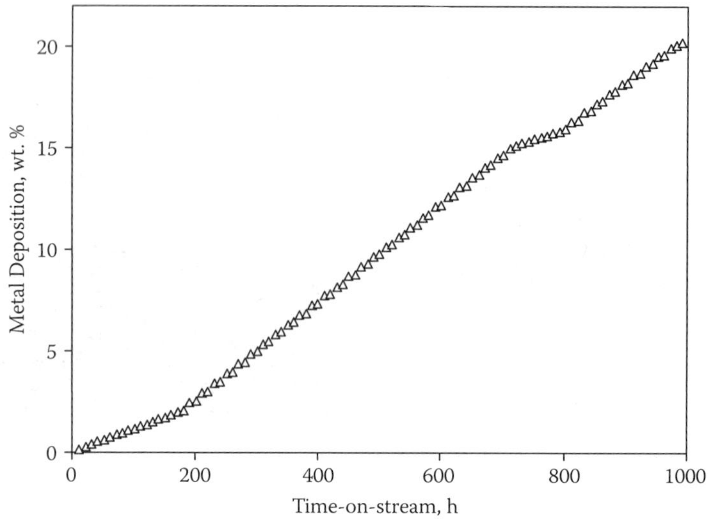

FIGURE 2.45 Ni + V metals deposition along with time-on-stream with $CoMo/Al_2O_3$-supported catalysts with pure Maya crude up to 1,000 h. (From Rana and Ancheyta, 2009. Unpublished result. With permission.)

the deposition of metal in solid catalysts is an irreversible process. Therefore, the contribution of catalytic sites with variation of time-on-stream is really unknown and a difficult task to identify, even in regenerated catalysts.

Asphaltenes are considered the major precursor of initial coke deposits due to adsorption on catalyst acid sites, subsequent reactions of the asphaltenes, and further deposition of metals and coke. They are not only deposited on the catalyst, but also precipitated through light paraffins in the reactor as well as on the refinery equipments, which generate large pressure drop in the reactor. Gray (2007) has reported the chemical structure of vacuum residue asphaltene and

TABLE 2.13
Spent Catalysts Composition and Reduction in Textural Properties Due to Deposited Species

Catalyst	Deposited species, wt%			Percentage of Reduction in Textural Properties	
	V	Ni	C	SSA	PV
CoMo/Mg-Al-0.01	1.5	0.4	10.12	64	56
CoMo/Mg-Al-0.1	3.1	0.6	8.23	47	52

Source: Trejo et al., 2008. *Catalysis Today* 130 (2–4): 327–336. With permission.

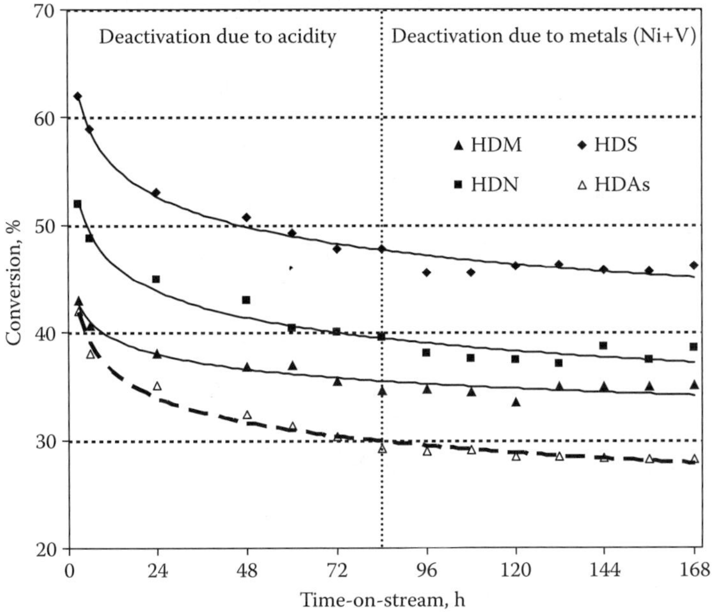

FIGURE 2.46 CoMo/(USY zeolite + alumina) supported Maya heavy oil hydroprocessing activity, relative to time-on-stream. (From Rana et al. 2008. *Catalysis Today* 130: 411–420. With permission.)

their implications for hydrotreating and hydroconversion, where the properties of certain species allow the application of powerful molecular simulation techniques to estimate thermo-physical properties, reaction pathways, adsorption onto surfaces, and aggregation in solution, with the aim of improving the catalysts and processes for hydroprocessing.

The role of an HDM catalyst is to reduce deactivation rate by an appropriate pore distribution. In general, large pore catalysts are preferred in order to avoid diffusion limitations into the porous and to the catalytic sites of NiMo or CoMo. The catalyst pore may be favored with bimodal pore size distribution to obtain a balance in HDS (smaller pore) and HDM activities (Rana et al., 2007e). The difference between the deactivation caused by coke and metal deposition is that the former is reversible while the latter is not. So, despite that metal concentration in the feed is generally much lower than other impurities, the deactivation by metals is much more serious. As it happens, contrary to coke deactivation, that produced by metal deposits does not stop when the original active sites on the catalyst surface are covered with deposited Ni and V (Sie, 2001). On the other hand, at high reaction temperatures, the carbonaceous deposits (mainly sludge) formed with acidic and bifunctional catalysts are due to the composition, which is practically independent of the reactant. The effect of temperature on the asphaltene reaction

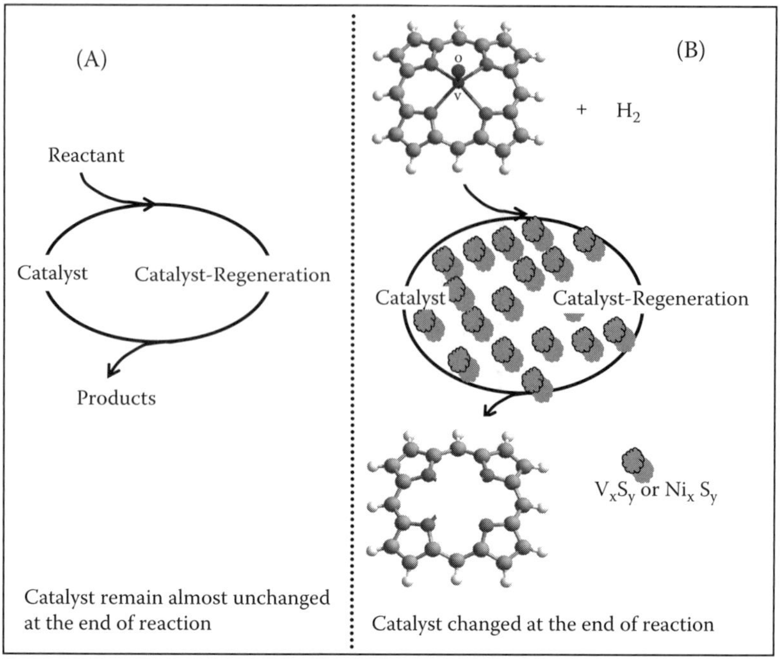

FIGURE 2.47 Life cycle of: (A) normal catalyst, and (B) catalysts for heavy oil hydroprocessing.

followed the same tendency than HDM reaction, which indicates that HDM and HDAs activities are controlled by temperature while HDN and HDS reactions are based on the number of catalytic sites. HDM and HDAs conversions follow similar behavior because most of the metals are associated with asphaltenes, which are located in the internal part of the molecules. The cracking results at high temperatures are confirmed by the light gaseous product formation, as shown in Figure 2.48, and relative increase in light gaseous product (380° to 420°C) selectivity, as observed in Figure 2.49 (Rana et al., 2008). Asphaltene cracking activities are depending on the paraffinic alkyl side chains of asphaltenes (Ancheyta et al., 2005b; Rana et al., 2007e). The gaseous paraffinic products also have an adverse effect on the unconverted asphaltenes, which favor precipitation at the end or exit line of the reactor because paraffins may cause asphaltene precipitation at room temperature (Michell and Speight, 1973), which further enhances reactor pressure drop and line plugging. Thus, in this study, the high temperature toward the end of the run leads to coke (sediment) formation and deactivation of the catalyst; the former is a solid material that is able to plug the inlet–outlet of the reactor. Sludge (sediment) formation usually occurs at high temperature and is due to changes in relative solubility of asphaltenes, maltene, and other paraffinic products. Andersen (1999) reported that sludge formation is owed to the critical solubility parameter of the solvent–precipitant mixture at which the least soluble asphaltene will precipitate or separate. Recently, Bartholdy and Andersen (2000)

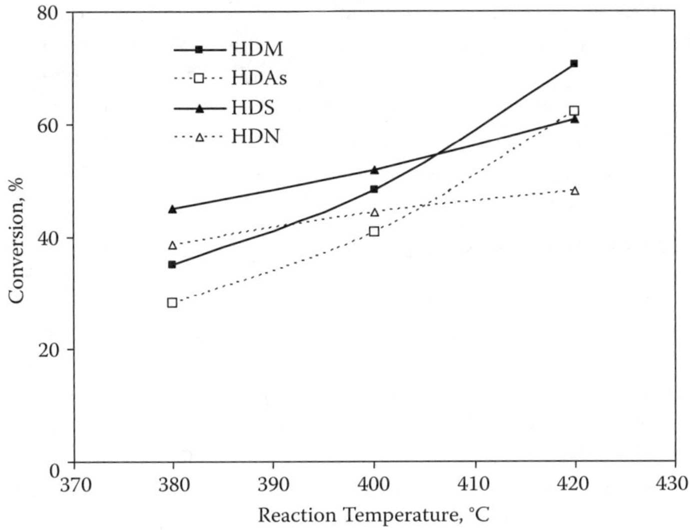

FIGURE 2.48 Effect of temperature on heavy oil conversions. (From Rana et al. 2008. *Catalysis Today* 130: 411–420. With permission.)

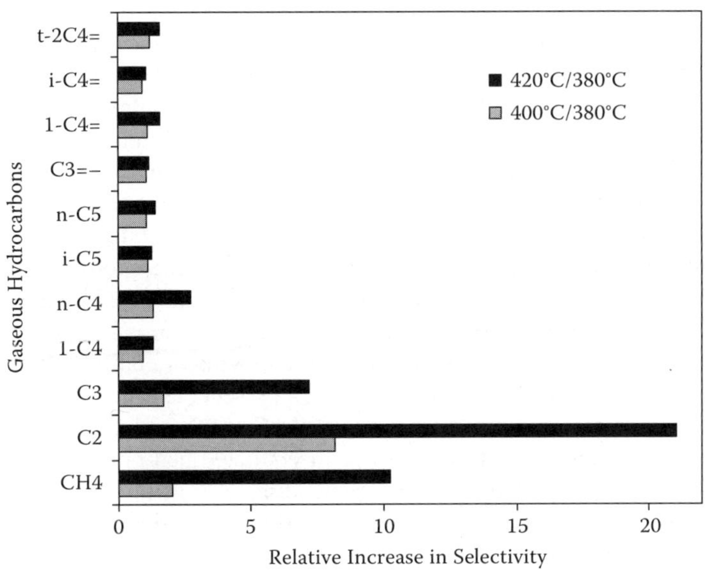

FIGURE 2.49 Relative increase in selectivity with respect to temperature. (From Rana et al. 2008. *Catalysis Today* 130: 411–420. With permission.)

reported that sludge formation at high temperature is mainly due to the chemistry of maltenes and asphaltenes. They have reported that as the temperature increased the critical solubility parameter also increased indicating an augmentation in the asphaltene solubility parameter, which agrees with the reduction in H/C ratios that are representative of a thicker structure like catacondensed polycyclic and/or pericondensed polycyclic aromatic hydrocarbons.

2.7 CONCLUDING REMARKS

Refining technologies are currently on the boom toward the heavy or extra heavy (bitumen) petroleum. The development of a catalytic process is required to have better understanding of (1) molecular structures of feedstock; (2) mechanisms of asphaltene conversion; (3) processes for removal of sulfur, nitrogen, and metals; (4) mechanisms of reacting molecules and the role of catalyst; and (5) mechanism of catalyst deactivation. The superior process technologies, catalyst formulations, and their combined effect on product selectivity and on clean fuel are the prime target for refiners to accomplish, which may compensate in some extent the effect of crude oil prices compared with alternative fuels.

The development of heavy oil process technologies for upgrading of heavy feeds will be necessary in view of the growing demand of lighter fraction and the increasing need to process heavy feedstocks with high contents of sulfur, nitrogen, metals, aromatics, and asphaltenes, which will have to give better or clean fuels that are the principal target to achieve. Hydroprocessing was found to play multiple roles and, also, it affects the market demand in a complicated and unique manner that is beyond the scope of the existing crude oil and residue quality. Thus, hydroprocessing conversion is considered to be the highly flexible and versatile process, which has demonstrated its great effect on product selectivity.

Apart from process aspects, a hydroprocessing catalyst requires an optimum level of macropore size distribution, which minimizes diffusion limitations of large molecules to the interior of pores. The place of catalyst loading in a multiple bed reactor system is also corroborated to these properties: the first catalyst bed must provide high HDM activity (macroporous catalyst) having large metal retention capacity, second bed provides high HDS and some HDM (macro or mesoporous catalyst), and the third bed must have adequate hydrocracking, HDS, and HDN activities.

In fixed-bed units, catalyst deactivation during the run is compensated by a progressive increase in bed temperature, while moving or ebullated bed reactors with a catalyst take out and replacement do not present the catalyst stability problem, but create safety and environmental problems associated to catalyst disposal. Nevertheless, it appears that moving bed or ebullated bed processes employing a suitable catalyst is an efficient way for handling petroleum bottoms and other heavy feeds. Also, fixed-bed technologies working at moderate conversion level with appropriate selection of type of feed and catalyst represent another good option for avoiding sediment formation problems and can upgrade heavy oils so that they can be transported or processed in conventional refineries.

REFERENCES

Absi-Halabi, M., Stanislaus, A., and Al-Zaid, H. 1993. Effect of acidic and basic vapors on pore size distribution of alumina under hydrothermal conditions. *Appl. Catal. A: Gen.* 101: 117–128.

Absi-Halabi, M., Stanislaus, A., and Trimm, D.L. 1991. Coke formation on catalysts during the hydroprocessing of heavy oils. *Appl. Catal. A: Gen.* 72: 193–215.

Afandizadeh, S. and Foumeny, E.A. 2001. Design of packed bed reactors: Guides to catalyst shape, size, and loading selection. *Appl. Themal Eng.* 21: 669–682.

Agrawal, R. and Wei, J. 1984a. Hydrodemetalation of nickel and vanadium porphyrins. 1. Intrinsic kinetics. *Ind. Eng. Chem. Proc. Des. Dev.* 23: 505–514.

Agrawal, R. and Wei, J. 1984b. Hydrodemetalation of nickel and vanadium porphyrins. 2. Intraparticle diffusion. *Ind. Eng. Chem. Proc. Des. Dev.* 23 (3): 515–522.

Ancheyta, J. Reactors for hydroprocessing, in *Hydroprocessing of Heavy Oil and Residua*, J. Ancheyta and J. G. Speight, Eds., Taylor & Francis: New York, 2007, Chap. 5.

Ancheyta, J. and Rana, M.S. 2008. Future technology in heavy oil processing, in *Encyclopedia of Life Support Systems (EOLSS)*, Developed under the Auspices of the UNESCO, Eolss Publishers, Oxford, U.K., at: http://www.eolss.net.

Ancheyta, J., Betancourt R.G., Marroquin, S.G.J., Centeno, N.G., Munoz, M.J.A., Alonso M.F. 2003. Method for the Catalytic Hydroprocessing of Heavy Petroleum Hydrocarbons, WO2005005581 (Appl. Number: WO2003MX00053 20030709).

Ancheyta, J., Rana, M.S., and Furimsky, E. 2005a. Hydroprocessing of heavy oil fraction. *Catalyst Today* 109: 1–3.

Ancheyta, J., Rana, M.S., and Furimsky, E. 2005b. Hydroprocessing of heavy petroleum feeds: Tutorial. *Catalyst Today* 109: 3–15.

Andersen, S.I. 1999. Flocculation onset titration of petroleum asphaltenes. *Energy Fuels* 13: 315–322.

Anderson, J.L. and Quinn J.A. 1974. Restricted transport in small pores: A model for steric exclusion and hindered particle motion. *Biophys. J.* 14: 130–150.

Araki, Y., Honna, K., and Shimada, H. 2002. Formation and catalytic properties of edge-bonded molybdenum sulfide catalysts on TiO_2. *J. Catal.* 207: 361–370.

Baltus, R.E. and Anderson, J.U. 1983. Hindered diffusion of asphaltenes through microporous membranes. *Chem. Eng. Sci.*, 38 (12): 1959–1969.

Bartholdy, J. and Cooper, B.H. 1993. Metal coke of resid hydroprocessing catalysts properties. *Am. Chem. Soc., Div. Petrol. Chem.* 38 (2): 386–390.

Bartholdy, J.S., and Andersen, I. 2000. Changes in asphaltene stability during hydrotreating. *Energy Fuels* 14: 52–55.

Bartholomew, C.H. 1994. Catalyst deactivation in hydrotreating of residua. In *Catalytic Hydroprocessing of Petroleum and Distillates*, Oballa, M.C. and Shih, S.S., Eds. CRC Press: USA, 1.

Beaton, E.I. and Bertolacini, R.J. 1991. Thermal conversion of PR spring bitumen-derived heavy oil in the presence of Na/alumina. *Catal. Rev. Sci. Eng.* 33 (3–4): 281.

Billon, A. and Bigeard, P.-H. 2001. Hydrocracking. In *Petroleum Refining*. P. Leprince, Ed., Technip Editions, Paris, 10.

Bonné, R.L.C., van Steenderen, P., and Moulijn, J.A. 2001. Hydrogenation of nickel and vanadyl tetraphenylporphyrin in absence of a catalyst: A kinetic study. *Appl. Catal., A: Gen.* 206: 171–181.

Breysse, M., Afanasiev, P., Geantet C., and Vrinat, M. 2003. Overview of support effects in hydrotreating catalysts. *Catalysis Today* 86: 5–16.

Bridge, A.G. 1990. Catalytic hydrodemetalation of heavy oils. In *Catalysts in Petroleum Refining*, 1989, Trimm, D.L., Akashah, S., Absi-Halabi, M., Eds., *Stud. Surf. Sc. and Catal.* Elsevier: Amsterdam 53: 363–383.

Caloch, B., Rana, M.S., and Ancheyta, J. 2004. Improved hydrogenolysis (C-S, C-M) function with basic supported hydrodesulfurization catalysts. *Catalysis Today* 98: 91–98.

Chen, H.J. and Massoth, F.E. 1988. Hydrodemetalation of vanadium and nickel porphyrins over sulfided cobalt-molybdenum/alumina catalyst. *Ind. Eng. Chem. Res.* 27 (9): 1629–1639.

Chuah, G.K., Jaenicke, S., and Xu, T.H. 2000. The effect of digestion on the surface area and porosity of alumina. *Micropor. Mesopor. Mat.* 37: 345–353.

Cooper, B.H., Donnis, B.B.L., and Moyse, B.M. 1986. Technology: Hydroprocessing conditions affect catalyst shape selection. *Oil Gas J.* Dec. 8: 39–44.

Daage, M. and Chianelli, R.R. 1994. Structure-function relations in molybdenum sulfide catalysts: The 'rim-edge' model. *J. Catal.* 149: 414–427.

de Boer, J.H. 1958. *The Structure and Properties of Porous Materials*, Butterworths: London, p. 68.

Dickenson, R.L., Biasca, F.E., Schulman, B.L., and Johnson H.E. 1997. Refiner options for converting and utilizing heavy fuel oil. *Hydrocarb. Proc.* February: 57–62.

Dickie, J.P. and Yen, T.F. 1967. Macrostructures of the asphaltic fractions by various instrumental methods. *Anal. Chem.* 39: 1847–1852.

EIA. (Energy Information Administration). 2007. http://www.eia.doe.gov/oil_gas/petroleum/info_glance/petroleum.html.

Eijsbouts, S., Heinemann, J.J.L., and Elzerman, H.J.W. 1993. MoS_2 structures in high activity hydrotreating catalysts. II. Evolution of the active phase during the catalyst life cycle deactivation model. *Appl. Catal. A: Gen.* 105: 69–82.

Fleisch, T. H., Meyers, B.L., Hall, J.B., and Ott, G.L. 1984. Multitechnique analysis of a deactivated resid demetallation catalyst. *J. Catal.* 86:147–57.

Fleischer, E.B. 1963. The structure of nickel etioporphyrin: I. *J. Am. Chem. Soc.* 85: 146–148.

Foumeny, E.A., Kulkarni, A., Roshani, S., and Vatani, A. 1996. Elucidation of pressure drop in packed-bed systems. *Appl. Thermal Eng.* 16: 195–202.

Francis, D.R. 2004. Has global oil production peaked? The Christian Science Monitor, January 29 edition.

Fukuyama, H., Terai, S., Uchida, M., Cano, J. L., and Ancheyta, J. 2004. Active carbon catalyst for heavy oil upgrading. *Catalyst Today* 98;, 207–215.

Furimsky, E. 1988. Selection of catalysts and reactor for hydroprocessing, *Appl. Catal. A: Gen.* 46: 177–206.

Furimsky, E. 1996. Spent refinery catalysts: Environment, safety and utilization, *Catalysis Today* 30: 223–286.

Furimsky, E. 2007. Catalyst for upgrading heavy petroleum feeds. In *Studies in Surface Science and Catalysis*, series no. 169, series ed. G. Centi, Elsevier: Amsterdam/New York.

Furimsky, E. and Massoth, F.E. 1999. Deactivation of hydroprocesssing catalysts. *Catalyst Today*, 52: 381–495.

Froment, G.F. 2001. Modeling of catalyst deactivation. *Appl. Catal. A: Gen.* 212: 117–128.

Gallezot, P. 2007. Process options for the catalytic conversion of renewables into bioproducts. In *Catalysis for Renewables from Feedstock to Energy Production,* 1. Eds. Centi, Gabriele, and van Santen, Rutger A., Wiley-VCH: Weinheim.

Gates, B.C. and Topsøe, H. 1997. Reactivities in deep catalytic hydrodesulfurization: Challenges, opportunities, and the importance of 4-methyldibenzothiophene and 4,6-dimethyldibenzothiophene. *Polyhedron* 16 (18): 3213–3217.

Gosselink, J.W. 1998. Sulfide catalysts in refineries. *CatTech.* 2: 127–144.

Gray, M.R. 2007. Chemical structure of vacuum residue components: Implications for hydrotreating and hydroconversion, Paper presented at the International Symposium on Advances in Hydroprocessing of Oil Fraction (ISAHOF), June 26–29, Morelia, Mexico.

Gray M.R. 1994. *Upgrading Petroleum Residues and Heavy Oils*, Marcel Dekker, New York.

Gregg, S.J. and Sing, K.S.W. 1982. *Adsorption, Surface Area and Porosity*, 2nd ed.; Academic Press: New York.

Hansford, R.C. 1964. Silica-zirconia-titania hydrocracking catalyst, U.S. Patent 3,159,588.

Hauser, A., Marafi, A., Stanislaus, A., and Al-Adwani, A. 2005. Relation between feed quality and coke formation in a three-stage atmospheric residue desulfurization (ARDS) process. *Energy Fuels* 19: 544–553.

Heck, R.H., Rankel, L.A., and DiGuiseppi, F.T. 1992. Conversion of petroleum resid from Maya crude: Effects of H-donors, hydrogen pressure and catalyst. *Fuel Proc. Tech.* 30: 69–81.

Huang, Y., White, A., Walpole, A., and Trimm, D.L. 1989. Control of porosity and surface area in alumina I. Effect of preparation conditions. *Appl. Catal.* 56: 177–186.

Huber, G.W., Iborra, S., and Corma, A. 2006. Synthesis of transportation fuels from biomass: Chemistry, catalysts, and engineering. *Chem. Rev.* 106, 4044–4098.

Jaffe, J. 1968. Co-precipitation method for making multi-component catalysts, U.S. Patent 3,401,125.

Janssens, J.P., Elst, G., Schrikkema, E.G., Van Langeveld, A.D., Sie, S.T., and Moulijn, J.A. 1996. Development of a mechanistic picture of the hydrodemetallization reaction of metallo-tetraphenylporphyrin on a molecular level. *Recl. Trav. Chim. Pay Bas.* 115 (11–12): 465–473.

Johnson, B.G., Massoth, F.E., and Bartholdy, J. 1986. Diffusion and catalytic activity studies on resid-deactivated HDS catalysts. *AIChE J.* 32: 1980–1987.

Johnson, M.F.L. and Mooi, J. 1968. The origin and types of pores in some alumina catalysts. *J. Catal.* 10. 342–354.

Kabe, T., Akamatsu, K., Ishihara, A., Otsuki, S., Godo, M., Zhang, Q., and Qian, W. 1997. Deep hydrodesulfurization of light gas oil. 1. Kinetics and mechanisms of dibenzothiophene hydrodesulfurization. *Ind. Eng. Chem. Res.* 36: 5146–5152.

Kaloidas, V., Thanos, A.M., Tsamatsoulis, D.C., and Papayannakos, N.G. 2000. Preparation of Al_2O_3 carriers in an integrated mini pilot unit. *Chem. Eng. Proc.* 39: 407–416.

Kamiya, Y. *Heavy Oil Processing Handbook*; RAROP; Japan, 1991.

Khorasheh, F., Rangwala, H., Gray, M.R., and Lana, I.G.D. 1989. Interactions between thermal and catalytic reactions in mild hydrocracking of gas oil. *Energy Fuels* 3: 716–722.

Kressmann, S., Morel, F., Harlé, V., and Kasztelan, S. 1998. Recent developments in fixed-bed catalytic residue upgrading. *Catalyst Today* 43: 203–215.

Landau, M.V., Berger, D., and Herskowitz, M. 1996. Hydrodesulfurization of methyl-substituted dibenzothiophenes: Fundamental study of routes to deep desulfurization. *J. Catal.* 158: 236–245.

Larson, O.A., and Beuther H. 1966. Processing aspects of vanadium and nickel in crude oils. *Prepr. Am. Chem. Soc., Div. Petrol. Chem.* 11 (2): B95–B103.

Lee, S.Y., Seader J.D., Tsai, C.H., and Massoth, F.E. 1991. Restrictive diffusion under catalytic hydroprocessing conditions. *Ind. Eng. Chem. Res.* 30 (1): 29–38.

Leliveld, R.G., van Dillen, A.J., Geus, J.W., and Koningsberger, D.C. 1997. The sulfidation of γ-alumina and titania supported (cobalt) molybdenum oxide catalysts monitored by EXAFS. *J. Catal.* 171: 115–129.

LePage, L.F., Chatila, S.G., and Davidson, M. 1992. *Resid and Heavy Oil Processing*. Technip; Paris.

Le Page, J.-F., Cosyns, J., Courty, P., Freund, E., Franck, J.-P., Jscquin, Y.J., Marcilly, B.C., Martino, G., Miquel, J., Montarnal, R., Sugier, A., and Landeghem, van H. 1987. Applied Heterogeneous Catalysis, Design, Manufacture, Use of Solid Catalysts. Technip: Paris.

Leyva, C., Rana, M.S., Trejo, F., and Ancheyta, J. 2007. On the use of acid-base supported catalyst for catalysts for hydroprocessing of petroleum. *Ind. Eng. Chem. Res.* 46 (23): 7448–7466.

Leyva, C., Rana, M.S., Trejo, F., Ancheyta, J. 2009. NiMo supported acidic catalysts for heavy oil hydroprocessing. *Catalyst Today* 141 (1–2): 168–175.

López-Salinas, E., Espinosa, J.G., Hernández-Cortez, J.G., Sánchez-Valente, J., and Nagira, J. 2005. Long-term evaluation of NiMo/alumina–carbon black composite catalysts in hydroconversion of Mexican 538°C+ vacuum residue. *Catalyst Today* 109: 69–75.

Luck, F. 1991. A review of support effects on the activity and selectivity of hydrotreating catalysts. *Bull. Soc. Chim. Belg.* 100 (11–12): 781–800.

Macias, M.J. and Ancheyta, J. 2004. Simulation of an isothermal hydrodesulfurization small reactor with different catalyst particle shapes. *Catalyst Today* 98: 243–250.

Maity, S.K., Ancheyta, J., Alonso, F., and Rana, M.S. 2004. Preparation, characterization and evaluation of Maya crude hydroprocessing catalysts. *Catalyst Today* 98: 193–199.

Maity, S.K., Ancheyta, J., Rana, M.S., and Rayo, P. 2006. Alumina-titania mixed oxide used as support for hydrotreating catalysts of Maya heavy crudes: Effect of support preparation methods. *Energy Fuels* 20: 427–431.

Maity, S.K., Ancheyta, J., Soberanis, L., Alonso, F., and Llanos, M.E. 2003. Alumina-titania binary mixed oxide used as support of catalysts for hydrotreating of Maya heavy crude. *Appl. Catal. A: Gen.* 244: 141–153.

Marafi, M. and Stanislaus, A. 1997. Effect of initial coking on hydrotreating catalyst functionalities and properties. *Appl. Catal. A: Gen.* 159: 259–267.

Marafi, M., Al-Omani, S., Al-Sheeha, H., Al-Barood, A., and Stanislaus, A. 2007. Utilization of metal-fouled spent residue hydroprocessing catalysts in the preparation of an active hydrodemetallization catalyst. *Ind. Eng. Chem. Res.* 46 (7): 1968–1974.

Marafi, A., Maruyama, F., Stanislaus, A., and Kam, E. 2008. Multicatalyst system testing methodology for upgrading residual oils. *Ind. Eng. Chem. Res.* 47 (3): 724–741.

Matsushita, K., Marafi, A., Stanislaus, A., and Hauser, A. 2004. Relation between relative solubility of asphaltenes in the product oil and coke deposition in residue hydroprocessing. *Fuel* 83 (11/12): 1669–1674.

Merdrignac, I and Espinat, D. 2007. Physicochemical characterization of petroleum fractions: The state of the art. *Oil Gas Sci. Technol. - Rev. IFP* 62 (1): 7–32.

Michell, D.L. and Speight, J.G. 1973. The solubility of asphaltenes in hydrocarbon solvents. *Fuel* 52: 149–152.

Miki, Y., Yamadaya, S., Oba, M., and Suggimoto, Y. 1983. Role of catalyst in hydrocracking of heavy oil. *J. Catal.* 83: 371–383.

Miller, J.T., Fisher, R.B., Thiyagarajan, P., Winans, R.E., and Hunt, J.E. 1998. Sub-fraction and characterization of Mayan asphaltenes. *Energy Fuels* 12 (6): 1290–1298.

Mills, K.J. 2004. Ceramics in catalytic reactors. *Hydrocarb. Eng.* 9 (9): 51–54.

Mojelsky, T.W., Ignasiak, T.M., Frakman, Z., McIntyre, D.D., Lown, E.M., Montgomery, D.S., and Strausz, O.P. 1992. Structural features of Alberta oil sand bitumen and heavy oil asphaltenes. *Energy Fuels* 6 (1): 83–96.

Mosby, J.F., Buttke, R.D., Cox, J.A., and Nickolaiclcs, C. 1986. Process characterization of expanded-bed reactors in series. *Chem. Eng. Sci.* 41 (4): 989–995.

Moulijn, J.A., van Leeuwen, P.V.N.M., and van Santen, R.A. 1993. An integrated approach to homogeneous, heterogeneous and industrial catalysis. *Stud. Surf. Sci. Catal.* 79, Elsevier: Amsterdam, V-Viii.

Mullins, O.C. and Groenzin, H. 1999. Petroleum asphaltene molecular size and structure. *J. Phys. Chem. A.* 103: 11237–11245.

Muralidhar, G., Rana, M.S., Maity, S.K., Srinivas, B.N., and Prasada Rao, T.S.R. 2000. Performance of Mo catalysts supported on TiO_2-based binary support for distillate fuel hydroprocessing. In *Chemistry of Diesel Fuels*. C. Song, S. Hsu, and I. Mochida, Eds., Taylor & Francis: New York, Chap. 8.

Muralidhar, G., Srinivas, B.N., Rana, M.S., Kumar, M., and Maity, S.K. 2003. Mixed oxide supported hydrodesulfurization catalysts-a review. *Catalysis Today* 86: 45–60.

Newson, E. 1975. Catalyst deactivation due to pore plugging by (residuum hydrodesulfurization) reaction products. *Ind. Eng. Chem. Proc. Des. Dev.* 14: 27–33.

Okamoto, Y., Breysse, M., Murali Dhar, G., and Song, C. 2003. Effect of support in hydrotreating catalysis for ultra clean fuels. *Catalysis Today* 86: 1–3.

Okamoto, Y., Maezawa, A., and Imanaka, T. 1989. Active sites of molybdenum sulfide catalysts supported on Al_2O_3 and TiO_2 for hydrodesulfurization and hydrogenation. *J. Catal.* 120: 29–45.

Ono, T., Ohguchi, Y., and Togari, O. 1983. Control of the pore structure of porous alumina, In Preparation of Catalyst III: Scientific bases for the preparation of heterogeneous catalysts, Grange, P., and Jacobs P.A. Eds. *Stud. Surf. Sci. Catal.* Elsevier: Amsterdam. 16: 631–642.

Papayannakos, N.G., Thanos, A.M., and Kaloidas, Y.E. 1993. Effect of seeding during precursor preparation on the pore structure of alumina catalyst supports. *Micropor. Mesopor. Mat.* 1: 423–430.

Penner, S.S., Benson, S.W., Camp, F.W., Clardy, J., Deutch, J., Kelley, A.E., Lewis, A.E., Mayer, F.X., Oblad, A.G., Sieg, R.P., Skinner, W.C. and Whitehurst, D.D. 1982. Assessment of research needs for oil recovery from heavy-oil sources and tar sands. *Energy* 7 (7): 567–602.

Phillips, G. and Liu, F. 2003. Invest in the future. *Hydrocarb. Eng.* 8 (9): 63–68.

Pratt, K.C., Sanders, J.V., and Christov, V. 1990. Morphology and activity of MoS_2 on various supports: Genesis of the active phase. *J. Catal.* 124: 416–432.

Quann, R.J., Ware, R.A., Hung, C.-W., and Wei, J. 1988. Catalytic hydrodemetallation of petroleum. *Adv. Chem. Eng.* 14: 95–259.

Rajagopalan, K., Lues, D. 1979. Influence of catalyst pore size on demetallation rate. *Ind. Eng. Chem. Process Des. Dev.* 28: 459–465.

Ramírez, J., Rana, M.S., and Ancheyta, J. 2007. Characteristics of heavy oil hydroprocessing catalysts. In *Hydroprocessing of Heavy Oil and Residua*, J. Ancheyta and J.G. Speight, Eds., Taylor & Francis: New York, Chap. 6.

Ramirez, J., Ruiz-Ramirez, L., Cedeno, L., Hale, V., Vrinat, M., and Breysse, M. 1993. Titania-alumina mixed oxides as supports for molybdenum hydrotreating catalysts. *Appl. Catal. A: Gen.* 93: 163–180.

Rana, M.S. and Ancheyta, J. 2009. Catalyst for guard-bed reactor. Unpublished results.

Rana, M.S., Srinivas, B.N., Maity, S.K., Murali Dhar, G., and Prasada Rao, T.S.R. 2000. Origin of cracking functionality of sulfided (Ni) $CoMo/SiO_2$-ZrO_2 catalysts. *J. Catal.* 195: 31–37.

Rana, M.S., Maity, S.K., Ancheyta, J., Murali Dhar, G., and Prasada Rao, T.S.R. 2003. TiO_2-SiO_2 supported hydrotreating catalysts: physico-chemical characterization and activities. *Appl. Catal. A: Gen.* 253: 165–176.

Rana, M.S., Ancheyta, J., Rayo, P., and Maity, S.K. 2004a. Effect of alumina preparation on hydrodemetallization and hydrodesulfurization of Maya crude, *Catalysis Today* 98: 151–160.

Rana, M.S., Maity, S.K., Ancheyta, J., Murali Dhar, G., Prasada Rao, T.S.R. 2004b. Cumene cracking functionalities on sulfided Co(Ni)Mo/TiO_2-SiO_2 catalysts. *Appl. Catal. A: Gen.* 258: 215–225.

Rana, M.S. and Ancheyta, J. 2009. Catalyst for guard-bed reactor. Unpublished results.

Rana, M.S., Navarro, R., and Leglise, J. 2004c. Competitive effects of nitrogen and sulfur content on activity of hydrotreating CoMo/Al_2O_3 catalysts: A batch reactor study. *Catalysis. Today* 98: 67–74.

Rana, M.S., Ancheyta, J., Maity, S.K., and Rayo, P. 2005a. Maya crude hydrodemetallization and hydrodesulfurization catalysts: An effect of TiO_2 incorporation in Al_2O_3. *Catalysis Today* 109: 61–68.

Rana, M.S., Ancheyta, J., Maity, S.K., and Rayo, P. 2005b. Characteristics of Maya crude hydrodemetallization and hydrodesulfurization. *Catalysis. Today* 104: 86–93.

Rana, M.S., Ancheyta, J., and Rayo, P. 2005c. A comparative study for heavy oil hydroprocessing catalysts at micro-flow and bench-scale reactors. *Catalysis Today* 109: 24–32.

Rana, M.S., Huidobro, M.L., Ancheyta, J., and Gomez, M.T. 2005d. Effect of support composition on hydrogenolysis of thiophene and Maya crude. *Catalysis Today* 107–108: 346–354.

Rana, M.S., Ancheyta, J., Maity, S.K., and Rayo, P. 2006. Efecto del Soporte y el diámetro de poro en la hidrodesmetalización de Crudo Maya. *Revista Mexicana de Ingeniería Química* 5 (3): 227–235.

Rana, M.S., Ancheyta, J., Maity, S.K., and Rayo, P. 2007a. Hydrotreating of Maya crude Oil: I. Effect of support composition and its pore-diameter on asphaltene conversion. *Petrol. Sci. Technol.* 25 (1–2): 187–200.

Rana, M.S., Ancheyta, J., Maity, S.K., and Rayo, P. 2007b. Hydrotreating of Maya crude oil: II. Generalized relationship between hydrogenolysis and hydrodeasphaltenization (HDAs). *Petrol. Sci. Technol.* 25 (1–2): 201–214.

Rana, M.S., Ancheyta, J., Rayo, P., and Maity, S.K. 2007c. Heavy oil hydroprocessing over supported NiMo sulfided catalyst: An inhibition effect by added H_2S. *Fuel* 86: 1263–1269.

Rana, M.S., Capitaine, E.M.R., Leyva, C., and Ancheyta, J. 2007d. Effect of catalyst preparation and support composition on hydrodesulfurization of dibenzothiophene and Maya crude oil. *Fuel* 86: 1254–1262.

Rana, M.S., Maity, S.K., and Ancheyta, J. 2007e Maya heavy crude oil hydroprocessing catalysts. In *Hydroprocessing of Heavy Oil and Residua.* J. Ancheyta and J.G. Speight, Eds., Taylor & Francis: New York, Chap. 7.

Rana, M.S., Ramírez, J., Gutierrez-Alejandre, A., Ancheyta, J., Cedeño, L., and Maity, S.K. 2007f. Support effects in CoMo hydrodesulfurization catalysts prepared with EDTA as a chelating agent. *J. Catal.* 246: 100–108.

Rana, M.S., Samano, V., Ancheyta, J., and Diaz, J.A.I. 2007g. A review of recent advances on process technologies for upgrading of heavy oils and residua. *Fuel* 86: 1216–1231.

Rana, M.S., Ancheyta, J., Maity, S.K., and Rayo, P. 2008. Maya heavy crude oil hydroprocessing: A zeolite based CoMo catalyst and its spent catalyst characterization. *Catalyst Today* 130: 411–420.

Rankel, L.A. 1981. Reactions of metalloporphyrins and petroporphyrins with H_2S and H_2. *Prepr. Am. Chem. Soc., Div. Pet. Chem.* 26: 689–698.

Rayo, P., Ancheyta, J., Ramírez, J., and G.-Alejandre, A. 2004. Hydrotreating of diluted Maya crude with NiMo/Al_2O_3-TiO_2 catalysts: Effect of diluent composition. *Catalysis Today* 98: 171–179.

Rayo, P., Ramírez, J., Ancheyta J., and Rana, M.S. 2007. HDS, HDN, HDM and HDAs of Maya crude over NiMo/Al_2O_3 modified with Ti and P. *Petrol. Sci. Technol.* 25: 215–229.

Renkin, E.M. 1954. Filtration, diffusion, and molecular sieving through porous cellulose membranes. *J. Gen. Physiol.* 38: 225–243.

Riazi, M.R. 1989. Characterization and Properties of Petroleum Fractions. 1st ed. ASTM International Manual Series, West Conshohocken, PA, 19428–2959, 156. manual series, PA, p. 156.

Richardson, J. T. 1989. *Principles of Catalyst Development*, Plennum Press: New York.

Saih, Y. and Segawa, K. 2003. Tailoring of alumina surfaces as supports for NiMo sulfide catalysts in the ultra deep HDS of gas oil: Case study of TiO_2-coated alumina prepared by chemical vapor deposition technique. *Catalysis Today* 86: 61–72.

Sanford, E.C., and Chung, K.H. 1991. The mechanism of pitch conversion during coking, hydrocracking and catalytic hydrocracking of Athabasca bitumen. *AOSTRA J. Res.* 7: 37–46.

Sakashita, Y., Araki, Y., and Shimada, H. 2001. Effects of surface orientation of alumina supports on the catalytic functionality of molybdenum sulfide catalysts. *Appl. Catal. A: Gen.* 215: 101–110.

Scherzer, J. and Gruia, A.J. (Eds.) 1996. *Hydrocracking Science and Technology*. Marcel Dekker: New York.

Schuetze, B. and Hofmann, H. 1997. How to upgrade heavy feeds. *Hydrocarb. Proc. Int. Ed.* 63 (2): 75–82.

Sharma, B.K., Sarowha, S.L.S., Bhagat, S.D., Tiwari, R.K., Gupta, S.K., and Venkataramani, P.S. 1999. Analysis of insolubles of petroleum vacuum residues using TLC-FID. *Petrol. Sci. Technol.* 17 (3–4): 319–332.

Sharma, B.K., Sharma, C.D., Bhagat, S.D., and Erhan S.Z. 2007a. Maltenes and asphaltenes of petroleum vacuum residues: Physico-chemical characterization. *Petrol. Sci. Technol.* 25 (1–2): 93–104.

Sharma, B.K., Sharma, C.D., Tyagi, O.S., Bhagat, S.D., and Erhan S.Z. 2007b. Structural characterization of asphaltenes and ethyl acetate insoluble fractions of petroleum vacuum residues. *Petrol. Sci. Technol.* 25 (1–2): 121–139.

Sharma, B.K., Stipanovic, A., and Sarowha, S.L.S. 2000. Size exclusion chromatographic analysis crude oils, petroleum residues and its soluble fractions. *ACS Div. Petrol. Chem. Inc. Preprints* 45 (4): 647–650.

Shbnura, M., Yoshlml, S., and Chlrato, T. 1986. Effect of catalyst pore structure on hydrotreating of heavy oil. *Ind. Eng. Chem. Fundament.* 25: 330–337

Shih, S.S., Mizahi, S., Green, L.A., and Sarli, M.S. 1992. Deep desulfurization of distillates. *Ind. Eng. Chem. Res.* 31: 1232–1235.

Shimada, H. 2003. Morphology and orientation of MoS_2 clusters on Al_2O_3 and TiO_2 supports and their effect on catalytic performance. *Catalyst Today* 86: 17–29.

Sie, S.T. 1980. Catalyst deactivation by pore plugging in petroleum processing. In *Catalyst Deactivation*. Delmon, B. and Froment G.F., Eds., Elsevier: Amsterdam, pp. 545–569.

Sie, S.T. 2001. Consequences of catalyst deactivation for process design and operation, *Appl. Catal. A: Gen.* 212: 129–151.

Smith, B.J. and Wei, J. 1991a. Deactivation in catalytic hydrodemetallation I. Model compound kinetic studies. *J. Catal.* 132: 1–20.

Smith, B.J. and Wei, J. 1991b. Deactivation in catalytic hydrodemetallation II. Catalyst characterization. *J. Catal.* 132: 21–40.

Smith, B.J. and Wei, J. 1991c. Deactivation in catalytic hydrodemetallation III. Random-spheres catalyst models. *J. Catal.* 132: 41–57.

Snel, R. 1984a. Control of the porous structure of amorphous silica-alumina. I. The effects of sodium ions and syneresis. *Appl. Catal.* 11: 271–280.

Snel, R. 1984b. Control of the porous structure of amorphous silica-alumina. II. The effects of pH and reactant concentration. *Appl. Catal.* 12: 189–200.

Snel, R. 1984c, Control of the porous structure of amorphous silica-alumina. III. The influence of pore regulating agents. *Appl. Catal.* 12: 347–357.

Snel, R. 1987. Control of the porous structure of amourphous silica-alumina: 4. Nitrogen bases as pore-regulating agents. *Appl. Catal.* 33 (2): 281–294.

Speight, J.G. 1999. *The Chemistry and Technology of Petroleum,* 3rd ed., Marcel Dekker: New York.

Srinivasan, S., Datye, A.K., and Peden, C.H.F. 1992. The morphology of oxide-supported MOS_2. *J. Catal.* 137: 513–522.

Tailleur, R.G. and Caprioli, L. 2005. Catalyst pore plugging effects on hydrocracking reactions in an ebullated bed reactor operation. *Catalysis Today* 109: 185–194.

Takahashi, T., Higashi, H., and Kai, T. 2005. Development of a new hydrodemetallization catalyst for deep desulfurization of atmospheric residue and the effect of reaction temperature on catalyst deactivation. *Catalysis Today* 104: 76–85.

Takuchi, C., Asaoka, S., Nakata, S.-I., and Shiroto, Y. 1985. Characteristics of residue hydrodemetallization catalysts. *Am. Chem. Soc. Prepr, Div. Petrol. Chem.* 30 (1): 96–107.

Tanaka, R., Hunt, J.E., Winans, R.E., Thiyagarajan, P., Sato, S., and Takanohashi, T. 2003. Aggregates structure analysis of petroleum asphaltenes with small-angle neutron scattering. *Energy Fuels* 17: 127–134.

Thakur, D.S. and Thomas, M. G. 1984. Catalyst deactivation during direct coal liquefaction: A review. *Ind. Eng. Chem. Prod. Res. Dev.* 23: 349–360.

Thakur, D.S. and Thomas, M. G. 1985. Catalyst deactivation in heavy petroleum and synthetic crude processing: A review. *Appl. Catal.* 15: 197–225.

Tischer, R.E. 1981. Preparation of bimodal aluminas and molybdena/alumina extrudates. *J. Catal.* 72: 255–265.

Tischer, R.E., Narain, N.K., Stiegel, G.J., and Cillo, D.L. 1985. Large-pore $Ni\text{-}Mo/Al_2O_3$ catalysts for coal-liquids upgrading. *J. Catal.* 95: 406–413.

Topsøe, H., Clausen, B.S., Candia, R., Wivel, C. and Mørup, S. 1981. *In situ* Mössbauer emission spectroscopy studies of unsupported and supported sulfided Co-Mo hydrodesulfurization catalysts: Evidence for and nature of a Co-Mo-S phase. *J. Catal.* 68: 433–452.

Topsøe, H., Clausen, B.S., and Massoth, F.E. 1996. *Hydrotreating Catalysis Science and Technology*. Springer-Verlag: New York, 1996.

Toulhoat, H., Plumail, J.C., Houpert, C., Szymanski, R., Bourseau, P., and Muratet, G. 1987. Modelling RDM catalyst deactivation by metal sulfides deposit: An original approach supported by HRTEM investigation and pilot test results. *Am. Chem. Soc., Div. Petrol. Chem.* 32 (2): 463–463.

Toulhoat, H., Szymanski, R., and Plumail, J. C. 1990. Interrelations between initial pore structure, morphology and distribution of accumulated deposits, and lifetimes of hydrodemetallisation catalysts. *Catalysis Today* 7: 531–568.

Toulhoat, H., Hudebine, D., Raybaud, P., Guillaume, D., and Kressmann, S. 2005. THERMIDOR: A new model for combined simulation of operations and optimization of catalysts in residues hydroprocessing units. *Catalysis Today* 109: 135–153.

Trejo, F., Rana, M.S., and Ancheyta, J. 2008. CoMo/MgO-Al_2O_3 supported catalysts: An alternative approach to prepare HDS catalysts. *Catalysis Today* 130 (2–4): 327–336.

Trimm D.L. and Stanislaus, A. 1986. The control of pore size in alumina catalyst supports. *Appl. Catal.* 21: 215–238.

Tynan, E.C. and Yen, T.F. 1970. General purpose computer program for exact ESR spectrum calculations with applications to vanadium chelates. *J. Mag. Res.* 3 (3): 327–335.

van Veen, A.R. 2002. Hydrocracking. *In Zeolite for Clean Technology,* M. Guisnet and J.-P. Gilson, Eds., Imperial College Press: UK, 131–152.

van Kessel, M.M., van Dongen, R.H., and Chevalier, G.M.A. 1987. Catalysts have large effect on refinery process economics. *Oil Gas* J. 85 (7): 55–59.

Walendziewski, J. and Trawczynski, J. 1993. Preparation of large pore alumina supports for hydrodesulfurization catalysts. *Appl. Catal. A: Gen.* 96: 163–174.

Walendziewski, J. and Trawczynski, J. 1994. Influence of the forming method on the pore structure of alumina supports. *Appl. Catal. A: Gen.* 119: 45–58.

Ware, R.A. and Wei, J. 1985a. Catalytic hydrodemetallation of nickel porphyrins: I. Porphyrin structure and reactivity. *J. Catal.* 93: 100–121.

Ware, R.A. and Wei, J. 1985b. Catalytic hydrodemetallation of nickel porphyrins. II. Effects of pyridine and of sulfiding. *J. Catal.* 93: 122–134.

Webb, P.A. and Orr, C. *Analytical Methods in Fine Particle Technology*. Micromeritics: Norcross, GA, 1997.

White, A., Walpole, H.Y., and Trimm, D.L. 1987. Control of porosity and surface area in alumina II. Alcohol and glycol additive. *Appl. Catal.* 56: 187–196.

Wilson, M.F., Fisher, I.P., and Kriz, J.F. 1986. Cetane improvement of middle distillates from Athabasca syncrudes by catalytic hydroprocessing. *Ind. Eng. Chem. Prod. Res. Dev.* 25: 505–511.

Yoshinaka, S. and Segawa, K. 1998. Hydrodesulfurization of dibenzothiophenes over molybdenum catalyst supported on TiO_2-Al_2O_3. *Catalysis Today* 45: 293–298.

Yui, S. 1999. The role of catalysts in syncrudés operation. *Can. Chem. News*, July/August 25–27.

3 Changes in Asphaltenes during Hydrotreating

3.1 INTRODUCTION

During hydroprocessing of heavy oils and residua, obtaining more high-quality valuable products is desirable. In this regard, asphaltenes play a very important role due to their nature and the contribution they play in producing such products. It is well known that asphaltenes are composed by aromatic rings and alkyl side chains. When asphaltenes are in a colloidal state in crude oil, they are surrounded by resins that stabilize those forming micelles. During upgrading reactions, different compounds turn into lighter and valuable products. Asphaltenes also experience changes because they are exposed to reactions that modify their structure. Asaoka et al. (1983) studied structural changes that asphaltenes from Athabasca bitumen, Boscan crude, and Khafji vacuum residue suffered, and concluded that, if asphaltenes contained a high content of vanadium, asphaltene aggregates in crude were broken as a consequence of the reaction. However, if vanadium is not present in high amounts, but sulfur is, then depolymerization was the predominant reaction because sulfur links different aromatic cores. Depolymerization occurs partially in aromatic rings linked weakly by sulfur bonds. Cracking reactions influence the changes of asphaltene structure during hydrotreating, but this also depends on the nature of the feedstock and the level of contaminants, such as sulfur and metals (mainly V and Ni).

Yoshida et al. (1982) studied the changes that asphaltenes experience after hydrogenolysis at 400°C and 22 MPa using red mud (Fe_2O_3 [~60%], Al_2O_3 [~20%], SiO_2 [10%] and other oxides) as the catalyst. After the reaction, it was observed that residual asphaltenes increased their hydrogen content and, at the same time, there was an increase of aliphatic carbon, which indicated that aromatics hydrogenation occurred. It seemed that aromatic systems with more than two rings were more easily hydrogenated. But, on the contrary, systems having only two aromatic rings were hardly hydrogenated. Changes in asphaltenes are mainly found in hydrogen to carbon (H/C) atomic ratio, aromaticity factor, molecular weight, and an average degree of cross linking. Regarding the degree of cross linking, hydrogenolysis is responsible for heteroatoms removal, which produces oil composed by two condensed aromatic rings. The remaining asphaltenes are formed by two or more condensed aromatic rings that cross link.

Differences in the asphaltenes' structure make them more or less reactive compounds because properties of asphaltenes and the way in which they are constituted guide their reactivity. For instance, asphaltenes are very prone to adsorb onto the catalyst surface when their main constituents undergo condensation, sulfonation, and oxidation reactions. The processability of asphaltenes depends on their molecular structure (Izquierdo et al., 1989) because asphaltenes having a higher degree of condensation are difficult to hydrogenate. Additionally, when the internal aromatic carbon-to-peripheral aromatic carbon ratio is higher than three, asphaltenes hardly react. When this ratio is between 1.5 and 2, asphaltenes react more easily. Rogel (2000) have shown that asphaltenes with low hydrogen content and high aromaticity undergo self-association at lower concentrations. On the other hand, asphaltenes undergo low self-association when they are less aromatic and have more hydrogen in their structures.

Zhang et al. (2007) also studied the changes that asphaltenes suffer after hydrotreating. The chemical structure of asphaltenes from a residue was investigated using ruthenium ions catalyzed oxidation (RICO). As products of the RICO reaction, various isomeric benzenepolycarboxylic acids from di- to hexacarboxylic acids were identified, which allows one to understand the chemical structure. Aromatic structures of asphaltenes are mainly constituted by catacondensed, pericondensed, and biphenyl-type structures; however, biphenyl is less abundant than cata- and pericondensed structures, whereas the most abundant one was the pericondensed structure. As severity of hydrotreating reaction increased, asphaltenes were constituted by large aromatic units being highly condensed structures with a pericondensed arrangement (Wang et al. 1997).

One of the most refractory heteroatoms present in asphaltene structure is nitrogen. Some authors have reported that its content remains almost constant during hydrotreating (Bartholdy and Andersen, 2000), but this trend is not always the same for all types of feedstocks. In this regard, Ancheyta et al. (2003) have found that nitrogen content in asphaltenes tends to increase, as hydrotreating proceeds due to nitrogen is not removed from aromatic rings. In addition, the content of nitrogen also depends on the concentration of metals (Ni + V), which increases proportionally in asphaltenes. Mochida et al. (1988) performed two-stage hydrotreating of a brown coal liquid vacuum residue and found that structure of the heavy polar fraction in asphaltenes was formed by remaining nitrogen compounds, which are the cause for obtaining lower reactivity toward denitrogenation and impact directly on catalyst deactivation. A very descriptive study of the changes that asphaltenes suffer after hydroprocessing and their influence in stability of hydroprocessed oil is reviewed by Gawel et al. (2005).

The main conclusion obtained from experimental data is that initial conversion of asphaltenes is a result of cracking of the relatively long aliphatic chains from the aromatic core. Aliphatic and aromatic bonds broken during initial conversion process are relatively easy to carry out, especially when aliphatic chains have one or two carbon atoms. (Gray et al. 1992).

3.2 ASPHALTENES CHARACTERIZATION AFTER HYDROTREATING

3.2.1 Elemental Analysis

Elemental analysis (or ultimate analysis) gives the percentage of carbon, hydrogen, oxygen, nitrogen, and sulfur (C, H, O, N, and S). Trejo et al. (2005) reported the changes in elemental composition of asphaltenes precipitated from hydrotreated heavy oil. In general, carbon, hydrogen, and sulfur contents tend to decrease when reaction conditions are more severe. Nitrogen content is very stable due to its location inside aromatic rings and its association with metal complexes, which confer high stability (Mitra-Kirtley et al., 1993a, 1993b), and its concentration in asphaltenes did not change significantly. Figure 3.1 shows the behavior of the H/C atomic ratio obtained by elemental analysis. When space velocity is reduced, the value of the H/C ratio decreases as a consequence of the more time asphaltenes last in the catalytic bed and conversion is higher. Alkyl chains are removed and aromatic carbon is more abundant. Lower values of H/C indicate that asphaltenes are more aromatic. When pressure is increased, the H/C ratio is slightly decreased. However, the most important changes in asphaltene composition deal with the increase of temperature. Temperature is responsible for the initiation of thermal reactions, which break alkyl chains forming free radicals. The higher the temperature, the higher the impact on alkyl chains rupture. At higher temperature, more aromatic asphaltenes are obtained. According to Whitehead (1994), at temperatures higher than 400°C, C-C bonds are easily broken and saturate hydrocarbons are cracked, which leads to polyaromatic rings to be more abundant.

3.2.2 Metals Content

Nickel content tends to increase as temperature and pressure are increased and space velocity is reduced. Trejo et al. (2005) stated that this behavior is due to alkyl chains removal because it is easier to remove aliphatic carbon than metals, which are located in the inner part of asphaltene molecules. Figure 3.2 shows an/in increase of nickel content as pressure and temperature increase and space velocity is reduced.

Vanadium is encrusted inside asphaltenes structure. During hydrotreating it is difficult to remove vanadium because it is mainly concentrated in asphaltenes as vanadyl porphyrins associated with large hydrocarbon molecules to form asphaltene micelles. For this reason, if asphaltenes are removed, vanadium is also eliminated. There are also some other types of vanadyl porphyrins concentrated in the nonasphaltenic portion of heavy oils, which are more easily removed. When a catalyst is used for hydrotreating, usually vanadium is deposited on the catalyst surface as vanadium sulfide. Similar to nickel, vanadium also follows almost the same tendency in the crude oil as well as in asphaltenes. Vanadium content is increased as temperature and pressure are increased and space velocity is reduced. Figure 3.3 shows the behavior of vanadium content in asphaltenes after hydrotreating. From Figure 3.2 and Figure 3.3, it is seen that vanadium content is always higher than nickel content, which is a common fact, not only for metals contained in asphaltenes, but also for most of crude oils. Figure 3.4 also shows the reduction of Ni + V content in crude as a function of asphaltene conversion. It is observed that metals removal follows a linear dependence with asphaltene removal. Thus, the higher the asphaltene conversion, the higher the removal of metals in crude.

It has been suggested by Asaoka et al. (1986) that removal of vanadium is responsible for decomposition of asphaltene micelles and residual vanadium

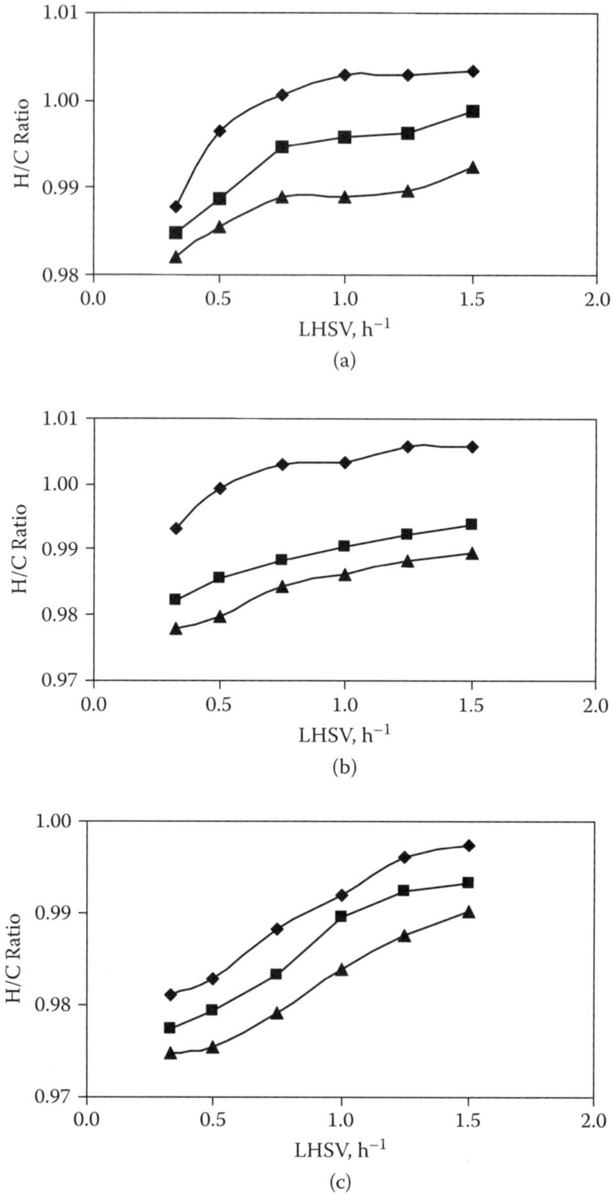

FIGURE 3.1 H/C atomic ratio of asphaltenes at different reaction conditions: (◆) 380°C; (■) 400°C; (▲) 420°C. (a) 70 kg/cm^2, (b) 85 kg/cm^2, (c) 100 kg/cm^2. (From Trejo et al. 2005. *Catalysis Today* 109 (1–4): 178–184. With permission.)

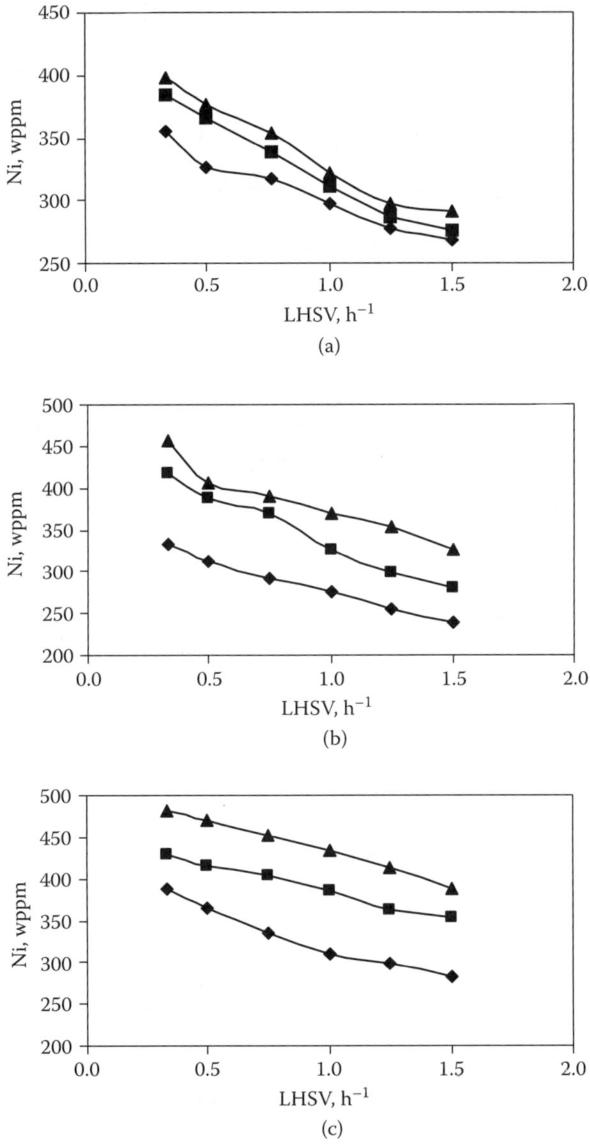

FIGURE 3.2 Nickel content in asphaltenes at different reaction conditions: (◆) 380°C; (■) 400°C; (▲) 420°C. (a) 70 kg/cm², (b) 85 kg/cm², (c) 100 kg/cm². (From Trejo et al. 2005. *Catalysis Today* 109 (1–4): 178–184. With permission.)

no longer forms additional micelles; however, remaining vanadium is combined with smaller molecules that are capable of reacting during hydrotreating, and which have a four-nitrogen coordination. Vanadyl complexes are also present in asphaltenes and they suffer changes during hydrotreating as well. The use

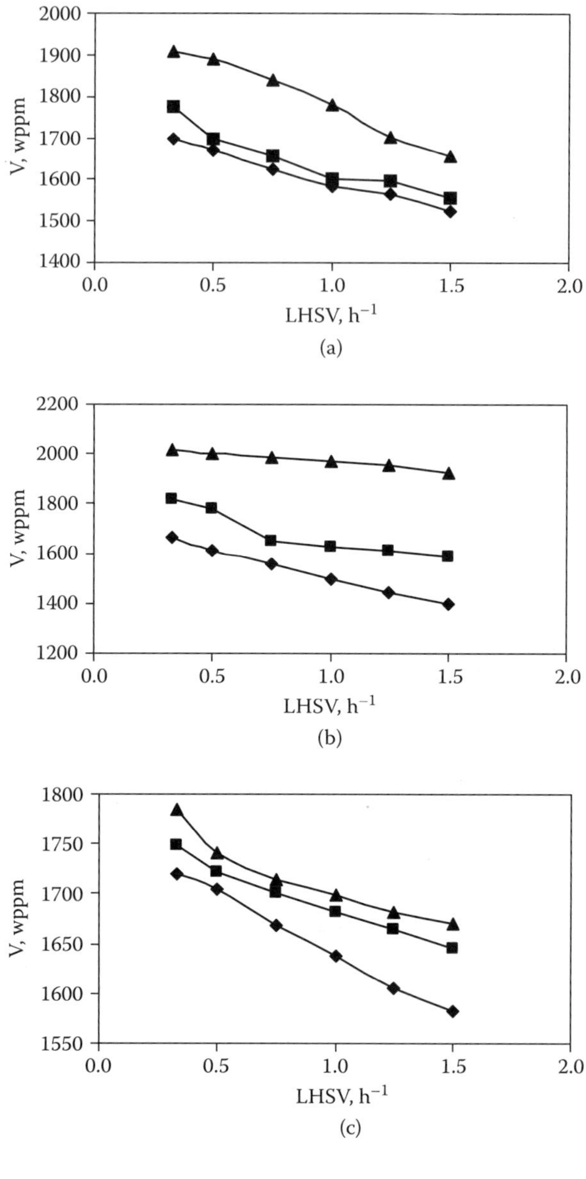

FIGURE 3.3 Vanadium content in asphaltenes at different reaction conditions: (◆) 380°C; (■) 400°C; (▲) 420°C. (a) 70 kg/cm^2, (b) 85 kg/cm^2, (c) 100 kg/cm^2. (From Trejo et al. 2005. *Catalysis Today* 109 (1–4): 178–184. With permission.)

of electron spin resonance (ESR) has made possible the study of these types of V-complexes. "Bound" vanadyl complexes are anisotropic, whereas "free" vanadyl compounds are isotropic as determined by ESR. It has been considered that vanadyl complexes are associated with large molecules of asphaltenes.

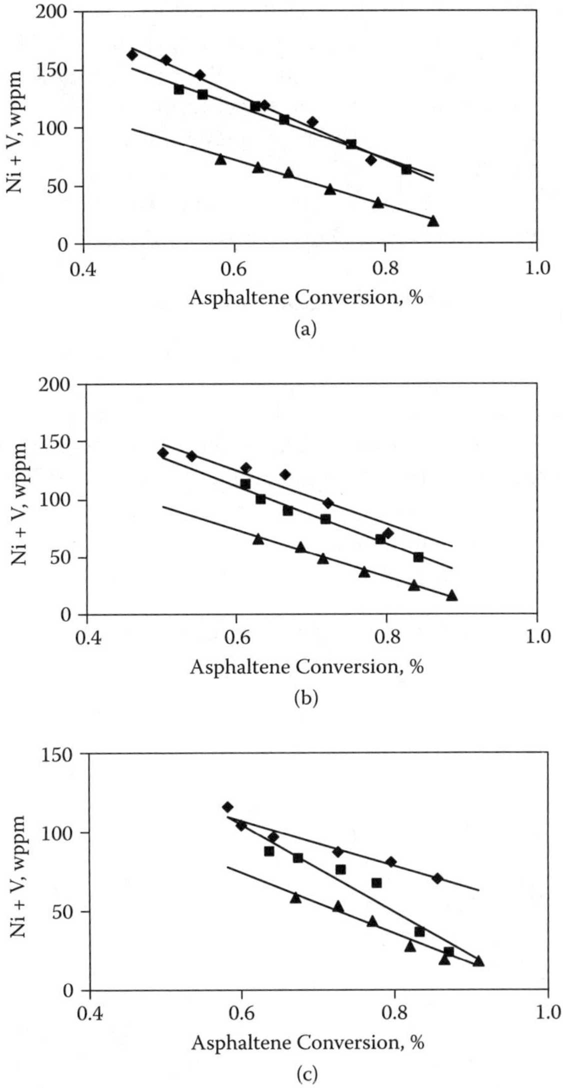

FIGURE 3.4 Metals content as function of asphaltene conversion: (◆) 380°C; (■) 400°C; (▲) 420°C. (a) 70 kg/cm^2, (b) 85 kg/cm^2, (c) 100 kg/cm^2. (From Trejo et al. 2005. *Catalysis Today* 109 (1–4): 178–184. With permission.)

As temperature is increased, vanadyl compounds change from being bound to free complexes. With this change measured by ESR and by using the Arrhenius equation, it is possible to calculate the dissociation energy of vanadyl present in asphaltenes before and after hydrotreating. Vanadyl in asphaltenes in the feedstock is considered to be in polymerized form forming complexes with some heteroatoms. Vanadyl in feedstock has a dissociation energy of 15 to 18 kcal/mol, whereas vanadyl in products has higher aromaticity due to the cleavage of weak linkages of heteroatoms; its dissociation energy is 8 to 10 kcal/mol. These values are in accordance of Tynan and Yen (1969) results. In this way, asphaltene micelles are destroyed during hydrotreating and vanadium is removed, but those remaining vanadyl complexes tend to form new asphaltenes and their aromatic layers are ready to be stacked. The presence of catalysts could help to process these new asphaltenes due to having smaller molecular weights by which they are able to enter inside the pores.

3.2.3 Changes in Molecular Weight and Size

The presence of aliphatic chains attached to aromatic cores in asphaltenes makes these structures able to break. Reduction of alkyl chains undoubtedly increases the aromaticity of asphaltenes. Many changes take place due to shortening of alkyl chains in asphaltenes, but depending on the catalyst type and reaction severity, asphaltenes can also suffer changes mainly due to hydrogenation of aromatic rings. Hydrogenation is a necessary step before cracking aromatic rings, and active sites along with textural properties of catalyst influence the course of reaction and, consequently, the asphaltene structure after reaction.

3.2.3.1 Molecular Weight by Vapor Pressure Osmometry (VPO)

During hydrotreating, Ancheyta et al. (2003) have reported that aggregate molecular weight obtained by VPO was reduced as temperature was increased indicating a reduction of size of the asphaltene molecule. Percentage of aromatic rings substituted with alkyl carbon was generally diminished and a drastic reduction of this value was observed during hydrotreating at 440°C. This could indicate a full removal of some alkyl chains and even a rearrangement of aromatic rings from catacondensed to pericondensed-like structures with reduction in the number of aromatic rings (R_a). Figure 3.5 shows the variation of percentage of substitution, average number of aromatic rings, and reduction of aggregate molecular weight of asphaltenes by VPO.

Based on size exclusion chromatography (SEC) analysis, Merdrignac et al. (2005) observed a reduction of the molecular weight of asphaltenes during hydrotreating at high reaction temperatures. Asphaltenes with higher molecular weight are easily cracked and preferentially converted into lighter molecules. Breaking of alkyl chains is also responsible for reducing the molecular weight. Seki and Kumata (2000) studied changes that asphaltenes and resins experience during hydrodemetallization and observed that asphaltene aromaticity increased at around 400°C as a consequence of the shortening of alkyl side chains. The authors have also concluded that asphaltene quality rather than quantity plays a key role in coke deactivation in catalyst.

Ali et al. (2006) stated that asphaltenes having the largest molecular weight show the highest aromaticity having high percentages of aromatic carbon in the bridge-head position and high amounts of heteroatoms. This is especially important when refining asphaltene-rich feedstocks. They concluded that low molecular

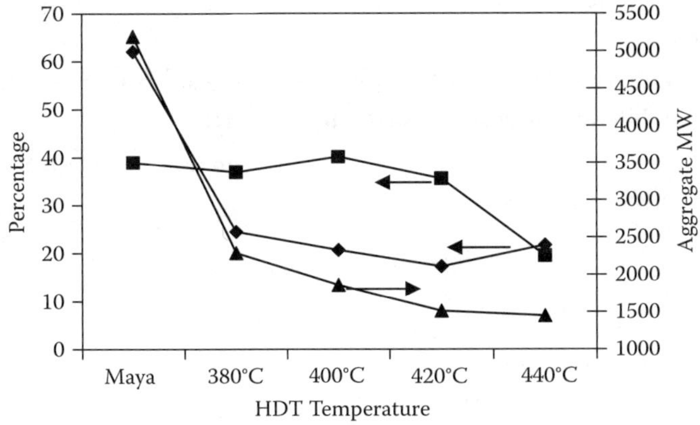

FIGURE 3.5 Changes in asphaltenes properties of hydrotreated products as function of reaction temperature compared with feedstock (Maya crude): (◆) Aromatic rings in the aggregates R_a; (■) percentage of substituted aromatic rings; (▲) aggregate molecular weight by VPO. (From Ancheyta et al. 2003. *Energy Fuels* 17 (5): 1233–1238. With permission.)

weight asphaltenes are converted into distillates, whereas high molecular weight asphaltenes are concentrated in the product oil. Despite the total amount of asphaltenes in the product oil being diminished, aromaticity of the remaining asphaltenes increased and the number of aliphatic chains decreased.

Michael et al. (2005) have stated that the main changes in asphaltenes involve alkyl chains and naphthenic rings. Asphaltene molecules in products after reaction are formed by cleavage of aliphatic carbon-to-carbon bonds at α, β, and γ positions from aromatic rings giving shorter aliphatic side chains and, as a consequence, there is an increase of aromaticity. The carbon skeleton of aromatic rings is highly condensed, but reaction did not lead to a significant fragmentation of these rings. The reduction in degree of substitution of these molecules after reaction is small, indicating that cracking of aliphatic chains from the aromatic rings is not complete.

Table 3.1 shows the main differences between asphaltenes from feed and product after thermal reaction. It is observed that hydrogen content is strongly reduced as well as the carbon content. However, according to these results, both types of asphaltenes preserve the aromatic core structure intact due to aromatic carbon (tertiary aromatic, quaternary, substituted, bridged, and nonbridged aromatic carbon) being almost the same. The number of S and N atoms per molecule also suffered a diminution, which impacts on the average asphaltene molecule as can be seen in the empirical formula in Table 3.1.

Merdrignac et al. (2006) have also reported changes of asphaltenes during hydroprocessing at different conditions. It was stated that when increasing the severity of reaction there is a reduction of alkyl chains length. The CH_2/CH_3 ratio is a parameter that gives an idea about dealkylation mechanism. The lower the CH_2/CH_3 ratio, the more dealkylation extent. It is more likely that a mechanism of β-scission is occurring during dealkylation, which increases aromaticity of

TABLE 3.1
Average Structural Data for Asphaltenes Separated from Feed and Product by Nuclear Magnetic Resonance (NMR)

	Atoms per Average Molecule	
Property	**Feed**	**Product**
Total hydrogen	163	57
Aromatic hydrogen	13	15
Aliphatic hydrogen	150	42
Aliphatic hydrogen in α position	15	13
Aliphatic hydrogen in β position	98	20
Aliphatic hydrogen in γ position	38	9
Total carbon	128	73
Aromatic carbon	59	55
Tertiary aromatic carbon	13	15
Quaternary aromatic carbon	47	40
Substituted aromatic carbon	7	5
Bridged aromatic carbon	35	32
Nonbridged aromatic carbon	25	23
Aliphatic carbon	69	18
Naphthenic carbon	14	3
n-alkyl carbon	43	10
Aliphatic carbon in methyl groups	11	5
Average number of carbon in aliphatic chains	10	3
Total number of aromatic ring per molecule	18	17
Total number of naphthenic rings per molecule	4	1
Aromaticity	0.46	0.76
Total sulfur	2.33	1.17
Total nitrogen	1.21	0.91
Empirical formula	$C_{128}H_{163}S_{2.33}N_{1.21}$	$C_{73}H_{57}S_{1.17}N_{0.91}$

Source: Michael et al. 2005. *Energy Fuels* 19 (4): 1598–1605. With permission.

remaining asphaltenes leaving only CH_3 groups attached to aromatic cores. Molecular weight (MW) of asphaltenes in hydrotreated products was estimated with scanning electron microscopy–mass spectrometry (SEC-MS) to be around 440 g/mol. An interesting comparison between changes in molecular weight of asphaltenes when hydroprocessing residue in fixed and ebullated-bed reactors has been made by these authors and their results indicated that, at lower conversions of residue (fixed bed), molecular weight is higher, corresponding to asphaltene aggregates. As conversion increases in the range of 30 to 50 wt%, molecular weight decreases significantly and at conversions higher than 50 wt% (that corresponds to more severe reaction conditions, such as ebullated-bed) there is a predominance of smaller molecular

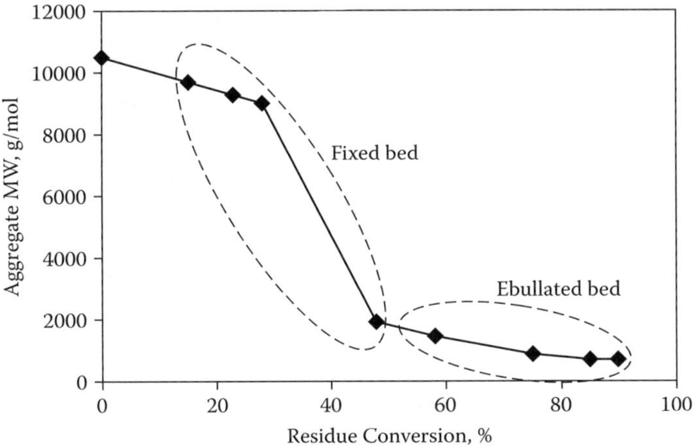

FIGURE 3.6 Qualitative evolution of molecular weight of asphaltene aggregates as a function of residue conversion. (From Merdrignac et al. 2006. *Energy Fuels* 20 (5): 2028–2036. With permission.)

weight asphaltenes. A qualitative representation of molecular weight of asphaltenes is given in Figure 3.6, in which it can be seen how MW reduction occurs as the function of residue conversion. Residue and asphaltene conversions are intimately linked so that at higher residue conversion more asphaltenes will be converted.

Heck and DiGuiseppi (1994) have worked with a stirred autoclave reactor and processing a 525°C$^+$ vacuum residue from Maya crude with high asphaltene content at 13.8 MPa of hydrogen pressure and concluded that in order to achieve high conversion of residue without significant coke formation, it is necessary to carry out the reaction initially at a low temperature in order to avoid a significant difference in the reaction rate between cracking and hydrogenation reactions. After some level of conversion has been carried out, temperature can be increased to complete the reaction. It was also reported a reduction of molecular weight of asphaltenes to be around two thirds of the initial molecular weight. Interestingly, they found that asphaltene molecular weight drops to less than a third of the feed value after 68% of residue conversion and remains at this level as conversion is increased up to 91% of residue conversion. The mean molecular size beyond 50% residue conversion is around 4 nm and only small additional changes occur at higher conversion. The tail corresponding to higher molecular weights is reduced, indicating that at around 80% of residue conversion the remaining asphaltenes have a molecular size of around 3.5 nm. Similar behavior is observed in the case of hydrocracked maltenes at 80% of residue conversion. At high residue conversion, maltenes size is around 2.5 nm in average. Reduction so drastic in asphaltenes size is due to cracking of aliphatic fragments leaving alone aromatic cores, which proliferate as conversion is increased; nonetheless, remaining smaller fragments are more difficult to remove or hydrogenate. Conversion of asphaltenes and reduction

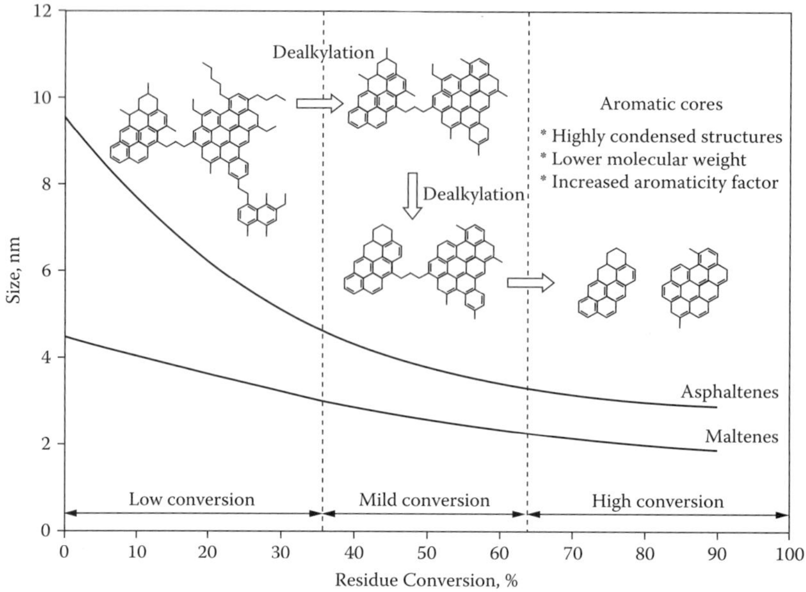

FIGURE 3.7 Proposed changes in asphaltenes contained in residue as a function of conversion.

of their molecular size are very fast in the beginning of reaction and tend to reach the steady state as residue conversion increases. At the same time, it is expected that asphaltenes aromaticity increases at the highest residue conversion. Based on aforementioned observations, it is possible to draw a scheme where differences in molecular size of asphaltenes and maltenes as a function of residue conversion are observed due mainly to dealkylation, which reduces considerably the molecular weight of asphaltenes as well. The final products, i.e., aromatic cores, have high aromaticity factor and are polycondensed structures. Only small fragments of alkyl chains are present at the highest conversion (Figure 3.7).

It is quite true that asphaltenes are concentrated in heavy cuts, i.e., residue. In this way, as residue was converted, asphaltenes also reacted, leading to significant changes in their structure and molecular weight. Gauthier et al. (2007, 2008) performed experiments for residue conversion in the range of 55 to 85% and asphaltenes conversion from 62 to 89% and summarized their proposed evolution mechanism of asphaltenes in three steps:

1. For residue conversion <40 wt%, only dissociation of asphaltene aggregates is the main mechanism.
2. For residue conversion in the range of 40 and 80 wt%, the main driving force is dealkylation of aliphatic chains linked to aromatic cores.
3. For residue conversion >80 wt%, asphaltenes become polycondensed structures with a minimal or null alkyl chains.

It is observed from Figure 3.7 that an asphaltene structure and its evolution are dependent on the conversion, which is directly related to reaction temperature, and archipelago structure of asphaltenes is gradually diminished. Firstly, a reduction of aggregates occurs at lower conversion followed by a dealkylation at moderate conversions. At this stage, labile alkyl bonds are broken and finally polycondensed aromatic rings are obtained at higher residue conversion obtaining a more compact aromatic core. Alkyl structures migrate to lighter fractions and they are converted to valuable products, whereas remaining asphaltenes are contained in nonconverted residue. However, the schematic representation does not take into account the role of metals within the asphaltene structure. It must be considered that multiple representations of asphaltenes are possible, thus a unique structure is not feasible.

3.2.3.2 Molecular Weight Distributions

SEC analysis with tetrahydrofuran (THF) as the mobile phase can be used for studying qualitative behavior of asphaltenes after hydrotreating. Trejo et al. (2007) analyzed different asphaltenes from hydrotreated Maya crude at different reaction conditions by SEC with THF as the solvent and found that, when pressure is varied at constant temperature and space velocity, molecular weight distribution did not exhibit a big difference with respect to asphaltenes without hydrotreating (Figure 3.8). On the contrary, when the temperature is increased from 380 to 420°C, molecular weight distribution showed a significant difference compared with nonhydrotreated asphaltenes. The smaller shoulder associated to the main peak is sometimes assigned to heavier molecular weight asphaltenes. In this case, temperature was responsible for reduction of those heavier asphaltenes due to the breaking of alkyl chains and linkages involving weak bonds with heteroatoms so that the higher the reaction temperature, the lower the molecular weight of asphaltenes. Bartholdy and Andersen (2000) have also observed this behavior before and it has been stated that at a lower temperature, hydrogenation is the preferred reaction, whereas at higher temperatures the reaction turns into cracking-dominated reaction. Transition temperature between both reactions has been established at around 380°C (Bartholdy et al., 2001). Space velocity also is responsible for changes in molecular weight distribution of asphaltenes. Seen in Figure 3.8, the displacement of chromatograms toward higher retention times indicates that lighter asphaltenes are obtained after hydrotreating.

One problem associated with THF as a mobile phase in SEC analysis is that chromatograms can tail toward smaller molecular weights and, virtually, become very long. For solving this issue, Herod et al. (1996) have recommended the usc of *N*-methyl-2-pyrrolidinone (NMP) as a solvent since significant changes in molecular weight distribution were observed, i.e., chromatograms shifted toward earlier elution times as the mobility of the fraction in planar chromatography decreased (Li et al., 2004). However, NMP can not fully dissolve asphaltenes (Ascanius et al., 2004; Strausz et al., 2002) because after hydrotreating, hydrogenation reactions make a greater proportion of the sample insoluble, along with the lack of solubility of aliphatics.

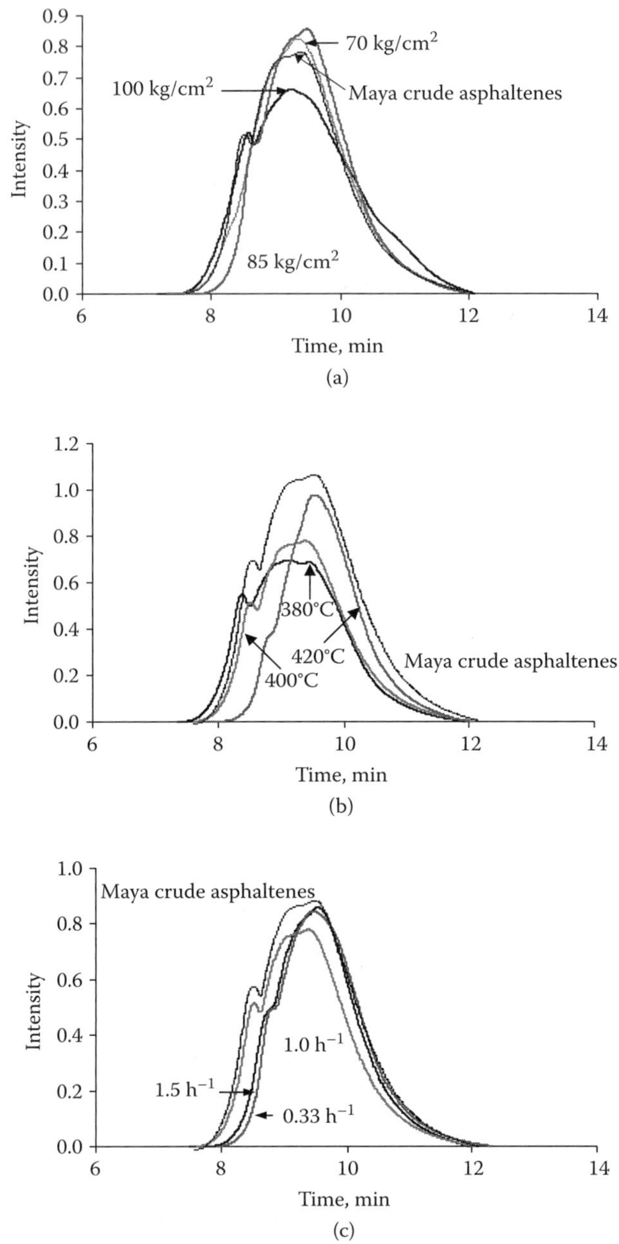

FIGURE 3.8 Size exclusion chromatography (SEC) profiles in tetrahydrofuran (THF) of asphaltenes from Maya crude oil and hydrotreated products. (A) effect of the pressure, (B) effect of the temperature, (C) effect of space velocity. (From Trejo et al. 2007. *Energy Fuels* 21 (4): 2121–2128. With permission.)

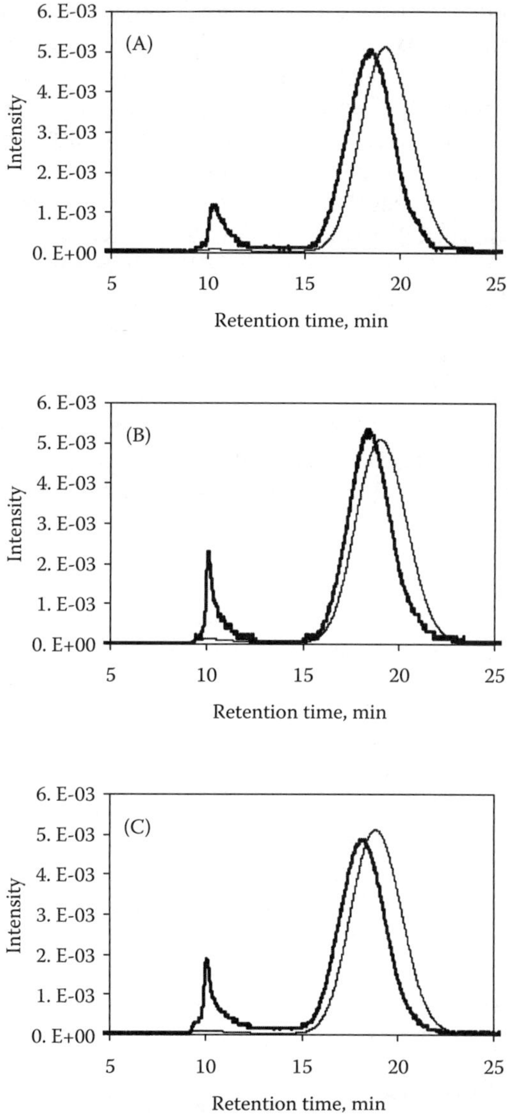

FIGURE 3.9 SEC profiles in *N*-methyl-2-pyrrolidinone (NMP) at 350 nm of asphaltenes: (A) 70 kg/cm^2, 400°C, LHSV = 1.0 h^{-1}; (B) 100 kg/cm^2, 400°C, LHSV = 0.33 h^{-1}; (C) 100 kg/cm^2, 420°C, LHSV = 1.0 h^{-1}. (From Trejo et al. 2007. *Energy Fuels* 21 (4): 2121–2128. With permission.)

However, an excluded peak is obtained between 8 and 13 min depending upon the sample, and a resolved peak is observed after 15 min in most cases (Figure 3.9). It is observed that, in all cases, it followed the same pattern in chromatograms. The first peak corresponds to higher molecular weight asphaltenes. The valley with a zero signal could be a consequence of a change in conformation of the sample going to three dimensional rather than a planar configuration (Karaca et al., 2004; Morgan et al., 2005). Calibration of columns with polystyrene indicates that at elution times of 15.5 and 16.1 min, the high-mass limits of the main peaks correspond to polystyrene masses of 6,000 and 3,000 g/mol, respectively. Differences were also observed between the UV-A (ultraviolet-absorption) and UV-F (ultraviolet fluorescence) chromatograms and only results at 350 nm are shown. On the other hand, when using the UV-A detector, the higher solvent power of NMP allowed for the detection of large-sized material eluting at the exclusion limit of the chromatographic column, which was observable from 6.5 min. The material corresponding to the excluded peak can be a consequence of very large molecular weights. The lack of fluorescence may be due to strong molecular interactions leading to quenching of fluorescence intensity (Ascanius et al., 2004). However, UV-A chromatograms show the narrowest distribution compared with UV-F chromatograms probably because of the smallest aromatics that have the strongest fluorescence. Suelves et al. (2003) have stated that molecular weights at 20 min correspond, in terms of polystyrene calibration, to asphaltenes having molecular weight ranging from 300 to 750 g/mol. This last value has been reported by fluorescence depolarization (Groenzin and Mullins, 2000).

LDMS (laser desorption mass spectrometry) and MALDI (matrix-assisted laser desorption/ionization) can also give a molecular weight distribution of asphaltenes. Trejo et al. (2007) reported the changes that molecular weight distribution of asphaltenes suffer after hydrotreating. Figure 3.10 shows both spectra (LDMS and MALDI) and the effects of pressure, temperature, and space velocity. At high pressure, the maximum ion abundances detected by LDMS and MALDI were *m/z* (mass to charge ratio) 1,500 and 2,000, respectively. However, bigger changes were observed with increasing temperature. At higher temperatures, asphaltene spectra are sharper indicating less polydispersity of asphaltenes as a consequence of reduction of bigger molecules of asphaltenes. The maximum ion abundances were *m/z* around 1,000 and 1,500 by LDMS and MALDI, respectively. At lower space velocity, the peak also is less polydisperse as in temperature. The maximum ion abundances were almost the same as in the case of temperature.

The upper mass limits and maximum ion abundances obtained by mass spectrometry compare reasonably well with those found by SEC using NMP for retained peaks. The two techniques agree within a factor of less than 2, but do not take into account the material of the excluded peaks of unknown mass (Millan et al., 2005).

3.2.4 Nuclear Magnetic Resonance

In previous sections, the importance of ^{13}C NMR analysis was stated in determining the average asphaltene structure. In this section, some structural changes that asphaltenes undergo after hydrotreating are analyzed using NMR data.

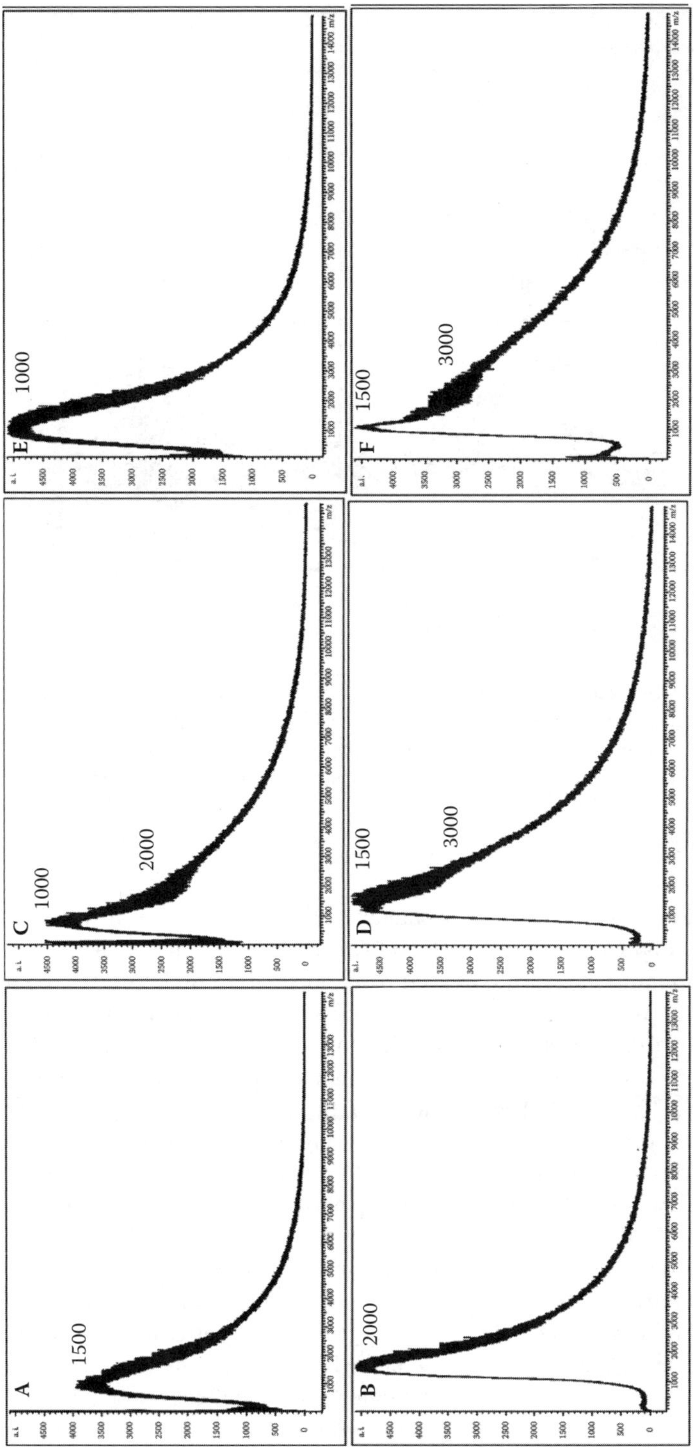

FIGURE 3.10A/F (A) LDMS and (B) MALDI spectra at LHSV = 1.0, T = 400°C, and P = 100 kg/cm^2; (C) LDMS and (D) MALDI spectra at LHSV = 1.0, T = 420°C, and P = 100 kg/cm^2; (E) LDMS and (F) MALDI spectra at LHSV = 0.33, P = 100 kg/cm^2, and T = 400°C. (From Trejo et al. 2007. *Energy Fuels* 21 (4): 2121–2128. With permission.)

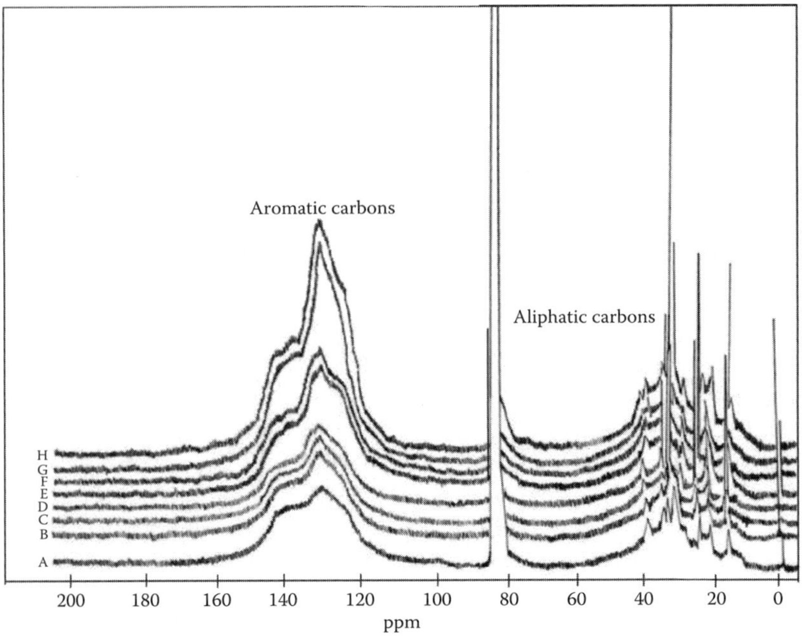

FIGURE 3.11 ^{13}C NRM spectra. (A) Pure Maya crude asphaltenes; (B) T = 400°, P = 70 kg/cm^2, LHSV = 1.0; (C) T = 400°C, P = 85 kg/cm^2, LHSV = 1.0; (D) T = 400°C, P = 100 kg/cm^2, LHSV = 1.5; (E) T = 380°C, P = 100 kg/cm^2, LHSV = 1.0; (F) T = 400°C, P = 100 kg/cm^2, LHSV = 1.0; (G) T = 420°C, P = 100 kg/cm^2, LHSV = 1.0; (H) T = 400°C, P = 100 kg/cm^2, LHSV = 0.33. (From Trejo et al. 2007. *Energy Fuels* 21 (4): 2121–2128. With permission.)

Trejo et al. (2007) characterized asphaltenes before and after hydrotreating by ^{13}C NMR. Figure 3.11 shows different spectra for pure and hydrotreated asphaltenes. Two main peaks are observed to be located around 15 to 40 ppm and 115 to 150 ppm, which correspond to aliphatic and aromatic carbons, respectively. It is seen qualitatively from spectra that, as reaction conditions are more severe, the peak of aromatic carbons increases indicating that aliphatic carbons are removed. Table 3.2 shows values of some structural parameters obtained from spectra, such as aromaticity factor (f_a), average length of alkyl carbon (n), percent of substitution of aromatic rings (A_s), and number of aromatic rings (R_a), calculated according to Equation (1.19) to Equation (1.22) seen in Chapter 1 (Calemma et al., 1995).

It is observed that the number of aromatic rings diminishes in general as temperature and pressure are increased, whereas space velocity is decreased. However, when pressure is increased, R_a changed only slightly. It is also seen that asphaltenes are less substituted, as indicated by A_s. This means that more aliphatic chains are removed from aromatic cores and asphaltenes probably tend to be catacondensed structures. Less aliphatic carbon and more aromatic carbon

TABLE 3.2
Structural Parameters by ^{13}C NMR for Pure and Hydrotreated Asphaltenes

Properties	Reaction Conditions		
		Pure Maya Crude asphaltenes	
R_a		62.0	
A_s		38.9	
n		6.80	
f_a		0.52	
		T = 400°C and LHSV = 1.0 h^{-1}	
	70 kg/cm²	85 kg/cm²	100 kg/cm²
R_a	28.32	27.86	25.90
A_s	33.89	34.29	32.70
n	6.07	5.90	5.22
f_a	0.55	0.55	0.58
		P = 100 kg/cm² and LHSV = 1.0 h^{-1}	
	380°C	400°C	420°C
R_a	40.18	25.90	31.35
A_s	36.61	32.70	26.95
n	6.01	5.22	3.34
f_a	0.55	0.58	0.72
		P = 100 kg/cm² and T = 400°C	
	1.5 h^{-1}	1.0 h^{-1}	0.33 h^{-1}
R_a	46.97	25.90	28.10
A_s	46.54	32.70	21.53
n	6.07	5.22	4.63
f_a	0.55	0.58	0.70

Source: Trejo et al. 2007. *Energy Fuels* 21 (4): 2121–2128. With permission.

increase the aromaticity factor as shown in Table 3.2. Since aliphatic chains are stripped from aromatic cores, the average length of alkyl carbon is reduced indicating that aromatic structures tend to prevail.

3.2.5 X-Ray Diffraction

When analyzing asphaltenes by XRD, two main peaks appeared in diffractograms, i.e., one around 20° and the other one around 26°. This last peak corresponds to the aromatic stacking of asphaltenes, which confers to solid asphaltenes some crystalline nature. However, one of the most important issues related to XRD diffractograms is baselines, which in most of the cases are distorted. When base lines are not well defined, the use of commercial software to fit the curves and reduce the noise is recommended. Curve fitting may include different models, among which the most commonly used are Gaussian or Lorentzian models. When hydrotreating asphaltenes, alkyl chains are reduced and more aromatic structures are obtained as a consequence of removal of aliphatic carbon and condensation reactions, which turns asphaltenes into smaller structures. When precipitating asphaltenes in solid state, they are ready to form piles of aromatic rings, which are able to form a crystallite. When severe conditions in hydrotreating are used, lighter asphaltenes are obtained and it is more feasible for small molecules to be stacked because of their small core and by the fact that their alkyl side chains are not too long, diminishing their interlayer distance once they are precipitated. It has been suggested by Tanaka et al. (2003) that asphaltenes aggregates change their morphology as the temperature increased, which is measured by small angle neutron scattering (SANS). At room temperature, asphaltenes aggregates are prolate ellipsoids, whereas at 350°C they have a sphere-like form. Therefore, at elevated temperatures, aggregates are not the predominant form in the crude. It is only after cooling and precipitation that asphaltenes would aggregate in crystalline structures.

Sakanishi et al. (2000) analyzed a vacuum residue and its fractions (maltenes and asphaltenes) by using a horizontal-type x-ray diffractometer by slow step scan mode at variable temperatures, in order to clarify the changes in molecular–aggregate structure during heating and hydrotreating. They found that the peak at around 26° in asphaltenes decreased by heating to 300°C. The addition of toluene was also effective for the reduction of the peak at around 26°. When using a NiMo/Al_2O_3 hydrotreating catalyst, it was observed that asphaltenes content diminished because they were transformed into maltenes at reaction temperatures below 400°C, whereas hydrotreating at 420°C increased the asphaltene aromaticity.

Michael et al. (2005) have characterized asphaltenes by XRD after thermal conversion to determine structural changes in asphaltene crystallite and reported an average distance between aromatic sheets (d_m) of 0.35 nm in both feedstock and products, and the average distance between aliphatic chains (d_γ) was also similar to 0.44 nm. However, the number of aromatic rings (R_a) increased in asphaltenes from products per aromatic core, whereas the number of stacked aromatic layers diminished from 4 to 3 in products, indicating a destruction of the piling of aromatic layers. Other authors have also studied the thermal decomposition of asphaltenes analyzed by XRD obtaining an increased aromaticity on asphaltenes after thermal treatment (Amerik, 1995). It was stated that asphaltenes are composed of rhombic crystal systems in the aromatic section (Korolev and Amerik, 1993).

Trejo et al. (2007) performed characterization of asphaltenes by XRD after and before hydrotreating, and the results are presented in Table 3.3, which shows quantitative measurements of crystallites of asphaltenes both in the aromatic and aliphatic stacking. It is observed that temperature favors more changes in asphaltene structure, i.e., there is a shortening of alkyl side chains. When increasing temperature from 380 to 420°C there is a diminution of stacking from 10.6 to

TABLE 3.3
Crystalline Parameters of Asphaltenes Before and After Hydrotreating of Maya Crude Oil

	Structural Parameters by XRD					
	d_m, Å		L_c, Å		M	
Variables	Aromatic Stacking	Alkyl Stacking	Aromatic Stacking	Alkyl Stacking	Aromatic Stacking	Alkyl Stacking
Pure Maya crude asphaltenes	3.53	4.98	69.5	21.3	20.7	5.3
Pressure, kg/cm^2	**T = 400°C and LHSV = 1.0 h^{-1}**					
70	3.58	4.72	35.6	19.9	10.6	5.2
85	3.51	4.87	33.5	19.0	10.4	4.9
100	3.53	4.87	33.5	16.3	10.5	4.3
Temperature, °C	**P = 100 kg/cm^2 and LHSV = 1.0 h^{-1}**					
380	3.60	4.83	32.5	19.3	10.6	5.0
400	3.53	4.87	33.5	16.3	10.5	4.3
420	3.53	4.72	26.3	16.6	8.3	4.5
LHSV, h^{-1}	**P = 100 kg/cm^2 and T = 400°C**					
1.5	3.60	4.82	34.9	18.8	10.7	4.9
1.0	3.53	4.87	33.5	16.3	10.5	4.3
0.33	3.48	4.80	31.7	17.9	10.1	4.7

Source: Trejo et al. 2007. *Energy Fuels* 21 (4): 2121–2128. With permission.

8.3 aromatic layers (M) in average. The interlayer distance (d_m) at the highest temperature reaction was 3.60 Å, whereas at the lowest temperature it decreased to 3.53 Å. Graphite-like structures have an interlayer spacing around 3.35 Å, whereas amorphous carbon has interlayer distance of 3.55 Å. According to these observations, asphaltenes from hydrotreated Maya crude at the most severe reaction conditions exhibit a very similar arrangement to amorphous carbon and in some cases their structures are semicoke-like (H/C atomic ratio of 0.75), which have an interlayer distance of 3.52 Å (Andersen et al., 2005). Asphaltenes hydrotreated at 85 and 100 kg/cm^2, 400 and 420°C, and space velocity of 1.0 and 0.33 h^{-1} are similar to semicoke-like structures. The average height of the stacking is reduced as the average number of aromatic layers is diminished. Space velocity also influences significantly crystalline parameters. There is a reduction in the average number of stacked molecules (M) from 10.7 to 10.1 and at the same time the average height of stacked molecules (L_c) diminishes from 34.9 to 31.7 at the lowest space velocity. There is also a reduction in the interlayer distance when space velocity is reduced.

The effect of pressure on stacked molecules (M) is less significant because there is not a big change in crystalline parameters. When increasing pressure from 70 to 100 kg/cm^2 there is only a small reduction in the average number of aromatic layers from 10.6 to 10.4 and, consequently, the average height (L_c) does not change because the number of aromatic layers is almost the same.

It was observed that the spacing between aromatic layers ranged from 3.48 to 3.60 Å at different reaction conditions, having around 10 aromatic layers in average. Sharma et al. (2002) found an average spacing of ~3.7 Å. XRD showed that stacking of asphaltene molecules after precipitation diminished when reaction conditions were more severe. Interlayer distances in crystallites correspond to amorphous carbon. Stacking of smaller molecules may occur during precipitation because their cores are smaller and alkyl chains are smaller as well. It is observed in all the cases that crystalline parameters of aliphatic chains remain almost constant. Only the L_c and M values diminished slightly at the most severe reaction conditions. Alkyl chains are the part of asphaltenes that suffers the major changes during hydrotreating due to the loss of aliphatic carbon. When alkyl chains pile up, they do so in an irregular way since chains are able to twist in the space. Stacking of aliphatic chains also depends on their length.

3.2.6 Microscopic Analysis

In spite of having different reports of asphaltenes characterization by microscopic techniques (SEM [scanning electron microscopy], TEM [transmission electron microscopy], STM [scanning tunneling microscopy], confocal laser-scanning), all of them are pertain to asphaltenes precipitated from heavy crude and, in some cases, asphaltenes are visualized in their natural medium in solid state, as in bitumen directly as reported when using AFM (atomic force microscopy). However, none of these reports have been applied to observing asphaltenes after hydroprocessing. In this regard, SEM and TEM emerge as useful techniques that allow for seeing how size and shape of asphaltenes are modified after reaction. During hydroprocessing, asphaltenes undergo several modifications that lead them to diminish their molecular weight, increase the aromaticity factor, and reduce the alkyl chains. These last changes modify the microscopic structure of asphaltenes and impact on their macroscopic behavior. At molecular level and viewed under TEM, asphaltenes present different morphologies depending on the severity of reaction. By SEM, some changes on the asphaltenes surface are notorious as explained in the following section.

3.2.6.1 SEM Analysis

Trejo et al. (2009) have analyzed by SEM asphaltenes from Maya crude, which were hydrotreated at different conditions. It has been reported that temperature is the most important variable that produces the most significant changes on asphaltenes when hydroprocessing heavy crudes (Trejo et al., 2005) and, for this reason, two temperatures during hydroprocessing of Maya crude were analyzed, i.e., 400 and 420°C. Figure 3.12 presents samples of asphaltenes coming from hydroprocessed Maya crude

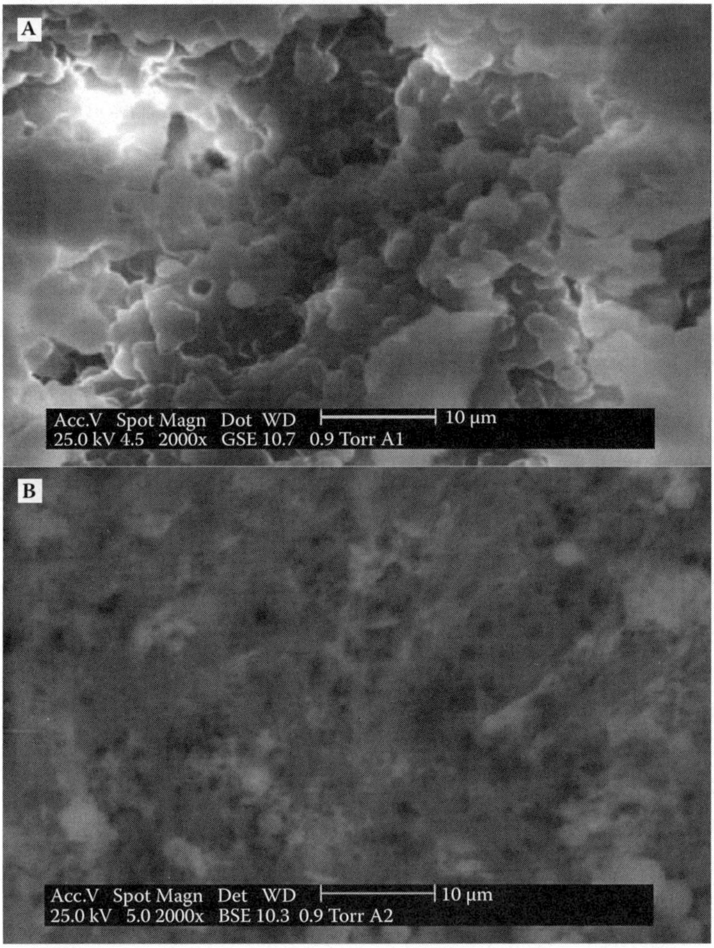

FIGURE 3.12 SEM images of asphaltenes from hydrotreated Maya crude at 400°C (A) and 420°C (B) keeping constant the pressure (100 kg/cm^2) and LHSV (1 h^{-1}). (Trejo et al. 2009, *Energy Fuels* 23(1): 429–439. With permission.)

at 400°C (case *a*) and 420°C (case *b*). Figure 3.12a shows cumulus of asphaltenes, whereas Figure 3.12b shows some pores on the asphaltenes matrix. Both asphaltenes samples were washed with a mixture of toluene and *n*-heptane (67/33 vol%) after precipitation. As a consequence of changing temperature during hydroprocessing, asphaltenes suffered modifications as well as mainly reducing their alkyl chains, which leads to a rearrangement at microscopic level and the forming of piles of aromatic layers, as reported later by TEM. Morphology of Figure 3.12a could be due to the combination of two effects: (1) hydroprocessing temperature is not enough to disrupt big cumulus of asphaltenes at all, and (2) solvents used for washing asphaltenes tend to form agglomerates of asphaltenes with similar properties. On the other hand, the porous structure showed in Figure 3.12b can be a consequence of trapped maltenes inside asphaltenes, which are released during reaction to transform them into valuable products leaving pores on the asphaltene matrix. The physical shape of asphaltenes in Figure 3.12a is spherical-like having ~2.1 μm of average diameter, whereas Figure 3.12b shows an irregular surface with void spaces of variable sizes. In the last case, it appears that higher hydrotreating temperature (420°C) is enough to break agglomerates of asphaltenes and transform more maltenes to lighter products. To compare changes suffered after hydroprocessing, asphaltenes from Maya crude were also washed with the same mixture of solvents. All samples were analyzed to obtain their elemental composition as shown in Table 3.4. The results correspond to the analyzed area by which it can be considered as local values. Nitrogen was not detected by this technique; however, it is seen that carbon content increased in asphaltenes as temperature was increased. In general, silicon also increased.

The presence of phosphorous is not commonly reported in asphaltenes; however, in the case of pure and hydrotreated asphaltenes from Maya crude, its presence is evident despite the highest hydrotreating temperature P amount being similar to that of nonhydrotreated asphaltenes. Sulfur diminished as hydrotreating temperature is increased indicating possible scission from aromatic rings and/or alkyl chains. The high sulfur content in nonhydrotreated asphaltenes is likely due to the analyzed area of the sample, which is very rich in this element. It is important to remember that this analysis must be carefully considered because the concentration and mapping of elements depend on the region where

TABLE 3.4
Elemental Composition of Asphaltenes and Fractions by Scanning Electron Microscopy (SEM)

	Elemental Composition, wt%		
Element	**Maya Crude Asphaltenes**	**HDT at 400°C**	**HDT at 420°C**
C	81.13	86.06	86.71
O	2.71	4.17	4.56
Si	0.17	0.94	0.34
P	0.42	0.26	0.43
S	14.99	7.96	7.23
V	0.41	0.28	0.15
Ni	0.08	0.21	0.43
Fe	0.08	0.11	0.15
S/C	0.0692	0.0347	0.0312
V/C (10^3)	1.1915	0.7671	0.4079
Ni/C (10^3)	0.2018	0.4994	1.0148

Source: Trejo et al. 2009. *Energy Fuels* 23(1):429–439.

asphaltenes are analyzed, e.g., Pérez-Hernández et al. (2003) in SEM characterization of asphaltenes did not find elements such as N, S, O, and metals. In our case, the presence of S and O was significant. Ni and Fe tend to concentrate as hydrotreating temperatures increase and a suitable explanation for this fact is that destruction of alkyl chains of asphaltenes leaves practically intact the core of micelle where porphyrins are located and the metals amount tends to concentrate. Table 3.4 also shows the S/C, V/C, and Ni/C atomic ratios where the behavior of these elements can be seen as a function of hydrotreating temperature.

3.2.6.2 TEM Analysis

TEM is a spectroscopic technique whereby a beam of electros is transmitted through an ultra thin specimen, interacting with asphaltenes in which an image is formed as a significantly higher resolve than SEM. It is possible using TEM to see structural modifications at molecular level when asphaltenes are hydrotreated at different reaction conditions. In this regard, temperature plays an important role since alkyl chains are de-attached from aromatic cores and let them pile up and form visible crystalline structures under microscopy. The stacking is possible because aromatic cores are flat and the height of the stacking is a function of the severity reaction conditions and the nature of the feedstock.

Figure 3.13 shows a photo from asphaltenes hydrotreated at 400°C and some type of rearrangement near the edge having an interlayer distance of around 0.355 nm in asphaltenes. This value is consistent with that reported for typical separation of aromatic structures of amorphous asphaltenes, but it is not the only rearrangement viewed in the sample. Graphite-like carbonaceous structures are also present having an interlayer separation around ~0.335 nm as a consequence of the very well-ordered aromatic sheets that do not have steric hindrance by the alkyl chains. These have been removed or reduced in average length during reaction obtaining a great number of stacked aromatic cores. By increasing the hydrotreating temperature at 420°C, asphaltenes suffered more significant changes (Figure 3.14) where poorly ordered structures are observed near the edge corresponding to amorphous asphaltenes (~0.355 nm), whereas, in the inner part of the sample, well-ordered structures at long extent are seen corresponding to graphite-like asphaltenes (~0.34 nm). This indicates that a rearrangement occurred due to the breaking of alkyl chains because the higher reaction temperature during hydrotreating favored the cracking of alkyl chains. Aromatic cores are preserved, being able to stack better because there are not alkyl chains impeding the asphaltenes layers from piling up. Aforementioned samples of Figure 3.13 and Figure 3.14 were previously washed with a mixture of toluene plus n-C_7 (67/33 vol%), the same as in the SEM analysis.

Elemental analysis of asphaltenes obtained by TEM is reported in Table 3.5 where nonhydrotreated asphaltenes were analyzed for comparing with hydrotreated samples. Carbon content increased as hydrotreating temperature was also increased and it is comparatively higher than that of the SEM analysis. Sulfur content increased slightly when the temperature was 400°C compared with nonhydrotreated

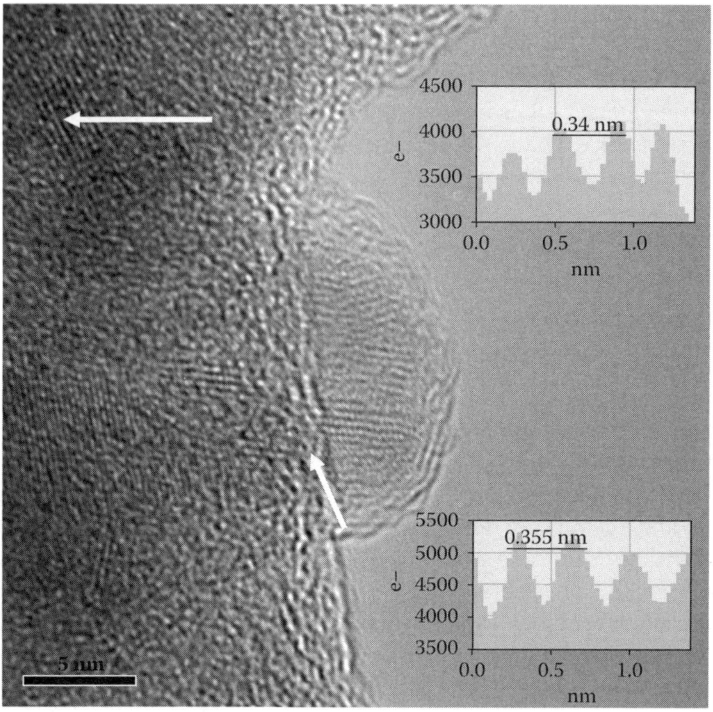

FIGURE 3.13 TEM image of asphaltenes from hydrotreated Maya crude at 400°C, 100 kg/cm^2, and LHSV = 1 h^{-1} washed with toluene (67 vol%) and *n*-heptane (33 vol%). (Trejo et al. 2009, *Energy Fuels* 23(1): 429–439. With permission.)

asphaltenes, but, in general, its content was lower at 420°C than at 400°C. Vanadium content also diminished as in SEM analysis. In this case, Ni content was only detected in traces because it is commonly present in lower amounts compared with V or Fe. When analyzing by SEM, Ni appeared, but in TEM it is not detected since the area of analysis in SEM is higher than that observed by TEM. Ni could be detected by TEM if other areas on the sample were focused. Fe also increased as hydrotreating temperature was higher. Furimsky (1978) has reported the presence of Fe in coke, which is not surprising that this element appears in asphaltenes because they are composed of polynuclear aromatics that are prone to generate coke. However, Sánchez-Berna et al. (2006) have observed the presence of Cr or P along with Fe, but, in our case, neither Cr nor P was detected.

These results demonstrate the importance of having microscopic analysis of asphaltenes because changes during hydroprocessing can be understood at molecular level. If other analytical techniques are applied to characterized asphaltenes after hydroprocessing, such as NMR or XRD, a better comprehension of structural rearrangements could be obtained.

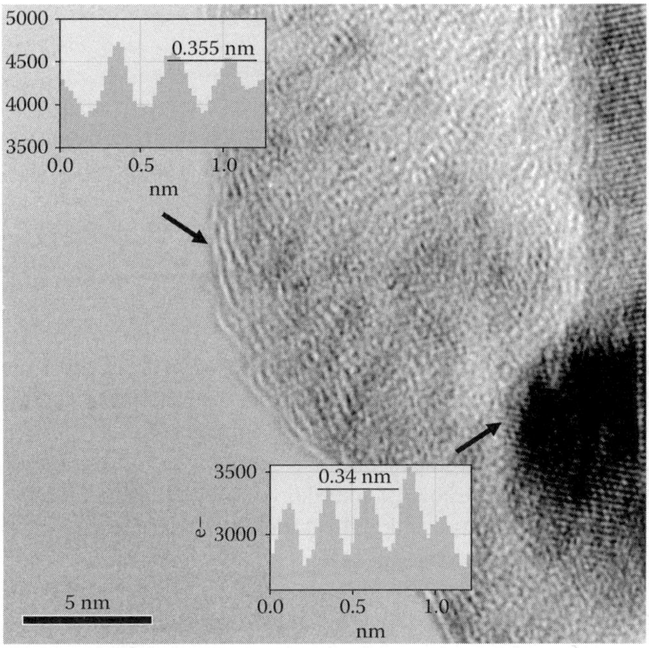

FIGURE 3.14 TEM image of asphaltenes from hydrotreated Maya crude at 420°C, 100 kg/cm^2, and LHSV = 1 h^{-1} washed with toluene (67 vol%) and *n*-heptane (33 vol%). (Trejo et al. 2009, *Energy Fuels* 23(1): 429–439. With permission.)

TABLE 3.5
Elemental Composition of Asphaltenes and Fractions by Transmission Electron Microscopy (TEM)

	Elemental Composition of Each Fraction, wt%		
Element	**Maya Crude Asphaltenes**	**HDT at 400°C**	**HDT at 420°C**
C	87.25	94.29	91.86
Al	0.00	0.00	0.20
N	7.42	0.00	0.00
Si	0.17	0.60	0.14
P	0.00	0.00	0.00
S	4.71	4.82	7.25
V	0.10	0.10	0.03
Ni	0.02	0.01	0.00
Fe	0.33	0.17	0.52

Source: Trejo et al. 2009. *Energy Fuels* 23(1): 429–439. With permission.

3.3 INFLUENCE OF REACTION CONDITIONS ON ASPHALTENE STRUCTURE

In view of the several changes that asphaltenes undergo during hydrotreating, it is useful to analyze separately the way in which each reaction variably influences the asphaltene structure. The purpose of this section is then to understand how temperature, space velocity, pressure, and catalysts guide the changes that asphaltenes experience during hydrotreating.

3.3.1 Influence of Temperature

Changes in asphaltene structure have been proposed by Seki and Kumata (2000) in respect to the increase of reaction temperature. Hydrodemetallization of Kuwait atmospheric residue was studied by varying temperature from 370 to 430°C at pressure of 14 MPa, space velocity of 0.5 h^{-1}, and H_2-to-oil ratio of 2,000 ft^3/bbl. It was observed that the number of internal carbons in asphaltenes increased up to 410°C and finally decreased suddenly at higher temperatures. External carbons showed the opposite trend. Skeleton suffers a rearrangement to a pericondensed-type up to 410°C and changed to a catacondensed-type at higher temperatures of the structure. The authors supported their observations in connection with the decrease of R_a and proposed that the shape of aromatic core of asphaltenes depends on reaction temperature without changing the average aromatic rings number; the change to catacondensed-type is suggested to be a potential precursor of sludge formation at higher temperatures.

Bartholdy and Andersen (2000) have stated that at low severity hydroprocessing, stability of products increases, but when reaction severity is increased, asphaltene stability is reduced. Asphaltenic fraction contains most of the heaviest components and, for this reason, sulfur content in asphaltenes represents less than 20% of the total sulfur content in the crude, but when hydrotreating crude, the sulfur content in asphaltenes is increased up to nearly 60% in respect of total sulfur content. This indicates that sulfur concentrates in asphaltenes, whereas sulfur in the rest of crude is converted and released as H_2S. It was also concluded that around 370°C, hydrotreating reactions change from being hydrogenation-dominated to hydrocracking-dominated, and at 380°C, asphaltenes suffer significant changes. Dealkylation makes more aromatic asphaltenes and more aliphatic maltenes. Asphaltenes are prone to precipitate in a more aliphatic medium by which hydrogenation at high extent will turn the medium into unfavorable for asphaltenes as pointed out by Mochida et al. (1990).

Buch et al. (2003) also concluded that higher temperatures during hydrotreating result in a reduction of asphaltenes size. Based on fluorescence depolarization, it was suggested that hydrogenation turns asphaltenes into smaller molecules with smaller ring systems. One important reason for instability of products, during hydrotreating of crude oils is not only condensation of molecules into large fused rings, but also the decrease in solubility caused by removal and reduction of alkyl chains. During hydrotreating, about 50% of original asphaltenes are converted to resins.

In general, temperature in hydrotreating reflects that from 380°C there is a change in the chemistry of reactions. Dealkylation is responsible for obtaining more condensed asphaltenic structures (Bartholdy and Andersen, 2000). Differences in temperature at which more prominent changes of asphaltenes take place observed by different authors may be due to the nature of the feedstock and reaction conditions (catalyst, pressure, temperature, space-velocity, etc.).

Ancheyta et al. (2003) have studied the changes that asphaltenes from Maya crude suffer during hydrotreating with changing temperature from 380 to 440°C and keeping constant the pressure at 70 kg/cm^2, H_2-to-oil ratio of 5,000 ft^3/bbl, and space velocity of 0.5 h^{-1}. It was observed that increasing asphaltenes conversion, heteroatoms conversion in crude also increased. HDS, HDNi, and HDV follow a linear trend; however, HDN increased at first and then it remained almost constant. Sulfur and metals, especially V, are converted almost at the same rate as asphaltenes. In the case of nitrogen, this compound is difficult to be removed because it forms part of very stable structures contained in aromatic rings. For this reason, their conversion is kept almost constant at higher asphaltene removal. Figure 3.15 shows the tendency of heteroatoms removal as a function of hydrodeasphaltenization (HDAs).

Ni is less converted than V. This is due to the nature of vanadyl porphyrins because V has a perpendicular oxygen atom linked to vanadium, whereas Ni does not have such a linkage (Ancheyta et al., 2002; Chen and Hsu, 1997; Kobayashi et al., 1987a). There is also a good correlation between metals and asphaltenes removal. Metals are present in asphaltenes as are porphyrins, so that the higher the asphaltene conversion, the higher the metals removed. Seki and Kumata (2000) observed an approximately linear correlation between hydrodemetallization and removal of asphaltenes. Asaoka et al. (1983) also observed a similar trend. Bartholdy and Andersen (2000) stated that asphaltenic metals are the least

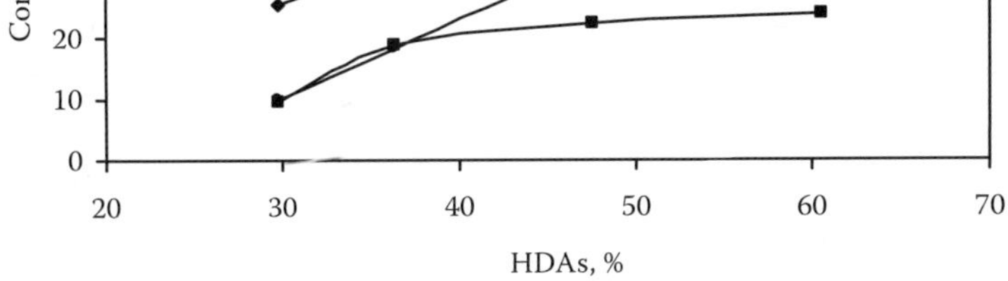

FIGURE 3.15 Conversion of contaminants as function of HDAs: (■) HDN; (●) HDNi; (♦) HDS; (▲) HDV. (From Ancheyta et al. 2003. *Energy Fuels* 17 (5): 1233–1238. With permission.)

reactive of all metals present in feedstock. They also observed that, as asphaltenes are removed, more metals are removed from the crude.

At higher temperatures asphaltenes hydrocracking is more pronounced and heteroatoms located at the outer edge of the asphaltene molecule are easily removed. Results obtained by Ancheyta et al. (2003) on asphaltenes characterization when varying temperature are summarized as:

- Sulfur content diminished as temperature increased due to breakage of alkyl chains where asphaltenic sulfur is located as sulfur bridges. Nitrogen and metals increased their content as temperature increased since these structures are present in the inner section of molecules and they do not undergo appreciable changes compared with the external part of molecules. Moreover, nitrogen is relatively located at stable aromatic rings by which its removal is difficult. Mitra-Kirtley et al. (1993a) reported that asphaltenic nitrogen is aromatic in nature and it is hardly removed when processing asphaltenes.
- There is a reduction of H/C atomic ratio of asphaltenes as temperature increased indicating that more aromatic structures are obtained (Figure 3.16) due to shortening of alkyl chains. Other atomic ratios are also seen in this figure. The results obtained by elemental analysis are supported by aromaticity factor (f_a) and average number of carbons in alkyl chains (n) obtained by ^{13}C NMR. In this case, f_a increased as temperature was reduced indicating that more aromatic structures are present after Maya crude hydrotreating. Shortening of alkyl chains along with possible dehydrogenation of napthenic rings are responsible

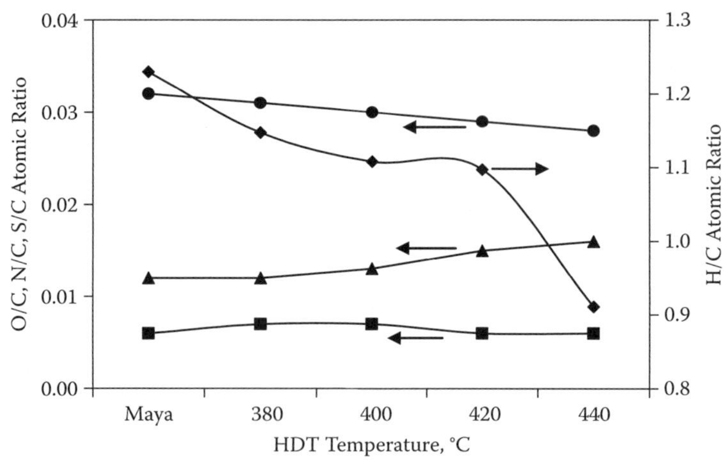

FIGURE 3.16 Changes in H/C and other atomic ratios in asphaltenes as a consequence of increasing the reaction temperature: (♦) H/C; (■) O/C; (▲) N/C; (●) S/C. (From Ancheyta et al. 2003. *Energy Fuels* 17 (5): 1233–1238. With permission.)

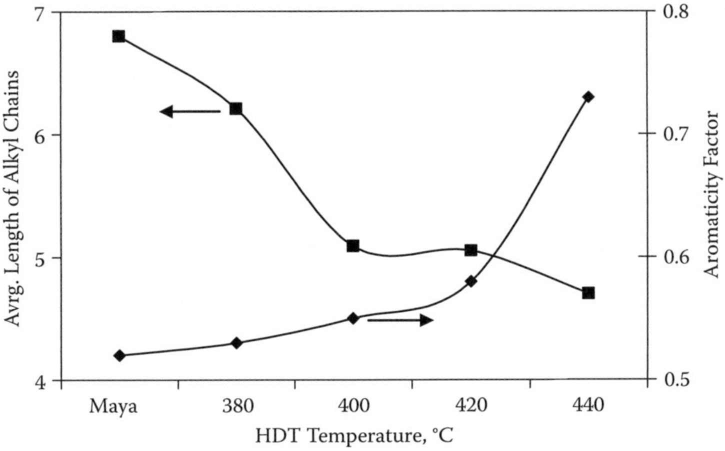

FIGURE 3.17 Change in aromaticity factor (♦) and average length of alkyl chains (■) as function of temperature. (From Ancheyta et al. 2003. *Energy Fuels* 17 (5): 1233–1238. With permission.)

for increasing f_a. A diminution of the average length of alkyl chains is demonstrated by the reduction of n going from 6.8 carbons in average in asphaltenes from nonhydrotreated Maya crude to 4.7 carbons after hydrotreating at 440°C (Figure 3.17).

- The number of aromatic rings (R_a) decreased in hydrotreated asphaltenes compared with Maya crude asphaltenes. According to this, hydrogenation of aromatic rings could be developed to some extent converting polycondensed aromatics to other fractions. Seki and Kumata (2000) found that the number of aromatic rings in asphaltenes was almost constant up to 410°C and after that a slight reduction of R_a at 430°C was observed, suggesting the conversion of aromatic rings to other species and those aromatic rings that remained unchanged are present in asphaltenes. The authors also found that f_a was almost unchanged up to 400°C and then steeply increased at higher temperatures.

Trejo et al. (2005) performed another series of experiments by changing the reaction conditions. Figure 3.18 shows the changes that asphaltenes suffered as reaction temperature was increased keeping constant space velocity and pressure. It was concluded that the higher the reaction temperature, the lower the length of alkyl chains. The smaller aliphatic carbons content makes the H/C atomic ratio increase as reaction temperature increases. Due to the increase of aromatic carbons content and alkyl chains, stripping of aromaticity factor also is higher as temperature was increased. Aggregate molecular weight (AMW) obtained by VPO indicated the rupture of asphaltene molecules as temperature increased. It

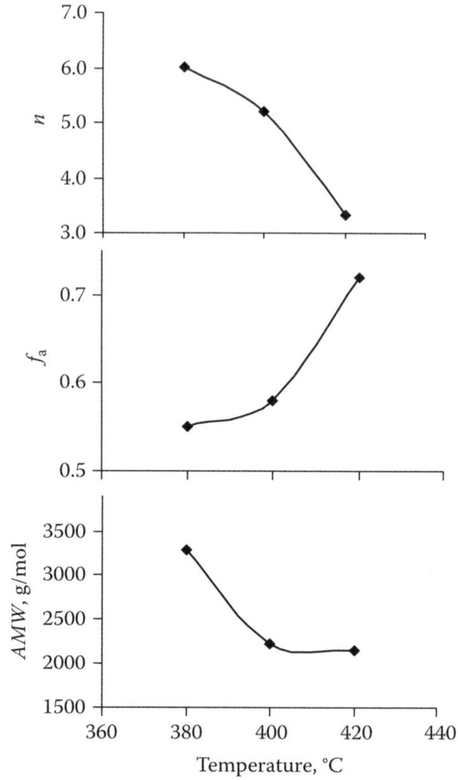

FIGURE 3.18 Effect of temperature on structural parameters of hydrotreated asphaltenes at P = 100 kg/cm^2 and LHSV = 1.0 h^{-1}. (From Trejo et al. 2005. *Catalysis Today* 109 (1–4): 178–184. With permission.)

can be also observed from Figure 3.18 that bigger changes in asphaltene structure occurred when reaction temperature is higher than 400°C.

Mosio-Mosiewski and Morawski (2005) also studied the effect of temperature on asphaltene conversion in vacuum residue of Ural crude keeping constant the catalyst concentration in the residue (1 wt%), pressure (16 MPa), and space velocity (0.5 h^{-1}). They observed that a nearly rectilinear increase in asphaltene conversion was carried out as temperature increased from 410 to 450°C.

3.3.2 Influence of Space Velocity

At lower space velocity, the contact time is higher and the reactants take more time in passing through the catalytic bed and, hence, higher conversion is also enhanced. Figure 3.19 shows the effect of space velocity on asphaltene structural parameters (Trejo et al., 2005). The average length of alkyl chains and aromaticity factor are reduced as space velocity is reduced. The most prominent changes

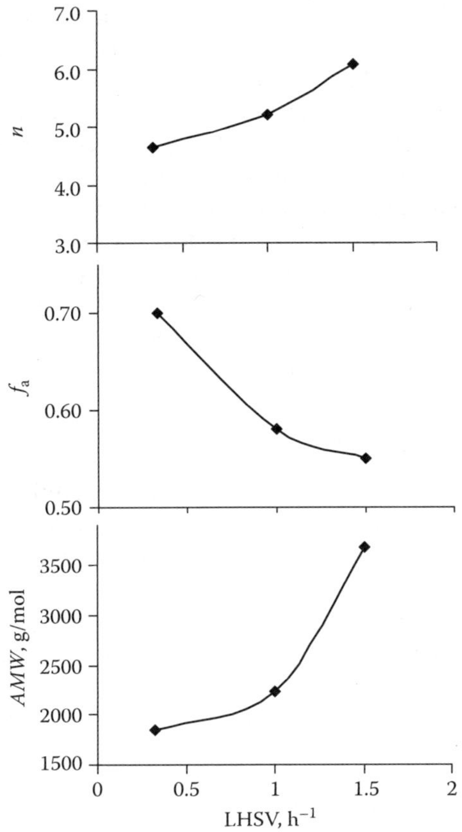

FIGURE 3.19 Effect of space velocity on structural parameters of hydrotreated asphaltenes at P = 100 kg/cm^2 and T = 400°C. (From Trejo et al. 2005. *Catalysis Today* 109 (1–4): 178–184. With permission.)

are observed when space velocity is 0.33 h^{-1}. Aggregate molecular weight is also diminished when space velocity is less than 1 h^{-1}.

The combination of temperature, pressure, and space velocity greatly influences asphaltenes structure. It was stated that hydrocracking of asphaltenes is more important at temperatures above 400°C, pressure greater than 85 kg/cm^2, and space velocity less than 1 h^{-1}. Hydrotreating temperature presents the major influence on shortening of alkyl chains. Average number of carbons per alkyl chain changed from 5.2 to 3.3 when temperature increased from 400 to 420°C. Changes over this same structural parameter are not as significant when pressure increases. As temperature is raised from 400 to 420°C, aromaticity factor also increases from 0.58 to 0.70 indicating that around 70% of total carbon is aromatic when space velocity decreases from 1 to 0.33 h^{-1}. When space velocity is reduced from 1.5 to 1.0 h^{-1} aggregate molecular weight diminishes from 3,666 to 2,230 g/mol. On the other hand, when pressure is increased from 380 to 400°C, aggregate molecular weight is reduced from 3,284 to 2,230 g/mol. It was also concluded that the effect of pressure is less significant than that of temperature or space velocity on changes suffered on the shortening of alkyl chains, aromaticity factor, and aggregate molecular weight.

The effect of space velocity on hydrotreating of vacuum residue was also studied by Morawski and Mosio-Mosiewski (2006), ranging from 0.75 to 0.25 h^{-1} keeping constant temperature, pressure, and catalyst concentration. Asphaltene conversion increased up to 80%, which evidences the great influence of space velocity to obtain high conversions. Several modifications were observed on asphaltene structure when space velocity was varied in the same proportion as temperature. In fact, similar conversions are obtained when changing temperature and space velocity, as shown in Figure 3.20.

3.3.3 Influence of Pressure

Pressure of hydrogen is a very important operating condition when dealing with hydroprocessing because the presence of hydrogen inhibits coke formation and, combined with a catalyst, hydrogen acts by capping free radicals diminishing retrograde reactions.

Figure 3.21 shows the effect of pressure on asphaltenes properties when keeping constant space velocity and temperature, according to Trejo et al. (2005). It is observed that alkyl chains are reduced as pressure is increased. At high pressure, the aromaticity factor increases and the H/C atomic ratio decreases as a consequence of breaking of aliphatic carbon in alkyl chains. The average length of alkyl chains is almost the same at low and moderate pressure, i.e., 70 and 85 kg/cm^2; however, at 100 kg/cm^2 the change is more relevant. The aromaticity factor is also higher at the highest pressure. Aggregate molecular weight diminished as pressure increased. It is observed that, in spite of having less drastic changes compared with increasing temperature, above 85 kg/cm^2 changes in asphaltene structure are slightly more notorious. The combined effect of pressure and temperature shows that more severe hydrocracking is occurring when pressure is above 85 kg/cm^2 and temperature is higher than 400°C.

When increasing pressure, asphaltene conversion is only slightly increased (Morawski and Mosio-Mosiewski, 2006). They also found a smaller effect of pressure on asphaltene conversion compared with those of temperature and space velocity, as seen also in Figure 3.20.

3.3.4 Influence of Catalyst

It has been indicated that the catalyst pore size and its distribution have a significant effect on activity when hydroprocessing heavy oils because pore diffusion plays an important role (Ohtsuka, 1977; Green and Broderick, 1981). Several reports have shown the importance of catalyst pore sizes in hydrotreating, especially in removing sulfur, metals, and asphaltenes (Shimura et al., 1986; Fischer and Angevine, 1986; Kobayashi et al., 1987b).

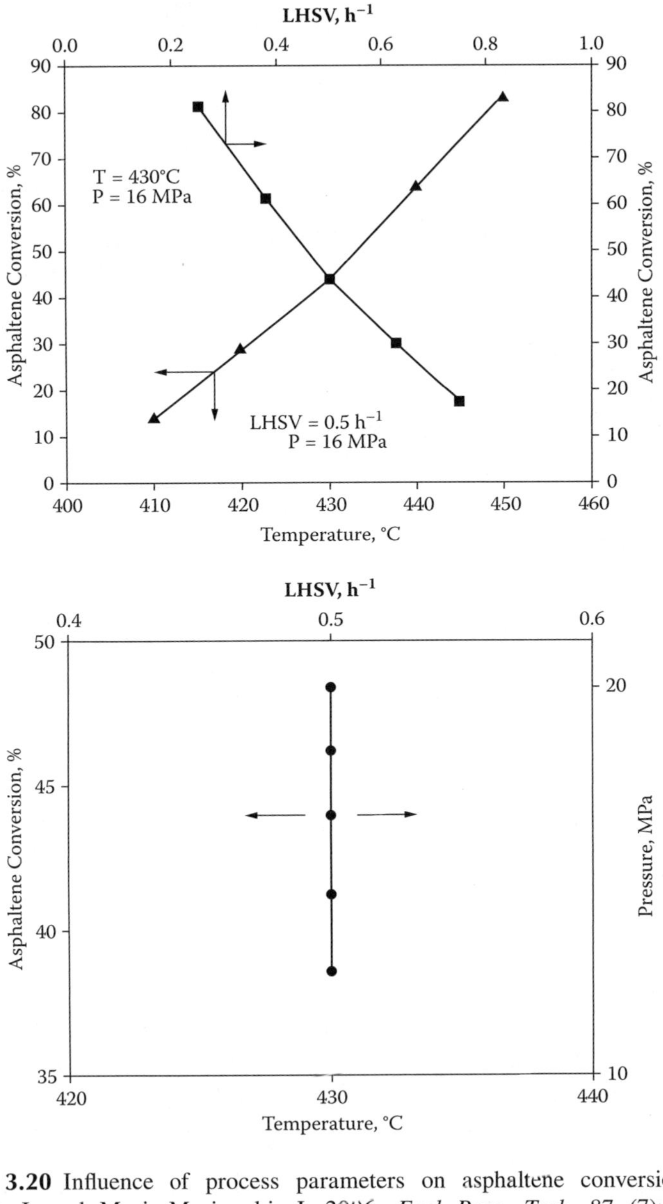

FIGURE 3.20 Influence of process parameters on asphaltene conversion. (From Morawski, I. and Mosio-Mosiewski, J. 2006. *Fuel Proc. Tech.* 87 (7): 659–669. With permission.)

Song et al. (1991) carried out a study to examine the influence of pore structure of Ni-Mo catalysts on their performance in asphaltene conversion studying mainly the effects of catalyst pore structure on asphaltenes conversion. Different NiMo/Al_2O_3 catalysts were prepared having unimodal average pore diameter. It was observed that after hydrocracking reaction the remaining asphaltenes have

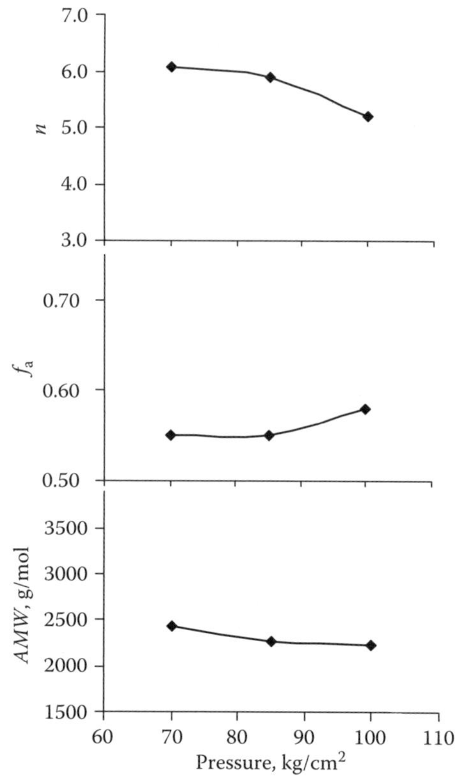

FIGURE 3.21 Effect of pressure on structural parameters of hydrotreated asphaltenes at T = 400°C and LHSV = 1.0 h^{-1}. (From Trejo et al. 2005. *Catalysis Today* 109 (1–4): 178–184. With permission.)

lower H/C ratios, higher aromaticity, and higher N and O contents, indicating that more aromatic and heteroaromatic compounds partially separated of their peripheral substituents remained as unconverted asphaltene molecules. In Figure 3.22, it is clear that apparent catalytic activity depends on pore structure, with the larger pore catalysts as the most efficient for converting asphaltenes. It is observed that optimum pore size is around 30 nm.

Stanislaus et al. (1996) also studied the effect of pore size of catalysts on asphaltenes conversion during hydroprocessing of Kuwait vacuum residue and analyzed the products after reaction at pressure of 12 MPa, space velocity 2 h^{-1}, temperature of 440°C, and H_2/oil ratio of 1,000 L/L. Different catalysts with pore distribution in the order of meso- and macropores were tested and a unimodal catalyst having around 60% of its pores in the mesopores range (10 to 20 nm) yielded the highest activity to asphaltenes conversion followed by another unimodal catalyst with 34% of its pores in the range of mesopores (10 to 20 nm). Residue

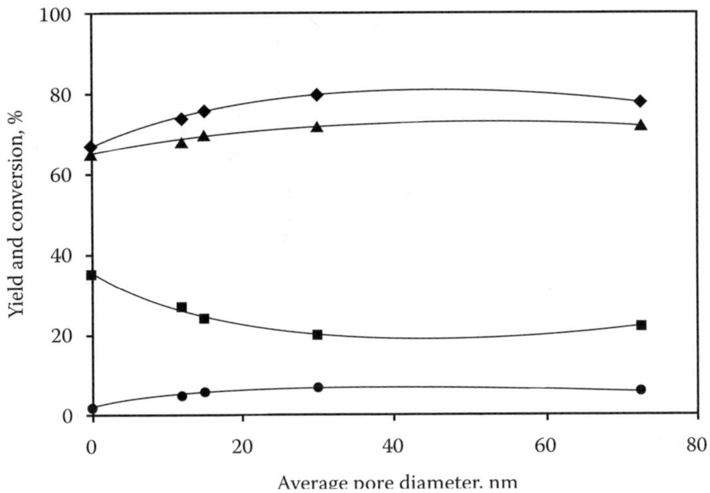

FIGURE 3.22 Effect of pore size of Ni-Mo/Al_2O_3 catalysts on hydrocracking of bitumen asphaltenes: (♦) total conversion; (▲) oil + gas; (■) asphaltenes; (●) coke. (From Song et al. 1991. *Ind. Eng. Chem. Res.* 30 (8): 1726–1734. With permission.)

conversion exhibited the same trend having the highest activity with the mesoporous catalyst (60% of its pores in the range of 10 to 20 nm). The catalyst with this pore size is able to promote asphaltene and residue conversion; however, more sediments (defined as toluene-insoluble matter) were generated with this catalyst (Figure 3.23).

When residual asphaltenes were analyzed after reaction, it was found that the lowest sulfur and vanadium content was obtained with the unimodal macroporous catalyst (Q). It seems that bimodal catalysts that have high proportion of mesopores and also macropores exhibit good activity for sulfur removal in asphaltenes, but removal of vanadium is not as good. The highest amount of sulfur and vanadium in asphaltenes after reaction was found in the highly mesoporous catalyst (60% in the range of 10 to 20 nm) even though this catalyst exhibited good asphaltene and residue conversions. It must be kept in mind that asphaltenes are polydisperse molecules that have different sizes. Bigger molecules of asphaltenes, which probably contain more sulfur and vanadium, cannot access easily to mesopores of catalysts and for this reason more sulfur, vanadium, and higher molecular weight are observed in residual asphaltenes. For large molecules, macropores are readily useful for converting asphaltenes. This is the reason why asphaltenes, after reaction, have less content of sulfur and vanadium when using macroporous catalysts.

Regarding pore size of catalysts, SANS analysis for determining the asphaltene size of asphaltene aggregates emerges as an important and useful technique because it gives an idea of the size and morphology of asphaltenes. By knowing

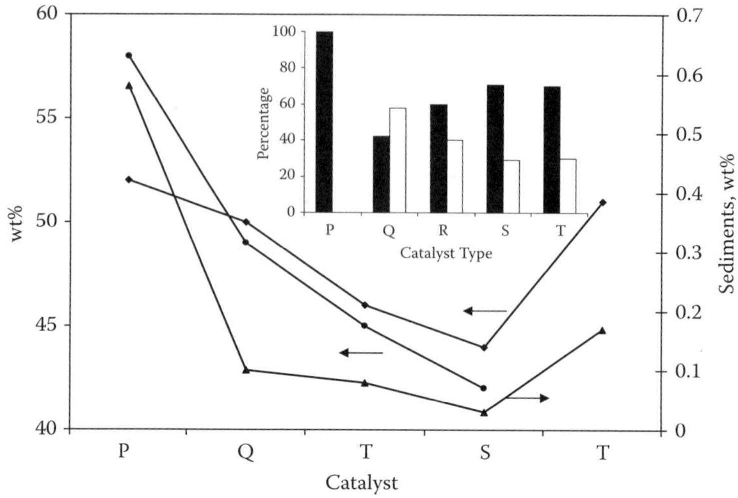

FIGURE 3.23 Influence of the catalyst pore size on: (●) asphaltenes conversion; (♦) distillate yield; (▲) sediments formation. Inset: (■) mesopores; (□) macropores. (Adapted from Stanislaus et al. 1996. *Catalysts in Petroleum Refining and Petrochemical Industries*, Elsevier: Amsterdam.)

the size of the aggregates, it is possible to design a better catalyst capable of processing even the most difficult and complex fractions in petroleum, such as asphaltenes. However, according to studies carried out with asphaltenes in organic solvents, the size and shape of asphaltenes change as temperature is increased. The size is reduced at higher temperatures (Thiyagarajan et al., 1995; Hunt and Winans, 1999) so that in order to design better catalysts these considerations also must be taken into account. An ideal catalyst would have meso- and macroporous in different ratio in which asphaltenes of variable size and morphology, including those which have suffered changes in their size as a result of temperature, are readily processed.

It is interesting to note that some reports have indicated that the radius of asphaltenes ranges from 2 to 15 nm (Baltus and Anderson, 1983; Quann et al., 1988). Hence, a catalyst having mesopores in the range of 10 to 20 nm is able to crack asphaltenes whose molecular radius is comprised within this interval of pores. The remaining asphaltenes could require bigger pores for being converted during reaction (Stanislaus et al., 1996). Morawski and Mosio-Mosiewski (2006) studied the catalyst concentration and its influence on asphaltene conversion and observed that by changing the amount of catalyst (1, 5, and 10 wt%), asphaltene conversion was practically the same indicating most probably that asphaltene conversion proceeded by means of free radicals when the catalyst employed was a $NiMo/Al_2O_3$ and temperature, pressure, and space velocity were kept constant at 430°C, 16 MPa and 0.5 h^{-1}, respectively.

Le Lannic et al. (2007) conducted tests for upgrading residue and analyzed the changes that asphaltenes undergo in a two-stage fixed-bed hydrotreating unit at the following conditions: space velocity of 1.0-0.1 h^{-1}, temperature of 380 to 400°C, pressure of 15 MPa, and Iraqi vacuum residue as feedstock. The first stage corresponds to a hydrodemetallization (HDM) process in which asphaltenes are exposed to severe reaction conditions. It can be observed that big aggregates of asphaltenes are converted into smaller ones with notorious molecular weight reduction as a consequence of thermal cracking. At this stage, the number of aromatic carbons increases at the expense of reduction of the number of alkyl carbons provoked by dealkylation of aromatic cores. More aromatic asphaltenes are obtained, which means that higher aromaticity factors (f_a) and lower H/C atomic ratios are observed in the remaining asphaltenes during HDM where higher temperature and lower space velocity are combined to achieve the highest conversion of asphaltenes as possible in the HDM stage. Guibard et al. (2005) have stated that before HDM sulfur is comprised in resins and asphaltenes, but after reaction, sulfur is concentrated in resins and aromatics. Important changes in asphaltenes have been observed at 400°C during hydrotreating of different feedstocks (Ancheyta et al., 2003; Hauser et al., 2005). At this temperature, asphaltenes of high molecular weight are converted into asphaltenes of lower molecular weight, which are less polydisperse. If the temperature is increased and space velocity is reduced, lighter asphaltenes but more aromatic ones are obtained, which can be prone to form coke with the consequent loss of catalyst activity (Hauser et al., 2005). In the second stage, hydrodesulfurization (HDS) is carried out, where smaller pores in the catalyst are required. Asphaltenes of lower molecular weight are readily reacted in the presence of an HDS catalyst. In this way, for obtaining asphaltenes of smaller size, it is necessary to increase the severity of reaction conditions in the HDM stage to assure that small molecules are obtained after this reaction and they readily go inside the pores in HDS catalysts. So, when residue is demetallized under severe reaction conditions, dealkylation of aromatic cores has already occurred by which no more alkyl chains are detached from the nucleus and remaining asphaltenes will not increase their aromaticity at this stage. However, in the HDS stage, asphaltenes will suffer additional conversion and concentration of remaining asphaltenes will diminish consequently.

HDM and HDS catalysts also influence the final shape of asphaltenes. For example, Seki and Yoshimoto (2001) studied the changes of asphaltenes under HDM followed by HDS when hydroprocessing Arabian light atmospheric residua. It was found that HDM reaction caused a decrease in the number of alkyl side chains without formation of naphthenic rings, which were mainly formed during HDS reaction, but their number decreased as HDS temperature increased, thus leading to an increase of aromaticity. Structural parameters allowed the authors to establish that asphaltenes at the end of a reaction have a coronene-type structure as a consequence of hydrotreating. On the other hand, when unpromoted catalysts are used, thermal cracking and condensation reactions may occur due to the lower hydrogenation function of catalysts (Marafi et al., 2004). Remaining asphaltenes are, as a result, highly condensed.

3.4 CONCLUDING REMARKS

Several experiments have shown that asphaltenes undergo a number of changes during hydroprocessing. These changes at molecular level are a consequence of the severity of the reaction. At the most severe reaction conditions, i.e., high temperature and pressure, and low space velocity, asphaltenes can undergo severe hydrocracking, which increases their conversion. Aromaticity increases because alkyl chains are removed. When increasing temperature, the most labile points to be broken are alkyl chains and rupture on C-C bonds forms free radicals. The broken aliphatic chains can experience cyclic reactions and dehydrogenation, which in the presence of condensation reactions, increases the aromaticity of residual asphaltenes after hydrotreating. Aromaticity is followed by a reduction of the H/C atomic ratio that is the result of removing aliphatic carbon. The reduction in length of alkyl chains leaves the aromatic core able to stack. This fact is observed when asphaltenes are precipitated and some sections of them present in crystalline arrangement as detected by XRD. Molecular weight is also reduced with removal of aliphatic chains and the molecular weight distribution also is expected to be diminished, becoming narrower. Asphaltenes conversion is directly dependent on crude oil residue conversion and at the highest conversion smaller asphaltene molecules are formed from bigger asphaltenes. Detailed characterization of asphaltenes could be especially important when designing catalysts, which must give evidence to all changes that asphaltenes suffer during hydrotreating. By this means, conversion of these high molecular weight molecules can be maximized in a suitable and profitable manner.

REFERENCES

Ali, F.A., Ghaloum, N., and Hauser, A. 2006. Structure representation of asphaltene GPC fractions derived from Kuwaiti residual oils. *Energy Fuel*s 20 (1): 231–238.

Amerik, Y.B. 1995. Manifestation of the macromolecular nature of asphaltenes in the reactions of destruction, polymerization, and polycondensation. *Neftekhimiya* 35 (3): 228–247.

Ancheyta, J., Betancourt, G., Marroquín, G., Centeno, G., Castañeda, L.C., Alonso, F., Muñoz, J.A., Gómez, M.T., and Rayo, P. 2002. Hydroprocessing of Maya heavy crude oil in two reaction stages. *Appl. Catal. A* 233 (1): 159–170.

Ancheyta, J., Centeno, G., Trejo, F., and Marroquín, G. 2003. Changes in asphaltene properties during hydrotreating of heavy crudes. *Energy Fuels* 17 (5): 1233–1238.

Andersen, S.I., Jensen, J.O., and Speight, J.G. 2005. X-ray diffraction of subfractions of petroleum asphaltenes. *Energy Fuels* 19 (6): 2371–2377.

Asaoka, S., Nakata, S., Shiroto, Y., and Takeuchi, C. 1983. Asphaltene cracking in catalytic hydrotreating of heavy oils. 2. Study of changes in asphaltene structure during catalytic hydroprocessing. *Ind. Eng. Chem. Proc. Des. Dev.* 22 (2): 242–248.

Asaoka, S., Nakata, S., Shiroto, Y., and Takeuchi, C. 1986. The characteristics of metal complexes before and after hydrotreating. *Am. Chem. Soc., Div. Petrol. Chem. – Prepr.* 31 (2): 597–605.

Ascanius, B.E., Merino-Garcia, D., and Andersen, S.I. 2004. Analysis of asphaltenes subfractionated by *N*-methyl-2-pyrrolidone. *Energy Fuels* 18 (6): 1827–1831.

Baltus, R.E. and Anderson, J.L. 1983. Hindered diffusion of asphaltenes through microporous membranes. *Chem. Eng. Sci.* 38 (12): 1959–1969.

Bartholdy, J. and Andersen, S.I. 2000. Changes in asphaltene stability during hydrotreating. *Energy Fuels* 14 (1): 52–55.

Bartholdy, J., Lauridsen, R., Mejlholm, M., and Andersen, S.I. 2001. Effect of hydrotreatment on product sludge stabiligy. *Energy Fuels* 15 (5): 1059–1062.

Buch, L., Groenzin, H., Buenrostro-Gonzalez, E., Andersen, S.I., Lira-Galeana, C., and Mullins, O.C. 2003. Molecular size of asphaltene fractions obtained from residuum hydrotreatment. *Fuel* 82 (9): 1075–1084.

Calemma, V., Iwanski, P., Nali, M., Scotti, R., and Montanari, L. 1995. Structural characterization of asphaltenes of different origins. *Energy Fuels* 9 (2): 225–230.

Chen, Y.-W. and Hsu, W.-C. 1997. Hydrodemetalation of residue oil over CoMo/alumina-aluminum phosphate catalysts in a trickle bed reactor. *Ind. Eng. Chem. Res.* 36 (7): 2526–2532.

Fischer, R.H. and Angevine, P.J. 1986. Dependence of resid processing selectivity on catalyst pore size distribution. *Appl. Catal. A* 27 (2): 275–283.

Furimsky, E. 1978. Chemical origin of coke deposited on catalyst surface. *Ind. Eng. Chem. Prod. Res. Dev.* 14 (4): 329–331.

Gauthier, T., Danial-Fortain, P., Merdrignac, I., Guibard, I., and Quoineaud, A.A. 2007. An attempt to characterize the evolution of asphaltene structure during hydroconversion conditions, in *Proceedings of International Symposium on Advances in Hydroprocessing of Oil Fractions*, Morelia, Mexico. June 26 to 29.

Gauthier, T., Danial-Fortain, P., Merdrignac, I., Guibard, I., and Quoineaud, A.-A. 2008. Studies on the evolution of asphaltene structure during hydroconversion of petroleum residues. *Catalyst Today* 130 (2–4): 429–438.

Gawel, I., Bociarska, D., and Biskupski, P. 2005. Effect of asphaltenes on hydroprocessing of heavy oils and residua. *Appl. Catal. A* 295 (1): 89–94.

Gray, M.R., Khorasheh, F., Wanke, S.E., Achia, U., Krzywicki, A., Sanford, E.C., Sy, O.K.Y., and Ternan, M. 1992. Role of catalyst in hydrocracking of residues from Alberta bitumens. *Energy Fuels* 6 (4): 478–485.

Green, D.C. and Broderick, D.H. 1981. Processing heavy crudes: Residuum processing in the 80's. *Chem. Eng. Prog.* 77 (12): 33–39.

Groenzin, H. and Mullins, O.C. 2000. Molecular size and structure of asphaltenes from various sources. *Energy Fuels* 14 (13): 677–684.

Guibard, I., Merdrignac, I., and Kressmann, S. 2005. Characterization of refractory sulfur compounds in petroleum residue, in *Heavy Hydrocarbon Resources: Characterization, Upgrading and Utilization*, Nomura, M., Rahimi, P.M., and Koseoglu, O.R., Eds., ACS Symposium Series 895, Washington, D.C., pp. 51–64.

Hauser, A., Marafi, A., Stanislaus, A., and Al-Adwani, A. 2005. Relation between feed quality and coke formation in a three-stage atmospheric residue desulfurization (ARDS) process. *Energy Fuels* 19 (2): 544–553.

Heck, R.H. and DiGuiseppi, F.T. 1994. Kinetic effects in resid hydrocracking. *Energy Fuels* 8 (3): 557–560.

Herod, A.A., Zhang, S.F., Johnson, B.R., Bartle, K.D., and Kandiyoti, R. 1996. Solubility limitations in the determination of molecular mass distributions of coal liquefaction and hydrocracking products: 1-methyl-2-pyrrolidinone as mobile phase in size exclusion chromatography. *Energy Fuels* 10 (3): 743–750.

Hunt, J.E. and Winans, R.E. 1999. An overview of resid characterization by mass spectrometry and small angle scattering techniques. Paper presented at 218th American Chemical Society National Meeting, New Orleans, LA, August 22 to 26.

Izquierdo, A., Carbognani, L., Leon, V., and Parisi, A. 1989. Characterization of Venezuelan heavy oil vacuum residua. *Fuel Sci. Tech. Int.* 7 (5–6): 561–570.

Karaca, F., Islas, C.A., Millan, M., Behrouzi, M., Morgan, T.J., Herod, A.A., and Kandiyoti, R. 2004. The calibration of size exclusion chromatography columns: Molecular mass distributions of heavy hydrocarbon liquids. *Energy Fuels* 18 (3): 778–788.

Kobayashi, S., Kushiyama, S., Aizawa, R., Koinuma, Y., Inoue, K., Shimizu, Y., and Egi, K. 1987a. Kinetic study on the hydrotreating of heavy oil. 1. Effect of catalyst pellet size in relation to pore size. *Ind. Eng. Chem. Res.* 26 (11): 2241–2245.

Kobayashi, S., Kushiyama, S., Aizawa, R., Koinuma, Y., Inoue, K., Shimizu, Y., and Egi, K. 1987b. Kinetic study on the hydrotreating of heavy oil. 2. Effect of catalyst pore size. *Ind. Eng. Chem. Res.* 26 (11): 2245–2250.

Korolev, Y.M. and Amerik, Y.B. 1993. X-ray diffraction study of crude oils and their components. *Petrol. Chem.* 33 (4): 338–344.

Le Lannic, K., Guibard, I., and Merdrignac, I. 2007. Behavior and role of asphaltenes in a two-stage fixed bed hydrotreating process. *Petrol. Sci. Tech.* 25 (1): 169–186.

Li, W., Morgan, T.J., Herod, A.A., and Kandiyoti, R. 2004. Thin-layer chromatography of pitch and a petroleum vacuum residue relation between mobility and molecular size shown by size-exclusion chromatography. *J. Chromatog. A* 1024 (1–2): 227–243.

Marafi, A., Hauser, A., and Stanislaus, A. 2004. Effect of catalyst type and operating severity on the products quality in hydrotreating Kuwait atmospheric residue. Paper presented at the Proceedings of 2004 AIChE Spring National Meeting, New Orleans, April 25–29, pp. 2220–2227.

Merdrignac, I., Quoineaud, A.-A., and Gauthier, T. 2006. Evolution of asphaltene structure during hydroconversion conditions. *Energy Fuels* 20 (5): 2028–2036.

Merdrignac, I., Truchy, C., Robert, E., and Kressmann, S. 2005. Size exclusion chromatography: Characterization of heavy petroleum residues. Application to resid desulfurization process. *Petrol. Sci. Tech.* 22 (7–8): 1003–1022.

Michael, G., Al-Siri, M., Khan, Z.H., and Ali, F.A. 2005. Differences in average chemical structures of asphaltene fractions separated from feed and product oils of a mild thermal processing reaction. *Energy Fuels* 19 (4): 1598–1605.

Millan, M., Behrouzi, M., Karaca, F., Morgan, T.J., Herod, A.A., and Kandiyoti, R. 2005. Characterising high mass materials in heavy oil fractions by size exclusion chromatography and MALDI-mass spectrometry. *Catalysis Today* 109 (1–4): 154–161.

Mitra-Kirtley, S., Mullins, O.C., Branthaver, J.F., and Cramer, S.P. 1993a. Nitrogen chemistry of kerogens and bitumens from x-ray absorption near-edge structure spectroscopy. *Energy Fuels* 7 (6): 1128–1134.

Mitra-Kirtley, S., Mullins, O.C., Elp, J.V., George, S.J., Chen, J., and Cramer, S.P. 1993b. Determination of the nitrogen chemical structures in petroleum asphaltenes using XANES spectroscopy. *J. Am. Chem. Soc.* 115 (1): 252–258.

Mochida, I., Xing, Z.Z, and Sakanishi, K. 1988. Structural analyses of the products obtained in the two-stage hydrotreatment of the asphaltene fraction of a brown coal liquid vacuum residue. *Nenryo Kyokaishi* 67 (5): 323–329.

Mochida, I., Zhao, X.Z., and Sakanishi, K. 1990. Catalytic two-stage hydrocracking of Arabian vacuum residue at a high conversion level without sludge formation. *Ind. Eng. Chem. Res.* 29 (3): 334–337.

Morawski, I. and Mosio-Mosiewski, J. 2006. Effects of parameters in Ni–Mo catalysed hydrocracking of vacuum residue on composition and quality of obtained products. *Fuel Proc. Tech.* 87 (7): 659–669.

Morgan, T.J., Millan, M., Behrouzi, M., Herod, A.A., and Kandiyoti, R. 2005. On the limitations of UV-fluorescence spectroscopy in the detection of high-mass hydrocarbon molecules. *Energy Fuels* 19 (1): 164–169.

Mosio-Mosiewski, J. and Morawski, I. 2005. Study on single-stage hydrocracking of vacuum residue in the suspension of Ni-Mo catalyst. *Appl. Catal. A* 283 (1–2): 147–155.

Ohtsuka, T. 1977. Catalyst for hydrodesulfurization of petroleum residua. *Catal. Rev.–Sci. Eng.* 16 (2): 291–325.

Pérez-Hernández, R., Mendoza-Anaya, D., Mondragón-Galicia, G., Espinosa, M.E., Rodríguez-Lugo, V., Lozada, M., and Arenas-Alatorre, J. 2003. Microstructural study of asphaltene precipitated with methylene chloride and *n*-hexane. *Fuel* 82 (8): 977–982.

Quann, R.J., Ware, R.A., Hung, C-W., and Wei, J. 1988. Catalytic hydrodemetallation of petroleum. *Adv. Chem. Eng.* 14 (1): 95–259.

Rogel, E. 2000. Simulation of interactions in asphaltenes aggregates. *Energy Fuels* 14 (3): 566–574.

Sakanishi, K., Manabe, T., Watanabe, I., and Mochida, I. 2000. Changes in molecular aggregate of vacuum residue fractions analyzed by slow step scan XRD. *J. Jpn. Petrol. Inst.* 43 (1): 15–16.

Sánchez-Berna, A.C., Camacho-Morán, V., Romero-Guzmán, E.T., and José-Yacamán, M. 2006. Asphaltene aggregation from vacuum residue and its content of inorganic particles. *Petrol. Sci. Tech.* 24 (9): 1055–1066.

Seki, H. and Kumata, F. 2000. Structural change of petroleum asphaltenes and resins by hydrodemetallization. *Energy Fuels* 14 (5): 980–985.

Seki, H. and Yoshimoto, M. 2001. Deactivation of hydrodesulfurization catalysts in two-stage resid desulfurization process (Part 5) characterization of asphaltenes after hydrodemetallization and subsequent hydrodesulfurization reactions. *J. Jpn. Petrol. Inst.* 44 (3): 154–162.

Sharma, A., Groenzin, H., Tomita, A., and Mullins, O.C. 2002. Probing order in asphaltenes and aromatic ring systems by HRTEM. *Energy Fuels* 16 (2): 490–496.

Shimura, M., Shiroto, Y., and Takeuchi, C. 1986. Effect of catalyst pore structure on hydrotreating of heavy oil. *Ind. Eng. Chem. Fund.* 25 (3): 330–337.

Song, C., Nihonmatsu., T., and Nomura, M. 1991. Effect of pore structure of Ni-Mo/$A1_20_3$ catalysts in hydrocracking of coal derived and oil sand derived asphaltenes. *Ind. Eng. Chem. Res.* 30 (8): 1726–1734.

Stanislaus, A., Absi-Halabi, M., and Khan, Z. 1996. Influence of catalysts pore size on asphaltenes conversion and coke-like sediments formation during catalytic hydrocracking of Kuwait vacuum residues, in *Catalysts in Petroleum Refining and Petrochemical Industries 1995,* Absi-Halabi, M., Beshara, J., Kabazard, H., and Stanislaus, A., Eds., Elsevier, Amsterdam, The Netherlands, pp. 189–197.

Strausz, O.P., Peng, P., and Murgich, J. 2002. About the colloidal nature of asphaltenes and the MW of covalent monomeric units. *Energy Fuels* 16 (4): 809–822.

Suelves, I., Islas, C.A., Millan, M., Galmes, C., Carter, J.F., Herod, A.A., and Kandiyoti, R. 2003. Chromatographic separations enabling the structural characterisation of heavy petroleum residues. *Fuel* 82 (1): 1–14.

Tanaka, R., Hunt, J.E., Winans, R.E., Thiyagarajan, P., Sato, S., and Takanohashi, T. 2003. Aggregates structure analysis of petroleum asphaltenes with small-angle neutron scattering. *Energy Fuels* 17 (1): 127–134.

Thiyagarajan, P., Hunt, J.E., Winans, R.E., Anderson, K.B., and Miller, J.T. 1995. Temperature-dependent structural changes of asphaltenes in 1-methylnapthalene. *Energy Fuels* 9 (5): 829–833.

Trejo, F., Ancheyta J., Centeno, G., and Marroquín, G. 2005. Effect of hydrotreating conditions on Maya asphaltenes composition and structural parameters. *Catalysis Today* 109 (1–4): 178–184.

Trejo, F., Ancheyta, J., Morgan, T.J., Herod, A.A., and Kandiyoti, R. 2007. Characterization of asphaltenes from hydrotreated products by SEC, LDMS, MALDI, NMR, and XRD. *Energy Fuels* 21 (4): 2121–2128.

Trejo, F., Rana, M.S., Ancheyta, J. 2009. Structural chracterization of asphaltenes obtained from hydroprocessed crude oils SEM and TEM *Energy Fuels.* 23 (1): 429–439.

Tynan, E.C. and Yen, T.F. 1969. Association of vanadium chelates in petroleum asphaltenes as studied by ESR. *Fuel* 43: 191–194.

Wang, Z., Liang, W., Que, G., and Qian, J. 1997. Study on molecular structure of fractions in Shengli vacuum residue by ruthenium ions catalyzed oxidation. *Acta Petrolei Sinica* (*Petrol. Proc. Sect.*) 13 (4): 8–9.

Whitehead, E.V. 1994. Fuel oil chemistry and asphaltenes, in *Asphaltenes and Asphalts. 1. Developments in Petroleum Science*, Yen, T.F. and Chilingarian, G.V, Eds., Elsevier, Amsterdam, The Netherlands, chap. 4.

Yoshida, R., Yoshida, Y., Bodily, D.M., and Takeya, G. 1982. Mechanism of high-pressure hydrogenolysis of Hokkaido coals (Japan) 3. Chemical structure changes in coal asphaltenes during hydrogenolysis. *Fuel Proc. Tech.* 6 (3): 225–234.

Zhang, H.-C., Yan, Y.-J., Cheng, Z.-Q., Sun, W.-F., and Guan, M.-H. 2007. Changes of asphaltene after hydrotreating by ruthenium ions catalyzed oxidation. *Acta Petrolei Sinica* (*Petrol. Proc. Sect.*) 23 (4): 32–38.

4 Catalyst Deactivation Due to Asphaltenes

4.1 INTRODUCTION

Catalyst deactivation is mainly due to carbon and metal deposition, which enhances with time-on-stream (TOS) leaving unutilized catalytic active sites. During hydrotreating, polynuclear aromatic (PNA) compounds and asphaltenes are the main element responsible for coke formation (Abotsi and Scaroni, 1989). Adsorption of PNAs and asphaltenes is mainly due to the presence of nitrogen compounds (Choi and Gray, 1988; Deng et al., 1997). Various authors have reported the presence of nitrogen (N) in coke (Furimsky, 1978; Satterfield et al., 1978; Speight, 1987; Choi and Gray, 1988). Furimsky concluded that adsorption of nitrogen compounds is stronger compared to other hydrocarbons, and adsorption is carried out by interaction of the unpaired electrons of N with Lewis sites. According to Speight (1987), nitrogen remains in coke as a nonvolatile compound after thermal reaction of asphaltenes, and Reynolds (1998) has pointed out that the majority of sulfur, nitrogen, and metals, which are the most important asphaltene constituents, are deposited on the coke when thermal decomposition of asphaltene is carried out.

The deposition of carbonaceous material can be classified as specific and nonspecific depending on the location of carbon deposits in respect to the catalytic active sites. Van Doorn and Moulijn (1990) reported the formation of filamentary carbon on Co and Ni supported on Al_2O_3 catalysts and stated that coke deposition on spent hydrotreating catalysts is a three-dimensional carbonaceous structure. Traditionally, it is considered that coke is primarily responsible for initial deactivation of the catalyst, i.e., during the first hours-on-stream, while continued deactivation has been attributed to deposition of metals (Tamm et al., 1981; Nielsen et al., 1981). Nielsen et al. have stated that distribution of coke and metals throughout the pore system is the main factor to determine the deactivation rate. According to these authors, there are optimal conditions; i.e., partial pressure of hydrogen, catalyst particle size, and quality of the feedstock, for maximum utilization of catalysts. However, the optimum catalytic performance does not depend only on one type of catalyst. It is desirable to have different types of catalysts during hydrotreating, particularly for heavy oil processing, where a special catalyst designed for metals removal is placed at the top bed or in the first reactors, and good catalysts for hydrodesulfurization (HDS) and/or hydrocracking are situated at the bottom of the bed or in the last reactors (Shimada et al., 2009).

Due to the enormous importance of asphaltene on the behavior of hydroprocessing catalysts, this chapter is devoted to studying the main aspects influencing catalyst deactivation, which are due to the presence of asphaltenes. Deactivation during thermal and catalytic processes is reviewed and different characterization techniques commonly used to analyze spent catalysts are described. A brief discussion concerning catalyst regeneration is also given.

4.2 ORIGIN OF COKE, MECHANISM OF ITS FORMATION, AND PROPERTIES

Coke is a carbonaceous black deposit, which remains intact after washing a catalyst with solvent (usually toluene) in order to remove light paraffins and adsorbed hydrocarbons. Coke structure is not defined and its formation is a consequence of thermal reactions and even catalytic reactions (Ho, 1992; de Jong, 1994) involving free radicals that lead to condensation reactions and dehydrogenation promoted by the catalyst at higher temperatures. It is thought that alkenes and aromatics are prone to generate coke, but, more specifically, asphaltenes are the main coke precursors due to their complex structure and physical properties, e.g., polar in nature. When a catalyst is present, polar aromatic compounds are readily adsorbed onto the catalyst surface, and, as a result, a significant amount of coke is deposited. Gray et al. (1999) reported that the yield of coke deposited onto the catalyst is proportional to asphaltenes content in the feedstock.

In this regard, Savage et al. (1988) have concluded that coke formation is due to the reduction of molecular weight of heavy fractions contained in petroleum. Takatsuka et al. (1989) have suggested that coke is formed by a polymerization process, while Wiehe (1992) assumed that both mechanisms can be present and the increase of the molecular weight involves the combination of two to five molecules of olefins, which form coke. Izquierdo et al. (1989) suggested that characterization of asphaltenes by nuclear magnetic resonance (NMR) could be feasible to establish the processability of any feedstock by measuring the ratio of percentage of bridging or internal aromatic carbons-to-percentage of nonbridging carbons and the value can correspond to the information about asphaltenes structure. High values of this ratio mean that the average molecule has more compact aromatic core that is difficult to process and prone to generate coke.

Coke formation is mainly due to big and complex molecules, i.e., asphaltenes, which contain polar species. Not only asphaltenes are coke precursors, but also every hydrocarbon species could be responsible for a part of the coke produced. In this sense, the origin of the coke from different sources has been studied by Furimsky (1978), who found that coke formed from heavy gas oil has more oxygen and nitrogen than that formed from bitumen and it is presumed that this trend is due to heterocyclic compounds present in heavy gas oil by which compounds that contain N and O in their structure can act as coke precursors. In the case of gas oil, Mochida et al. (1977) demonstrated that compounds like dibenzofuran, carbazole, and dibenzothiophene could be converted to coke in the presence of a Lewis acid. For dibenzofuran and carbazole, content of N and O in coke was higher than in the case of sulfur from dibenzothiophene. The explanation to this observation is related to the nature and strength of bonds; i.e., the C–S (carbon-sulfur) bond is weaker and easier to break undergo when it is present in heterorings compared with C–N (carbon-nitrogen) and C–O (carbon-oxygen). Heterorings are strongly adsorbed on the catalyst surface due to their polarity, and undergo hydrocarbon decomposition, which leads to coke formation.

The separation of vacuum residue into fractions according to their polarity is useful to track the propensity to coke formation of each fraction present in heavy crude or residue. Examination by hot-stage microscopy of Hamaca vacuum residue fractions showed that, after thermal reaction at 440°C and 5.3 MPa under nitrogen atmosphere, the propensity to coke formation in each fraction was in the following order (Rahimi et al., 1998):

Amphoteric > Basic > Acidic > Neutral > Aromatic fraction

The formation of coke is likely due to the presence of amphoteric, basic, and acid fractions that are prone to the generation of coke. Amphoteric compounds also were found to be the heaviest fraction.

Regarding the propensity of coke formation of each fraction, Wiehe (1994) explained the heavy crude upgrading in terms of the pendant–core model in which the core is formed by large aromatics and tends to precipitate when the limit of solubility is reached, that enough changes in the medium have been carried out and asphaltenes are not soluble any more by maltenes. The apparition of another phase in the medium is known as mesophase, in which aromatic molecules array into a well-ordered and stacked structure and form a separate liquid phase, which is finally converted into a solid phase called *coke*. The important reactions that contribute to coke formation are cracking of alkyl chains linked to aromatic cores. If the temperature is high enough, then dehydrogenation of naphthenes is likely to take place to form aromatics, condensation of aliphatic structures, and dehydrogenation to form aromatic structures. Condensation of aromatics to form bigger polycondensed structures and dimerization or oligomerization reactions are mainly determined to be coke precursors. If hydrogen atmosphere where petroleum is being upgraded is not enough, cleavage of alkyl chains with posterior condensation of alkyl fragments and dehydrogenation will occur to form coke (Speight, 1998).

Wiehe (1992) was the pioneer in analyzing the thermal products from a residue and its fractions (volatiles, saturates, aromatics, resins, asphaltenes, coke) obtained by thermal decomposition by means of elemental analysis and molecular weight. Coke formation was reported to be a consequence of asphaltenes decomposition; however, asphaltenes not only form coke, but also resins, aromatics, saturated, and volatiles in the following sequence of products:

$$\text{Saturates} \rightarrow \text{Volatiles}$$

$$\text{Resins} \leftarrow \text{Aromatics} \rightarrow \text{Saturates} + \text{Volatiles}$$

$$\text{Asphaltenes} \leftarrow \text{Resins} \rightarrow \text{Aromatics} + \text{Saturates} + \text{Volatiles}$$

$$\text{Coke} \leftarrow \text{Asphaltenes} \rightarrow \text{Resins} + \text{Aromatics} + \text{Saturates} + \text{Volatiles}$$

In each case, thermal conversion of a fraction formed the following fraction being more aromatic and having higher molecular weight. As can be seen from Table 4.1 and the above reactions, saturates generate volatiles by fragmentation, aromatics form saturates plus volatiles by scission reactions, and resins by condensation; resins give lighter products, such as aromatics, saturates, and volatiles due to dealkylation and fragmentation, whereas asphaltenes generate resins, aromatics, saturates plus volatiles as a result of scission of alkyl chains. However, they

TABLE 4.1
Thermal Conversion of Residue Fractions

Reactants	Product	Yield, wt%	H/C Product	VPO MW Products, g/mol
Saturates + Aromatics	Volatiles	29.5		
	Saturates	32.1	1.74	694
	Aromatics	38.4	1.50	645
	Resins	19.4	1.29	839
Resins	Volatiles	10.8		
	Saturates	5.7	1.70	670
	Aromatics	30.8	1.46	442
	Resins	30.6	1.23	804
	Asphaltenes	22.1	0.95	1841
Asphaltenes	Volatiles	10.4		
	Saturates	2.6	1.79	
	Aromatics	14.2	1.51	422
	Resins	12.4	1.18	622
	Asphaltenes	21.0	0.90	1557
	Coke	39.4	0.80	7525

Source: Adapted from several literature-based data.

also form coke by condensation involving free radicals. Byproducts including asphaltenes and resins were lighter in products compared to reactants. It can also be seen that the hydrogen to carbon (H/C) atomic ratio for saturates and aromatics are almost constant after thermal decomposition. Resins were formed from saturates plus aromatics and asphaltenes; resins have almost the same H/C as in reactants, but with lower molecular weight. As shown in Table 4.1, asphaltenes formed from resins, and asphaltenes without reacting (which are those formed from asphaltenes after reaction), have similar H/C atomic ratios compared with asphaltenes as reactants (0.95, 0.90, and 0.91, respectively).

According to the author, resins can be formed from aromatics by either molecular weight growth or by decreasing the hydrogen content, which includes alkyl chains removal or aromatization of naphthenes. Asphaltenes can be formed in the same way either by molecular weight growing of resins or by decreasing their hydrogen content. Coke is probably formed due to the molecular weight growing of partially converted asphaltenes as well as by decreasing the hydrogen content.

However, it must be remembered that not only heavier fractions of petroleum are responsible for coke formation, but also low molecular weight and polar species, such as resins, play an important role. Resins and asphaltenes are polar in nature, which tend to originate coke at high temperatures. Resins and asphaltenes may produce between 25 and 60 wt% of coke or even more. Asphaltenes themselves can produce from 35 to 65 wt% of coke when thermal decomposition is taking place and the remaining percent is constituted by volatile thermal products, which can vary from condensable liquids to gases (Speight, 2004).

Wang et al. (1998) stated that resins with larger pericondensed aromatic units and shorter alkyl chains are converted into asphaltenes whereas highly pericondensed asphaltenes with small alkyl chains are transformed to toluene insolubles (coke) due to exceeding the solubility limit. On the other hand, Liu et al. (1998) concluded that the yield of coke in feedstock linearly depends on asphaltenes, resins, and aromatics concentration in hydrocracked residue. The authors upgraded a Chinese vacuum residue rich in asphaltenes and the following relationship was obtained by fitting the data with a correlation coefficient of 0.988.

$$\text{Benzene} - \text{insoluble}\,(\text{wt\%}) = 4.76\left[\frac{\text{Asphaltenes (wt\%)}}{\text{Aromatics (wt\%)} + \text{Resins (wt\%)}}\right] - 0.1 \tag{4.1}$$

Toluene insolubles also can be correlated with asphaltenes and resins content in heavy fractions. For this case, Liu et al. (1999) have studied 40 samples of Chinese, Oman, and Saudi Arabia resids and data fitting by linear regression gave a correlation coefficient of 0.875.

$$\text{Toluene} - \text{insoluble}\,(\text{wt\%}) = 7.12\left[\frac{\text{Asphaltenes (wt\%)}}{\text{Aromatics (wt\%)} + \text{Resins (wt\%)}}\right] - 1.3 \tag{4.2}$$

These authors also reported the following correlation for cracking conversion of residue in terms of H/C, wt% of S, N, and molecular weight at 350°C using data of 40 samples, with a correlation coefficient of 0.874:

$$\text{Conversion}_{350^\circ\text{C}} = 74.2[(\text{H/C})^{0.257}(\text{S\%})^{0.158}(\text{N\%})^{0.184}(\text{MW})^{-0.151}] + 0.3 \quad (4.3)$$

The above correlation indicates that the higher the H/C, S%, and N%, the higher the cracking conversion.

Conversion was also correlated to saturate, aromatics, resin, and asphaltenic (SARA) fractions as (correlation coefficient of 0.885):

$$\text{Conversion}_{350^\circ\text{C}} = 29.4\left[\begin{array}{l}(\text{Saturates})^{0.026}(\text{Aromatics})^{0.008}(\text{Resins})^{0.380} \\ (\text{S\%})^{0.170}(\text{MW})^{-0.230}\end{array}\right] + 0.6 \quad (4.4)$$

The quality of coke and asphaltenes depends mainly on temperature, pressure of hydrogen, type of solvent, and catalysts. Savage et al. (1988) conducted a set of reactions, such as pyrolysis and hydropyrolysis, with and without solvents and catalytic hydroprocessing over a $CoMo/Al_2O_3$ hydrotreating catalyst. The results allowed the discrimination of a reaction network for thermal and catalytic pathways and the main conclusions were

- Pyrolysis at different reaction conditions indicated that coke did not appear at 350°C even after 2 h of reaction; however, the conversion of asphaltenes was around 100% after 30 min of reaction at 450°C. During pyrolysis reaction, the following atomic ratios were found:
  - For asphaltenes: H/C diminished from 1.24 to 1.02. S/C diminished from 0.038 to 0.030. N/C increased from 0.022 to 0.026.
  - For coke: H/C changed from 0.90 to 0.59. S/C decreased slightly from 0.029 to 0.026. N/C increased from 0.033 to 0.037.
- Hydropyrolysis (pyrolysis in presence of hydrogen) at 400°C gave an asphaltene conversion around 85% at 120 min of reaction and high content of coke was obtained (>70 wt%). The use of H_2 at 6.9 MPa yielded a small amount of gases even at 450°C being less than 2.5 wt%. Major gaseous products were H_2S, methane, ethane, and propane. Maltenes obtained at 400°C were constituted by alkylcyclohexanes, alkylbenzenes, alkylnaphthalenes, alkylthiophenes, benzothiophenes, and substituted indenes and indans. The qualities of asphaltenes and coke in terms of their atomic ratios are:
  - For asphaltenes: H/C diminished from 1.24 to 0.91. S/C diminished from 0.038 to 0.028. N/C increased from 0.022 to 0.034.
  - For coke: H/C was reduced from 0.87 to 0.58. S/C diminished from 0.029 to 0.023. N/C increased from 0.036 to 0.038.

From the aforementioned results, it is observed that coke is highly aromatic with increasing content of nitrogen. The differences in coke properties are due to the cross-linkage during coking and physical appearance is also a consequence of this. For example, Nandi et al. (1978) identified two types of coke. Asphaltenes yielded a fine mosaic grain structure and, with increasing temperature, it turned into a coarse grain structure. On the other hand, deasphalted heavy oil gave a flow-type structure. The authors concluded that both types of coke are independent from the other.

Depending on the type of catalyst and reactor severity, the cycle length of a hydroprocessing unit is typically between a few months to years. In fixed-bed units, catalyst deactivation during the run is compensated by a progressive increase in catalytic bed temperature (Figure 4.1), up to a certain value dictated by metallurgical constraints or product qualities of the process. The end-of-cycle is usually low activity and product specifications along with reactor high pressure drop, so that the unit shutdown is required. As reported before, deactivation is usually due to three main causes: carbon laydown (step 1), active phase sintering (step 2), and diffusion limitation into the pores (step 3). The detrimental effects of coke and metal deposition are a reduction of support porosity (more than 50%), leading to diffusional limitations and blocking the access to active sites. These pores can be effectively recovered by regeneration of carbon, but metal deposition is permanent. Fe, Na, Mg, and Ca are metal contaminants often found in spent catalysts, but these impurities have a rather low catalyst deactivating effect and come essentially from corrosion of upstream, desalter, etc., and are usually found at the top bed catalyst in a reactor (Rana et al., 2008).

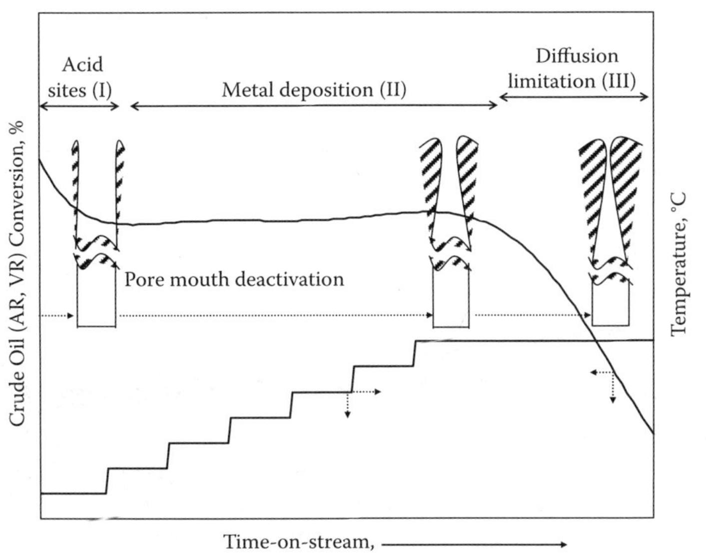

FIGURE 4.1 Proposed mechanism for heavy oil hydroprocessing catalyst deactivation with time-on-stream at pore mouth due to carbon and metal depositions with variation of temperature and uphold constant conversion of crude oil and its derived residues (AR, atmospheric residue and VR, vacuum residue).

TABLE 4.2
Summary of Deactivation Mechanisms of Heavy Oil Hydroprocessing Catalysts

Reaction Parameter	Phenomenon in Which the Structure and State of the Catalyst Change	Problem	Remedies
Thermal	Sintering Alloying Support changes Metal/metal oxide-support interactions Metal volatilization	Loss of catalytic sites	Use low operation temperatures
Chemical	Poisoning: irreversible adsorption or reaction on/with the surface Physical/chemical blockage of support pore structure	Block the number of catalytic sites Reconstruction of catalytic surfaces	Decrease basic nitrogen and H_2S concentration
Fouling	Coke formation (carbon deposits) Metal deposition Catalyst bed plugging	Diffusion limitation Decrease number of catalytic sites	Graded bed loading
Mechanical	Thermal shock Attrition Physical breakage	Loss of catalytic material Reactor plugging Pressure drop	

The reaction parameters that generally affect catalyst deactivation are shown in Table 4.2.

4.3 THE ROLE OF THE ASPHALTENES IN COKE FORMATION

The complex structures of asphaltenes are responsible for many reactions when upgrading heavy oils, residue, or bitumen. It is to be remembered that heavy fractions containing asphaltenes can undergo thermal or catalytic reactions; however, when upgrading heavy cuts, asphaltenes undergo a number of reactions, one of them being a retrograde reaction that tends to produce coke along with several other petroleum products, which can be separated by using different method of fractionation, as shown in Figure 4.2. The influence of thermal and catalytic paths of coke formation reactions mechanism from asphaltenes is discussed in detail in the following sections.

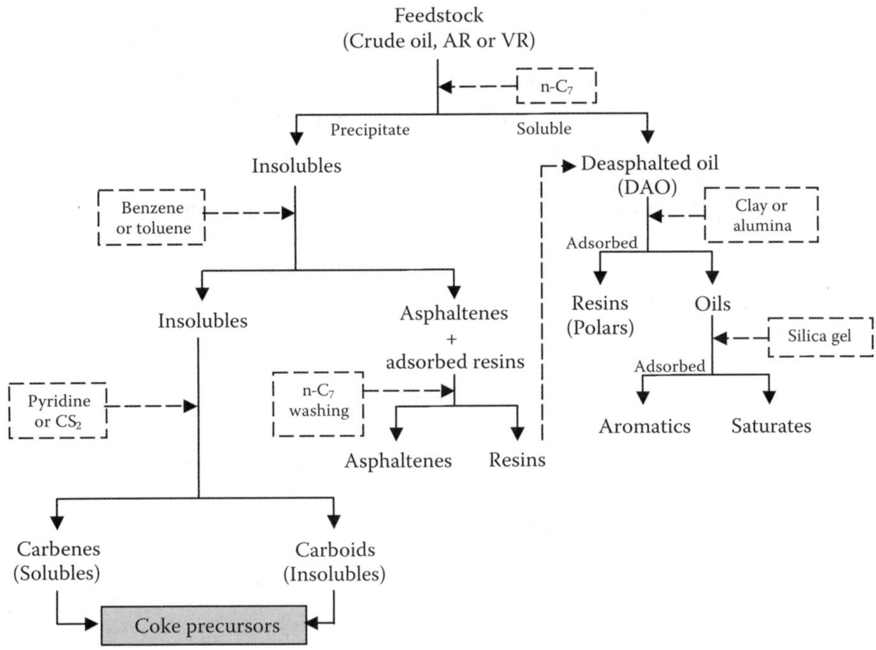

FIGURE 4.2 Fractions of crude oil and their component effect on catalyst deactivation.

4.3.1 Thermal Reaction

For heavy oil processing, it is usually accepted that coke formation is associated with asphaltene cracking. The coke formation in a thermal process is due to the dehydrogenation of paraffins and naphthens into olefins and cycloolifins, respectively. These olefins further condense and form coke, which is time-dependent, as reported by Del Bianco et al. (1993). Some studies to analyze the direct influence of asphaltenes toward coke formation have been carried out by Banerjee et al. (1986), who fractionated Athabasca bitumen, Topped Athabasca bitumen, Light Arabian vacuum bottoms, and Cold Lake heavy oil into asphaltenes, hard and soft resins, aromatics, and saturates. Kinetic studies of coke formation were performed. Soft and hard resins were obtained by column fractionation with methyl ethyl ketone (MEK) and tetrahydrofuran (THF), respectively. It was observed that aromatic fractions produced coke, but at a lower velocity than that of asphaltenes. The rate of coke formation from saturates is low because saturates need to convert into several products (olefins) before producing coke. The reaction order was supposed to be unity and activation energy of coke formation for different fractions was

- Asphaltenes (41 to 47 kJ mol^{-1})
- Soft resins (36 to 40 kJ mol^{-1})
- Hard resins (34 to 40 kJ mol^{-1})
- Aromatics (52 to 58 kJ mol^{-1})

The influence of temperature is one of the most important parameters in coke formation. Regarding this, studies involving residue with a high content of asphaltenes have been carried out by Zhang et al. (2005), who have established a link between coking, colloidal stability, and molecular properties of petroleum residue during thermal reactions. A parameter called *colloidal stability* was defined based on the mass of added *n*-heptane-to-mass of residue; it was observed that during the induction period (time necessary to coke appears as a separated phase) for different residue samples analyzed, coke formation was not seen, but it was formed very fast after reaching the induction period. At the same time, the colloidal stability decreases during the induction period and almost remains unchanged after it. A summary of the main observations is

- Asphaltenes concentration in residue increases during the induction period reaching a maximum at the end of the induction period and after that there is a smooth decrease as the time progresses indicating that aggregation of asphaltenes occurs and they turn into coke. At the same time, maltenes amount decreases sharply during induction period and after that the decrease is very smooth.
- The diminution of the amount of maltenes reduces the resins content along with colloidal stability of asphaltenes during the induction period, which leads to saturation of asphaltenes in the colloidal medium making asphaltenes prone to aggregate and transform into coke.
- Molecular weight of maltenes diminishes as thermal reaction progresses. Molecular weight of asphaltenes increases during induction period having a maximum due to aggregation reactions, but after that period, asphaltenes undergo a reduction of their molecular weight.
- Dipole moment of residue increases as thermal reaction progresses and its maximum value is reached at the end of the induction period and then decreased. This means that asphaltenes with a high dipole moment tend to aggregate during the induction period and after that they form coke.

A schematic representation of the variation of the colloidal stability parameter and coke formation is shown in Figure 4.3. As thermal reaction advances, asphaltenes and resin ratio will tend to vary with temperature and accelerate the deterioration of colloidal stability. Thus, asphaltenes having large conglomerating parameters prefer to aggregate into coke. It is well known that less than one resin/asphaltene ratio favors sediment formation (Chapter 5).

The induction period also has been referred to as the time necessary for the mesophase to become visible under microscopy, being bright under cross-polarized light. Mesophase appears when precursors, such as asphaltenes, exceed their solubility limit and depend on the solvent power of maltenes present in heavy crude, which stabilize asphaltenes. However, maltenes change their solvent power due to the extent of thermal reactions that modify colloidal stability of

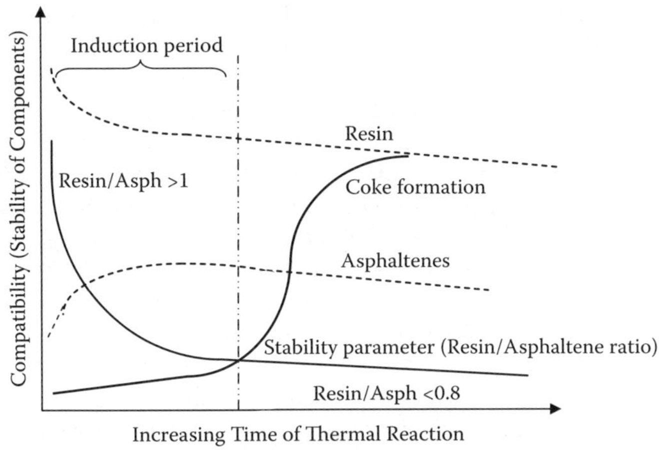

FIGURE 4.3 Schematic representation of changes occurring during coke formation (Asph = asphaltene).

the medium (Gentzis et al., 2001). Conversion of maltenes during reaction leaves asphaltenes unprotected and more likely to agglomerate and precipitate. At one point, the physical appearance of the medium changes as a result of the progress of thermal reactions provoking the initial appearance, i.e., asphaltenes in maltenes, changes to another phase (maltenes in asphaltenes), as reported by Gimaev et al. (1980). This phenomenon is known as phase inversion and was observed earlier by Brooks and Taylor (1965). Gimaev et al. (1980) concluded that:

- As thermal reaction proceeds, free radicals are formed and condensation reactions make those products accumulate in the system with a consequent phase transition in the liquid medium.
- Asphaltenes experience a gradual separation of phases because of intermolecular attractions and by attraction of unpaired electrons of free radicals originating in the formation of a mesophase considered as anisotropic.
- Accumulation of aromatic structures produces the formation of a second liquid phase, and van der Waals and dipole–dipole forces do this phase to orient the molecules of the mesophase sphere.
- When condensation occurs within mesophase, then solid crystallites of coke appear.
- Phase separation is mainly promoted by incompatibility of the medium with asphaltenes due to their shortening of alkyl chains when processed at high temperature. Asphaltenes and resins are not thermally stable and they begin to suffer changes when increasing temperature.

Hence, the main reasons for incompatibility of asphaltenes and the medium are cracking and polymerization reactions, as stated by Wiehe (1993). He also reported that coke formation is very dependent on the behavior of asphaltenes. As conversion increases, asphaltene cores become more aromatic until the solubility limit is exceeded with the consequent phase separation forming a second liquid phase, which has less hydrogen to extract and stabilize free radicals. This makes recombination of asphaltene cores feasible and can give a higher yield of coke. In this respect, Wiehe (1994) created the concept of the pendant–core building model for explaining that macromolecules can be approached by a combination of two blocks linked by means of weak bonds. The pendant unit is a volatile unit, whereas the core is nonvolatile in nature. Cracking of heavy fractions leaves free the pendant units and cores tend to concentrate. According to this model, all fractions can be obtained by a combination of pendants and cores units. Further cracking permits the recombination of cores only and the apparition of coke is imminent, whereas pendant units form all distillates. A schematic view of the pendant–core model is shown in Figure 4.4.

Rahmani et al. (2003) observed the phase inversion in some sections of coke produced from asphaltenes. During phase inversion, coke is the major phase and oil is dispersed in it, which was shown by scanning electron microscopy (SEM). The coke is separated as spheres or as coalesced structures from the continuous oil phase, whereas coke is the continuous phase. Le Page et al. (1987) have also explained the mesophase formation during thermal treatment and proposed that at the beginning asphaltenes have some void spaces in their structures; however, during the heating asphaltenes turn into more compact structures, which cannot be dissolved even in a high medium of resin. The core aggregation indicated that asphaltenes have a tridimensional structure before to the apparition of the mesophase. At this

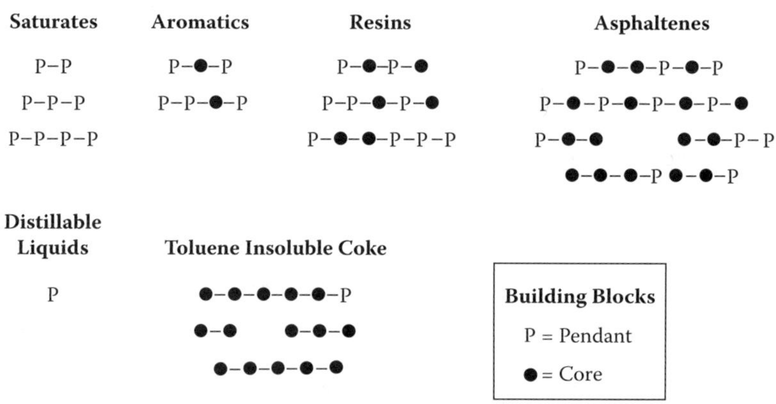

FIGURE 4.4 Distribution of species with the pendant–core building block model. (From Wiehe, I.A. 1994. *Energy Fuels* 8 (3): 536–544. With permission.)

point, a bigger rearrangement occurs and mesophase appears with a disc-like shape. Moreover, Storm et al. (1996) explained rheological changes of the petroleum as a consequence of the formation of dispersed droplets of the liquid phase. It can be concluded that asphaltene changes depend on temperature directly, which leads asphaltenes to reorganize the asphaltic micelles during the coke induction period in which large two-dimensional particles are formed. The induction period is the main cause of changes of asphaltenes, i.e., heptane-soluble material is turned into pentane insolubles and, at the same time, pentane insolubles are converted to heptane insolubles. These changes are responsible for macroscopic phase separation, which has been supported by small angle x-ray scattering (SAXS) analyses. It has been stated that the size of original micelles is 0.004 μm, and after reorganization of particles their characteristic dimension is between 0.02 and 0.03 μm (Storm et al., 1998).

4.3.2 Catalytic Reaction

To define the best method for conversion of bottoms of barrel into valuable products, detailed feedstock analysis is required. One of the most important components of the feedstock for this selection is asphaltenes. This macromolecule is closely related to other crude oil properties, such as viscosity and API (American Petroleum Institute) gravity among others. Asphaltenes are commonly defined by their solubility in toluene and insolubility in *n*-heptane, and are a highly aromatic, polydisperse mixture consisting of the heaviest and most polar fraction of crude oil. It is well known that asphaltenes contain in their molecule metal, sulfur, and nitrogen along with phosphorous as shown by the scanning transmission electron microscopy (STEM) mapping of virgin Maya crude asphaltenes (Figure 4.5). The form in which phosphorous exists in crude oil asphaltenes is still undefined; it may be just a substitution of nitrogen or may have some structural formula. The presence of phosphorous is not limited to virgin Maya crude oil asphaltenes, but it also is present in asphaltenes precipitated from hydrotreated oils, as shown in Figure 4.6 and Figure 4.7. Phosphorous has also been detected deposited along with other metals (Ni, V, Fe, etc.) on the spent heavy oil hydroprocessing catalyst (Figure 4.8); this also confirms phosphorous presence in the feedstock. Since crude oil composition widely varies with its origin, it may be possible that the detected concentration of phosphorous is only available in Maya crude oil.

To design a catalyst for hydroprocessing of heavy oils, and particularly to maximize asphaltenes conversion, two distinct properties are required: (1) number of acid sites and (2) large pore diameter, which allow large molecules for easy diffusion into the pores (or catalytic sites). Acid sites are essential to enhance hydrocracking selectivity (Delmon, 1990); however, coke formation is also favored by these catalytic acid sites and, consequently, a balance between acidity and textural properties of catalysts is necessary.

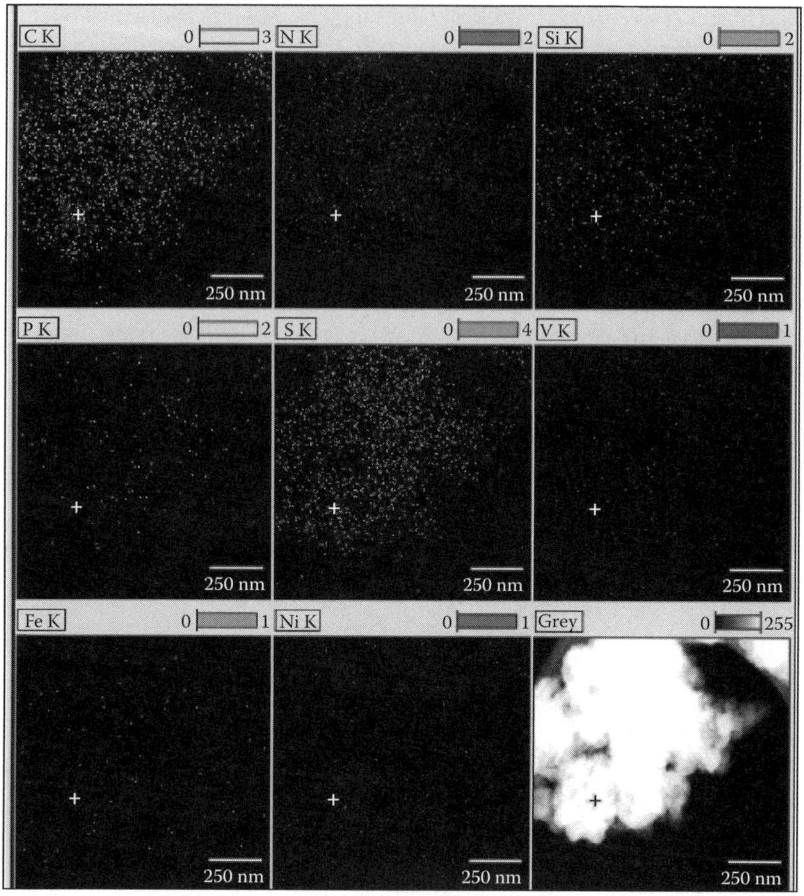

FIGURE 4.5 STEM images of the pure Maya crude precipitated asphaltene and its elemental composition mapping of a selected area. (From Rana, unpublished results.)

The function of the catalyst is to remove contaminants, such as sulfur, nitrogen, and metals present in the feedstock. A catalyst also hydrogenates cracked fragments and increases the H/C atomic ratio of products. However, the catalyst suffers from deactivation mainly due to coke and metal deposition (Ni + V) (see Figure 4.1). Catalysts with narrower pores and high surface area possess high hydrotreating activity, but they are deactivated very quickly (Bartholdy and Cooper, 1993; Absi-Halabi et al., 1991; and Tahkur and Thomas, 1985: Rana et al., 2004). Because asphaltenes are the aggregate of big molecules present in crude, their molecular size makes them difficult to enter into the pores. Increasing the pore size of catalysts improves their diffusion; however, diffusivity is also

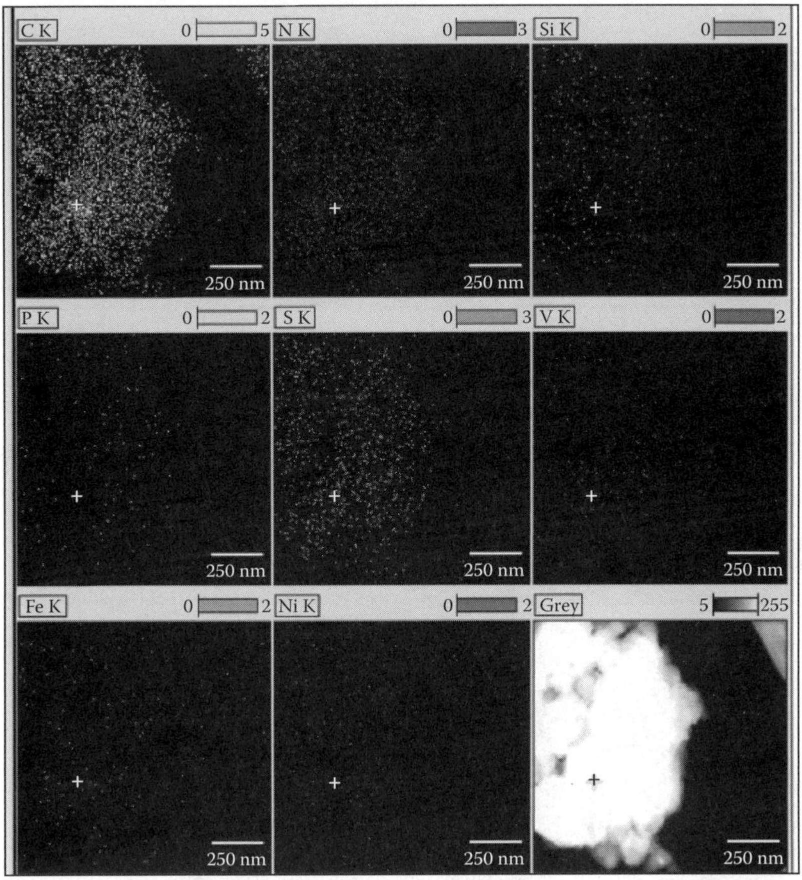

FIGURE 4.6 STEM of hydrotreated Maya crude asphaltene at 400°C and precipitated with 33/67 v/v, *n*-C_7/toluene. (From Rana, unpublished results.)

controlled by catalyst particle and pore structure. That is why it would be desirable that the catalyst has the maximum possible surface area to minimize diffusion restrictions. Metals are also linked to asphaltenes as metalloporphyrins. Large unimodal pore catalysts with pores ranging from 100 to 300 nm exhibited the highest activity for vanadium removal for Kuwait vacuum residue, as reported by Absi-Halabi et al. (1995). Nitrogen removal is achieved mainly in meso- and macropores because their structures have been related to porphyrinic compounds as well.

A rough relationship between catalyst activities and the effect of average pore diameter of a catalyst is shown in Figure 4.9. Because the deactivation through pore diffusion will depend on the catalyst pore diameter as well as the composition of

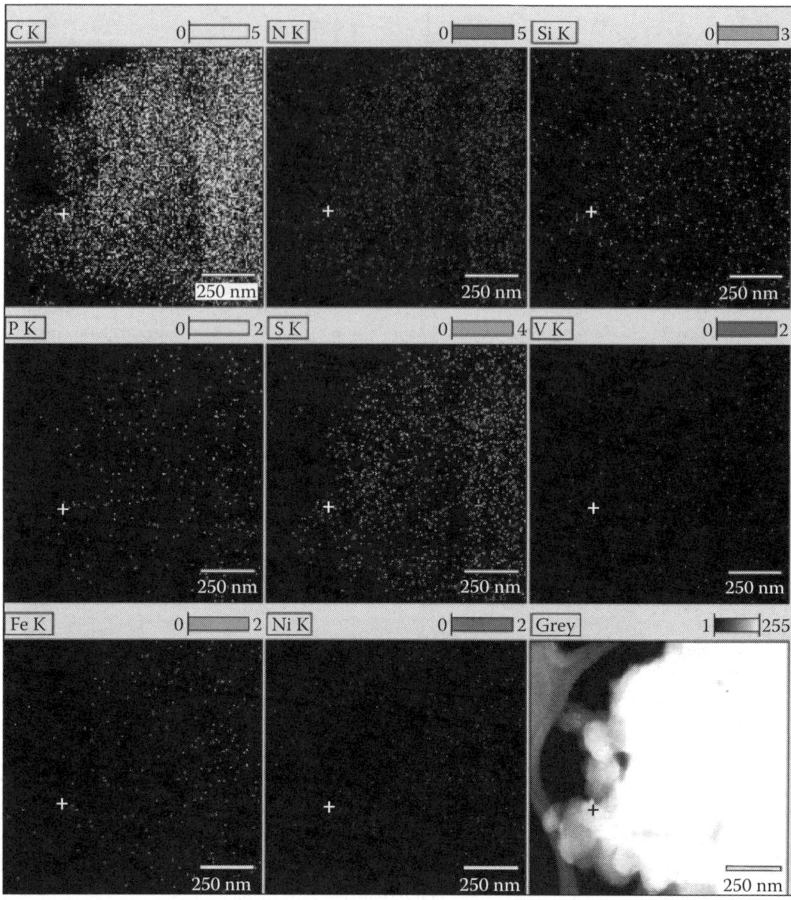

FIGURE 4.7 STEM of hydrotreated Maya crude asphaltene at 420°C and precipitated with 33/67 v/v, *n*-C_7/toluene. (From Rana, unpublished results.)

feedstock, the external surface of catalyst particles and the diffusion into the catalyst interior becomes a rate-limiting factor, especially for the molecule of complex metals (Ni, V) and asphaltene. This can be followed by the textural properties and HDS, hydrodemetallization (HDM), and hydrodeasphaltenization (HDAs) conversions, as shown in Figure 4.10a and Figure 4.10b. This behavior is observed for different average pore diameters (APD) of alumina-supported Co-Mo catalysts, which contain similar amounts of Co and Mo. The variation in pore diameter can be obtained by changing the preparation method, as reported in Table 4.3. In general, HDM and HDAs activities increase with pore diameter, while HDS activity decreases. These results further confirm diffusion limitations against the large molecule into the pores or catalytic sites, i.e., Coordinated Unsaturated Sites (CUS).

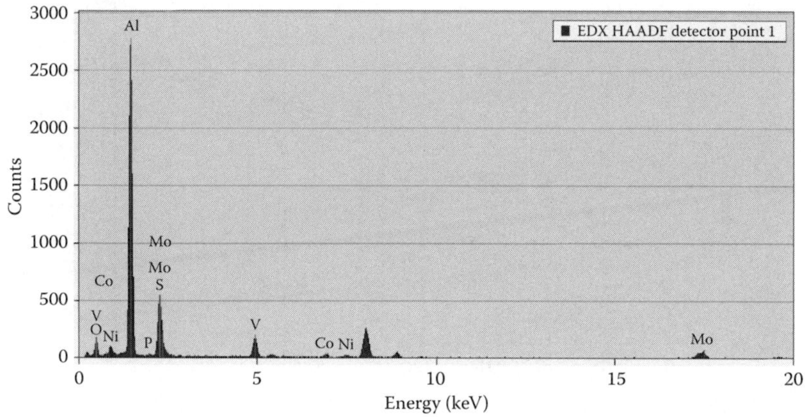

FIGURE 4.8 STEM energy disperse (EDX) spectrum of spent CoMo/Al_2O_3 catalyst after 200 h time-on-stream (TOS). (From Rana, unpublished results.)

Coke deposition over the catalyst was related to pore size and it was concluded that big molecules, such as asphaltenes, are prone to decompose and form coke, spend more time in narrower pores accomplishing with the reaction diffusion-limited (Ancheyta et al., 2005). Asphaltenes that can go inside the small pores experience over-cracking in the pores due to the fact that they move slowly and for this reason more carbon is generated. Catalysts with high content of macropores

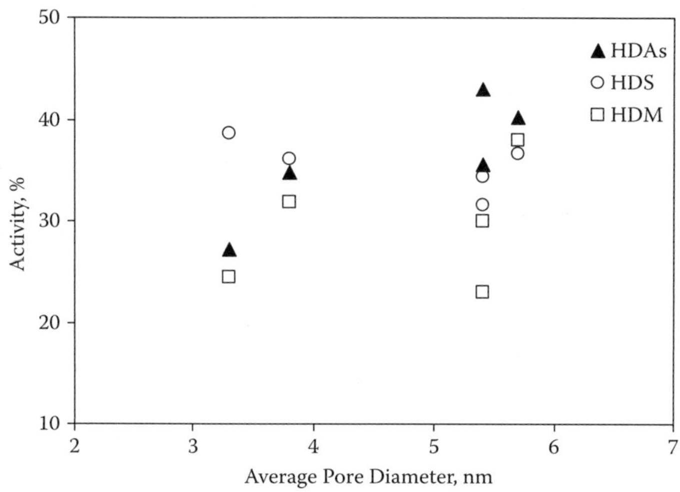

FIGURE 4.9 Effect of average pore diameter (APD) on HDS, HDM, and HDAs conversions. (From Rayo et al. 2007. *Petrol. Sci. Tech.* 25 (1–2): 215–229. With permission.)

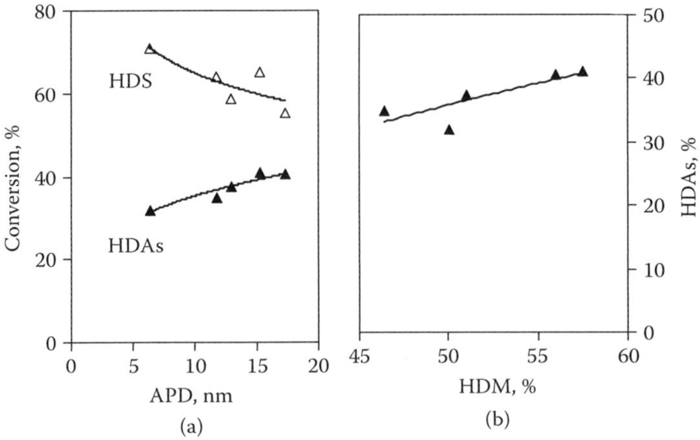

FIGURE 4.10 (A) Effect of pore diameter on the HDS, HDAs, and (B) HDAs selectivity against HDM (results after 120 h TOS). (From Rana et al. 2007d. *Petrol. Sci. Tech.* 25 (1–2): 201–214. With permission.)

TABLE 4.3
Effect of Alumina Support Preparation on the Textural Properties

Catalyst	Preparation Method of Al_2O_3 Support	Textural Properties		
		SSA, m²/g	PV, ml/g	APD, nm
CoMo/γ-Al_2O_3-u	Precipitated with urea at 95°C for 10 h	136	0.39	9.4
CoMo/γ-Al_2O_3-acs	Precipitated with NH_4CO_3 using swing method (variation in pH) and final pH was *ca.* 8	160	0.47	11.2
CoMo/γ-Al_2O_3-ac	Precipitated with NH_4CO_3 and reflux for 4 h	201	0.37	6.0
CoMo/γ-Al_2O_3-am	Precipitated with ammonia (NH_4OH) using 10% aqueous solution at room temperature and pH *ca.* 8	169	0.27	7.0

Note: APD = average pore diameter, u = urea, ac = ammonium carbonate, acs = ammonium carbonate swing method, am = ammonia.

Source: Rana, et al. 2004. *Catalysis Today* 98 (1–2): 151–160. With permission.

assist asphaltenes to enter deeper inside and go out more easily, thus spending less time and producing less carbon. Deactivation can also be carried out by metals, e.g., metalloporphyrins cannot enter very deep inside the small pores where molecules have restricted access to internal surface being that demetallization be carried out at the mouth of the pores.

During the upgrading of heavy crude oils, heavy and light fractions are converted at different extents. Lighter fractions have smaller size and lower aromaticity and polarity than heavy fractions, which can turn these molecules into pentane soluble during catalytic hydrocracking. However, the highly condensed aromatic core is a potential precursor of coke (Zhao et al., 2001). In order to prevent coke formation, different types of treatments are available; one of them is the deasphalting with solvents. Since asphaltenes are coke precursors, if they are removed, then the tendency to form coke is reduced. The hydrodemetallization of feedstock is also recommended in a guard reactor. The catalyst employed for HDM must possess bigger pores for capturing metals, which usually depend on the combination of textural properties of the catalyst, as shown in Figure 4.11. However, apart from the reaction conditions, the deposition of metals also depends on the metal content in the feedstock, as illustrated in Figure 4.12. These results indicated that metal deposition affects more or less at a similar magnitude the conversion of HDAs and HDM, while the HDS activity is less affected. The plausible explanation for

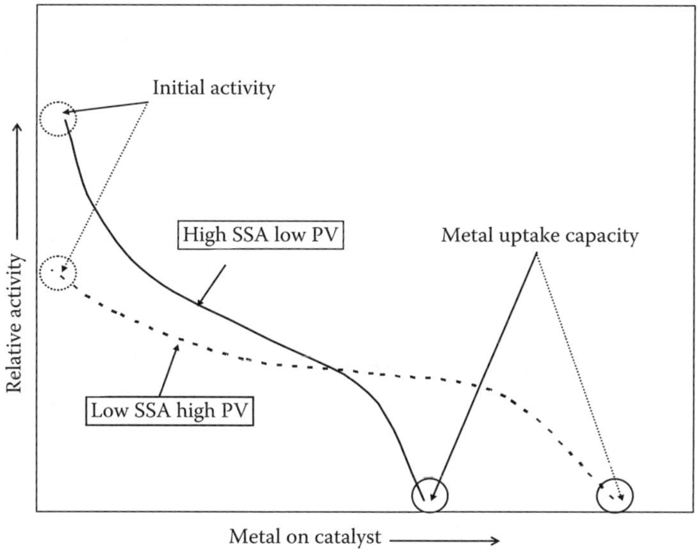

FIGURE 4.11 Effect of pore volume (PV), specific surface area (SSA), and their relationship between catalyst activity and metal accumulation.

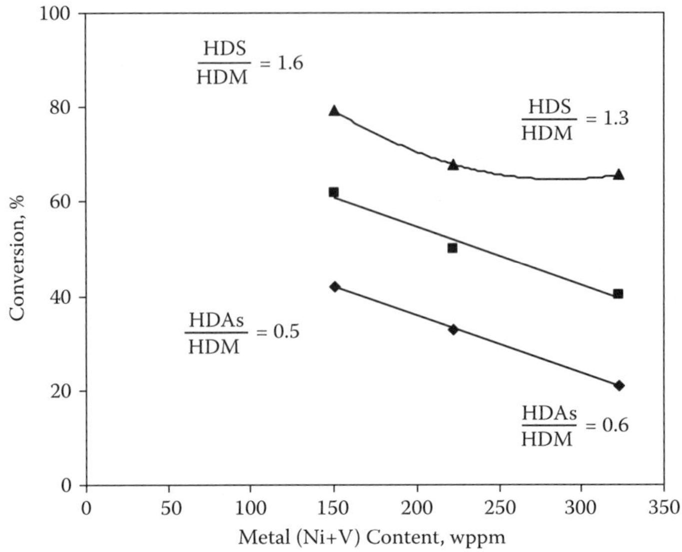

FIGURE 4.12 Effect of metal (V + Ni) content in feedstock on HDS, HDAs, and HDM. (From Rana et al. 2007d. *Petrol. Sci. Tech.* 25 (1–2): 201–214. With permission.)

this behavior is the smaller molecules of sulfur compound compared with those of the asphaltene, as well as Ni and V porphyrin molecules, which can be easily diffused into the pores. The activity results are well supported by the textural properties of spent catalysts in which pore volume and specific surface area are reduced by 40 and 25%, respectively, after 120 h TOS (Rana et al., 2007e). Pore size distribution analysis indicated that coke and metals in spent catalysts were deposited predominantly in the narrow pores (i.e., <15 nm) of the catalyst, as shown in Figure 4.13. An increase in the spent catalyst absolute isotherm area is due to the reduction of pore mouth by the deposition of metals (Ni and V) and carbon during the reaction. Thus, the existence of ink-bottle types of pores possibly decreases or slows down the desorption of adsorbed nitrogen, and, as a result, an increase is expected, which can be a semiquantitative way to calculate the pore mouth plugging compared with the fresh catalyst (Ancheyta et al., 2005; Rana et al., 2005a).

The presence of hydrogen during thermal reactions initially limits the hydrogen donation ability of asphaltenes preserving their structures, especially the naphthenic and aliphatic bonds, which can be further cracked. Because the content of hydrogen in asphaltenes is still high at the beginning of thermal reactions, asphaltenes conversion is slower and coke is also slowly formed. In absence of an effective hydrogen donor, the rate of coke formation is faster; however, by using a good hydrogen donor, the reaction proceeds slowly, but increases the selectivity

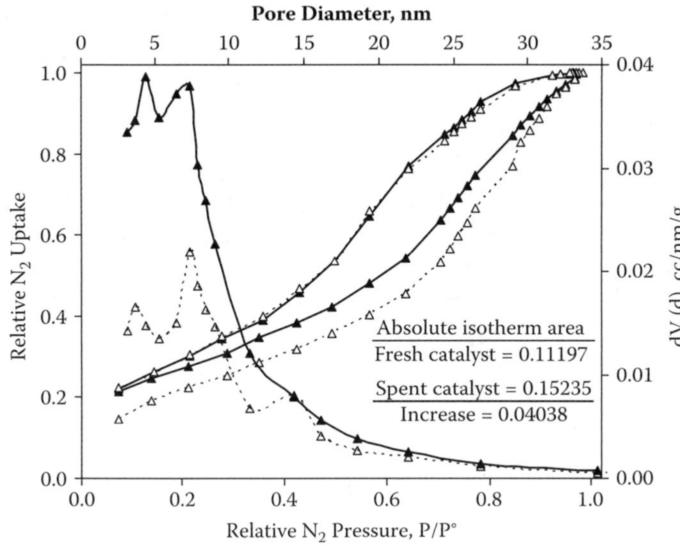

FIGURE 4.13 Adsorption–desorption isotherms and variation in pore diameter of (▲) fresh and (Δ) spent $CoMo/Al_2O_3$ catalysts. (From Rana et al. 2007d. *Petrol. Sci. Tech.* 25 (1–2): 201–214. With permission.)

toward maltenes. The presence of a catalyst keeps high selectivity to maltenes at a intermediate reaction rate. The catalyst efficiently uses the hydrogen dissolved in the organic medium and hydrogenates part of residual asphaltenes (Savage et al., 1988). When hydrogen is not present in the reaction, a deficiency of hydrogen in asphaltenes is created producing free radicals, which can condensate and/or polymerize and, in consequence, coke is readily formed. By using moderate hydroprocessing conditions, asphaltenes catalytic conversion is limited (less than 50%), but some literature has reported initial conversion as very high by using acidic catalysts (Figure 4.14). On the other hand, a detailed study of the diluent effect (hydrodesulfurized naphtha and diesel) on Maya crude hydrotreating has indicated that diluent has an important role in HDS and HDAs conversions (Rayo et al., 2004), as shown in Figure 4.15 and Figure 4.16, over $NiMo/TiO_2$-Al_2O_3-supported catalysts. The authors conclude from their results that naphtha diluent leads to an increase in the poisoning of the catalyst surface by carbon deposition originated by the insolubility of asphaltene in light components present in naphtha, which indicates that careful attention must be paid to the type of diluent used when experimenting with heavy crude oil. However, dilutions of crude with diesel evade such a insolubility problem or minimize.

During hydroprocessing in a fixed-bed reactor, Guibard et al. (2005) have pointed out that the presence of asphaltenes in the HDM section limits the HDS of

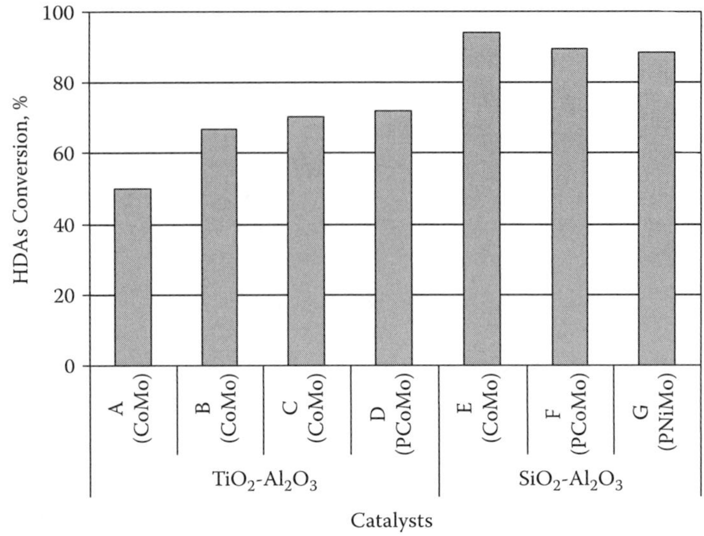

FIGURE 4.14 Effect of support preparation and active metal components on asphaltene conversion. (Adapted from Maity et al. 2003a. *Appl. Catal. A* 244 (1): 141–153. With permission; Maity et al. 2003b. *Appl. Catal. A* 250 (2): 231–238. With permission; Maity et al. 2003c. *Appl. Catal. A* 253 (1): 125–134. With permission.)

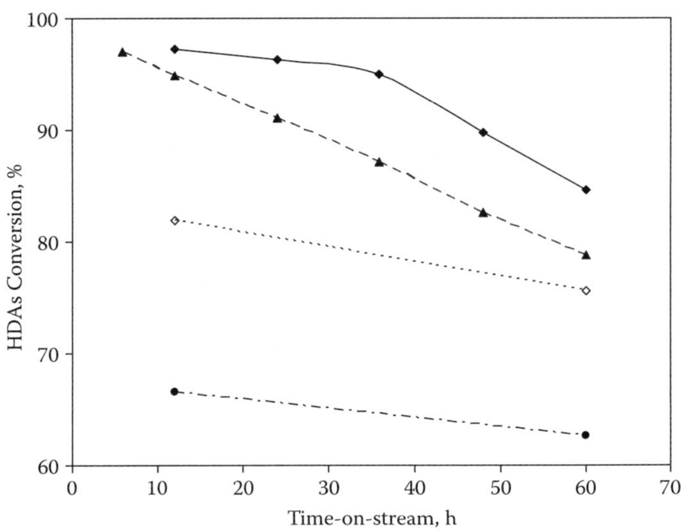

FIGURE 4.15 Effect of diluent (HDS naphtha or HDS diesel) or feed composition on HDS conversion of Maya crude oil. (▲) NiMo/Al_2O_3 (Maya HDT + naphtha), (◆) NiMo/Al_2O_3-TiO_2 (Maya HDT + naphtha); (◇) NiMo/Al_2O_3-TiO_2 (Maya HDT + diesel); (●) NiMo/Al_2O_3-TiO_2 (Maya + diesel). (From Rayo et al. 2004. *Catalysis Today* 98 (1–2): 171–179. With permission.)

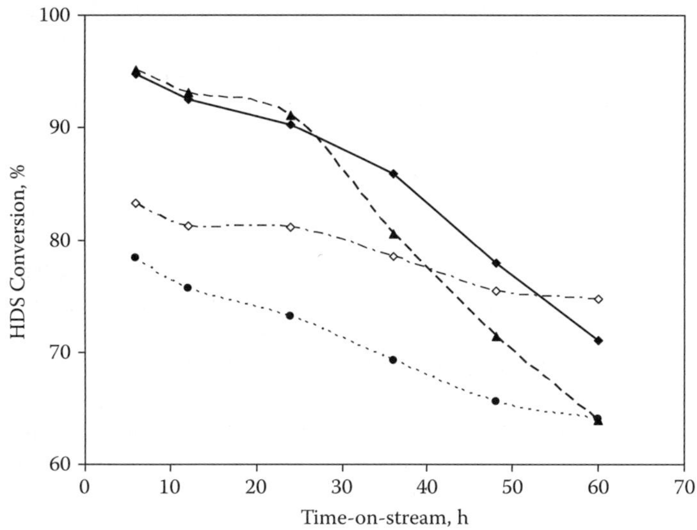

FIGURE 4.16 Effect of diluent on the HDAs conversion of Maya crude oil. (▲) NiMo/Al_2O_3 (Maya HDT + naphtha); (◆) NiMo/Al_2O_3-TiO_2 (Maya HDT + naphtha); (◇) NiMo/Al_2O_3-TiO_2 (Maya HDT + diesel); (●) NiMo/Al_2O_3-TiO_2 (Maya + diesel). (From Rayo et al. 2004. *Catalysis Today* 98 (1–2): 171–179. With permission.)

resins and aromatics. Asphaltene aggregates that go to the HDS section proceeding from the HDM section are still large so that they cannot easily enter inside the pores limiting the HDS reaction. To establish the influence that asphaltenes have during HDS, Le Lannic et al. (2007) carried out experiments in which demetallized and deasphalted feedstocks were submitted to HDS. Resins and aromatics were desulfurized at around 100% when asphaltenes were previously removed. Compared with asphaltenic feedstocks, HDS was slower due to the presence of asphaltenes that block and retard the access to the active sites. Thus, it is important to remember that asphaltenes, resins, aromatics, and saturates are present in residue in a delicate balance. However, during hydroprocessing, destabilization of asphaltenes occurs when resins are converted leaving asphaltenes unprotected to the point in which they are no longer soluble causing instability and incompatibility. Higher temperature during hydroprocessing turns the medium more incompatible due to competition of thermal and cracking reactions and the difference in reactivity of distinct components. Aromatics and resins are less complex structures than asphaltenes and react more easily, but due to resins being under hydrogenating conditions they modify their structure and at the same time asphaltenes experience cracking of alkyl chains that leaves the aromatic core with a minimum number of branched chains. Asphaltenes are not soluble in hydrogenated resins and precipitate enhancing coke formation on the catalyst, as reported by Matsushita et al. (2004). They defined a new relationship that takes into account the H/C atomic ratio of asphaltenes and maltenes from the HDM of Kuwaiti atmospheric residue:

$$\text{Relative solubility index (RSI)} = \frac{\text{H/C of asphaltenes}}{\text{H/C of maltenes}} \tag{4.5}$$

The relationship may give information about the solubility of asphaltenes and its influence on coke formation. Conclusions drawn from different relative solubility index values indicate that the higher the relative solubility index, the lesser the coke formation. The quality of coke deposited onto the catalyst also can be explained in terms of the relative solubility index (RSI), i.e., at lower RSI, coke deposited will have less H/C (more aromatic coke). Thus, the lower the RSI, the higher the amount of coke on the catalyst. The proper selection of catalyst and reaction conditions is the major parameter to improve the quality of liquid products with reduced coke deposited.

When using a catalyst for cracking, e.g., zeolite-based catalyst, which is common to upgrade residue, fast growing of coke produced by influence of temperature is capable of causing damage in zeolite-supported catalysts by blockage or encapsulation before zeolite develops its cracking capability. The cleavage of alkyl chains will favor the generation of free radicals, which, in turn, will combine together to condensate and create bigger molecules that precipitate from the crude and deposit as coke. As reaction progresses, a second type of coke is observed to be present on the catalyst, which is formed by associated molecules of asphaltenes and maltenes. In the case of asphaltenes, the presence of nitrogen in their structures makes them basic in nature being preferentially adsorbed even in weakly acidic surfaces. Desorption is not as easy to carry out and molecules tend to cross link giving a product with higher molecular weight. As time goes on, more cross linking is achieved and formation of coke is imminent, but, in this case, having a catalytic influence. When using zeolite-supported catalysts, not all molecules are prone to be adsorbed into the pores because large molecules still remain. Big molecules, such asphaltenes, will deposit at the entrance of the pores blocking the access to its interior and forming coke. Only when molecules have smaller size than pores, will the catalyst be really useful and exert its cracking activity. However, it has to be considered that at this stage many pores could be blocked by coke, as reported by Ho (1992), who has proposed a scheme by which coke can be formed according to the aforementioned discussion (Figure 4.17).

de Jong (1994) has also stated that coke formation is a parallel reaction that involves thermal and catalytic reactions. Thermal reactions enhance the formation of free radicals especially in big molecules, such as asphaltenes. At this point two cases are possible: (1) free radicals recombine to form a condensate

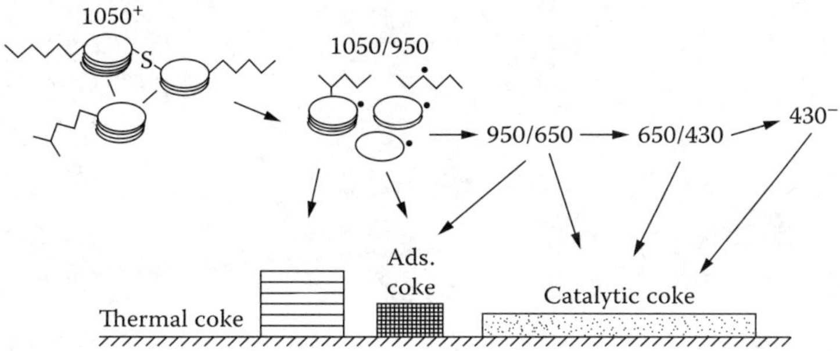

FIGURE 4.17 Schematic diagram showing the formation of different types of coke, where asphaltene particles are shown as oligomers of unit sheets with alkyl side chains, naphthenic moieties, and/or heteroatom-containing moieties. (From Ho, T.C. 1992. *Ind. Eng. Chem. Res.* 31 (20): 2281–2286. With permission.)

molecule, which eventually grows up to form coke, and (2) free radicals could be neutralized by action of activated hydrogen (hydrogen radicals). Activated hydrogen is formed by catalytic action, and, in the case of hydroprocessing catalysts, it is due to presence of molybdenum even at low concentrations. However, at high temperatures, dehydrogenation of naphthenes is thermodynamically favored by the catalyst, which could imply that naphthenes linked to aromatic rings dehydrogenate increasing the aromaticity of molecules. The continuously repeated dehydrogenation would be responsible for forming more polynuclear aromatic rings.

Images of asphaltenes obtained by SEM magnified at around 2000× are shown in Figure 4.18. Fractionated asphaltenes from hydrotreated crude at reaction temperatures of 400 and 420°C along with asphaltenes from virgin Maya crude oil are shown in this figure with two different magnifications for each sample for better comparison. Asphaltenes were separated by using *n*-heptane and washed with a mixture of solvents (67/33 toluene/n-C_7). SEM micrographs clearly indicated the aggregates of asphaltenes at the two reaction temperatures. This change in morphology can be attributed to the temperature at which reaction was carried out and significant differences in precipitated asphaltenes are observed particularly when reaction temperature changes in this range (from 400 to 420°C). It appears from the figure that the virgin asphaltenes have small porosity and structure as fragile aggregates, while in the case of hydrotreated asphaltenes, structure became stronger or defined, which is further magnified with the hydrotreating (HDT) reaction temperature at 420°C. This structural information also indicated that with increasing reaction temperature only side and fragile structure converted into the more solid and highly aromatic in nature, which is more difficult to crack catalytically.

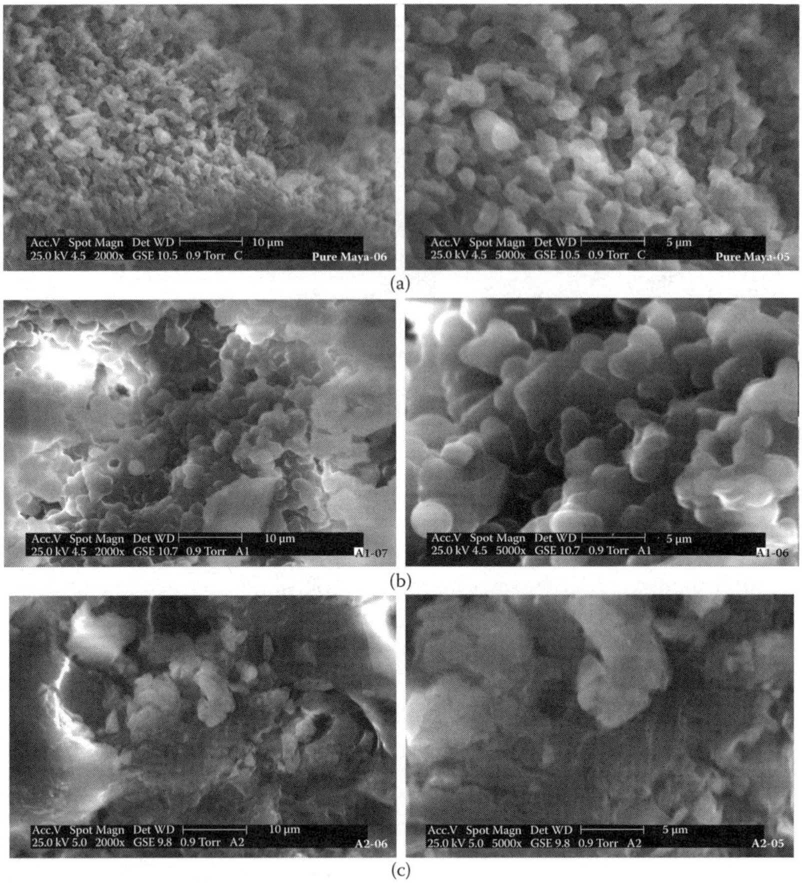

FIGURE 4.18 SEM images of purified asphaltenes: (a) asphaltenes from virgin Maya crude oil; (b) asphaltenes of hydrotreated Maya crude at 100 kg/cm^2, 400°C, and LHSV=1 h^{-1}; (c) asphaltenes of hydrotreated Maya crude at 100 kg/cm^2, 420°C, and LHSV=1 h^{-1}.

Similar types of results have been reported by Tanaka et al. (2003), who have shown that variation of structure of asphaltenes is just depending on the temperature.

4.4 CATALYST DEACTIVATION IN FIXED-BED REACTORS

Catalytic hydroprocessing is a common way to transform heavy and extra heavy crude into more valuable products. It is useful to reduce the molecular size of large molecules by hydrocracking and eliminate heteroatoms, such as sulfur, nitrogen, and metals. The elimination of these contaminants can be as gases (H_2S, NH_3) or by metal deposition on the catalyst surface. Catalytic hydroprocessing is carried out at industrial scale with temperatures between 375 and 430°C and pressures of 3 to 14 MPa. In general, during hydroprocessing, it is observed that even 50% of the specific surface area of the catalyst is lost during the first hours of the run. Initial deactivation occurs through the acid site while later deactivation is originated by deposition of metals in sulfide form (Figure 4.1). Molybdenum sulfide migrates through such deposits to the external surface and interactions between Ni, V, and molybdenum sulfides result in the formation of a deposit (Trimm, 1996).

Hydroprocessing of bottom of barrel (residue) is well documented in the literature (Kressmann, 1998; Rana et al., 2007b), but direct crude oil treating (heavy or extra heavy) is relatively new for catalytic processing. Virgin crude oil hydroprocessing has emerged as a new approach for refiners to upgrade heavy petroleum in order to improve API gravity and remove impurities before distillation. Usually, processing of such a kind of feedstock is carried out in fixed-, moving- or ebullated-bed reactors (Ancheyta, 2007; Gosslink, 1998). Due to product quality requirements and investment cost, a fixed-bed reactor operating at moderate reaction severity is preferred. Because heavy feeds contain very high amounts of metal and asphaltenes, catalysts are very sensitive to deactivation, which represents decreased life cycle. Catalyst deactivation inside the reactor also generates pressure drop (ΔP), which can be diminished by a proper selection of catalyst size (graded bed loading) as well as by using different types of catalyst loading, as shown in Figure 4.19. Thus, to determine what option one must use in a particular hydroprocessing operation, an integrated approach that takes into account size, shape, pore size, and catalyst loading must be considered (Cooper et al., 1986).

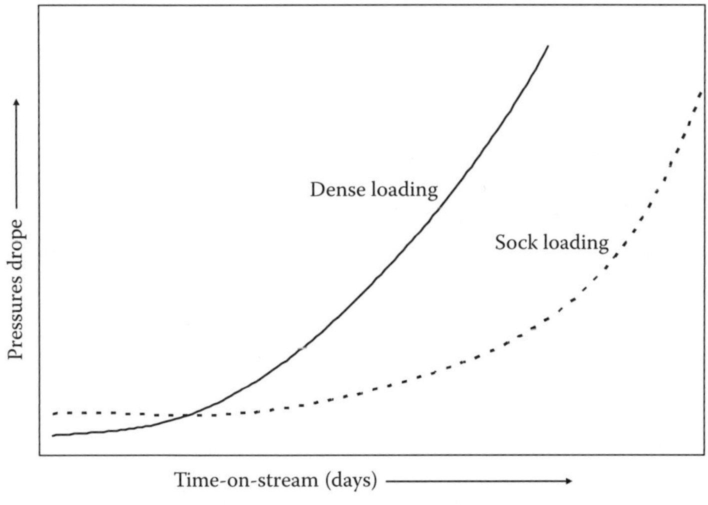

FIGURE 4.19 Effect of type of catalyst loading on the reactor pressure drop along with time-on-stream (TOS).

The way in which asphaltenes originate coke during catalytic hydroprocessing is by their adsorption onto acid sites of the catalyst, which enhances as a function of the TOS (Absi-Halabi et al., 1991; Marafi and Stanislaus, 2001; LePage et al., 1992). Two main contaminants are responsible of the activity decrease: metals deposition and coke. Metals are present in both resins and asphaltenes; however, the demetallization rate is not the same since metals in resins are faster removed than in asphaltenes (Dolbear et al., 1987). As mentioned before, it is common practice at the industrial level to increase the temperature to compensate for the loss of catalyst activity, therefore, the deactivation is also faster. At the initial period of the run, there is a high increase in the amount of deposited coke on the catalyst surface. However, after a period of time, more stable coke deposition is achieved. The specific surface area of the catalyst decreases with increasing coke and metal deposition (Figure 4.20). During hydroprocessing in fixed-bed reactors, it has been observed that surface area and pore volume losses are higher at the top-bed catalyst. At the same time, more metals are deposited at the entrance of the reactor. The reason is that catalysts are mainly designed to catch as much large metals content as possible in the first stages. Deactivation by metals is gradually decreased with the depth of the bed. The maximum metals depositions have been found at around 20% of reactor loading, which indicates that the metal porphyrin at the beginning requires hydrogenating the ring (molecular induction period) and subsequently promoting HDM. On the other hand, less coke is found

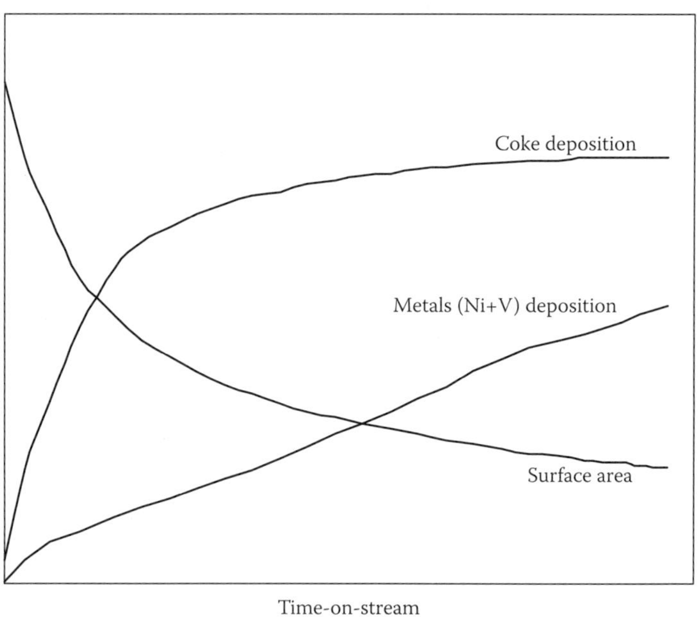

FIGURE 4.20 Effect of time-on-stream on metals, coke deposition over the catalyst, and the observed decrease in specific surface area of the catalyst.

at the top bed, while more coke is deposited at the exit of the reactor. The possible explanations for this tendency are:

1. At the entrance of the reactor, the feedstock may contain low reactive polynuclear aromatic hydrocarbon and highly condensed asphaltenes, which are less reactive and prone to coke generation. With increasing TOS, the amount of these components increases at the end of the reactor depositing coke on the catalyst.
2. The number of acid sites remains constant, thus coke deposition increases relatively at the end of the reactor.
3. The increase on H_2S concentration (depending on HDS rate) and the reduction of hydrogen partial pressure at the end of the reactor can also influence the deposition of coke (Al-Dalama and Stanislaus, 2006) due to the lower hydrogen amount in this part of the reactor and more acid sites (-SH) (Rana et al., 2007a).

During the hydroprocessing of heavy oils, it has been found that the amount of vanadium and nickel deposited on the catalyst is higher in the first reactor and it diminishes toward the last reactor. On the contrary, carbon content on the catalyst increases toward the last reactor, and the lowest content is found in the first reactor. This fact makes it possible that, when regenerating catalysts, those contained in the last reactor will recover a considerably higher surface area and pore volume compared with catalysts in the first reactors because coke is predominant on the catalyst in the last one, which can be easily regenerated. Thus, a catalyst in the last reactor has more coke deposition, which is the main cause for losses of pore volume and surface area, whereas metals deposition is the reason why a catalyst loses its porosity in the first sections.

Higher deposition of metals at the top of fixed-bed occurs because the catalyst is macroporous and designed to retain the higher amount of metals as possible (Rana et al., 2007e). Asphaltenes are big molecules that cannot be properly processed as the reaction time goes forward because the catalyst activity on the top of the bed is decaying so that when the temperature is increased at the end of the run to compensate for the loss of activity, unreacted asphaltenes would suffer fragmentation, but their size still would be too large to enter into the pores in the last stage of the catalytic bed being that they deposit on the catalyst forming the coke. Additionally, asphaltenes would precipitate from the medium since resins are converted, leaving asphaltenes unstable, which are prone to precipitate on the catalyst surface. It was also affirmed that the presence of unstable radicals increment the coke formation due to the presence of H_2S, which reduces hydrogen partial pressure and, as a result, the capping of radicals cannot be achieved by hydrogen.

4.5 INFLUENCE OF COKE AND METALS DEPOSITION ON CATALYST DEACTIVATION DURING HYDROPROCESSING

Deactivation in fixed-bed catalysts is commonly observed in three stages (Figure 4.1), which is well documented in the literature (Tamm et al., 1981; Hannerup and Jacobsen, 1983; Kam et al., 2005). It has been reported that coke deposition reaches up to 25 wt% while metals are up to 20 wt% of the fresh catalyst weight. In the following sections, descriptions of each deactivation stage are discussed.

4.5.1 Initial Catalyst Deactivation

Furimsky (1978) has stated that deposits of carbon are highly enriched with nitrogen, which could mean that nitrogen is a coke precursor. Several other authors have detected important contributions of sulfur and nitrogen products (or intermediate species) in the initial period of deactivation (Wiwel et al., 1991). When using a catalyst, polynuclear aromatics, such as asphaltenes, adsorb on the surface even with low or moderate acid sites.

4.5.1.1 Initial Coke Deposition

Being catalysts porous solids, different coke deposition modes can be carried out depending on the nature of the pores and operating conditions (Richardson et al., 1996; Muegge and Massoth, 1991). The modes can be distinguished in: (1) uniform surface deposition, (2) pore-mouth plugging, and (3) bulk phase deposition. In the case of uniform surface deposition, coke is homogeneously distributed in the inner surface of the pore, most probably on acid sites. When pore-mouth plugging occurs, coke blocks the access to the inner pores leaving a part of the surface area unused. Bulk phase deposition causes plugging of pores in the outer catalyst surface and, as a consequence, molecules cannot access inside the pores. Due to such phenomena, a major loss in specific surface area can occur, but the later two stages will be affected with time-on-stream (Figure 4.1), with coke and metal deposition around the pores. As an effect of catalyst, coke formation is highly dependent on the balance function of hydrogenation and cracking of asphaltenes especially in the later stages of coking. Richardson et al. (1996) stated that, in the initial coking using a NiMo/alumina catalyst, the coke layer is formed on the support surface as a result of cracking reactions, whereas the active sites, where hydrogenation is carried out, are not deactivated in that stage. In general, at the initial stage of coking, there are two possible mechanisms in which coke formation proceeds either by parallel coking or by a series coking mechanism as represented below:

$$\left.\begin{array}{l} \mathrm{A} \rightarrow \mathrm{B} \\ \mathrm{A} \rightarrow \mathrm{C(coke)} \end{array}\right\} \text{parallel coking}$$
$$\mathrm{A} \rightarrow \mathrm{B} \rightarrow C\mathrm{(coke)}\} \text{ in series coking} \tag{4.6}$$

Other coking mechanisms could explain coke formation when processing a heavy feedstock, such as vacuum residue, as reported by Del Bianco et al. (1993).

They suggested that coke formation could be due to the presence of an intermediate as shown:

$$\begin{array}{c} \text{Vacuum residue unconverted} \rightarrow \text{Distilliates} \\ \downarrow \\ \text{Intermediate} \rightarrow \text{Coke} \end{array} \tag{4.7}$$

However, in the model proposed by Del Bianco et al. (1993), gases are not included into the kinetics, where they have reported activation energy to coke formation of 63.9 kcal/mol. In another study, a four-lump kinetic model has been reported by Sawarkar et al. (2007), which takes into account the production of gases:

$$\begin{array}{ccc} & \text{Vacuum residue} & \\ \swarrow & \downarrow & \searrow \\ \text{Gases} & \text{Distillates} & \text{Coke} \end{array} \tag{4.8}$$

They have also reported that coke formation starts almost at the beginning of the reaction and proceeds very quickly at higher temperatures. The activation energy for coke formation was 62.65 kcal/mol, which is very similar to the one reported by Del Bianco et al. (1993). These high values of activation energy indicate that reactions involved in coke formation are relatively more important at higher temperatures. Sometimes the parallel or series mechanisms commonly used to explain coke deposition in fixed-bed reactors are not capable of representing the behavior of coke formation on catalysts during hydrotreating. For instance, S-shape profiles have been obtained by Amemiya et al. (2000) and, in this case, neither in series nor parallel models can explain adequately this trend in coke deposition.

Thakur and Thomas (1984) studies have reported that up to 20 to 30 wt% of coke is deposited on the surface of catalysts during the initial stage of deactivation. The rapid deposition of coke is achieved during the first hours of hydroprocessing and, at the same time, a loss of specific surface area of the catalyst occurs reaching, in some cases, 50 to 60 wt%. Other authors (Marafi and Stanislaus, 1997) have reported that rapid deposition of coke as high as 10 wt% is deposited onto the catalyst surface during the first 3 h of a run diminishing around 40% on the specific surface area of the catalyst. Coke is mainly deposited into narrow pores (<5 nm).

Coke deposition during the first hours of TOS up to 120 h has been investigated by Hauser et al. (2005). They used a Kuwaiti atmospheric residue as feedstock and observed that coke deposition reaches around 40% of the total amount during the first 12 h of TOS. The use of solvents, such as THF, to wash spent catalyst after different TOS allows for removing small aromatic compounds highly alkyl-substituted. The solubility or insolubility of some compounds in toluene and THF lets one better understand the constituents and nature of coke. At the beginning, H/C atomic ratio of initially deposited coke is very similar to the feedstock.

Rather, at the initial stage, the compounds deposited on the catalyst surface are coke precursors. These authors have proposed some structures based on characterization data and showed that coke deposited after 1 h of TOS is a less condensed structure compared with asphaltenes. However, after 12 h, the number of methyl groups has reduced significantly. S and N are present in coke structures and the degree of condensation is almost the same even up to 120 h of TOS, whereas the number of tertiary carbons increases indicating some rearrangement of the structure. According to the changes in structure from characterization data, different structures were proposed to represent the changes of coke during hydrotreating in the early stages of TOS (Figure 4.21).

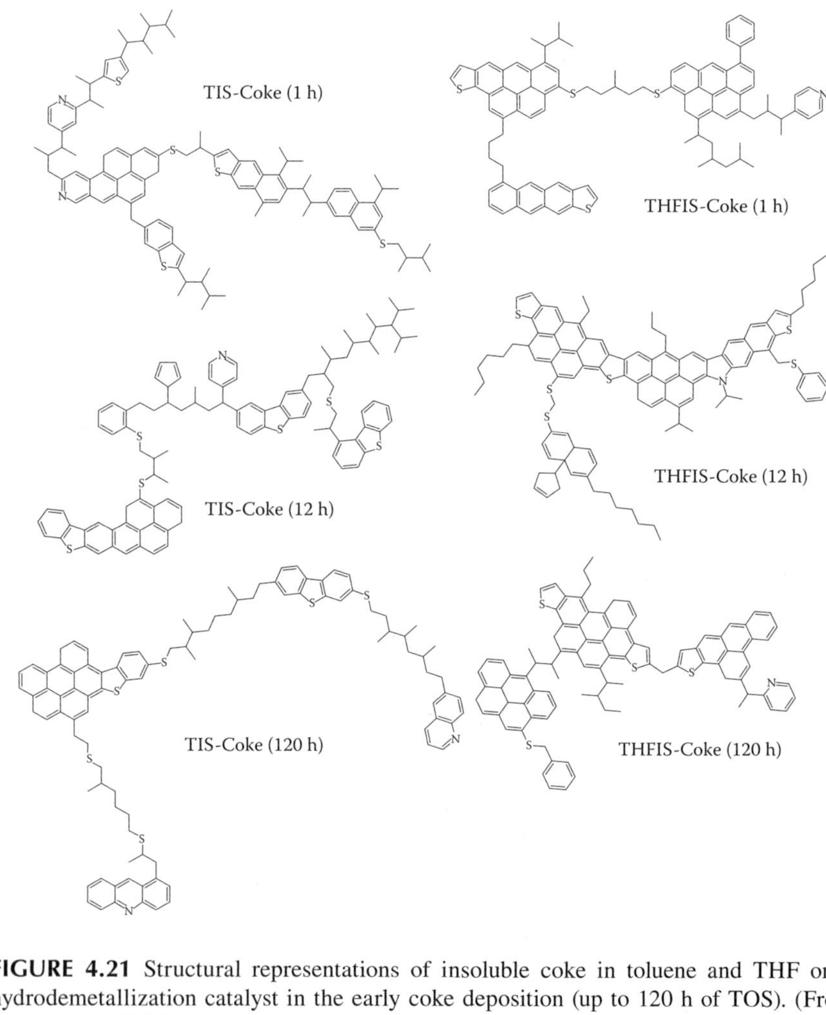

FIGURE 4.21 Structural representations of insoluble coke in toluene and THF on a hydrodemetallization catalyst in the early coke deposition (up to 120 h of TOS). (From Hauser et al. 2005. *Fuel* 84 (2–3): 259–269. With permission.)

It can be observed that after 1 h of TOS, insoluble coke in toluene (TIS-coke) remains highly branched compared with insoluble coke in THF (THFIS-coke). However, after 12 h of TOS, toluene insoluble coke is less branched. THFIS-coke after 12 h is a more compact structure even after some long ramification. TIS-coke after 120 h has the aspect of an archipelago structure with long alkyl chains. On the other hand, THFIS-coke is a more compact and highly condensed structure with a more aromatic core.

4.5.1.2 Initial Metals Deposition

Furimsky (1998) and Beaton and Bertolacini (1991) stated that with a low metals content feedstock, the catalyst will suffer of deactivation initially by coke and, subsequently, by metals deposition, by increasing metals content feedstock deactivation will be stronger in the second stage. Deactivation continues up to where the catalyst is inactive and needs to be substituted or it will regenerate.

Metal deposits build up monotonically as the metals sulfides (V_3S_4, Ni_xS_y, etc.) monolayer covers the support as well as the existing catalytic sites (Mo and Ni or Co). An estimation of the amount of vanadium sulfur deposited as the monolayer in a catalyst having a surface area of 200 $m^2/mL_{catalyst}$ is around 0.076 to 0.176 g $V/mL_{catalyst}$ (Tamm et al., 1981).

4.5.2 Steady State Deactivation

It is observed that a steady state period follows after the first deactivation period, in which coke deposition appears to be constant. However, this is only apparent because it is the result of some compensating factors (Tamm et al., 1981). As metals deposit on the catalyst, there is an increase in the catalyt density and for this reason it appears that the percentage of coke remains constant even as coke amount indeed increases. At this stage, smaller pores (micropores) are likely to reduce or be blocked during the initial deactivation, and acidic sites are not available to adsorb further.

4.5.2.1 Intermediate Coke Deposition

Once the initial coke deposition has taken place, a period of relative steady state deactivation comes over the catalyst. In this stage, the build up of coke is steady and it could be one of the most important reasons for catalyst deactivation, as reported by Tamm et al. (1981). When characterization of catalysts is carried out, it is seen that there is a gradual decreasing in the catalyst surface area and pore volume as the amount of coke increases affecting mainly the smaller pores. After micropores or small mesopores are affected, larger pores tend to undergo pore blockage. When using fixed-bed reactors for processing heavy crudes, coke deposition can be well represented in most of the cases by series or parallel mechanisms, but different patterns are obtained depending on whether or not diffusion limitations are taking place. According to the series model, coke deposition increases as the depth of catalytic bed increases and higher deposition will occur inside the pores (Koyama et al.,

1996). On the other hand, the parallel model establishes that the amount of coke deposited on the catalyst diminishes with distance from the reactor inlet, as reported by Chang et al. (1982) and coke deposition will be more remarkable at the catalyst's outer surface. But when strong diffusion limitations occur, then most of the coke is deposited on the outer surface of the catalyst. Zhang (2006) summarized these typical profiles of coke deposition and stated that fast deactivation due to coke deposition is more prominent during the first hours of TOS. However, as aforementioned, coke continues to be deposited on the catalyst surface. Ancheyta et al. (2003) have reported that during hydroprocessing of Maya crude, deposited coke after 490 h of TOS was 18.3 wt%, whereas at 1,120 h of TOS deposited coke was almost the same (18.5 wt%) in spite of increasing temperatures. At the same time, metals deposition was more notorious at this stage and 25 wt% of metals were deposited on the catalyst (fresh basis).

4.5.2.2 Intermediate Metals Deposition (HDM Reaction Mechanism)

In the steady state stage, metals deposition takes place with different patterns for Ni and V. In the case of vanadium, its deposition is in the exterior of the pellet whereas nickel deposition is located inside the pellet (Thakur and Thomas, 1984). It is accepted that intraparticle diffusion is present when hydroprocessing residue has a significant amount of asphaltenes and can play an important role during reaction (Adkins and Limmer, 1990). Removal of Ni and V present in residue is diffusion-limited. Ni and V are deposited on the catalyst following a specific pattern in which maximum metal deposition can be present on the surface or inside of the pellet catalyst (Tamm et al., 1981). Bartholdy and Hannerup (1990) concluded that metals removal occurs as a sequential reaction and metalo-porphyrins can react with H_2S to form metal sulfides on the catalyst. The reaction is essentially first-order and follows the network:

$$A \xrightarrow{k_1} B \xrightarrow{k_2} C \tag{4.9}$$

The first step corresponds to hydrogenation of the metal-bearing porphyrins, which turns them into hydrogenated metal porphyrins. The second step is the reaction between hydrogenated porphyrins and H_2S to form a deposited metal sulfide on the catalyst surface. k_1 depends on the partial pressure of hydrogen (because the first step is a hydrogenation step) and k_2 depends on the partial pressure of H_2S. In the case of Ni, it has been suggested by using model compounds such as Ni-etioporphyrin, Ni-tetraphenylporphyrin, and Ni-tetra(3-methylphenyl) porphyrin that a partially hydrogenated intermediate is involved in the reaction, as reported by Ware and Wei (1985a, 1985b). The porphyrins react via a sequential mechanism, first involving hydrogenation of peripheral double bonds, followed by hydrogenolysis step, which fragments the ring and removes the metal. The enhanced metal-deposition activity of the sulfided catalyst compared with the oxide catalyst was steeper with more U-shape profiles and less deposits in the center of the pellets. In general, the reaction of Ni porphyrin can be represented as:

$$Ni\text{-}R + H_2 \rightarrow Ni\text{-}RH + H_2 \rightarrow Ni\ deposit\ (Ni_xS_y) \tag{4.10}$$

The pattern of deposition of Ni is not the same along the reactor. At the end of the reactor, Ni deposition occurs closer to the edge of the pellet. (Trimm, 1996). Two-step reaction mechanism is involved in the case of vanadium porphyrins where in the first step an addition of a proton to the metal–prophyrin complex is carried out and in the second stage hydrogenation of the ring occurs. Removal of vanadium has been already established as a sequential reaction (Bartholdy and Hannerup, 1991). The evolution of H_2S during hydrotreating inhibits mainly the HDS reaction, but it has been reported that, at the same time, elimination of vanadium is favored. Even removal of metals is carried out at high H_2 and H_2S partial pressures without a catalyst (Bonné et al., 2001). Dautzenberg and de Deken (1985) have stated that sulfur can be strongly coordinated to metals like vanadium and nickel. They have also suggested that H_2S coordinates to metals in porphyrins and weakens the covalent metal–nitrogen bonds in the molecule. Porphyrins are prone to successive hydrogenation and the final step is the ring fragmentation. It is not clear where the metals in porphyrins go after demetallization without using a catalyst; however, in the presence of a catalyst, metals are deposited as metal sulfides on the catalyst surface.

Recently, Rana et al. (2007a) have proposed that the presence of a catalyst enhances the final step of hydrodemetallization (HDM) of the metal–prophyrin molecule and carries out the hydrogenolysis of metal–nitrogen. Hydrogenation and hydrogenolysis take place in different catalytic sites favored by adsorption of H_2S (saturated sites, -SH group), which replaces the vacancies with increasing the H_2S partial pressure. The first step is the adsorption of the metal–porphyrin ring to saturated sites with the consequent hydrogenation of the porphyrin leading to a chlorine metal structure and, finally, the demetallization is accomplished. Because these studies have been carried out with a real feedstock, several other parameters are also affecting the reaction mechanism. Moreover, one should consider that presumption of the HDM mechanism is not so easy because there are several parameters that need to be established, such as optimum pore diameter of the catalysts and the nature of the feedstock, which may vary the diffusion limitation and, thus, the rate of reaction.

Toulhoat and Plumail (1990) have reported that removal of vanadium and asphaltenes has a limit when increasing the pore diameter of a catalyst. Mass transfer limits diffusion of asphaltenes and vanadium complexes inside the pore in catalysts having a pore diameter of around 20 nm. Determination of asphaltene size is a hard task and, as reported in the Chapter 1, size exclusion chromatography (SEC) analysis gives an indication about the size of asphaltenes. In this regard, Sughrue et al. (1991) used this technique to quantify the size of vanadium associated in complex molecules and determined a Gaussian distribution of molecular size of vanadium complexes to be in the range of 8 to 10 nm where most of the molecules are located. It is possible now to compare the size of vanadium compounds with diffusion limitations in small pores and it is easy to infer why vanadium is mainly deposited onto the catalyst surface instead of going deeper inside the pores (Trimm, 1996). Bridge (1990) has also concluded that HDM is diffusion-limited, but HDS is not. In addition, the maximum deposition profile of metals is near the external surface when the pore diameter of the catalyst is diminished. Thus, pore blockage is continuously increasing due to coke and metals deposition. Original pores in the fresh catalyst suffer changes after deactivation, as reported by Ancheyta et al. (2002). It was found that pores having a diameter of 50 nm or higher are partially covered by poisons contributing to increasing the percentage of mesopores in the range of 25 to 50 nm. Pores located in this region also experience blockage and can form smaller pores. Pores having diameters of less than 10 nm are the result of blockage of mesopores in the region of 10 to 25 nm. A schematic view based on this observation is seen in Figure 4.22.

Although catalyst deactivation is taking place due to metals and coke deposition, the catalyst exhibits subsequent activity. This catalytic behavior can be

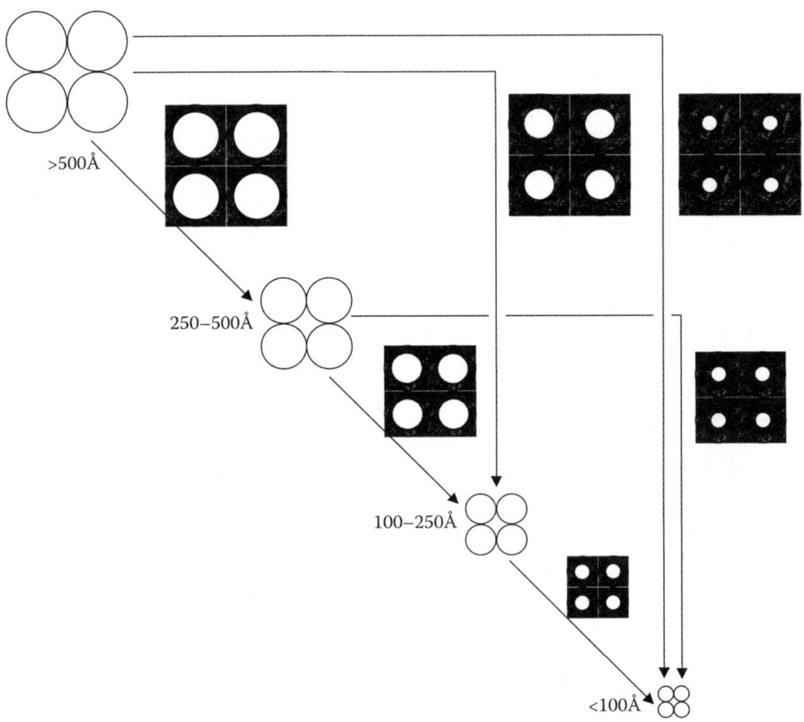

FIGURE 4.22 Possible blockage mechanism of pores during hydroprocessing where void circles represent unblocked pores and black crowns indicate covered pores. (From Ancheyta et al. 2002. *Energy Fuels* 16 (6): 1438–1443. With permission.)

related to the migration of molybdenum as sulfide to the catalyst surface, which represents around 10% of the total molybdenum content (Simpson, 1991) and, along with vanadium, nickel sulfides on the surface are responsible for catalytic activity. V_3S_4 and Ni_3S_2 or Ni_2S have been detected by Fleisch et al. (1984) by means of x-ray photoelectron spectroscopy (XPS) and electron microprobe extrudated cross-section analyses in the catalyst surface. Once no further catalytic activity is developed, the ultimate catalyst deactivation occurs and the unit shutdown is necessary.

4.5.3 Final Catalyst Deactivation

The last stage of deactivation is due to severe pore plugging and diffusion limitations. At this stage, the only option is the replacement of the catalyst or unloading the catalyst because most of the catalytic sites will be buried under the deposited metal and/or carbon. The amount of metals and coke deposited is a function of a combined effect of catalyst properties, such as pore size distribution, activity, acidity, and operating conditions (Morel et al., 1997).

The final result of severe deactivation is the end of the catalytic activity. Deposited coke possesses different properties depending on the type of catalysts, i.e., coke on NiW spent catalysts was more aromatic than that of the CoMo catalyst (Yoshimura et al., 1994). Hauser et al. (2005) have proposed some structures of deposited coke at the last stage of deactivation. After 6,500 h of TOS, coke shows a compact and highly condensed structure. The structure of coke consists of a number of aromatic rings in which nitrogen and sulfur appear as part of coke in the form of condensed low hydrogen containing rings. As reported earlier, there are several factors that are responsible for carbon deposition, and a principal one is pore diameter, which stands in linear relationship between the average pore diameter (APD) and coke deposition, as shown in Figure 4.23.

On the other hand, metal deposition estimation of the deposited vanadium sulfide can be obtained by using an HDM conversion as a function of time, which also represents a linear relationship with TOS. This deposit obviously diminished the pore diameter of the catalyst and showed increasing tendency with increasing average pore diameter. Attempts have been made in order to study the deactivation models (semiquantitative) based on a pore plugging mechanism (Rana et al., 2005a). For deactivation studies, different kinetic models have been reported, for instance, first-order has been used to fit demetallization reactions (Dautzenberg et al., 1978; Rajagopalan and Luss, 1979), while second-order kinetics has been employed as well, as reported by Nitta et al. (1979).

At the end of the run, very high amounts of metals and carbon are expected to be deposited on the surface of the catalyst. This amount depends on the nature of feedstock and reaction conditions (Absi-Halabi et al., 1991). According to Tamm et al. (1981), nickel is distributed more evenly throughout the pellet and vanadium and iron are deposited near the surface. The removal of nickel and vanadium is diffusion controlled, whereas the iron removal is reaction

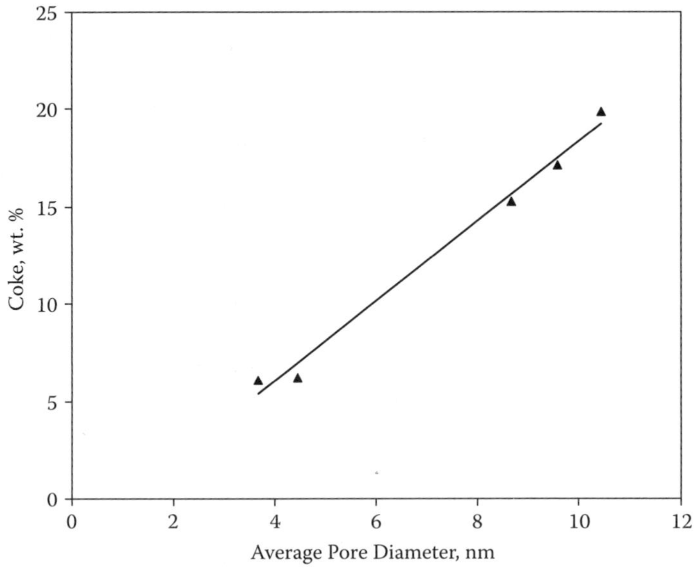

FIGURE 4.23 Relationship between average pore diameter of fresh catalysts and coke formation (60 h, TOS) during hydrotreating of Maya crude diluted with naphtha. (From Maity et al. 2003a. *Appl. Catal. A* 244 (1): 141–153. With permission.)

controlled. Furimsky (1978) showed that the amount of Fe compared with V and Ni is small; however, during catalytic hydrotreating, it is likely that Fe is converted to sulfide, which is catalytically active. In a fixed-bed reactor, after the catalyst has been exposed to the long cycle length, patterns of metal deposition observed in each section of the bed are different. Simpson (1991) used electron microprobe line scans applied to a demetallization catalyst from Unocal® to show the profile of vanadium, nickel, and iron deposition. Both metals and sulfur deposit in large amounts at the top of the bed while decreasing at the bottom. The apparent stoichiometric molar ratio of metal to sulfur varied from 1:1 with an average formula of MS (metal sulfide) at the top to 1:2 at the bottom of the bed with an average formula of MS_2, where M is the metal in the following ratios: 2/3 V + 1/6 Ni + 1/6 Fe. Usually Ni and V deposits are found in every section of catalytic bed; however, Fe deposit is found at the top and middle sections of the bed. The main precursor of these metals and coke is asphaltenes, which is available in the crude oil in abundance, particularly in heavy or extra heavy crude oils, as shown in Figure 4.24. Thus, its concentration in feedstock provokes more coke and metal deposition on the catalyst, seen in Figure 4.25.

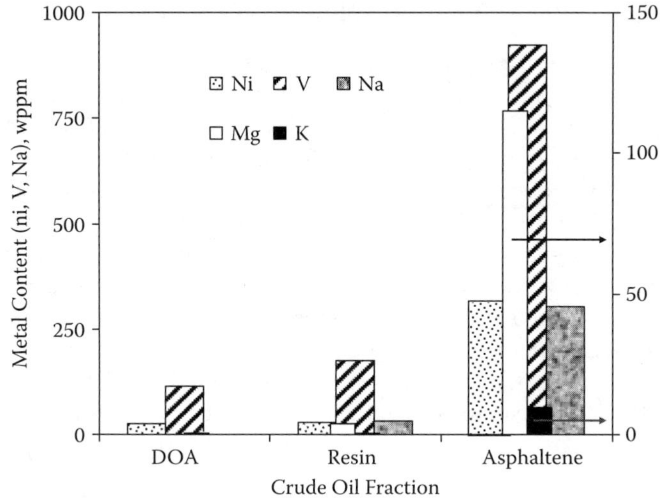

FIGURE 4.24 Fractions of heavy crude oil (API gravity = 13) and their metal contaminants distribution. (From Rana, unpublished results.)

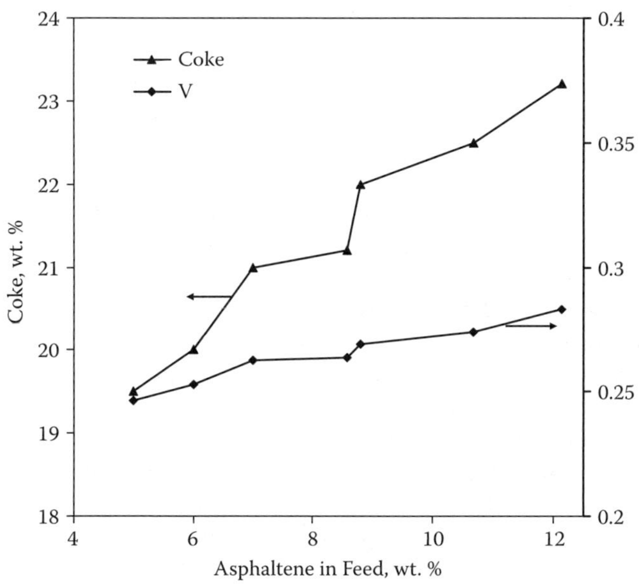

FIGURE 4.25 Effect of asphaltenes content on metal and coke deposition. (From Maity et al. 2007. Petroleum Science and Technology 25(1): 241–249. With permission.)

4.6 CHARACTERIZATION OF SPENT CATALYSTS

Previously, to characterize the spent catalysts after crude oil hydrotreating, it was necessary to wash them by using toluene refluxing in Soxhlet in order to remove the physically adsorbed hydrocarbon on the catalyst. Saito and Kumata (1998) used Soxhlet washing of spent catalysts with toluene followed by LGO (light gas oil). The use of LGO allows for removing of some carbonaceous compounds (25 to 45%) deposited on the catalyst surface. During this treatment, the coke scrub-down is known as *soft coke* while the remaining coke is called *hard coke*. It was found that hard coke was mainly deposited in larger pores. Soft coke can be considered as a temporal poison that exists on the catalyst surface. Thus, the removal of light carbonaceous compounds by LGO reflux permits the characterization of spent catalysts in a better way. There are different analyses that can be applied to characterize spent catalyst. Some of the main techniques are described in the following sections.

4.6.1 Temperature Programmed Oxidation (TPO)

Characterization of spent catalyst by TPO is used to obtain information for oxidation of coke at different temperatures along with the oxidation of deposited S and N species that are present in coke. Because sulfur can be released from the metal sulfides (Mo, Co, Ni, V, Fe, etc.), it is difficult to quantify the total sulfur obtained as SO_2. Al-Dalama and Stanislaus (2006) carried out TPO analyses for CO_2, SO_2, and NO_2 on spent catalysts obtained from different beds in a fixed-bed reactor. In the case of TPO for CO_2 analysis, it was observed that in the first reactors, whose catalysts are more poisoned with vanadium, the peak for CO_2 ranged from 200 to 450°C. On the other hand, in the last reactors whose catalysts have less vanadium deposited, the peak for CO_2 began around 250°C and it was extended up to 550°C. Spent catalysts from the first reactors exhibited a lower temperature range for coke oxidation, which can be due to the presence of vanadium that may catalyze the oxidation (Massoth, 1981; Zeuthen et al., 1995). The TPO profiles for SO_2 exhibited three peaks. The first peak between 200 and 250°C has been suggested to correspond to the oxidation of metal sulfides (Yoshimura et al., 1991). The second peak is broader and ranged from 300 to 500°C, which may correspond to CO_2 profiles and can be due to sulfur in coke that is oxidized along with carbon (Al-Dalama and Stanislaus, 2006). The third peak is located in the interval of 600 to 700°C and could be assigned to decomposition of sulfates to SO_2 (Clark et al., 1993). TPO of NO_2 profiles showed three peaks according to Al-Dalama and Stanislaus (2006). The first peak coincides with that of the SO_2 and could be due to the adsorption of basic nitrogen in the sulfide phase by which nitrogen species could be oxidized along with sulfur species. The second peak coincides with CO_2 at around 380°C. This could imply a relationship between C and N in coke. Some authors have reported the presence of N in coke (Speight, 1987). The third peak around 700°C is not fully identified; however, it could be due to

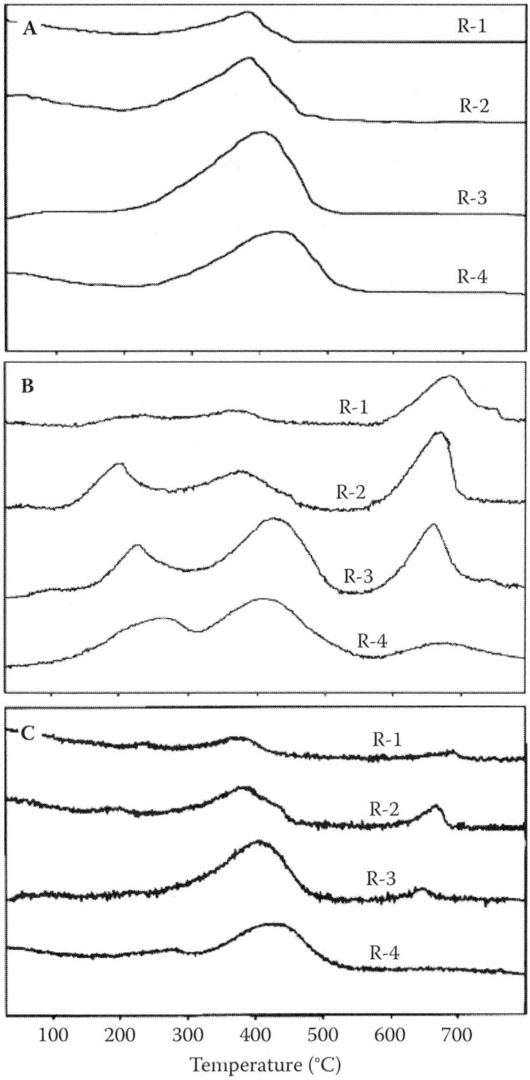

FIGURE 4.26 TPO profiles for: (a) CO_2, (b) SO_2, (c) NO_2 of spent catalysts in a series of four reactors (R-1 to R-4). (From Al-Dalama, K. and Stanislaus, A. 2006. *Chem. Eng. J.* 120 (1–2): 33–42. With permission.)

the presence of metal nitrates along with metal sulfates. Figure 4.26 shows typical TPO profiles obtained by Al-Dalama and Stanislaus (2006). In order to perceive soft coke and hard coke by TPO, soft coke burns at lower temperature (low temperature CO_2 peak) while hard coke at higher temperature (high temperature CO_2 peak), particularly coke formed due to the asphaltenes precipitation or due to sediment formation.

4.6.2 Nuclear Magnetic Resonance (NMR)

Generally, coke has a high content of carbon atoms and very few hydrogen atoms. In order to identify the nature of carbon and its surrounding atoms connectivity, ^{13}C NMR is the best technique to determine its structure. Usually coke deposition increases along the reaction and its nature also becomes more aromatic at the end of reaction, which has less alkyl chains linked. Structural parameters can be identified easily by using the well-established techniques of high power decoupling, magic angle spinning (MAS) to remove chemical shift anisotropy (approximately 100 ppm for aromatic carbons), and cross-polarization (CP) in which magnetization is transferred from abundant ^{1}H to dilute ^{13}C spins to improve sensitivity and thus avoid long relaxation delays. By using solid state ^{13}C NMR, it has been reported that coke represents different types of structures (tertiary and quaternary carbons, other carbon types present), carbon groups (aliphatic or aromatic), which are usually affected by the nature of feedstock, catalyst type, and reaction conditions (Al-Dalama and Stanislaus, 2006). Egiebor et al. (1989) have characterized organic residues on spent hydroprocessing catalysts by NMR and it was confirmed that coke contained a significant amount of aliphatic, hydroaromatic, and aromatic compounds. Snape et al. (1999) have defined the existence of two kinds of coke depending on the percent of deposited carbon on the catalyst and type of solvent used to extract coke. The "soft" coke is chloroform-extractable and accounts for about 30% of the total carbonaceous deposit on the catalyst being highly polar with significant aliphatic nature. "Hard" coke is not solvent-extractable (5 to 7 wt% of carbon content) and it has an aromaticity around 50%. When longer periods of deactivation by coke are present, carbon content is around 15 wt% or higher and it seems to be very aromatic (90% or even more). ^{13}C NMR analysis indicated that "hard" coke is highly aromatic and "soft" coke has an aliphatic peak and, for these reasons, this kind of coke is chloroform-extractable. For deep analysis of the hard coke, different experiments have been carried out in order to identify clearly the contribution of different groups, such as ^{13}C SPE/MAS (single pulse excitation/magic angle spinning) and CP/MAS (cross-polarization/magic angle spinning). NMR experiments were employed to quantitatively measure insoluble coke deposits, as can be seen in Figure 4.27 (Sahoo et al., 2004). The authors have also tried to determine coke composition from MAS NMR along with other analytical techniques (soluble and insoluble components) that are extracted from spent hydrocracking catalyst. It was also reported that soft coke is related more to the feedstock composition, whereas the insoluble coke or hard coke formation may depend on the nature of the catalyst and its functional reactions.

4.6.3 Raman Spectrometry

Characterization of spent catalysts by Raman spectrometry has been used by several authors in order to explain coke structure on supported catalysts (Digne et al., 2007; Amemiya et al., 2003; Chua and Stair, 2003; Espinat et al., 1985).

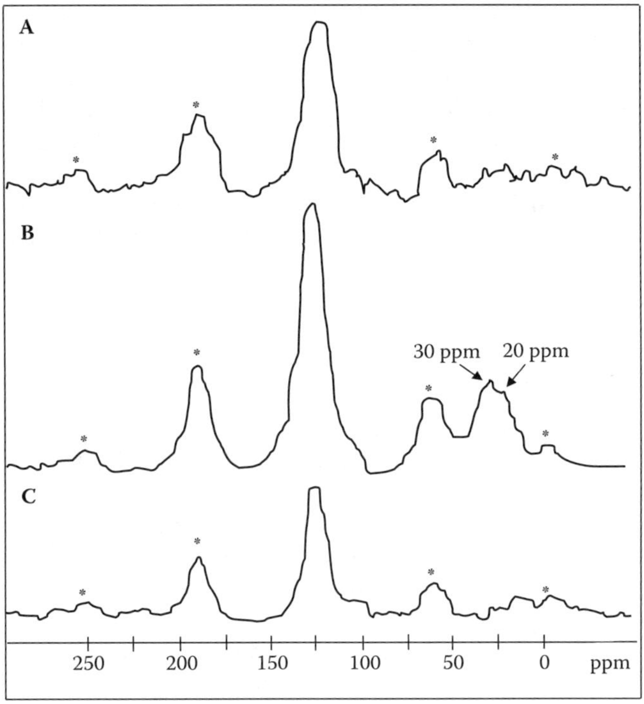

FIGURE 4.27 Solid state ^{13}CCP MSA NMR spectra of hard coke (insoluble coke) of the deposits on spent hydrocracking catalysts (Cat-HY); (a) SPE/MAS, (b) CP/MAS, and (c) CP–DD/MAS 13C NMR spectra. (From Sahoo et al. 2004. *Appl. Catal. A* 278 (1): 6–91. With permission.)

Classical characteristic bands of MoS_2 phase at 384 and 406 cm-1 were identified, which were also reported by Brown et al. (1977). Apart from MoS_2 two more peaks were detected, one of them between 1,550 and 1,650 cm^{-1} and the second one around 1,250 and 1,450 cm^{-1}. The coke spectrum appearance has similarities with graphite spectrum, which exhibits two peaks at 1,585 cm^{-1} (G mode) and 1,355 cm^{-1} (D mode) that correspond to the ring stretching of polynuclear aromatic rings. However, when using different feedstocks, such as gas oil, vacuum distillate, and vacuum gas oil over the same catalyst in a pilot plant test, supplementary differences were found. The signal at around 1,600 to 1,605 cm^{-1} remained almost without changes, but when vacuum distillate was used, two peaks at 1,585 and 1,618 cm^{-1} appeared that indicate the presence of intermediate species. The peak around 1,250 and 1,450 cm^{-1} presents some displacements and, in the case of using vacuum distillate and vacuum gas oil, it is broader. Amemiya et al. (2003) have also reported the presence of peaks at 1,600 and 1,350 cm^{-1} when characterizing spent catalyst for vacuum gas oil hydrotreating operating at 8 MPa and 360 to 400°C for one year in a fixed-bed reactor. These two peaks are stronger as the depth of catalytic bed increases. Carbon deposition is slightly more remarkable on the edge and fewer on the center of the catalyst particle. Both signals at 1,600 and 1,350 cm^{-1} showed this trend, i.e., higher intensity near the edge, intermediate intensity in the middle, and lower intensity in the center of the catalyst particle. Calcination of spent catalyst increased the intensity of the bands mainly in the catalyst located at the top of the catalytic bed. The catalyst located in the lower section of the catalytic bed exhibited partially graphitized carbon. Interpretation of Raman spectra of coke is not easy; however, this technique is useful for characterizing not only fresh catalysts, but also deposited coke.

4.6.4 SEM-EDX Analysis

The traces of deposited species can be identified by using different characterization techniques, such as energy dispersive XRD. With this technique, different metals, such as Ni, V, Fe, and Na deposited on the catalyst surface have been detected (Figure 4.28). Because most of the metalloporhyrins are associated with the asphaltene molecules, it is likely that these complex molecules of asphaltene do not diffuse into the catalyst pore. Consequently, their conversion took place at the entrance of the pore mouth due to pore diffusion limitation. Thus, normally the distribution of vanadium and carbon deposition as a function of catalyst particle (extrudate) adopts a U-shape pattern with higher deposition near the edge of the particle, as seen in Figure 4.29. Usually, heavy oil catalysts are loaded as graded bed approach in which the catalyst loaded at top bed that has

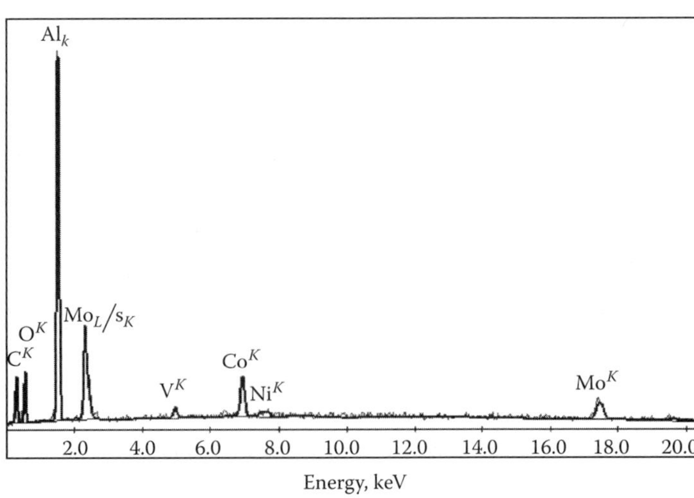

FIGURE 4.28 Elemental composition of spent catalyst ($CoMo/Al_2O_3$) using SEM-EDX analysis. (From Rana et al. 2007c. *Petrol. Sci. Tech.* 25 (1–2): 187–200. With permission.)

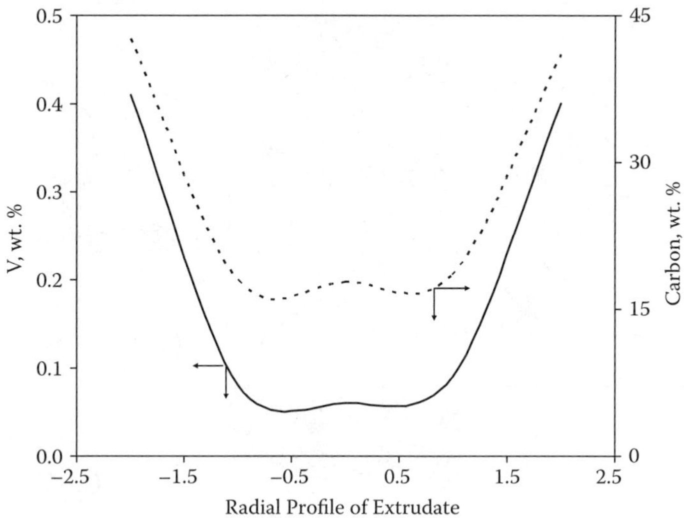

FIGURE 4.29 Deposited V and carbon concentrations profiles obtained in radial distribution of cylindrical extrudate. (From Rana et al. 2007c. *Petrol. Sci. Tech.* 25 (1–2): 187–200. With permission.)

big pores for high metal retention and its activity is low. Metals are transported deeper inside the pores and their deposition is homogeneous across the pellet. On the other hand, catalysts employed in the last reactors are more active having smaller pores by which the diffusion rate will be low. Molecules have a tendency to react as they enter inside the pores and metal deposition will occur near the pore mouths. As a result, nonhomogeneous deposition of vanadium will be obtained in the pellets of the last reactors having maximum concentrations in the edges.

Ancheyta et al. (2003) have performed the hydrotreating of Maya crude in a fixed-bed reactor with using a $NiMo/Al_2O_3$ tetra-lobe commercial catalyst and found (by SEM analysis) that deposition profiles of V along the longest and the shortest distance through the lobes were almost similar. TOS was 1,120 h, and temperature was increased (400°, 420°, and 430°C) for compensating catalyst deactivation. The deposition of V metal as a function of extrudated radial distribution zones, which shows an enrichment of V in the superficial region of the catalyst particle and its radial analyses, is seen in Figure 4.30. These patterns differ from those obtained at lower TOS (490 h) and at a constant temperature (400°C) reported elsewhere (Ancheyta et al., 2002), where the longest distance through the lobes presented more deposited vanadium. Differences between both cases are due to the TOS and the increase of temperature. Qualitative deposition of Ni and V on spent catalysts also can be carried out by scanning transmission electron microscopy (STEM). Trejo et al. (2008) have shown that the distribution of Ni

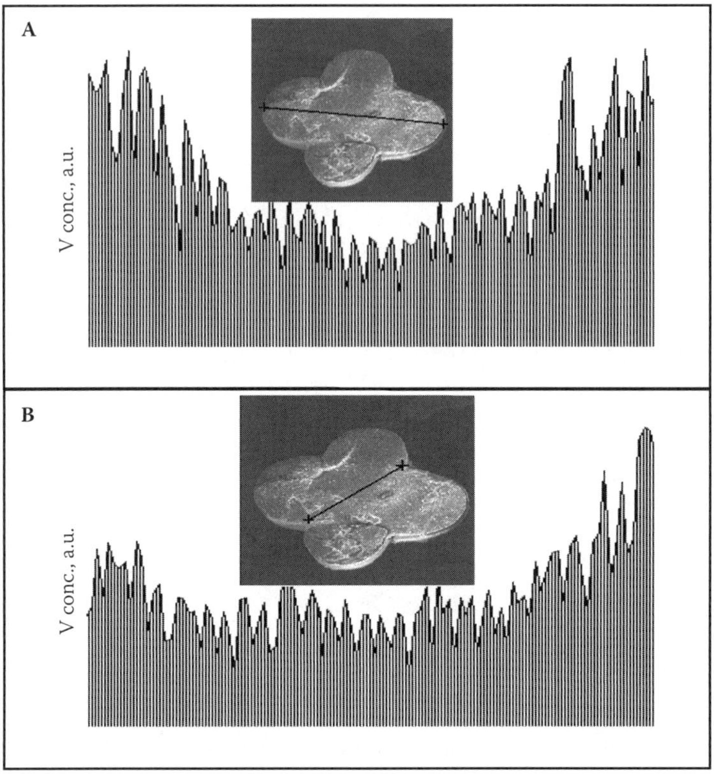

FIGURE 4.30 Intraparticle SEM-EDAX analysis of V profiles in tetra lobe (A) longest and (B) shortest extremes of spent commercial catalyst. (From Ancheyta et al. 2003. *Energy Fuels* 17 (2): 462–467. With permission; Ancheyta et al. 2002. *Energy Fuels* 16 (6): 1438–1443. With permission.)

and V over CoMo/Al_2O_3-MgO spent catalysts after 204 h of TOS exhibited a homogeneous deposition of metals on the catalyst particle. However, the catalyst with larger pore diameter (CoMo/Al-Mg-0.1) had a higher amount of metals due to its wider pores. On the other hand, the smaller pore diameter catalyst has more coke deposition (Figure 4.31). The less coke formation could be due to the higher MgO content in the catalyst that may prevent coke formation due to its basic nature. The distribution of vanadium is in the outer surface of the catalyst while Ni distribution is more uniform throughout the catalyst particle (Figure 4.32). Diffusion of nickel porphyrin reactant molecule inside the pores is higher and for this reason nickel is uniformly deposited across the pellet. It has been established that vanadium is more reactive than nickel due to the presence of a double bond in vanadium, which makes it more prone to react (Quann et al., 1988). Some authors have reported that reactivity of vanadium compounds is three to five times higher than reactivity of nickel. However, the effective diffusivity of nickel is 10 times higher than vanadium (Sato et al., 1971).

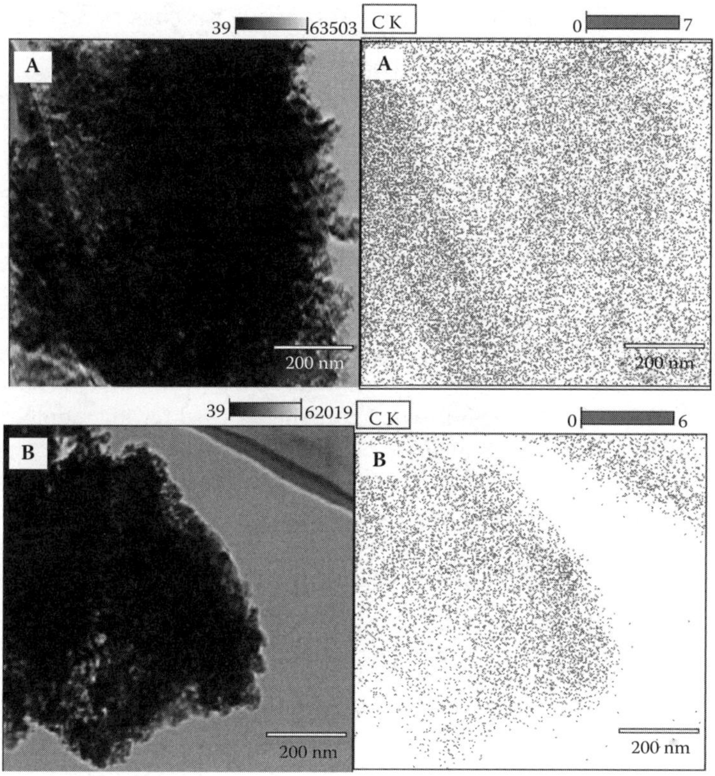

FIGURE 4.31 Scanning transmission electron micrographs (STEM) of carbon (CK) distribution over two different catalysts (CoMo/MgO-Al_2O_3) having different MgO content in the support: (A) 1.5 wt% MgO (7.4 nm, APD) and (B) 9.4 wt% MgO (9.5 nm, APD). (From Trejo et al., *Catalysis Today* 130 (2–4): 327–336. With permission.)

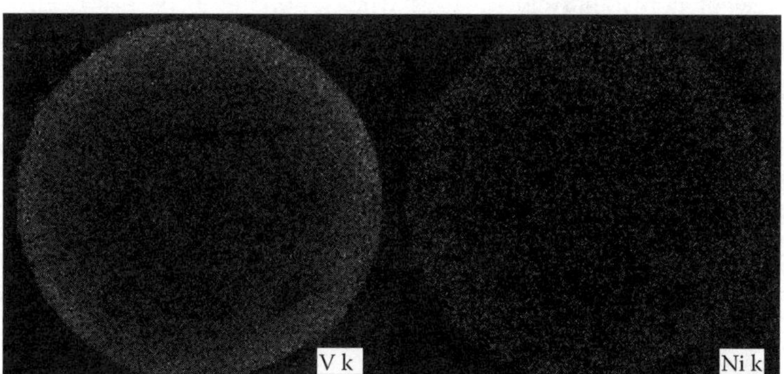

FIGURE 4.32 Radial distribution of vanadium (VKα) and nickel (NiKα) along the extrudate (CoMo/Zeolite-Alumina) supported catalyst. (From Rana et al., *Catalysis Today* 130 (2–4): 411–420. With permission.)

4.6.5 Thermogravimetric Analysis (TGA) and Kinetics of Coking

Thermogravimetric analyses of coke not only identify the type of coke, but also are a useful tool to estimate kinetic parameters, which can be obtained from the pyrolysis of asphaltenes. Reaction order in coking is frequently assumed to be one. A number of kinetic studies have been reported using TGA data during the coking reactions (Cotte and Calderon, 1981; Schucker, 1983; Jiménez-Mateos et al., 1996; Dalla-Lana et al., 1997; Li and Yue, 2003; Reynolds et al., 1995; Coats and Redfern, 1964.). This nonisothermal technique is helpful because a short time is required and problems due to the initial heat-up period that accompanies the isothermal experiments are avoided (Shih and Sohn, 1980). The influence of asphaltenes in coke formation in different atmospheric residues was analyzed by Gonçalves et al. (2007). They observed that not only are asphaltenes responsible for coke formation when thermal cracking is used for upgrading residues, but other heavy components also are coke precursors, such as polar aromatics. In one test, atmospheric residue was thermally cracked and the amount of coke was measured. In another test, atmospheric residue without having asphaltenes, which were precipitated from residue by using the IP-143 standard method, was also cracked and the amount of coke was compared with that obtained when asphaltenes were present in residue. In spite of removing asphaltenes by precipitation, a small amount of coke was formed in each case. When SARA fractionation and TGA are combined, it is possible to establish the propensity of each fraction to form coke, as reported by Jiménez-Mateos et al. (1996). They fractionated different heavy products into saturates, aromatics, resins, and asphaltenes and each fraction along with the entire bitumen were analyzed by TGA to show quantitatively that asphaltenes are responsible for 50 wt% of carbonaceous residue, at least during pyrolysis, followed by resins, which are capable of forming up to 20 wt%. The whole bitumen, as such, is responsible for around 20% of coke during pyrolysis, whereas the smallest fractions prone to form coke were aromatics and saturates. In all cases, the loss of weight was almost constant beyond 500°C.

The use of nonisothermal kinetics when applying TGA has been reviewed at great extent by Flynn and Wall (1966). The variation of heating rate was also discussed and reliable measurements were obtained. In this case, the method was applied to polymers, but the mathematical treatment is not exclusive to these materials and thermogravimetry of heavy fractions of petroleum and residue can be analyzed in the same way. Different methods of analysis have been reported, which include:

- Integral method
- Differential method
- Difference-differential method
- Initial reaction rates
- Nonlinear or cyclic heating rate methods

Trejo et al. (2009) used a nonisothermal technique with various heating rates applied to the determination of kinetics of coking of asphaltenes from a 13°API crude, which were purified under Soxhlet reflux with n-C_7 overnight in order to remove adsorbed resins. Approximately 10 mg of purified sample was analyzed at four different heating rates, i.e., 4, 8, 12, 16°C/min, ranging from room temperature up to 900°C with nitrogen flow of 50 mL/min. TGA data were analyzed by using Friedman's procedure (Friedman, 1964) and kinetics of coking was well-fitted to first order. It was assumed that asphaltenes are decomposed thermally according to the equation (Schucker, 1983):

$$A \rightarrow aC + (1-a)V \tag{4.11}$$

where A is asphaltene, C is the coke product, V is the volatile fraction, and a is the stoichiometric coefficient. Using this mechanism, it is possible to have a kinetic expression for volatiles formation according to:

$$\frac{1}{Vo}\frac{dv}{dt} = Ae^{-E/RT}\left[1 - \frac{V}{Vo}\right] \tag{4.12}$$

where Vo is the total amount of volatilized material. Taking the logarithm of both sides and equating $V/Vo=x$, the following expression is obtained, as also reported by Li and Yue (2003):

$$\ln\frac{dx}{dt} = \ln[A(1-x)] - \frac{E}{RT} \tag{4.13}$$

By applying a linear regression to this equation, a series of activation energies (E) and preexponential factors (A) are obtained as function of volatilized asphaltenes (x). The left side of the equation is the rate of change of weight and, as such, was directly obtained from the time derivative on the TGA. Analyzed asphaltenes are reponsible for around 47 wt% of coke generated when suffering thermal decomposition in this case. Thermograms for asphaltenes are shown in Figure 4.33 at four different heating rates. It is observed that a more prominent weight loss is obtained at higher velocities. According to pyrolysis, the obtained results can be divided into three stages when analyzing asphaltenes. The first stage corresponds to a minimal loss of weight indicating that chemical reactions are negligible at this point where temperature increases from ambient to around 350°C; the second stage corresponds to an increase of temperature from approximately 350 to 450°C. In this stage, there is a loss of weight due to volatiles; however, evolved gas is not high. The reduction of weight is due to that in the second stage, intermolecular associations and weaker chemical bonds are destroyed accompanied by rupture of alkyl chains to generate small gaseous molecules. The third stage is normally located above 450°C where more volatiles are produced constantly up until no more gases are released. At this point, the stronger bonds are broken and the asphaltene molecule is fully destroyed and, as a consequence, gaseous molecules are obtained until coke prevails as the final product of asphaltenes pyrolysis, as also reported by Dong et al. (2005). It is also observed that at 4 and 8°C/min there are small shoulders that could correspond to resins or aromatics.

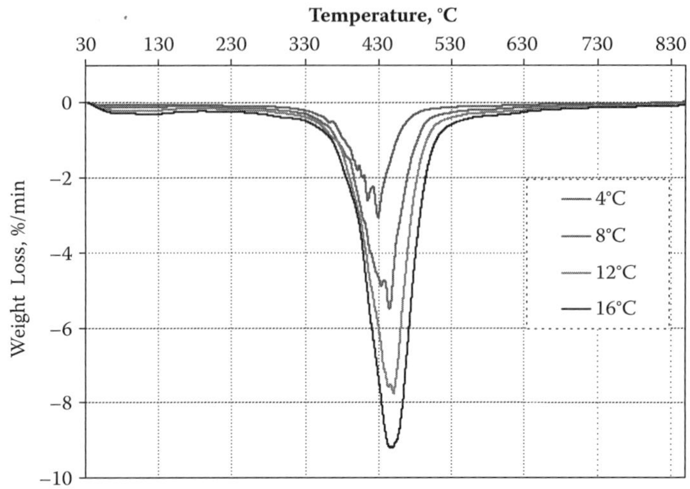

FIGURE 4.33 Thermal decomposition of asphaltenes at different heating rates. (From Trejo et al., 2009. *Catalysis Today*. Submitted. With permission.)

Figure 4.34 shows the volatilization of asphaltenes at different heating rates where asymptotic curves are obtained at higher temperatures indicating that the coke amount is constant. However, values of 90% volatilization are uncertain due to the difficulty for obtaining accurate values at this conversion. Friedman's method

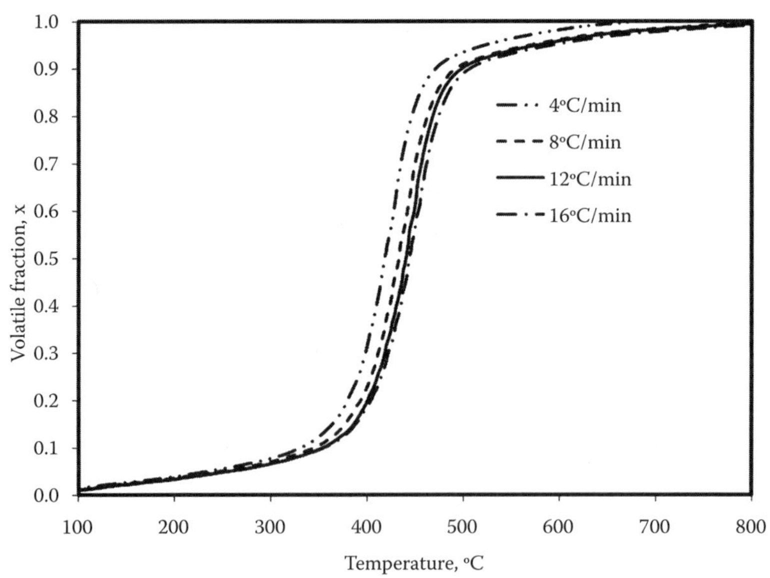

FIGURE 4.34 Volatilization of asphaltenes from a 13° API crude as a function of temperature. (From Trejo et al., 2009. *Catalysis Today*. Submitted. With permission.)

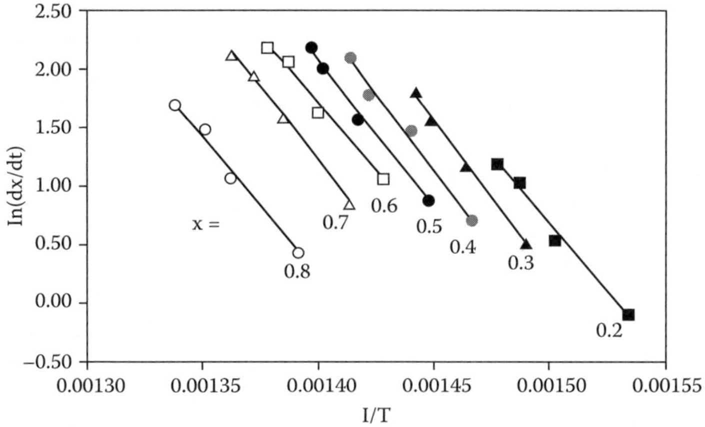

FIGURE 4.35 Determination of kinetic parameters at different heating rates. (From Trejo et al., 2009. *Catalysis Today.* Submitted. With permission.)

gives simultaneously the reaction order, activation energy, and preexponential factor, which are calculated without the knowledge of the rate dependence on conversion. Once that volatilization curves are obtained, linear regression is applied to obtain the activation energy and preexponential factor at different volatilization ratio (Figure 4.35). It is seen that lines are not completely parallel, which indicates that activation energy presents different values as a function of the degree of volatilization of asphaltenes. Results of activation energy and preexponential factor are summarized in Table 4.4 where the differences are clearly observed. A good correlation

TABLE 4.4
Calculation of Activation Energy and Preexponential Factors

V/Vo = x	E_A, kcal/mol	A, min-1	r
0.1	39.9	6.20×10^{13}	0.996
0.2	46.1	3.23×10^{15}	0.996
0.3	50.9	8.67×10^{16}	0.998
0.4	52.2	1.79×10^{17}	0.999
0.5	49.9	3.03×10^{16}	0.996
0.6	45.5	1.13×10^{15}	0.997
0.7	49.6	1.65×10^{16}	0.998
0.8	48.8	5.04×10^{15}	0.994
0.9	43.3	1.68×10^{13}	0.980

Source: Trejo et al., 2009. *Catalysis Today.* Submitted. With permission.

coefficient (r) was obtained in all the cases. Schuker (1983) also obtained a similar behavior on activation energy values. These differences in activation energy likely reflect changes in the strength of bonds, which is intimately related to a change in mechanism reaction during coking. This could mean that kinetics of coking is not always well represented by simple kinetic mechanisms having only one activation energy, as reported by Miura and Maki (1998) as well. Obtained results are also in agreement with those reported by Shih and Sohn (1980).

Other attempts to study the pyrolysis reaction are based on the distributed activation energy models that consider that activation energy is not a unique value; on the contrary, in complex reactions, such as pyrolysis, a distribution of activation energy is required. Shih and Sohn (1980) and Braun and Burnham (1987) have stated that activation energy dependent on the volatilization obtained from the Friedman method is a consequence of having a distribution of activation energy instead of having only a unique value. However, parameter estimation requires big computational efforts and an approximation of analytical solutions to time–temperature integrals is necessary. To simplify the solution, Lakshmanan and White (1994) have used the Weibull distribution to represent the kinetics of complex thermal transformations. Miura and coworkers (Miura, 1995; Maki et al., 1997, Miura and Maki, 1998) have reported an approach and simple method for estimating the distribution of activation energies $f(E_A)$ and preexponential factor $k_o(E_A)$ from three sets of experiments carried out at different heating rates without assuming any functional form for $f(E_A)$ and $k_o(E_A)$. The following equation is based on Miura's method:

$$\mathrm{Ln}\left(\frac{a}{T^2}\right) = \mathrm{Ln}\left(\frac{k_o R}{E_A}\right) + 0.6075 - \frac{E_A}{R}\frac{1}{T} \tag{4.14}$$

where a is the heating rate, T the temperature in Kelvin, and R the ideal gases constant. Dong et al. (2005) have used the distributed activation energy model (DAEM) based on Miura's procedure to analyze the pyrolysis kinetics and observed that the loss of weight curves was completely predicted by this method.

4.7 EFFECT OF REACTION CONDITIONS ON CATALYST DEACTIVATION

The influence of process variables in coke and metal deposition on catalysts is explained in terms of temperature, pressure of hydrogen, feedstock, and catalyst properties in the following sections.

4.7.1 Influence of Temperature

Increasing the temperature is a common practice for compensating activity with time-on-stream (TOS) of a catalyst and keeping constant a level of conversion. During this process, an increasing temperature induces the reactant molecule and provokes metal and carbon deposition on the catalyst surface. Thus, it is observed that more vanadium and carbon deposits occur on the edge of the pellet because the reaction takes place at the surface of the catalyst and not on the active sites. Dong et al. (1997) studied the influence of N-compounds in VGO along with the increase of temperature and observed that the higher temperature, the higher the coke deposition. However, the presence of carbazole and acridine in VGO considerably increased the tendency to form coke on a $CoMo/Al_2O_3$ catalyst. Working with fixed-bed reactors, Takahashi et al. (2005) concluded that the increase of temperature resulted in an accelerated decomposition rate of maltenes and asphaltenes with the consequence of deposition of metals and coke in the guard reactor, which was responsible for the rapid catalyst deactivation. The transformation of maltenes is the main factor for asphaltenes being insoluble in the medium. Additionally, rupture of alkyl carbons of asphaltenes makes them less soluble and aromatic cores are very prone to form coke. These reactions are increased at higher temperature and it is to be expected that larger amounts of coke are formed at higher temperature.

4.7.2 Influence of Hydrogen Pressure

Experiments for studying the influence of the pressure have been carried out by Tamm et al. (1981). They studied the effect of two partial pressures (12.5 and 18.6 MPa) keeping constant the temperature at 370°C using Arabian heavy atmospheric residue as feedstock. When increasing partial pressure, more vanadium is deposited in the edge of the pellet and less deposition in the center is observed. The opposite trend was observed at the lowest pressure, i.e., a slight increase of vanadium deposit is found at the center of the catalyst pellet. With respect to coke deposition, it has been reported by Richardson et al. (1996) that initial coke deposition onto the catalyst surface diminishes slightly when increasing hydrogen pressure.

4.7.3 Influence of the Feedstock

Metals deposition is very dependent on the nature of feedstock. The deposition profile of vanadium is more prominent with feedstocks containing more asphaltenes and Ni + V. In this case, the deposition in the edge is more obvious than light feedstock (Tamm et al., 1981). It has been stated that the kinetics of HDM of heavy feedstocks is not fully represented by a first-order power law. Instead, demetallization follows a second-order kinetics because not all molecules have the same reactivity during the reaction. A simple kinetic representation of the reaction does not match the metal deposition profile in an integral reactor operated as fixed-bed and a better representation of the reaction could include parallel reactions. Marafi et al. (2008a) have stated that the catalyst deactivation rate is more than five times faster for a high metal content feedstock (e.g., Boscan feed with 1308 wppm Ni + V) than a conventional Kuwaiti straight run atmospheric residue of Kuwait export crude (84 wppm Ni + V).

4.7.4 Influence of the Catalyst and Its Properties

The catalyst is the main component of any catalytic upgrading processes for heavy crude and residue, which influences the conversion as well as selectivity of reaction. Different aspects, such as catalyst/feedstock ratio or catalyst properties, are important to choose the best alternative for improving the crude oil quality. Regarding the influence of catalyst amount in coke formation, Richardson et al. (1996) studied the hydroprocessing of Athabasca bitumen in a batch reactor and in a continually stirred tank reactor (CSTR). They found that coke deposition was almost constant when a ratio of feedstock-to-catalyst between 15 and 20 $g_{feedstock}/g_{catalyst}$ was used. In the case of the CSTR, the deposited coke on the surface catalyst was almost constant from 30 $g_{feedstock}/g_{catalyst}$. Thus, modifications of catalyst support or preparation method of supports are other ways to reduce the deactivation tendency. This is especially important when the feedstock is highly contaminated with metals and asphaltenes. The pore size distribution and average pore diameter influence notably the catalytic activities toward HDM and HDAs (Rana et al., 2005a). Maity et al. (2005a) reported differences in deactivation behavior for CoMo catalysts supported on alumina, alumina-silica, and alumina-titania (Table 4.5). Higher coke deposition was observed on an alumina-silica-supported catalyst that may be due to the stronger acid sites provided by the support. An alumina-titania-supported catalyst exhibited a better catalytic behavior in most of the hydroprocessing reactions during the initial hours of stream. Rana et al. (2005b) stated that TiO_2 modified the nature of active metal interaction with support

TABLE 4.5
Physical Properties and Coke and Vanadium Deposits on Catalyst

	$CoMo/Al_2O_3$ (A)		$CoMo/Al_2O_3$-SiO_2 (B)		$CoMo/Al_2O_3$-TiO_2 (C)	
Properties	**Fresh**	**Spent**	**Fresh**	**Spent**	**Fresh**	**Spent**
SSA, m^2/g	241	178	218	119	257	195
TPV, mL/g	0.75	0.30	0.3	0.2	0.62	0.32
APD, nm	12.0	6.7	5.6	4.8	9.6	6.3
PSD, vol%						
>100 nm	2.9	0	0	0	0	0
100–50 nm	11.1	5.8	0	0	0	0
50–20 nm	26.7	24.3	50.1	1.8	1.4	0.8
20–10 nm	30.2	27.5	5.3	3.4	38.6	36.4
10–5 nm	25.6	18.8	41.5	22.6	53.8	49.7
5–1.7 nm	3.5	23.6	3.1	72.2	6.2	13.1
C, wt%		19		9.6		6.0
V, wt%		0.87		0.45		0.16

Source: Maity et al. 2005a. *Energy Fuels* 19 (2): 343–347. With permission.

TABLE 4.6
Atomic Absorption Analysis of Spent Catalysts

	Textural Properties		Weight %						S/Mo,
Catalysts	**SSA**	**TPV**	**Mo**	**Co**	**Ni**	**V**	**C**	**S**	**mol/mol**
AT-1 (Fresh)	230	0.43	5.5	2.1	–	–	–	–	–
AT-1-WW	82	0.09	2.8	1.5	0.106	0.563	11.4	3.7	3.9
AT-1	115	0.11	3.0	1.1	0.116	0.521	10.1	3.1	3.1
AT-1-R	158	0.23	5.4	1.8	0.152	0.556	0.01	–	–
AT-2	88	0.27	3.5	1.5	0.110	0.501	9.54	3.5	3.0
AT-3	82	0.30	4.6	1.3	0.112	0.513	9.99	3.9	2.5
AT-4	102	0.09	3.5	1.3	0.11	0.44	9.88	3.5	3
AT-5	107	0.17	4.7	1.3	0.142	0.368	9.10	3.4	2.2
AT-6	106	0.17	4.6	1.3	0.195	0.42	9.30	3.6	2.4

Note: SSA, m^2/g; TPV, ml/g; AT-1-WW: catalyst without washing; AT-1-R: catalyst regenerated.
Source: Rana et al. 2005b. *Catalysis Today* 109 (1–4): 61–68. With permission.

facilitating the dispersion of active phases on the support surface. Table 4.6 shows the way in which catalyst properties are modified after initial deactivation. In some cases, textural properties decrease nearly 50%. Carbon deposition on an alumina-titania-supported catalyst was the lowest and the hydrodeasphaltenization (HDA) for homogeneity reaction showed a more or less constant conversion around 70% after 60 h of TOS. The addition of phosphorous to the support also has a benefit in the diminution of coke deposition. Maity et al. (2005b) reported that less deposition of coke on $CoMo/Al_2O_3-TiO_2$ modified with P was observed. The addition of P to the support has been explained in terms of a reduction in the number of strong acid sites in the support (Stanislaus et al., 1988). Another possible explanation given by Kushiyama et al. (1990) indicates that P interacts with V in the feedstock retarding the catalyst deactivation.

The catalytic activities (HDS, HDM, and HDN) along with a H/C ratio (HYD) decrease almost in similar magnitude with TOS, as shown in Figure 4.36. The hydrogenation function is much lower than hydrogenolysis (HDS, HDM) activities, which is a well-known fact for a CoMo catalyst. The changes in H/C ratio (HYD) are in agreement with asphaltene conversion (HDAs) measured as *n*-heptane precipitation (Figure 4.37), in which the H/C ratio increases when increasing the asphaltene conversion, which confirms a relationship between the two different analyses. Hydrogen content against HDAs (calculated by *n*-heptane precipitation) indicated that HDAs conversion is proportional to the percentage increase in elemental hydrogen. These results point out that asphaltene conversion is partially carried out with hydrogen. The remaining conversion of asphalene is more likely due to the cracking of complex asphaltene molecules into the lower molecular weight of asphaltene, aromatics, naphthenic, and aliphatic chains. On the other

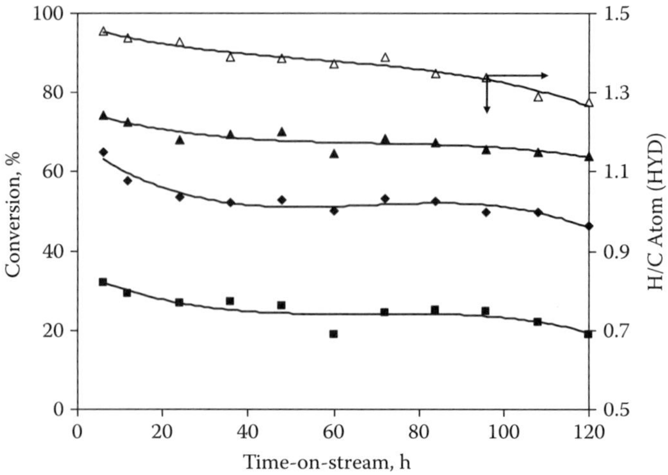

FIGURE 4.36 Catalytic activities on micro-flow reactor (CoMo/γ-Al_2O_3) with time-on-stream. (◆) HDM, (▲) HDS, (■) HDN, (Δ) H/C ratio (HYD) (Rana et al., 2007d *Petroleum Science Technology* 25 (1–2): 2021–214. With permission).

hand, variation of support can also have an important effect on the crude oil processing since the basic nature of supports is less prone to deactivation. However, this catalyst has lower hydrogenation activity or asphaltenes conversion, while acidic support has a beneficial effect on asphaltenes conversion. Recently Trejo

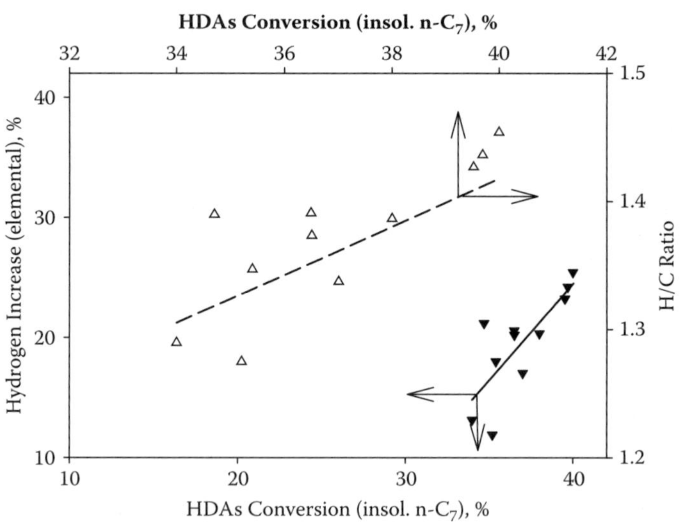

FIGURE 4.37 Relationship between hydrodeasphaltenization (HDAs) and an increase in elemental hydrogen in hydrotreated product for CoMo/CA catalyst. (From Rana et al. 2007d. *Petrol Sci. Tech.* 25 (1–2): 201–214. With permission.)

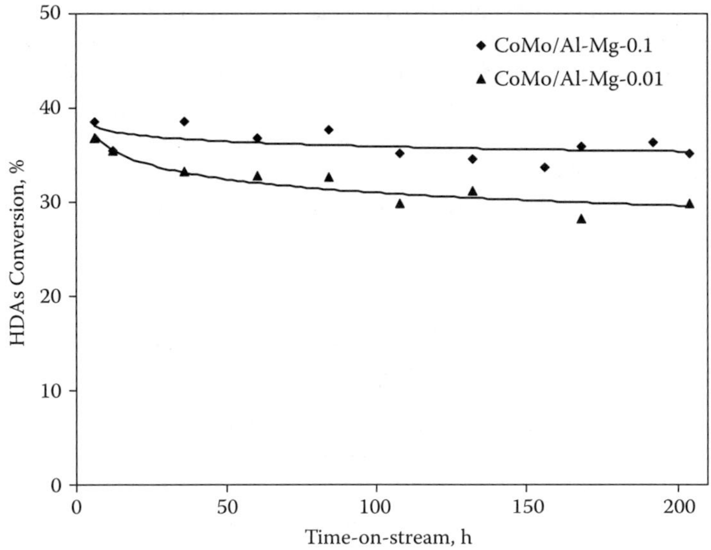

FIGURE 4.38 Activity during HDAsph of Maya crude with CoMo/MgO-Al_2O_3-1 (1 wt% MgO) and CoMo/ MgO-Al_2O_3-10 (10 wt% MgO) catalysts. (From Trejo et al. 2008. *Catalysis Today* 130 (2–4): 327–336. With permission.)

et al. (2008) intended to find out an effect of support composition (MgO-Al_2O_3) and catalyst functionalities, and observed that, after 204 h of TOS, conversion of asphaltenes was almost constant over a CoMo/Al_2O_3-MgO catalyst. The amount of MgO varied in the support; e.g., 1 wt% (CoMo/Mg-Al-1) and 10 wt% (CoMo/Mg–Al-10), and better conversion of asphaltenes was achieved on the CoMo/Mg-Al-10 catalyst and is seen in Figure 4.38, where asphaltenes conversion almost remained constant at around 40%.

The relatively high MgO content in the catalyst and larger pore diameter promote diffusion of large asphaltene molecules inside the pores making the catalyst highly active and more stable. The effect of pore size is very important in catalyst deactivation, as reported by Ancheyta et al. (2004), who studied three different catalysts with different pore size distribution. The combination surface area and textural properties indicated that asphaltenes conversion depends on the optimum pore diameter of the catalyst. Moreover, removal of heteroatoms from the asphaltenes depends on the localization of the heteroatoms in the asphaltenes molecule, thus, the catalytic activities are severely affected by the reaction temperature. The first catalyst (C-1) had most of its pores in the region of 10 to 25 nm, whereas most of the pores in the second catalyst (C-2) were mesopores, and the third catalyst (C-3) had pores between 5 to 10 nm and higher than 20 nm. It was observed that C-3, which has the highest specific surface area, showed very low asphaltene conversion due to the poor activity of this catalyst. C-1, which has its pores on the mesoporous region, showed the highest conversion of asphaltenes during hydrotreating. C-2 showed an intermediate conversion. The catalysts were analyzed after reaction and it was observed that the average pore volume of C-1 and C-2 decreased in the 10 to 25 nm regions, whereas this region increased in the C-3 spent catalyst. Initial porosity followed the same trend in all catalysts. When asphaltenes were analyzed after precipitation of hydrotreated crude, it was demonstrated that they were more hydrogenated with C-1. More sulfur was removed from asphaltenes with using C-3 due to its higher specific surface area (SSA) and pore size distribution in the mesoporous region.

4.8 CATALYST FORMULATION AND ITS EFFECT ON ASPHALTENES CONVERSION

A number of literature reports describe the importance and application of different aspects of catalyst formulation for light feedstock hydrotreating, but only very few works deal with hydroprocessing of heavy or extra heavy crude oils (Leyva et al., 2007). The acidity of hydroprocessing/hydrocracking catalysts is one of the most important parameters to be taken into account because the amount of coke formed depends on it. If the feedstock is asphaltene-rich, then it would be necessary to regulate the acidity in the catalyst to avoid a fast deactivation. Marafi and Stanislaus (1997) prepared a series of $NiMo/Al_2O_3$ catalysts having different degrees of acidity by using different amounts of sodium or fluoride and found that acidity decreased in the following order: Ni-Mo-F(5%) > Ni-Mo-F(2%) > Ni-Mo > Ni-Mo-Na(1%) > Ni-Mo-Na (5%). The presence of sodium decreased the sulfur and nitrogen conversion and produced more coke likely as a consequence of a strong interaction of molybdenum and sodium, which could suppress hydrogenation activity of the catalyst. It is well known that asphaltenes present in residue are very hard to crack because of their size, stacking, and metals present inside their molecule. For this reason, small pores cannot be used to process big molecules, such as asphaltenes, and when heavy, extra heavy oils, or residue are upgraded, asphaltenes suffer the effect of temperature, begin to crack, and free radical combination leads them to increase their degree of polycondensation turning them into more aromatic structures. Only maltenes can be upgraded to smaller pores when being hydrogenated, whereas asphaltenes are increasingly more aromatic. At some extent of the reaction, there is a significant incompatibility between maltenes and asphaltenes, and coke and sediments deposition occurs (Inoue et al., 1998). Studies carried out by Absi-Halabi et al. (1998) evaluated the performance of alumina-supported NiMoW catalysts and found that the W addition enhances the hydrogenation function of the catalyst. In the case of asphaltenes conversion, it was found that there was an initial conversion 10% higher than that of NiMo catalysts; however, after 250 h of TOS, asphaltenes conversion on NiMo and NiMoW/alumina approached almost the same value. The increasing activity in HDS was in the order CoMo > NiMo > NiW, while hydrogenation followed the order NiW > NiMo > CoMo. The very contrasting activity is due to the difference in the nature of active sites. Nickel-promoted catalysts have stacks of MoS_2 slabs forming five to six layers (Eijsbouts et al., 1993), whereas, in cobalt-promoted catalysts, the MoS_2 slabs are distributed as a monolayer.

Few other supports like sepiolite have been reported by Takeuchi et al. (1983) in order to prevent coke formation, promote asphaltene conversion, and metals removal. The synergistic effect of sepiolite and alumina as supports makes the activity of the resulting catalyst to be twice that of the original sepiolite catalyst, as reported by Inoue et al. (1998). Mesoporous silica materials are also reported for their adsorption and reaction of large molecules, e.g., asphaltenes. The use of mesoporous materials allows the asphaltenes micelles to access to catalytic active sites and enhances the production of lighter distillates with minimal coke formation, according to Byambajav and Ohtsuka (2003a). These authors employed Ni and Fe/SBA-15 catalysts and found that, in the case of 10% of Fe catalysts, asphaltene conversion showed a linear behavior as a function of the pore size up to 12 nm reaching 70% of conversion, being constant at higher pore diameters. Maltene yield is increased up to around 40% with the same pore diameter. From their study, catalysts having pores in the interval of 12 to 15 nm are suitable for asphaltene conversion turning them into maltenes. Asphaltene conversion and maltene yields are shown in Figure 4.39. The variation of metal loading (Ni and Fe) on the support was observed to influence the conversion of asphaltenes and maltene yield, which corresponds to the better hydrogenation function carried out by the Ni/SBA-15 catalyst (Byambajav and Ohtsuka 2003a, 2003b).

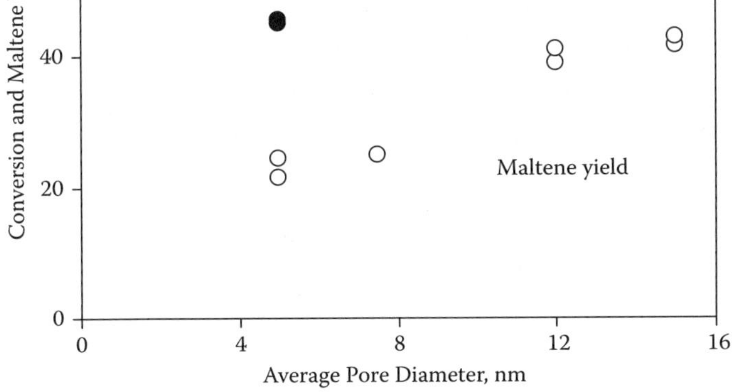

FIGURE 4.39 Performances of 10% Fe/SBA-15 with different pore diameters in hydrocracking of asphaltene. (From Byambajav, E. and Ohtsuka, Y. 2003a. *Appl. Catal.* A 252 (1): 193–204. With permission.)

The higher maltene yield was obtained at 10% of Ni or Fe with average pore diameter of 12 nm. In the case of Fe/SBA-15, it promotes higher asphaltene conversion and fewer maltene yields while with Ni/SBA-15 less conversion of asphaltenes is achieved and higher maltene yield is obtained. A possible explanation of this is the enhanced hydrogenation function of the Ni/SBA-15 catalyst compared with the Fe/SBA-15 catalyst. However, both catalysts demonstrated to be stable. In addition, characterization by XRD and TPO showed that Fe-S and Ni-S may be catalytically active species. Clay-like supported catalysts, such as montmorillonite (natural clay) and laponite (synthetic clay), have also been used for hydrocracking. The catalysts were pillared with active metals (Sn, Cr, Al, etc.) as well as double layered using polyoxo-vanadate and polyoxo-molybdate (Bodman et al., 2002, 2003). The catalysts were evaluated and compared with conventional NiMo/alumina and Mo(Co)6 dispersed catalysts. Especially, chromium pillared montmorillonite calcined at 500°C and tin laponite catalysts showed improved behavior compared with conventional catalysts. This kind of catalysts is useful for improving conversion, reduction of molecular weight, heteroatom removal, hydrogenation of polynuclear aromatic rings, such as asphaltenes, and improving the resistance to fast deactivation. The high performance of the pillared clays is attributed to their large galleries, which are available to the adsorption of asphaltenes and other large molecules. An important feature of the pillared clays is their high thermal stability and their ability to crack heavy molecules due to their strong acid sites. However, the applicability of new materials either in batch reactor or in very primary experimental stages has to be confirmed with detailed deactivation studies with real feeds at reaction conditions close to the industrial practice.

In summary, all of these catalysts could be taken into consideration when selecting the best catalytic option for upgrading heavy crude with high amounts of asphaltenes to reach higher conversions.

4.9 ALTERNATIVES APPROACHES FOR PREVENTING COKE DEPOSITION

Some additional remedies can be applied during heavy crude upgrading during hydroprocessing of heavy petroleum fractions or bitumen, which include the addition of fine solids to avoid the coalescence of the mesophase into bigger particles that eventually precipitate forming coke deposition. Another alternative is the removal of asphaltenes by solvent deasphalting before processing the heavy crude or residue. An explanation of the use and application of these alternatives is given below.

4.9.1 Use of Fine Solids

The presence of fine solids can make that the emerging new phase during induction period interacts with the solids by means of nucleation or dispersion of coke. Tanabe and Gray (1997) have used fine solids, such as clays, to inhibit coke formation. The mechanism by which fine solids act in preventing coke formation is due to clays

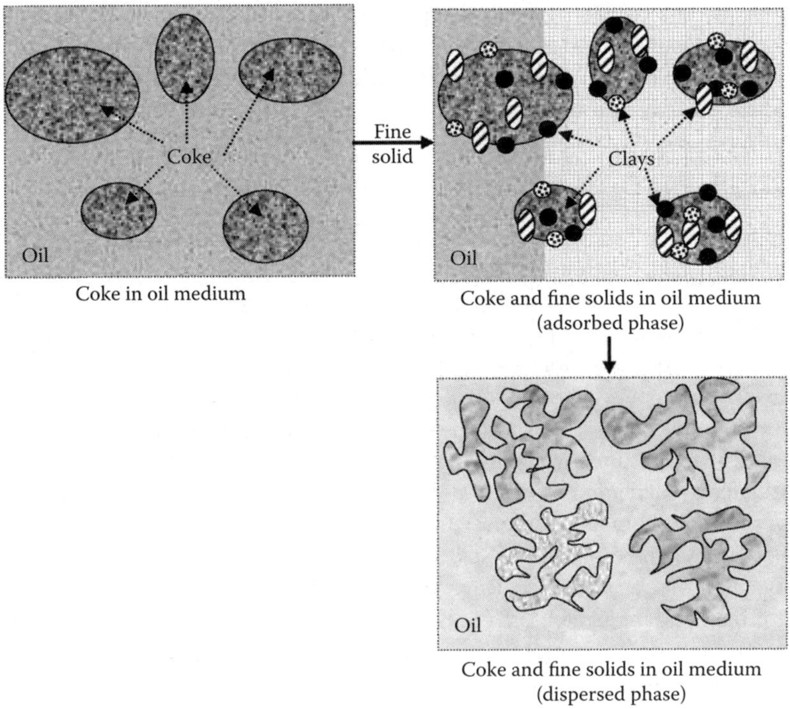

FIGURE 4.40 Effect of fine solid particles into the coke and its interfaces (adsorption and dispersed) in oil medium.

sticking to the external surface of coke particles at the coke/oil interface avoiding coalescence of small coke particles into larger ones (Figure 4.40). It can be observed that clays are added to the surface of coke and eventually retard their coalescence. Figueiras et al. (1998) showed that carbonaceous material insoluble in quinoline tend to avoid the growing and coalescence of mesophase spheres. Carbon black also has been used in preventing coke formation. Kanno et al. (1997) used carbon black in different amounts and they observed that, with 5 wt% of carbon black added, it was sufficient to suppress the mesophase formation during pyrolysis at lower temperatures. Sakanishi et al. (1999) also used different carbon adsorbents for preventing mesophase growing. They showed that asphaltenes are adsorbed onto carbon black, and mesoporous carbons, and that even carbon with low specific surface area (150 m^2/g) was able to remove around 500 wppm of V and 200 wppm of Ni from a vacuum residue. Thus, mesoporous carbon particles can be used in selective removal of asphaltenes and metals from heavy and extra heavy oils. Mochida (1999) also reported that some carbon materials are outstanding entrants for selective adsorbents for asphaltenes; the only problem with carbon material is the mechanical strength and porosity, which require further consideration of researchers. Carbon also has a good capacity for removing metals and asphaltenes at high extent, thus, it can be used as support for the catalyst for hydroprocessing. However, the carbon-supported catalyst requires the use of hydrogen donors because it has low activity for ring hydrogenation. The main issue with carbon black-supported catalysts is that strong adsorption of asphaltenes is carried out into the support and some modifications of support are necessary to reduce its surface polarity.

4.9.2 Solvent Deasphalting

Hydroprocessing is a hydrogen addition process which reduces coke formation by enhancing hydrogenation function either by catalytic site or by hydrogen partial pressure or by some solvent that provides hydrogen (Kubo et al., 1994). In all cases, hydrogen addition processes require very high investment. If asphaltenes are mainly responsible for coke deposition, then why not remove them before crude oil processing? The answer is "yes" because that will improve feed quality but at industrial scale it may not be simple because several technical and economical issues are involved when removing asphaltenes e.g. what to do with solid material (asphaltenes) separated from crude oil, its transportation, solvent recovery process etc. Nevertheless the application of solvent deasphalting (SDA) coupled with some other processes, such as hydrotreating and gasification, could be advantageous, as shown by Rana et al. (2007b) in Figure 4.41.

Elimination of asphaltenes from the crude or from residues (AR/VR) improves the quality of crude because many contaminants, such as metals and sulfur, are associated to asphaltenes. In this regard, the use of solvents for removing asphaltenes has been practiced with good results. Brons and Yu (1995) reported that separation

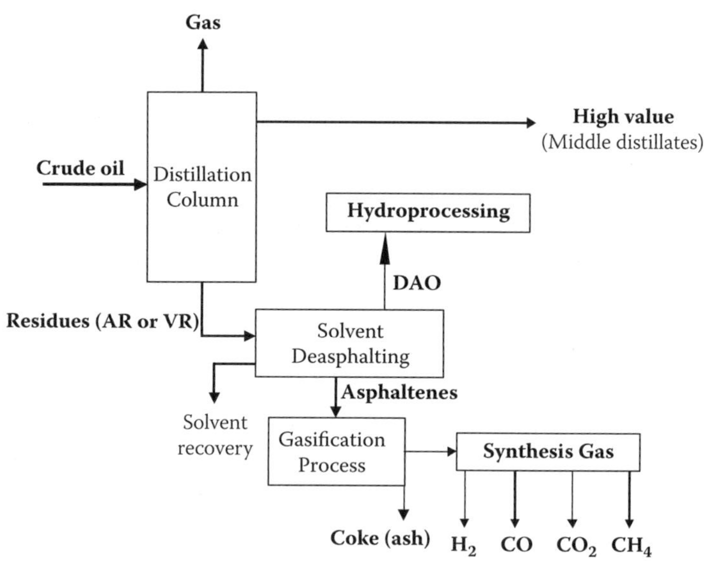

FIGURE 4.41 Heavy oil upgrading integrated process combining solvent deasphalting (SDA), thermal process, and hydroprocessing. (From Rana et al. 2007d. *Petrol Sci. Tech.* 25 (1–2): 201–214. With permission.)

of asphaltenes from Cold Lake bitumen increased the API gravity of the deasphalted oil (DAO). As more deasphalting occurs, the oil is richer in hydrogen and poorer in metals, sulfur, nitrogen, and Conradson carbon residue (CCR). It was also found that thiophenic sulfur is present in asphaltenes and that sulfides act as cross links, which confer higher molecular weight to asphaltenes being that the heaviest asphaltenes are responsible for changes in viscosity of the oil. There are different purposes for removing asphaltenes from crude or residue and the most important are to make commercial asphalts from the bottoms and to increase the yield of gas oil, which can be further upgraded into a fluid catalytic cracking (FCC) unit. It is possible to recover gas oil from vacuum distillation towers using SDA.

The fixed-bed processes for upgrading atmospheric and vacuum residue can be coupled along with solvent deasphalting with the objective of producing as much middle distillates as possible. One type of configuration that allows increasing the conversion is by adding a SDA unit after hydrodesulfurization of vacuum residue. DAO can be mixed with vacuum gas oil to meet the specifications for resid catalytic cracking (RCC) feedstocks. Asphaltenes obtained after deasphalting can be partially oxidized by means of gasification to obtain hydrogen as proposed by Kressmann et al. (1998). Another option is to use a semidirect scheme, which is more profitable to the preparation of feedstocks for an FCC unit. In this case, hydrodemetallization of residue is a previous stage of SDA. DAO can be integrated to a vacuum gas oil stream to be hydrotreated. The hydrotreated product is able to be used as FCC feedstock. This process offers the possibility of reducing the CCR content from 25 to 35 wt% in the feedstock to 1.5 wt% with final sulfur content around 0.15 wt%. Rana et al. (2007b) have proposed a scheme in which solvent deasphalting is used after distillation of crude. DAO is hydrotreated in order to meet specifications for further refining processes. The remaining asphaltenes are sent to gasification wherein CH_4, H_2, and CO are obtained. Hydrogen can be used in hydrotreating reactors or can be used along with CO as synthesis gas for obtaining valuable products, such as kerosene or gas oil, by means of Fischer–Tropsch synthesis and hydroisomerization.

SDA requires a significant amount of heating to separate or recycle the solvent. The gasification process produces the necessary heat to be employed in solvent recovering. The heat is used to strip the solvent from the oil and asphaltenes so that solvent can be recovered and reused in the process. Thus, the products of SDA and gasification also can be integrated. The DAO requires hydrotreating and catalytic cracking to produce diesel. Hydrogen is required for doing this, but it is obtained from gasification eliminating an external supply of gas. In addition, sulfur compounds are present in asphaltenes, but, when they are gasified, sulfur is converted to hydrogen sulfide, which is further removed from the synthesis gas by means of conventional acid gas absorption and then converted to elemental sulfur (Claus process). The benefits of integrated gasification with SDA are (Wallace et al., 1998):

- Beneficial use of asphaltenes
- Internal consumption of heat
- Production of hydrogen for DAO hydrotreating
- Recovery of hydrogen from hydrotreating purge gas

Penrose et al. (1999) have shown the benefits when integrated gasification with SDA, gasification, hydrotreating, cogeneration plant, and sulfur recovering is carried out. Coupling of SDA with gasification and hydrotreating has been also patented by Wallace and Johnson (2002). This process uses the SDA for recovering maltenes and leaving asphaltenes alone, which are further gasified and the released hydrogen is used in hydrotreating units. Ushio et al. (2005) have proposed some reaction schemes based on bench reaction results that include DAO as feedstock to be hydrocracked at mild and severe conditions. At mild hydrocracking, the process would consist of deasphalted vacuum residue as feedstock at 50% of 560°C+ conversion giving gasoline, middle distillates, and VGO. Around 65 wt% of VGO and uncracked bottoms are sent to an FCC unit. When severe hydrocracking is selected, deasphalted vacuum residue can be converted up to 70% of 560°C+ conversion. Around 30 wt% of VGO and uncracked oil are sent to FCC. In this case, a recycle of VGO and bottoms from a fractionator are sent to the reactor. DAO hydrocracking conversion can be improved if zeolite-supported catalysts are used in order to obtain a high yield of middle distillates. Synthesizing mesoporous, zeolite-based, hydrocracking catalysts with titanium obtains high cracking activity compared with amorphous catalysts. Regarding these zeolite-based catalysts, Honna et al. (2003) have prepared titanium-modified USY zeolites, which gave higher selectivity to middle distillates than zeolite-based catalysts without titanium. Ti-modified zeolites have larger mesopore volumes with diameter of pore higher than 10 nm and less acidity than the initial ultrastable Y-type (USY) zeolite. The low metal (Ni and V) in feedstock, particularly vanadium, is less sensitive to the zeolite structural decomposition during the regeneration through vanadic acid formation. Thus, titanium may act as a scavenger to zeolitic framework, and TiO_2 also promotes the activity of the sulfided phase. Experimental tests involving SDA coupled to mild thermal cracking at continuous bench scale have been carried out to improve the quality of extra-heavy crudes. This process (mild cracking solvent deasphalting) was reported to produce more DAO with lower metals content and CCR than solvent deasphalting alone. The hydrogen losses obtained as a consequence of the cracking are significantly reduced. Elimination of metals is highly improved with this process as well. On the other hand, one must keep in mind that SDA is a physical separation that cannot modify the distribution of contaminants in heavy crudes. For this reason, it is necessary to introduce a reaction which can displace contaminant heteroatoms present in heavy crude oil toward asphaltenes, leaving the maltenes with fewer amounts of impurities; one of the preferred reactions is thermal cracking (Chen et al., 1994).

Deasphalting for upgrading heavy fractions is recognized as a useful process and for this reason some plants of SDA are operating worldwide, as shown in Table 4.7 The importance of SDA could grow in the coming years since feed quality is decreasing every day and its processing is becoming more difficult. However, SDA should be integrated to other upgrading processes in order to obtain better benefits and to meet the feedstock quality specifications. There are a few reports that explain precipitation of asphaltene before exploration (underground

TABLE 4. 7
Worldwide Solvent Deasphalting Residue Process Technologies

Country or Area	Units in Operation[a]
North and South America	15
Europe and Japan	3
Others	6
Total	24

[a] March 2003.

treatment) introducing heptane/pentane into the well, these solvents precipitate asphaltene and light deasphalted oil can be easily taken out. In this way, the disposal of asphaltene is not a problem at onsite exploration.

4.10 REGENERATION AND REJUVENATION OF SPENT CATALYST

As shown in Figure 4.1, when further increases of temperature at the end of the run are not enough to keep the conversion at the desired level, the catalyst then needs to be regenerated or replaced. Some catalytic activity can be recovered from spent catalysts when contaminants are eliminated by different procedures. For catalysts that are not highly fouled by metals, it is recommendable to burn off the coke, which is done either *in situ* or off-site; the former is suitable for fixed-bed reactors, whereas the catalyst from ebullated-bed reactors must be regenerated off-site. If metal fouling is considerably high, especially when hydroprocessing heavy fractions, then the most appropriate way to reestablish catalytic activity is by regenerating as well as leaching, in which acids or bases are used and remove at least some extent of deposited metals, i.e., Ni, V, Fe, Na, etc. Moreover, catalysts with high metals may not be economic after regeneration and such catalysts may be suitable for metal reclamation; if that also is not possible, an environmentally safe method for catalyst disposal must be found (Marafi and Stanislaus, 2008).

4.10.1 Regeneration by Calcination

To remove deposited coke on the catalyst, an oxidative (reductive) regeneration can be applied by slowly increasing the temperature to 600°C. There are several basic phenomena that occur during oxidative regeneration of hydroprocessing catalysts. When studying the removal of carbon and sulfur as a function of temperature, during this period redispersion of the active metal–sulfide phase occurs, which is sensitive to the temperature and other regenerating compositions, such as air (oxygen) and nitrogen (inert gas). Calcination under oxidative atmosphere is a common way in which spent catalysts recover catalytic activity by burning the carbonaceous deposits. Burning of coke is a noncatalytic reaction and the most probable set of reactions that occurs can be written as (Furimsky and Yoshimura, 1987):

$$\begin{aligned} &C+\frac{1}{2}O_2 \rightarrow CO \\ &C+O_2 \rightarrow CO_2 \\ &H+\frac{1}{4}O_2 \rightarrow \frac{1}{2}H_2O \end{aligned} \tag{4.15}$$

When regenerating hydrotreating catalysts, sulfides are converted to oxides according to:

$$\begin{aligned} &MoS_2\ (WS_2) \xrightarrow{O_2} MoO_3\ (WO_3) \\ &Co_9S_8\ (Ni_3S_2) \xrightarrow{O_2} CoO\ (NiO) \end{aligned} \tag{4.16}$$

Not only carbon is present when regenerating the catalyst, but also organic sulfur and nitrogen species are present and suffer from oxidation:

$$\begin{aligned} &S_{org} \xrightarrow{O_2} SO_2 \\ &N_{org} \xrightarrow{O_2} NO_x \end{aligned} \tag{4.17}$$

The heat released by these reactions contributes to the overall heat during burning. In fact, TPO studies revealed that oxidative regeneration proceeds by previous removal of sulfidic S as SO_2 and then the removal of carbon as CO and CO_2. There is a small amount of SO_2 along with carbon oxidation, which could exist due to the presence of organic sulfur and residual sulfides (Yoshimura et al., 1994). When NiW catalysts are regenerated, more severe oxidation conditions need to be employed compared with spent CoMo catalysts. Bogdanor and Rase (1986) reported the presence of two types of coke, one being burnt below 380°C and the other one higher than 380°C. If regeneration is carried out at a temperature, e.g., 380°C, only partial regeneration of the catalyst is achieved especially near the outer surface, and around 0.1 to 0.2 mm from the particle edge would be removed by calcination. In addition, sulfur is not completely removed and only the so-called "fixed sulfur" remains on the catalyst, whereas the "active sulfur," which is bound to Mo, is fully eliminated. Whole sulfur is removed at higher temperatures. Calcination temperatures above 500°C also totally remove the deposited coke burning carbonaceous deposits from the external surface to the center of the pellet. Three main region losses of weight of coke were observed by TGA. The first region where coke loses weight ranged from 25° to 150°C, which corresponds to water evaporation; the second region in the interval of 360° to 380°C in which coke is easily oxidized; and the third region in the range of 430° to 450°C corresponding to coke that is more difficult to oxidize. Burning coke permits reutilization of the catalyst, which can be sulfided and used in hydrotreating reactions again. However, when calcination was carried out at higher temperature (i.e., 560°C), there is a migration of Mo toward the pellet surface reducing the catalytic activity by agglomeration of Mo, which diminishes the hydrogenation function.

Sajkowski et al. (1989) have also concluded that Mo sulfide agglomerated during hydrotreating, but Mo sulfide phases redisperse during oxidative regeneration as MoO_3 on alumina support. A similar finding was reported by Arteaga et al. (1986, 1987) who stated that Mo and Co phases in a $CoMo/Al_2O_3$ spent catalyst are redispersed after regeneration and were better distributed on the catalyst surface, but at high temperature, e.g., 500°C, some of Co reacted with the alumina support to form $CoAl_2O_4$ diminishing the active sulfided sites. When sulfiding regenerated the catalyst, it is possible to control the desired profile by increasing the sulfidation temperature (Jepsen and Rase, 1981). In this case, the Mo/Ni ratio shifted from the particle center to become a maximum at the particle edge when sulfidation temperature increased from 232° to 371°C. A comparison of reductive regeneration with oxidative regenerations of a $NiMo/Al_2O_3$ deactivated hydrotreating catalyst showed that activity was partially recovered by using a diluted oxygen concentration in a mixture of gases (1.6 vol% O_2) compared with a reductive treatment (5 vol% H_2) (Texeira da Silva et al., 1998). The initial activity of fresh catalyst was not reached due to the probable formation of Ni spinels ($NiAl_2O_4$) during burning of coke.

Sometimes spent catalysts can be reused after decoking and acid leaching, which allows using them in the preparation of a new catalyst (Marafi and Stanislaus, 2008). When mixing the regenerated catalyst with variable amounts of boehmite, the resultant catalyst had high specific surface area and pore volume that gave to the catalyst substantially higher HDM and HDS activities compared with a fresh catalyst. The combination of decoking with mild leaching or hydrothermal treatment is capable of removing contaminants and leaving the catalyst able to be used in further applications in spite of having V and Ni deposited on its surface. Other studies carried out by Gardner and Kidd (1989) were focused on the preparation of hydrodesulfurization catalysts from spent catalysts, which were mixed with alumina to form extrudates. Spent catalysts have been also milled, sintered at high temperature to obtain wider pore size, and mixed with additives in order to form new hydrotreating catalysts, as reported by de Boer (2000). These studies reveal the usefulness of recycling spent catalysts as an alternative, thus aiding in reducing environmental damage.

4.10.2 Rejuvenation and Metals Removal by Leaching

Usually heavy, extra heavy, and residues (AR/VR) hydroprocessing deposited very high amounts of metals contaminants, which also depended on the reaction condition, time-on-stream, and quality of feedstock. Thus, such spent catalysts (carbon, sulfur, vanadium, iron, and nickel) are the main contaminants deposited. However, when the catalyst is not able to be regenerated because of its high degree of deactivation, and burning of coke is not enough for recovering the activity, then it is recommended to extract the metals by leaching since spent hydrotreating catalysts are classified as hazardous solid waste by the U.S. Environmental Protection Agency (EPA). For this reason, it is mandatory to find out an economical solution to prevent pollution by catalyst disposal. Leaching is a common way in which metals are recovered. A schematic representation of metals leaching has been given by Marafi et al. (2008b) according to:

$$\text{A(fluid)} + \text{B(solid)} \rightarrow \text{Soluble and insoluble product} \tag{4.18}$$

It has been established that the reaction proceeds through diffusion of molecules (A) from the liquid to the solid and diffusion of species (A) through a layer of porous solids to the surface where reaction takes place on the solid (B). Main diffusion limitations are present when deposits of metals block the access of the species (A) into the pores. Different options for rejuvenating spent catalysts have been studied. A preliminary, but detailed, study for proposing a catalyst rejuvenation process to fouled catalysts generated during hydroprocessing of residue has been carried out. Process design was based on a capacity of 6,000 tons of spent catalyst per year and different options for processing were developed, as seen in Figure 4.42 (Marafi et al., 2008b). Each option depends on the level of fouling in catalysts. For example, in Option II, where leaching is applied to a medium fouled catalyst, the best leaching agent consisted of 6% of oxalic acid and 8% of ferric nitrate in water at 40°C. Catalyst properties showed a significant recuperation after removal of metals and compared with fresh catalysts, rejuvenated catalysts exhibited almost the same activity for HDS. Different chemical agents can be used in leaching depending on the desired metal to be removed and reaction conditions. For instance, Marafi et al. (1994) used oxalic acid with ferric nitrate that allowed them to remove vanadium in high amounts and, as a result, the specific surface area, pore volume, and catalytic activity were recovered. Tartaric acid did not exhibit the same effectiveness as oxalic acid. In both cases, poor acid activity for leaching was observed when ferric nitrate was not used. Ferric nitrate probably works favoring oxidizing and complexing reactions. A preliminary economical study was performed on the recovering of V and Mo from hydrotreating spent catalysts by basic leaching. It was found that the best pH for recovering Mo was 8.5, and pH of 9 for V after 12 and 8 h of stirring, respectively, with NaOH (10 wt%) as a leaching agent. The economical benefit was estimated to be between 0.79 and 0.86 US$/barrel of processed feedstock, but this benefit is strongly dependent on the metals market and does not include facilities and production costs. However, it demonstrates the advantage when metal recuperation is carried out (Alonso et al., 2008).

Other authors have reported that a two-stage leaching process is useful for recovering high amount of metals (Sun et al., 2001). Concentrated ammonia used in the first stage is capable of removing up to 83% (w/v) of Mo. Sulfuric acid is used in the second stage to recover up to 77% of Co and 4% of the remaining Mo. Roasting is another process by which metals in spent catalysts

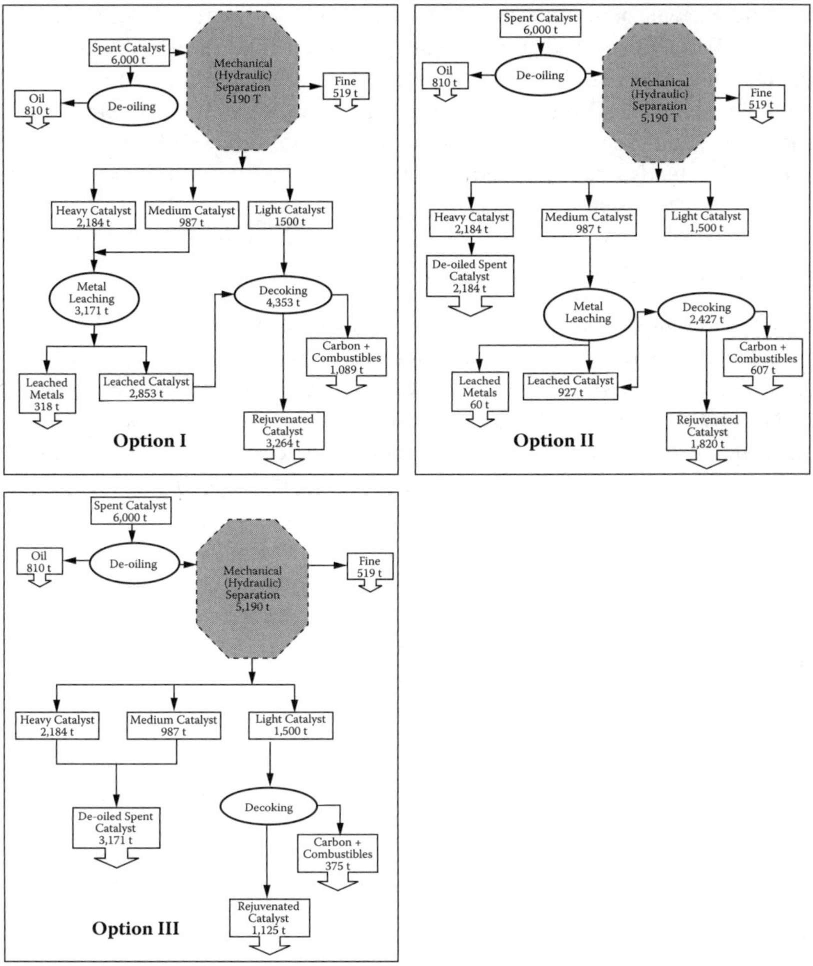

FIGURE 4.42 Rejuvenation and leaching processes. (From Marafi et al. 2008b. *b. J. Environ. Manag.* 86 (4): 665–681. With permission.)

can be recovered at high yields. Chen et al. (2006a) have developed a procedure to recover, in high amounts, metals present in fouled catalysts. The yield of recovered metals during roasting is around 90% for molybdenum and vanadium by using a ratio of sodium carbonate-to-spent catalyst of 0.15 at 750°C during 45 min. Purification with water at 80° to 90°C was required after roasting and then molybdenum and vanadium were extracted with trialkylamine (20 vol%) and octyl alcohol (10 vol%) in sulfonated kerosene. With a pH of 7 to 8.5 (given by NH_4NO_3) over 99% of vanadium is recuperated as ammonium metavanadate. Molybdenum is obtained as ammonium polymolybdate (~98%) when HNO_3 is used to adjust the pH to around 1.5 to 2.5. Finally, MoO_3 and V_2O_5 are obtained from calcinations of ammonium molybdate and ammonium metavanadate at 500° to 550°C. Chen et al. (2006b) also reported that alumina is obtained as $Al(OH)_3$ with purity higher than 99% by using sodium aluminate and further calcination. Thus, leaching is an optional way to reutilize metals extracted from spent catalysts reducing the environmental contamination, especially when highly fouled feedstocks are used.

4.11 CONCLUDING REMARKS

Catalyst deactivation is a result of a number of unwanted chemical and physical changes in the catalyst textural properties as well as in its composition. The causes of deactivation are classically divided into three categories: chemical, thermal, and mechanical. Deactivation is a function of temperature, which is usually increased to maintain the conversion of different reactions, such as HDS, HDM, etc., as well as API gravity of hydroprocessed crude or residue. The main reasons for deactivation of catalyst in heavy oil processing are coke and metals (Ni + V) depositions on pore mouth and blocking of the pores of the catalyst.

REFERENCES

Abotsi, G.M.K. and Scaroni, A.W. 1989. A review of carbon-supported hydrodesulfurization catalysts. *Fuel Proc. Tech.* 22 (2): 107–133.

Absi-Halabi, M., Stanislaus, A., and Al-Dolama, K. 1998. Performance comparison of alumina-supported Ni-Mo, Ni-W and Ni-Mo-W catalysts in hydrotreating vacuum residue. *Fuel* 77 (7): 787–790.

Absi-Halabi, M., Stanislaus, A., Al-Mughni, T., Khan, S., and Qamra, A. 1995. Hydroprocessing of vacuum residues: Relation between catalyst activity, deactivation and pore size distribution. *Fuel* 74 (8): 1211–1215.

Absi-Halabi, M., Stanislaus, A., and Trimm, D.L. 1991. Coke formation on catalysts during the hydroprocessing of heavy oils. *Appl. Catal. A* 72 (2): 193–215.

Adkins, B.D. and Limmer, K.R. 1990. A simple diffusion-reaction model for resid hydroprocessing catalysts. *Am. Chem. Soc., Div. Petr. Chem.—Prepr.* 35 (4): 601–603.

Al-Dalama, K. and Stanislaus, A. 2006. Comparison between deactivation pattern of catalysts in fixed-bed and ebullating-bed residue hydroprocessing units. *Chem. Eng. J.* 120 (1–2): 33–42.

Alonso, F., Ramírez, S., Ancheyta, J., and Mavil, M. 2008. Alternatives for recovering of metals from heavy hydrocarbons spent hydrotreating catalysts —A case of study. *Rev. Int. Contam. Ambient.* 24 (2): 55–69.

Amemiya, M., Korai, Y., and Mochida, I. 2003. Catalyst deactivation in distillate hydrotreating (Part 2): Raman analysis of carbon deposited on hydrotreating catalyst for vacuum gas oil. *J. Jpn. Petrol. Inst.* 46 (2): 99–104.

Amemiya, M., Suzuka, T., Korai, Y., and Mochida, I. 2000. Catalyst deactivation in distillate hydrotreating (Part 1): Catalyst deactivation in the commercial vacuum gas oil hydrotreating unit. *Sekiyu Gakkaishi* 43 (1): 52–58.

Ancheyta, J. 2007. Reactors for hydroprocessing, in *Hydroprocessing of Heavy Oil and Residua,* Ancheyta, J. and Speight, J.G., Eds., Taylor & Francis, New York, Chap. 5.

Ancheyta, J., Betancourt, G., Centeno, G., Marroquín, G., Alonso, F., and Garciafigueroa, E. 2002. Catalyst deactivation during hydroprocessing of Maya heavy crude oil. 1. Evaluation at constant operating conditions. *Energy Fuels* 16 (6): 1438–1443.

Ancheyta, J., Betancourt, G., Centeno, G., and Marroquín, G. 2003. Catalyst deactivation during hydroprocessing of Maya heavy crude oil. (II) Effect of temperature during time-on-stream. *Energy Fuels* 17 (2): 462–467.

Ancheyta, J., Centeno, G., and Trejo, F. 2004. Effects of catalyst properties on asphaltenes composition during hydrotreating of heavy oils. *Petrol. Sci. Tech.* 22 (1–2): 219–225.

Ancheyta, J., Rana, M.S., and Furimsky, E. 2005. Hydroprocessing of heavy petroleum feeds: Tutorial. *Catalysis Today*, 109*:* 3–15.

Arteaga, A., Fierro, J.L.G., Delannay, F., and Delmon, B. 1986. Simulated deactivation and regeneration of an industrial CoMo/γ–Al_2O_3 hydrodesulphurization catalyst. *Appl. Catal. A* 26 (1–2): 227–249.

Arteaga, A., Fierro, J.L.G., Grange, P., and Delmon, B. 1987. Simulated regeneration of an industrial CoMo/CoMo/γ–Al_2O_3 catalyst. Influence of the regeneration temperature. *Appl. Catal. A* 34 (1–2): 89–107.

Banerjee, D.K., Laidler, K.J., Nandi, B.N., and Patmore, D.J. 1986. Kinetic studies of coke formation in hydrocarbon fractions of heavy crudes. *Fuel* 65 (4): 480–484.

Bartholdy, J. and Cooper, B.H. 1993. Metal and coke deactivation of resid hydroprocessing catalysts. *Am. Chem. Soc., Div. Petr. Chem.–Prepr.* 38 (2): 386–390.

Bartholdy, J. and Hannerup, P.N. 1990. Hydrodemetallation in resid hydroprocessing. *Am. Chem. Soc., Div. Petr. Chem.–Prepr.* 35 (4): 619–625.

Beaton, W.I. and Bertolacini, R.J. 1991. Resid hydroprocessing at Amoco. *Catal. Rev.–Sci. Eng*. 33 (3–4): 281–317.

Bodman, S.D., McWhinnie, W.R., Begon, V., Millan, M., Suelves, I., Lazaro, M-J., Herod, A.A., and Kandiyoti, R. 2003. Metal-ion pillared clays as hydrocracking catalysts (II): Effect of contact time on products from coal extracts and petroleum distillation residues. *Fuel* 82 (18): 2309–2321.

Bodman, S.D., McWhinnie, W.R., Begon, V., Suelves, I., Lazaro, M-J., Morgan, T.J., Herod, A.A., and Kandiyoti, R. 2002. Metal-ion pillared clays as hydrocracking catalysts (I): Catalyst preparation and assessment of performance at short contact times. *Fuel* 81 (4): 449–459.

Bogdanor, J.M. and Rase, H.F. 1986. Characteristics of commercially aged NiMo/Al_2O_3 hydrotreating catalyst: Component distribution, coke characteristics, and effects of regeneration. *Ind. Eng. Chem. Prod. Res. Dev.* 25 (2): 220–230.

Bonné, R.L.C., van Steenderen, P., and Moulijn, J.A. 2001. Hydrogenation of nickel and vanadyl tetraphenilporphyrin in absence of a catalyst. A kinetic study. *Appl. Catal. A* 206 (2): 171–181.

Braun, R.L. and Burnham, R.L. 1987. Analysis of chemical reaction kinetics using a distribution of activation energies and simpler models. *Energy Fuels* 1 (2): 153–161.

Bridge, A.G. 1990. Catalytic hydrodemetalation of heavy oils, in *Catalysis in Petroleum Refining 1989*, Trimm, D.L., Akashah, S., Absi-Halabi, M., and Bishara, A, Eds., Elsevier, Amsterdam, 363.

Brons, G. and Yu, J.M. 1995. Solvent deasphalting effects on whole Cold Lake bitumen. *Energy Fuels* 9 (4): 641–647.

Brooks, J.D. and Taylor, G.H. 1965. The formation of graphitizing carbons from the liquid phase. *Carbon* 3 (2): 185–186.

Brown, F.R., Makovsky, L.E., and Rhee, K.H. 1977. Raman spectra of supported molybdena catalysts: II. Sulfided, used, and regenerated catalysts. *J. Catal.* 50 (3): 385–389.

Byambajav, E. and Ohtsuka, Y. 2003a. Hydrocracking of asphaltene with metal catalysts supported on SBA-15. *Appl. Catal. A* 252 (1): 193–204.

Byambajav, E. and Ohtsuka, Y. 2003b. Cracking behavior of asphaltene in the presence of iron catalysts supported on mesoporous molecular sieve with different pore diameters. *Fuel* 82 (13): 1571–1577.

Chang, H.J., Seapan, M., and Crynes, B.L. 1982. *Catalyst decay during hydrotreatment of a heavy coal oil. ACS Symposium Series* 196, 309–320, Washington, D.C.

Chen, S-L., Jia, S-S., Luo, Y-H., and Zhao, S-Q. 1994. Mild cracking solvent deasphalting: A new method for upgrading petroleum residue. *Fuel* 73 (3): 439–442.

Chen, Y., Feng, Q., Shao, Y., Zhang, G., Ou, L., and Lu, Y. 2006a. Investigations on the extraction of molybdenum and vanadium from ammonia leaching residue of spent catalyst. *Int. J. Mineral Proc.* 79 (1): 42–48.

Chen, Y., Feng, Q., Shao, Y., Zhang, G., Ou, L., and Lu, Y. 2006b. Research on the recycling of valuable metals in spent Al_2O_3-based catalyst. *Minerals Eng.* 19 (1): 94–97.

Choi, J.H.K. and Gray, M.R. 1988. Structural analysis of extracts from spent hydroprocessing catalysts. *Ind. Eng. Chem. Res.* 27 (9): 1587–1595.

Chua, Y.T. and Stair, P.C. 2003. An ultraviolet Raman spectroscopic study of coke formation in methanol to hydrocarbons conversion over zeolite H-MFI. *J. Catal.* 213 (1): 39–46.

Clark, F.T., Springman, M.C., Willcox, D., and Wachs, I.E. 1993. Interactions in alumina-based iron oxide-vanadium oxide catalysts under high temperature calcination and SO_2 oxidation conditions. *J. Catal.* 139 (1): 1–18.

Coats, A.W. and Redfern, J.P. 1964. Kinetic parameters from thermogravimetric data. *Nature* 201 (4914): 68–69.

Cooper, B.H., Donnis, B.B.L., and Moyse, B.M. 1986. Hydroprocessing conditions affect catalyst shaper selection. *Oil Gas J.* 84 (49): 39–44.

Cotte, E.A. and Calderon, J.L. 1981. Pyrolysis of Boscan asphaltenes: Process description and nature of the products. *Am. Chem. Soc., Div. Petrol. Chem.–Prepr.* 26 (2): 538–547.

Dalla Lana, I.G., Fiedorow, R., Gaworski, B., and Song, X. 1997. Thermogravimetric análisis of the catalytic hydrogenation of a coal-derived liquid. *Fuel* 76 (14–15): 1503–1508.

Dautzenberg, F.M. and de Deken, J.C. 1985. Modes of operation in hydrodemetallization. *Am. Chem. Soc., Div. Petrol. Chem.–Prepr.* 30 (1): 8–20.

Dautzenberg, F.M., Van Klinken, J., Pronk, K.M.A., Sie, S.T., and Wijffels, J.B. 1978. *Catalyst deactivation through pore mouth plugging during residue desulfurization*. ACS Symposium Series 65, 254–267, Washington, D.C.

de Boer, M. 2000. Process for preparing a large pore hydroprocessing catalyst. U.S. patent 6,030,915.

de Jong, K.P. 1994. Effects of vapor-liquid equilibria on coke deposition in trickle-bed reactors during heavy oil processing. 1. Experimental results. *Ind. Eng. Chem. Res.* 33 (4): 821–824.

Del Bianco, A., Panariti, N., Anelli, M., Beltrame, P.L., and Carniti, P. 1993. Thermal cracking of petroleum residues. 1. Kinetic analysis of the reaction. *Fuel* 72 (1): 75–85.

Delmon, B. 1990. Advances in hydropurification catalysts and catalysis, in *Catalysts in Petroleum Refining 1989*, Trimm, D.L., Akashah, S., Absi-Halabi, M., and Bishara, H., Eds., Elsevier, Amsterdam, The Netherlands, 1–40.

Digne, M., Marchand, K., and Bourges, P. 2007. Monitoring hydrotreating catalysts synthesis and deactivation using Raman spectrometry. *Oil Gas Sci. Tech.–Rev. IFP* 62 (1): 91–99.

Dolbear, G.E., Tang, A., and Moorehead, E.L. 1987. Upgrading studies with Californian, Mexican and Middle Eastern heavy oils. *Fuel* 66 (2): 267–270.

Dong, D., Jeong, S., and Massoth, F.E. 1997. Effect of nitrogen compounds on deactivation of hydrotreating catalysts by coke. *Catalysis Today* 37 (3): 267–275.

Dong, X-G., Lei, Q-F., Fang, W-J., and Yu, Q-S. 2005. Thermogravimetric analysis of petroleum asphaltenes along with estimation of average chemical structure by nuclear magnetic resonance spectroscopy. *Thermochim. Acta* 427 (1–2): 149–153.

Egiebor, N.O., Gray, M.R., and Cyr, N. 1989. ^{13}C-NMR of solid organic deposits on spent hydroprocessing catalysts. *Chem. Eng. Comm.* 77 (1): 125–133.

Eijsbouts, S., Heinerman, J.J., and Elzerman, H.J.W. 1993. MoS_2 structures in high activity hydrotreating catalysts. II. Evolution of the active phase during the catalyst life cycle. Deactivation model. *Appl. Catal. A* 105 (1): 69–82.

Espinat, D., Dexpert, D., Freund, E., Martino, G., Couzi, M., Lespade, P., and Cruege, F. 1985. Characterization of the coke formed on reforming catalysts by laser Raman spectrometry. *Appl. Catal. A* 16 (3): 343–354.

Figueiras, A., Granda, M., Casal, E., Bermejo, J., Bonhomme, J., and Menéndez, R. 1998. Influence of primary QI on pitch pyrolysis with reference to unidirectional C/C composites. *Carbon* 36 (7–8): 883–891.

Fleisch, T.H., Meyers, B.L., Hall, J.B., and Ott, G.L. 1984. Multitechnique analysis of a deactivated resid demetallation catalyst. *J. Catal.* 86 (1): 147–157.

Flynn, J.H. and Wall, L.A. 1966. General treatment of the thermogravimetry of polymers. *J. Res. Natl. Bur. Stand. A* 70A (6): 487–523.

Friedman, H.L. 1964. Kinetics of thermal degradation of char-forming plastics from thermogravimetry. *J. Polymer Sci. C* 6: 183–195.

Furimsky, E. 1978. Chemical origin of coke deposited on catalyst surface. *Ind. Eng. Chem. Prod. Res. Dev.* 17 (4): 329–331.

Furimsky, E. 1998. Selection of catalysts and reactors for hydroprocessing. *Appl. Catal. A* 171 (2): 177–206.

Furimsky, E. and Yoshimura, Y. 1987. Mechanism of oxidative regeneration of molybdate catalysts. *Ind. Eng. Chem. Res.* 26 (4): 657–662.

Gardner, L.E. and Kidd, D.R. 1989. Preparation of hydrotreating catalysts from spent catalysts. U.S. patent 4,888,316.

Gentzis, T., Rahimi, P., Malhotra, R., and Hirschon, A.S. 2001. The effect of carbon additives on the mesophase induction period of Athabasca bitumen. *Fuel Proc. Tech.* 69 (3): 191–203.

Gimaev, R.N., Gubaidullin, V.Z., Rogacheva, O.V., Davydov, G.F., and Danil'yan, T.D. 1980. Mechanism of coke formation from liquid phase. *Chem. Tech. Fuels Oils* 16 (3): 196–199.

Gonçalves, M.L.A., Ribeiro, D.A., Teixeira, A.M.R.F, Teixeria, M.A.G. 2007. Influence of asphaltenes on coke formation during the thermal cracking of different Brazilian distillation residues. *Fuel* 86 (4): 619–623.

Gosselink, J.W. 1998. Sulfide catalysts in refineries. *CatTech.* 2 (2): 127–144.

Gray, M.R., Zhao, Y., McKnight, C.M., Komar, D.A., and Carruthers, J.D. 1999. Coking of hydroprocessing catalyst by residue fractions of bitumen. *Energy Fuels* 13 (5): 1037–1045.

Guibard, I., Merdrignac, I., and Kressmann, S. 2005. *Characterization of refractory sulfur compounds in petroleum residue*. ACS Symposium Series 895, 51–64, Washington, D.C.

Hannerup, P.N. and Jacobsen, A. C. 1983. A model for the deactivation of residues hydrodesulfurization catalysts. *Am. Chem. Soc., Div. Petr. Chem.–Prepr.* 28 (3): 576–599.

Hauser, A., Stanislaus, A., Marafi, A., and Al-Adwani, A. 2005. Initial coke deposition on hydrotreating catalysts. Part II. Structure elucidation of initial coke on hydrodemetallation catalysts. *Fuel* 84 (2–3): 259–269.

Ho, T.C. 1992. Study of coke formation in resid catalytic cracking. *Ind. Eng. Chem. Res.* 31 (10): 2281–2286.

Honna, K., Araki, Y., Enomoto, T., Yoshimoto, M., and Shimada, H. 2003. Titanium modified USY zeolite-based catalysts for hydrocracking residual oil (Part 1). Preparation and activity test of molybdenum supported catalysts. *J. Jpn. Petrol. Inst.* 46 (4): 249–258.

Inoue, S., Takatsuka, T., Wada, Y., Nakata, S., and Ono, T. 1998. A new concept for catalysts of asphaltene conversion. *Catalysis Today* 43 (3): 225–232.

Izquierdo, A., Carbognani, L., León, V., and Parisi, A. 1989. Characterization of Venezuelan heavy oil vacuum residua. *Petrol. Sci. Tech.* 7 (5–6): 561–570.

Jepsen, J.S. and Rase, H.F. 1981. Effect of sulfiding temperature on dispersion and chemical states of the components of Co-Mo and Ni-Mo. *Ind. Eng. Chem. Prod. Res. Dev.* 20 (3): 467–474.

Jiménez-Mateos, J.M., Quintero, L.C., and Rial, C. 1996. Characterization of petroleum bitumens and their fractions by thermogravimetric analysis and differential scanning calorimetry. *Fuel* 75 (15): 1691–1700.

Kam, E.K.T., Al-Shamali, M., Juraidan, M., and Qabazard, H. 2005. A hydroprocessing multicatalyst deactivation and reactor performance model-pilot-plant life test applications. *Energy Fuels* 19(3), 753–764.

Kanno, K., Fernandez, J.J., Fortin, F., Korai, Y., and Mochida, I. 1997. Modifications to carbonization of mesophase pitch by addition of carbon blacks. *Carbon* 35 (10–11): 1627–1637.

Koyama, H., Nagai, E., and Kumagai, H. 1996. *Catalyst deactivation in commercial residue hydrodesulfurization*. ACS Symposium Series 634, Chap. 17, Washington, D.C.

Kressmann, S., Morel, F., Harlé, V., and Kasztelan, S. 1998. Recent developments in fixed-bed catalytic residue upgrading. *Catalysis Today* 43 (3–4): 203–215.

Kubo, J., Miyagawa, R., and Takahashi, S. 1994. Radical scavenging material obtained from petroleum. *Energy Fuels* 8 (3): 804–805.

Kushiyama, S., Aizawa, R., Kobayashi, S., Koinuma, Y., and Uemasu, I. 1990. Effect of addition of sulphur and phosphorus on heavy oil hydrotreatment with dispersed molybdenum-based catalysts. *Appl. Catal. A* 63 (2): 279–292.

Lakshmanan, C.C. and White, N. 1994. A new distributed activation energy model using Weibull distribution for the representation of complex kinetics. *Energy Fuels* 8 (6): 1158–1167.

Le Lannic, K., Guibard, I., and Merdrignac, I. 2007. Behavior and role of asphaltenes in a two-stage fixed bed hydrotreating process. *Petrol. Sci. Tech.* 25 (1–2) 169–186.

Le Page, J.F., Morel, F., Trassard, A.M., and Bousquet, J. 1987. Thermal cracking under hydrogen pressure: Preliminary step to the conversion of heavy oils and residues. *Am. Chem. Soc., Div. Petr. Chem.–Prepr.* 32 (2): 470–476.

Le Page, J-F., Chatila, S.G., and Davidson, M. 1992. *Resid and Heavy Oil Processing*, Editions Technip: Paris, 169–176.

Leyva, C., Rana, M.S., Trejo, F., and Ancheyta, J. 2007. On the use of acid-base supported catalyst for catalysts for hydroprocessing of petroleum. *Ind. Eng. Chem. Res.* 46 (23): 7448–7466.

Li, S. and Yue, C. 2003. Study of different kinetic models for oil shale pyrolysis. *Fuel Proc. Tech.* 85 (1): 51–61.

Liu, C., Zhu, C., Jin, L., Shen, R., and Liang, W. 1998. Relationship between thermal reactivities and chemical compositions of vacuum residues and their SFEF asphalts. *Am. Chem. Soc., Div. Petr. Chem.–Prepr* 43 (1): 160–168.

Liu, C., Zhu, C., Jin, L., Shen, R., and Liang, W. 1999. Step by step modeling for thermal reactivities and chemical and chemical compositions of vacuum residues and their SFEF asphalts. *Fuel Proc. Tech.* 59 (1): 51–67.

Maity, S.K., Ancheyta, J., Soberanis, L., Alonso, F., and Llanos, M.E. 2003a. Alumina-titania binary mixed oxide used as support of catalysts for hydrotreating of Maya heavy crude. *Appl. Catal. A* 244 (1): 141–153.

Maity, S.K., Ancheyta, J., Soberanis, L., and Alonso, F. 2003b. Alumina-silica binary mixed oxide used as support of catalysts for hydrotreating of Maya heavy crude. *Appl. Catal. A* 250 (2): 231–238.

Maity, S.K., Ancheyta, J., Soberanis, L., and Alonso, F. 2003c. Catalysts for hydroprocessing of Maya heavy crude. *Appl. Catal. A* 253 (1): 125–134.

Maity, S.K., Ancheyta, J., and Rana, M.S. 2005a. Support effects on hydroprocessing of Maya heavy crude. *Energy Fuels* 19 (2): 343–347.

Maity, S.K., Ancheyta, J., Rana, M.S., and Rayo, P. 2005b. Effect of phosphorous on activity of hydrotreating catalyst of Maya heavy crude. *Catalysis Today* 109: 42–48.

Maity, S.K., Perez, V.H., Ancheyta, J., Rana, M.S., and Centeno, G. 2007. Effect of asphaltene contained in feed on deactivation of Maya crude hydrotreating catalyst. *Petroleum Science and Technology* 25 (1): 241–249.

Maki, T., Takatsuno, A., and Miura, K. 1997. Analysis of pyrolysis reactions of various coals including Argonne premium coals using a new distributed activation energy model. *Energy Fuels* 11 (5): 972–977.

Marafi, M. and Stanislaus, A. 1997. Effect of initial coking on hydrotreating catalyst functionalities and properties. *Appl. Catal. A* 159 (1–2): 259–267.

Marafi, M. and Stanislaus, A. 2001. Influence of catalyst acidity and feedstock quality on hydrotreating catalyst deactivation by coke deposition. *Petrol. Sci. Tech.* 19 (5–6): 697–710.

Marafi, M. and Stanislaus, A. 2008. Preparation of heavy oil hydrotreating catalyst from spent residue hydroprocessing catalysts. *Catal. Today* 130 (2–4): 421–428.

Marafi, M., Stanislaus, A., and Absi-Halabi, M. 1994. Heavy oil hydrotreating catalyst rejuvenation by leaching of foulant metals with ferric nitrate-organic acid mixed reagents. *Appl. Catal. B* 4 (1): 19–27.

Marafi, A., Almarri, M., and Stanislaus, A. 2008a. The usage of high metal feedstock for the determination of metal capacity of ARDS catalyst system by accelerated aging tests. *Catalysis Today* 130 (2–4): 395–404.

Marafi, M., Stanislaus, A., and Kam, E. 2008b. A preliminary process design and economic assessment of a catalyst rejuvenation process for waste disposal of refinery spent catalysts. *J. Environ. Manag.* 86 (4): 665–681.

Massoth, F.E. 1981. Characterization of coke on coal catalysts by an oxidation technique. *Fuel Proc. Tech.* 4 (1): 63–71.

Matsushita, K., Marafi, A., Hauser, A., and Stanislaus, A. 2004. Relation between relative solubility of asphaltenes in the product oil and coke deposition in residue hydroprocessing. *Fuel* 83 (11–12): 1669–1674.

Miura, K. 1995. A new simple method to estimate f(E) and k_0(E) in the distributed activation energy model from three sets of experimental data. *Energy Fuels* 9 (2): 302–307.

Miura, K. and Maki, T. 1998. A simple method for estimating f(E) and k_0(E) in the distributed activation energy model. *Energy Fuels* 12 (5): 864–869.

Mochida, I. 1999. Structural insights to heavy resid and coal for designing upgrading. *Am. Chem. Soc., Div. Fuel Chem.* 44 (4): 791–795.

Mochida, I., Inoue, S-I., Maeda, K., and Takeshita, K. 1977. Carbonization of aromatic hydrocarbons—VI: Carbonization of heterocyclic compounds catalyzed by aluminum chloride. *Carbon* 15 (1): 9–16.

Morel, F., Kressmann, S., Harlé, V., and Kasztelan, S. 1997. Processes and catalysts for hydrocracking of heavy oil and residues, in *Hydrotreatment and Hydrocracking of Oil Fractions*, Froment, G.F., Delmon, B., and Grange, P., Eds., Elsevier, Amsterdam, The Netherlands, 1–16.

Muegge, B.D. and Massoth, F.E. 1991. Basic studies of deactivation of hydrogenation catalysts with anthracene. *Fuel Proc. Tech.* 29 (1–2): 19–30.

Nandi, B.N., Belinko, K., Ciavaglia, L.A., and Pruden, B.B. 1978. Formation of coke during thermal hydrocracking of Athabasca bitumen. *Fuel* 57 (5): 265–268.

Nielsen, A., Cooper, B.H., and Jacobsen, A.C. 1981. Composite catalyst beds for hydroprocessing of heavy residua. *Am. Chem. Soc., Div. Petrol. Chem.–Prepr.* 26 (2): 440–455.

Nitta, H., Takatsuka, T., Kodama, S., and Yokoyama, T. 1979. Deactivation model for residual hydrodesulfurization catalysts, Paper 34E, presented at the 86th AIChE Annual Meeting, Houston, TX.

Penrose, C.F., Wallace, P.S., Kasbaum, J.L., Anderson, M.K., and Preston, W.E. 1999. Enhancing refinery profitability by gasification, hydroprocessing and power generation. Gasification Technologies Conference. San Francisco:http://www.gasification.org/Docs/1999_Papers/GTC99270.pdf.

Rahimi, P., Gentzis, T., and Cotte, E. 1998. Observation of optical behavior of heavy oil fractions obtained by ion exchange chromatography. *Am. Chem. Soc., Div. Petrol. Chem.–Prepr.* 43 (4): 619–622.

Rahmani, S., McCaffrey, W., Elliott, J.A.W., and Gray, M.R. 2003. Liquid-phase behavior during the cracking of asphaltenes. *Ind. Eng. Chem. Res.* 42 (17): 4101–4108.

Rajagopalan, K. and Luss, D. 1979. Influence of catalyst pore size on demetallation rate. *Ind. Eng. Chem. Proc. Des. Dev.* 18 (3): 459–465.

Rana, M., Ancheyta, J., Rayo, P., and Maity, S.K. 2004. Effect of alumina preparation on hydrodemetallization and hydrodesulfurization of Maya crude. *Catalysis Today* 98 (1–2): 151–160.

Rana, M.S., Ancheyta, J., Maity, S.K., and Rayo, P. 2005a. Characteristics of Maya crude hydrodemetallization and hydrodesulfurization catalysts. *Catalsis Today* 104 (1): 86–93.

Rana, M.S., Anchetya, J., Maity, S.K., and Rayo, P. 2005b. Maya crude hydrodemetallization and hydrodesulfurization catalysts: an effect of TiO_2 incorporation in Al_2O_3. *Catalsis Today* 109 (1–4): 61–68.

Rana, M.S., Ancheyta, J., Rayo, P., and Maity, S.K. 2007a. Heavy oil hydroprocessing over supported NiMo sulfided catalyst: an inhibition effect by added H_2S. *Fuel* 86 (9): 1263–1269.

Rana, M.S., Sámano, V., Ancheyta, J., and Diaz, J.A.I. 2007b. A review of recent advances on process technologies for upgrading of heavy oils and residua. *Fuel* 86 (9): 1216–1231.

Rana, M.S., Ancheyta, J., Maity, S.K., and Rayo, P. 2007c. Hydrotreating of Maya crude oil: I. Effect of support composition and its pore-diameter on asphaltene conversion. *Petrol. Sci. Tech.* 25 (1–2): 187–200.

Rana, M.S., Ancheyta, J., Maity, S.K., and Rayo, P. 2007d. Hydrotreating of Maya crude Oil: II. Generalized relationship between hydrogenolysis and hydrodeasphaltenization (HDAs). *Petrol. Sci. Tech.* 25 (1–2): 201–214.

Rana, M.S., Anchetya, J., Rayo, P., and Maity, S.K. 2007e. Catalyst for a first-step hydrodemetallization in a multi-reactor hydroprocessing system for upgrading of heavy and extra-heavy crude oils. Mexican Patent applied, IMPI, Mx/a/2007/009504.

Rana, M.S., Ancheyta, J., Maity, S.K., and Rayo, P. 2008. Maya heavy crude oil hydroprocessing: A zeolite based CoMo catalysts and its spent catalyst characterization. *Catalysis Today*, 130 (2–4): 411–420.

Rayo, P., Ancheyta, J. Ramírez, J., and Gutiérrez-Alejandre, A. 2004. Hydrotreating of diluted Maya crude with NiMo/Al_2O_3-TiO_2 catalysts: Effect of diluent composition. *Catalysis Today* 98 (1–2): 171–179.

Rayo, P., Ramírez, J., Ancheyta J., and Rana, M.S. 2007. HDS, HDN, HDM and HDAs of Maya crude over NiMo/Al_2O_3 modified with Ti and P. *Petrol. Sci. Tech.,* 25 (1–2): 215–229.

Reynolds, J.G. 1998. Metals and heteroelements, in *Heavy Crude Oils. Petroleum Chemistry and Refining*, Speight, J.G., Ed., Taylor & Francis, Washington, D.C., Chap. 3.

Reynolds, J.G., Burnham, A.K., and Mitchell, T.O. 1995. Kinetic analysis of California petroleum source rocks by programmed temperature micropyrolysis. *Org. Geochem.* 23 (2): 109–120.

Richardson, S.M., Nagaishi, H., and Gray, M.R. 1996. Initial coke deposition on a NiMo/γ-Al_2O_3 bitumen hydroprocessing catalyst. *Ind. Eng. Chem. Res.* 35 (11): 3940–3950.

Sahoo, S.K., Ray, S.S., and Singh, I.D. 2004. Structural characterization of coke on spent hydroprocessing catalysts used for processing of vacuum gas oils. *Appl. Catal. A* 278 (1): 83–91

Saito, T. and Kumata, F. 1998. Bench scale plant tests of three kinds of different pore size HDS catalyst and characteristics of coked catalysts: Effect of extraction using LGO. *Am. Chem. Soc., Div. Petr. Chem.–Prepr.* 43 (4): 628–633.

Sajkowski, D.J., Roth, S.A., Iton, L.E., Meyers, B.L., Marshall, C.L., Fleisch, T.H., and Delgass, W.N. 1989. X-ray absorption study of vanadium on regenerated catalytic-cracking catalysts. *Appl. Catal. A* 51 (1): 255–262.

Sakanishi, K., Manabe, T., Watanabe, I., and Mochida, I. 1999. Characterization and adsorption treatment of vacuum residue fractions with carbons. *Am. Chem. Soc., Div. Fuel Chem.* 44 (4): 759–761.

Sato, M., Takayama, N., Kurita, S., and Kwan, T. 1971. Vanadium and nickel deposits distribution in desulfurization catalysts. *Nippon Kagaku Zasshi* 92 (10): 834.

Satterfield, C.N., Modell, M., Hites, R.A., and Declerck, C.J. 1978. Intermediate reactions in the catalytic hydrodenitrogenation of quinoline. *Ind. Eng. Chem. Prod. Res. Dev.* 17 (2): 141–148.

Savage, P.E., Klein, M.T., and Kukes, S.G. 1988. Asphaltene reaction pathways. 3. Effect of reaction environment. *Energy Fuels* 2 (5): 619–628.

Sawarkar, A.N., Pandit, A.B., and Joshi, J.B. 2007. Studies in coking of Arabian mix vacuum residue. *Chem. Eng. Res. Des.* 85 (A4): 481–491.

Schabron, J.F., Pauli, A.T., and Rovani, J.F. 2003. Coke predictability maps: Stability and compatibility of heavy oils and residua. *Am. Chem. Soc., Div. Petr. Chem.–Prepr.* 48 (1): 1–5.

Schucker, R.C. 1983. Thermogravimetric determination of the coking kinetics of Arab heavy vacuum residuum. *Ind. Eng. Chem. Proc. Des. Dev.* 22 (4): 615–619.

Shih, S-M. and Sohn, H.Y. 1980. Nonisothermal determination of the intrinsic kinetics of oils generation from oil shale. *Ind. Eng. Chem. Proc. Des. Dev.* 19 (3): 420–426.

Shimada, H., Sato, K., Honna, K., Enomoto, T., and Ohshio, N. 2009. Design and development of Ti-modified zeolite-based catalyst for hydrocracking heavy petroleum. *Catalysis Today* 141 (1–2): 43–51.

Simpson, H.D. 1991. Implications of analyses of used resid detemetallation catalysts, in *Catalyst Deactivation 1991*, Bartholomew, C.H., and Butt, J.B., Eds., Elsevier, Amsterdam, The Netherlands, 265–272.

Snape, C.E., Tyagi, Y.R., Castro-Díaz, M., Hall, P.J., Murray, I.P., and Martin, S.C. 1999. Characterisation of coke on deactivated hydrodesulfurisation catalysts and a novel approach to catalyst regeneration. *Am. Chem. Soc., Div. Fuel Chem.* 44 (4): 832–836.

Speight, J.G. 1987. Initial reactions in the coking of residua. *Am. Chem. Soc., Div. Petr. Chem.–Prepr.* 32 (2): 413–418.

Speight, J.G. 1998. The chemistry and physics of coking. *Korean J. Chem. Eng.* 15 (1): 1–8.

Speight, J.G. 2004. Petroleum asphaltenes. Part 2. The effect of asphaltenes and resins constituents on recovery and refining processes. *Oil Gas Sci. Tech.–Rev. IFP* 59 (5): 479–488.

Stanislaus, A., Absi-Halabi, M., and Al-Dolama, K. 1988. Effect of phosphorus on the acidity of γ-alumina and on the thermal stability of γ-alumina supported nickel-molybdenum hydrotreating catalysts. *Appl. Catal. A* 39 (1–2): 239–253

Storm, D.A., Barresi, R.J., and Sheu, E.Y. 1996. Flocculation of asphaltenes in heavy oil at elevated temperatures. *Petrol. Sci. Tech.* 14 (1–2): 243–260.

Storm, D.A., Barresi, R.J., Sheu, E.Y., Bhattacharya, A.K., and DeRosa, T.F. 1998. Microphase behavior of asphaltic micelles during catalytic and thermal upgrading. *Energy Fuels* 12 (1): 120–128.

Sughrue, E.L., Adarme, R., Johnson, M.M., Lord, C.J., and Phillips, M.D. 1991. Demetallization of asphaltenes: Modeling feed and product molecular size distributions and metal distribution parameters, in *Catalyst Deactivation 1991*, Bartholomew, C.H., and Butt, J.B., Eds., Elsevier, Amsterdam, The Netherlands, 281–288.

Sun, D.D., Tay, J.H., Cheong, H.K., Leung, D.L.K., and Qian, G.R. 2001. Recovery of heavy metals and stabilization of spent hydrotreating catalyst using a glass–ceramic matrix. *J. Hazardous Materials* 87 (1–3): 213–223.

Takahashi, T., Higashi, H., and Kai, T. 2005. Development of a new hydrodemetallization catalyst for deep desulfurization of atmospheric residue and the effect of reaction temperature on catalyst deactivation. *Catalysis Today* 104 (1): 76–85.

Takatsuka, T., Kajiyama, R., Hashimoto, H., Matsuo, I., and Miwa, S. 1989. A practical model of thermal cracking of residual oil. *J. Chem. Eng. Jpn.* 22 (3): 304–310.

Takeuchi, C., Fukui, Y., Nakamura, M., and Shiroto, Y. 1983. Asphaltene cracking in catalytic hydrotreating of heavy oils. 1. Processing of heavy oils by catalytic hydroprocessing and solvent deasphatting. *Ind. Eng. Chem. Proc. Des. Dev.* 22 (2): 236–242.

Tamm, P.W., Harnsberger, H.F., and Bridge, A.G. 1981. Effects of feed metals on catalyst aging in hydroprocessing residuum. *Ind. Eng. Chem. Proc. Des. Dev.* 20(2), 262–273.

Tanabe, K. and Gray, M.R. 1997. Role of fine solids in the coking of vacuum residues. *Energy Fuels* 11 (5): 1040–1043.

Tanaka, R., Hunt, J.E., Winans, R.E., Sato, P.T., and Takahashi, T. 2003. Aggregates structure analysis of petroleum asphaltenes with small-angle neutron scattering. *Energy Fuels* 17 (1): 127–134.

Teixeira da Silva, V.L.S., Lima, F.P., Dieguez, L.C., and Schmal, M. 1998. Regeneration of a deactivated hydrotreating catalyst. *Ind. Eng. Chem. Res.* 37 (3): 882–886.

Thakur, D.S. and Thomas, M.G. 1984. Catalyst deactivation during direct coal liquefaction: A review. *Ind. Eng. Chem. Proc. Des. Dev.* 23 (3): 349–360.

Thakur, D.S. and Thomas, M.G. 1985. Catalyst deactivation during coal liquefaction: A review. *Appl. Catal. A* 45 (2): 197–225.

Toulhoat, H. and Plumail, J.C. 1990. Upgrading heavy ends into marketable distillates: New concepts and new catalyst for two key stages, HDM and HDN, in *Catalysis in Petroleum Refining 1989*, Trimm, D.L., Akashah, S., Absi-Halabi, M., and Bishara, A., Eds., Elsevier, Amsterdam, The Netherlands, 463.

Trejo, F., Rana, M.S., and Ancheyta, J. 2008. CoMo/MgO–Al_2O_3 supported catalysts: An alternative approach to prepare HDS catalysts. *Catalysis Today* 130 (2–4): 327–336.

Trimm, D.L. 1996. Catalytic deactivation, In *Catalyst in Petroleum Refining and Petrochemical Industries 1996*, Absi-Halabi, M., Beshara, J., Qabazard, H., and Stanislaus, A., Eds., Elsevier: Amsterdam, The Netherlands, 65–76.

Ushio, M., Futigami, J., Kuroda, R., Ida, T., Hayashida, K., and Ishihara, H. 2005. Study on DAO hydrocracking with zeolitic catalyst. *Am. Chem. Soc., Div. Petr. Chem.–Prepr.* 50 (4): 347–350.

Van Doorn, J. and Moulijn, J.A. 1990. Carbon deposition on catalysts. *Catalysis Today* 7 (2): 257–266.

Wallace, P.S., Anderson, M.K., Rodarte, A.I., and Preston, W.E. 1998. Heavy oil upgrading by the separation and gasification of asphaltenes. Paper presented at the Gasification Technologies Conference. San Francisco.

Wallace, P.S. and Johnson, K.A. 2002. Integration of solvent deasphalting, gasification, and hydrotreating. U.S. Patent 6,409,912.

Wang, Z., Que, G., Liang, W., and Qian, J. 1998. Chemical structure analysis of Shengli crude vacuum residuum related to thermal conversion. *Am. Chem. Soc., Div. Petr. Chem.–Prepr.* 43 (1): 151–154.

Ware, R.A. and Wei, J. 1985a. Catalytic hydrodemetallation of nickel porphyrins I. Porphyrin structure and reactivity. *J. Catal.* 93 (1): 100–121.

Ware, R.A. and Wei, J. 1985b. Catalytic hydrodemetallation of nickel porphyrins II. Effects of pyridine and of sulfiding. *J. Catal.* 93 (1): 122–134.

Wiehe, I.A. 1992. A solvent-resid phase diagram for tracking resid conversion. *Ind. Eng. Chem. Res.* 31 (2): 530–536.

Wiehe, I.A. 1993. A phase-separation kinetic model for coke formation. *Ind. Eng. Chem. Res.* 32 (11): 2447–2454.

Wiehe, I.A. 1994. The pendant-core building block model of petroleum residua. *Energy Fuels* 8 (3): 536–544.

Wiwel, P., Zeuthen, P., and Jacobsen, A.C. 1991. Initial coking and deactivation of hydreotreating catalysts by real feeds, in *Catalyst Deactivation 1991,* Bartholomew, C.H. and Butt, J.B., Eds., Elsevier, Amsterdam, The Netherlands, 257–264.

Yoshimura, Y., Matsubayashi, N., Yokokawa, H., Sato, T., Shimada, H., and Nishijima, A. 1991. Temperature-programmed oxidation of sulfided cobalt-molybdate/alumina catalysts. *Ind. Eng. Chem. Res* 30 (6): 1092–1099.

Yoshimura, Y., Sato, T., Shimada, H., Matsubayashi, N., Imamura, M., Nishijima, A., Yoshitomi, S., Kameoka, T., and Yanase, H. 1994. Oxidative regeneration of spent molybdate and tungstate hydrotreating catalysts. *Energy Fuels* 8 (2): 435–445.

Zeuthen, P., Bartholdy, J., and Massoth, F.E. 1995. Temperature-programmed oxidation studies of aged V-containing hydroprocessing catalysts. *Appl. Catal. A* 129 (1): 43–55.

Zhang, L., Yang, G., and Que, G. 2005. The conglomerating characteristics of asphaltenes from residue during thermal reaction. *Fuel* 84 (7–8): 1023–1026.

Zhang, L., Yang, G., Que, G., Zhang, Q., and Yang, P. 2006. Colloidal stability variation of petroleum residue during thermal reaction. *Energy Fuels* 20 (5): 2008–2012.

Zhang, X. 2006. The impact of multiphase behavior on coke deposition in heavy oil hydroprocessing catalysts. PhD diss., University of Alberta, Edmonton, Canada.

Zhao, S., Kotlyar, L.S., Woods, J.R., Sparks, B.D., Hardacre, K., and Chung, K.H. 2001. Molecular transformation of Athabasca bitumen end-cuts during coking and hydrocracking. *Fuel* 80 (8): 1155–1163.

5 Sediments Formation

5.1 INTRODUCTION

The complex and heaviest part of petroleum is the asphaltene, which is, in large amount, in heavy or extra heavy crude oils. The asphaltene has been defined first by the French scientist J. B. Boussingault (1837) when he published his celebrated journal on the *Composition of Bitumen* in the year 1837. Later, in 1859, crude oil had been treated more extensively in Pennsylvania in order to distillate and discover commercial applicability of different fractions. Since then refiners and technology developers have been working to figure out what to do with the rest of the barrel. The importance of crude oil began in the late 1990s, when the price of bottom of barrel increased due to the decline in the light feedstock. Since that time, steady progress was made over the past several years. The crude oil processing generally starts with desalting and, subsequently, atmospheric and vacuum distillations; after that, due to the large amount of bottom of barrel, atmospheric residue (AR) and/or vacuum residue (VR) are converted into lighter fractions with hydrogen addition (catalytic) or carbon rejection (thermal) processes. Direct hydroprocessing of heavy crude oils to produce upgraded oil has also been recently proposed (Ancheyta and Rana, 2008c).

The term *heavy oil* refers to heavy or extra heavy crudes, oil sands bitumen, bottom of barrel, and residua that are left after the distillation process. Heavy oils contain a substantial quantity of high boiling fractions that include a significant quantity of asphaltenes. When processing heavy oils, these asphaltenes have detrimental effects on the supported catalyst due to the formation of coke, sludge, and sediments, particularly in the hydroprocessing operations. Moreover, crude oil contains other impurities, e.g., metals (Ni, V), sulfur, carbon residue, etc., which contribute to coke formation, catalyst deactivation, and product degradation in refinery. The amount of these impurities vary depending upon the particular crude oil origin.

The word *asphaltenes* means a heavy polar fraction and is the heaviest molecule present in petroleum, which concentration increases after the distillation. Asphaltenes from vacuum residue are generally characterized by Conradson or Ramsbottom carbon residue of 15 to 90 wt% that has a hydrogen-to-carbon (H/C) atomic ratio of 0.5 to 1.5. Generally, asphaltenes are cages of metal porphyrins in which metal contains from 0.05 wt% to over 0.5 wt% vanadium and from 0.02 wt% to over 0.2 wt% nickel. The sulfur and nitrogen concentration in asphaltenes can be from 100 to 350% greater than its content in the crude oil.

Usually, before hydroprocessing of this kind of heavy oil feedstock, a dispersant or antifoulant is used, particularly for crude oils where the ratio of resin to asphaltene is not high enough to prevent asphaltene flocculation and, as a result,

TABLE 5.1
Different Crude Oils and Their Asphaltene and Resin Composition

Country, Crude	Resin	Asphaltene
Canada, Athabasca	14	15
Venezuela, Boscan	29.4	17.2
Canada, Cold Lake	25	13
Mexico, Panucon	26	12.5
Iraq, Qayarah	36.1	20.4
Canada, Lloyminster	38.4	12.9
USA, MS, Baxterville	8.9	17.2
Russia, Balachany	6	0.5
Russia, Bibi-Eibat	9	0.3
USA, TX, Mexia	5	1.3
Iraq, Kirkuk	15.5	1.3
Mexico, Isthmus	8.1	1.3
USA, Tonkawa	2.5	0.2
France, Lagrave	7.5	4
Algeria, Hassi Messaoud	3.3	0.15

heavy organic depositon carried out known as *sediments* (LePage et al., 1992; Tojima et al., 1998). The most important part of heavy oil feedstocks is the resin content, which acts as a peptizing agent of the asphaltene molecules. The content of asphaltenes and resin depends on the source of crude, as shown in Table 5.1. Crude oils with higher asphaltene-to-resin ratios are more prone to heavy organics deposition or sediment formation; the ratio is also affected by the API (American Petroleum Institute) gravity of the crude oil, as shown in Figure 5.1. Thus, a crude oil that has low API gravity or a high asphaltene/resin ratio is a crude with frequent deposition problems, e.g., Boscan crude. However, one should not consider only this ratio as a factor behind sediment formation, but polydispersivity and polarities of asphaltenes and resins also play an important role. Moreover, the presence of compounds like paraffins, wax, and organometallics and their electrokinetics behavior are also responsible for coke deposition. In order to quantify all of these factors, one has to develop a predictive model in which all of these effects are incorporated (Mansoori, 1996). The high temperature during hydroprocessing reactions can lead to the formation of insoluble material that is also known as sludge products. Therefore, stability of the asphaltenes is a fine balance between intermolecular associations and solvent–solute interactions.

The compatibilities of these fractions mostly depend on the physical properties of the various components, such as dipole moment of asphaltenes and resins. The determination of asphaltene and resin dipole moments is important, which allows a valuable database for studying the effect of resins on precipitation and the modeling

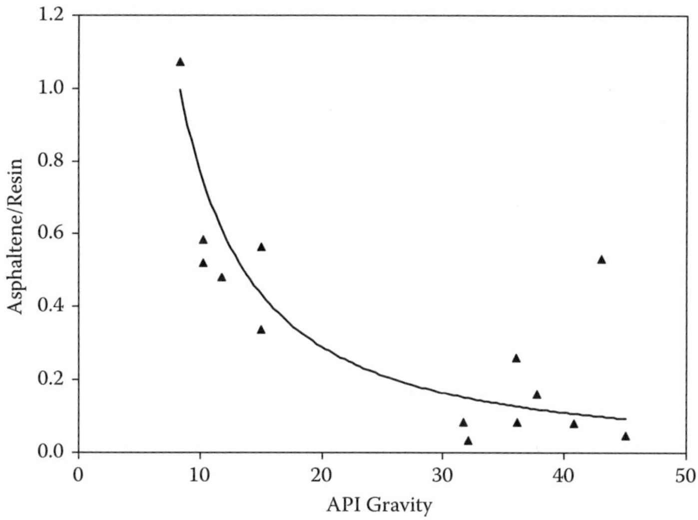

FIGURE 5.1 Effect of API gravity on asphaltene and resin obtained from virgin crude oil.

of asphaltene–asphaltene and asphaltene–resin interactions (Goual and Firoozabadi, 2002). Physical properties of various crude oils are reported in Table 5.2, in which contents of oils, resins, and asphaltene are compared for a given petroleum. Asphaltenes have higher dipole moments than resins; however, resins from certain petroleum can have higher dipole moments than asphaltenes obtained from another petroleum. Thus, sediment deposition over the catalyst, as well as the reactor, can be avoided due to the balance between different hydrocarbon compositions.

In general, sediments are undesired hydrocarbon, whose effluence can persist over a long time and can have direct toxic effects on human health. Because there is no beneficiary effect from sediments, it is just a heavy hydrocarbon that has a very high amount of metals. Recently, Wu et al. (2008) investigated the possibility of the conversion of thermally treated sediment into zeolite by adopting the fusion method. On the other hand, solid sediments can be considered as a solid coke that is relatively better than coke, particularly in countries like China, Brazil, The Netherlands, and India where the industrial boom is expected. To give an example, the United States exports about half (46%) of its coal to Canada and Brazil. In 2007, Canada received roughly the same amount of coal as in 2006, while Brazil's import increased dramatically, which was almost 50% more than in the previous year (EIA, 2008). Moreover, sediments have high amounts of metal, which is another benefit for the steel industries.

Apart from the catalyst, there are organic surfactants (i.e., additives) or hydrogen donor solvents that have been used to reduce the interfacial tension and change the physical properties of crude oil or residue medium (Kubo et al., 1994, 1996; Ovalles et al., 2001; Rodolfo et al., 1988). The additives either instantly improve transportation of crude oil or help with the miscibility of asphaltene into

TABLE 5.2
Physical Properties of Crude Oil at Ambient Conditions (Dielectric Constant, Refractive Index, Density, and Molecular Weight of n-C_5 Asphaltenes, Resins, and Oils

Properties		Crude Oil H	TE	C	TA	TK	S	U	B
API gravity		9	10	33	37	35	32	34	56
Density at 20°C, g/cm³		1.009	0.999	0.859	0.837	0.852	0.863	0.855	0.755
n-C_5	$n^{20°C}$	1.647	1.707	1.638	1.676	1.706	1.668	1.685	1.65
Asphaltene	ε_r	16.2	18.4	15.2	9.2	10.9	10.1	8.2	5.5
Insolubles	$P_{20°C}$, g/cm³	1.2	1.2	1.2	1.2	1.2	1.2	1.2	1.2
	M, g/mol	900	900	900	900	900	900	900	900
Resins	$n^{20°C}$	1.576	1.587	1.606	1.585	1.608	1.608	1.595	1.606
	ε_r	3.9	3.8	5.1	4.8	4.5	3.9	3.8	4.7
	$P_{20°C}$, g/cm³	1.0	1.0	1.0	1.0	1.0	1.0	1.0	1.0
	M, g/mol	700	700	700	700	700	700	700	700
Oils	$n^{20°C}$	1.527	1.522	1.487	1.497	1.496	1.498	1.496	1.472
	ε_r	2.6	2.5	2.2	2.3	2.4	2.4	2.4	2.1
	$P_{20°C}$, g/cm³	0.935	0.935	0.875	0.876	0.876	0.87	0.881	0.83
	M, g/mol	371	389	269	287	300	290	320	196

Note: Dielectric constant (E_r) = C/C_0, where C and C_0 are the capacitance of the condenser filled with the dielectric and air, respectively.

Source: Goual, L. and Firoozabadi, A. 2002. *AIChE J.* 48 (11): 2646–2663. Reprinted with permission of John Wiley & Sons, Inc.

the oil and resin of the system. Thus, during hydroprocessing, additives prevent the condensations of decomposed asphaltene molecules and this fraction remains in stable or in micelle form. In addition, the surfactant will be adsorbed on the asphaltene particles and prevent the agglomeration of coke-like solid material.

Generally, sediment is a common contaminant of fuels and its presence provokes coke formation over hydroprocessing catalysts. It usually consists of rust and other insoluble impurities like metals. To address this problem, hydroprocessed products have been analyzed by using standard methods to remove sediments as solid material.

5.2 SEDIMENTS ANALYSIS OR SEDIMENT TESTS

There are various methods that are used for sediment analysis that relate to different aspects of compatibility between the aliphatic, aromatic, and asphaltene compositions. The most common method used is specified as SASTM D4870-04 (standard test method for determination of total sediment in residual fuel), and quantifies sludge and sediments formation in hydroprocessed products. This

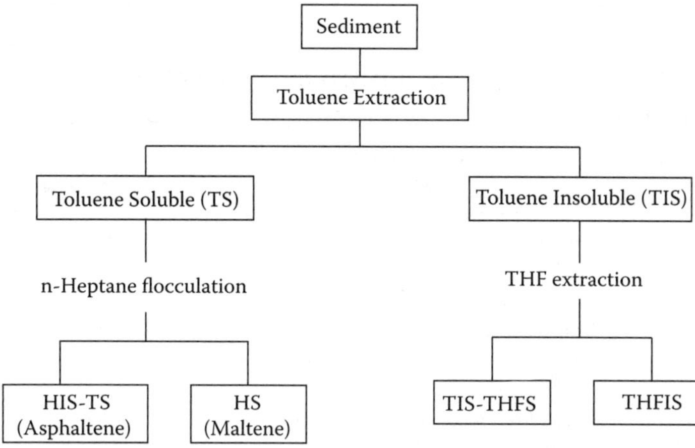

FIGURE 5.2 Extraction of sediments. (From Stanislaus et al. 2005. *Catalysis Today* 109 (1/4): 167–177. With permission.)

method is based on or similar to the IP 375, which was established by Institut Français du Petrole (IFP) (*Sedimenti totali in oli combustibili da residui—Parte 1: Determinazione mediante filtrazione a caldo*, IP 375-99). Determination of total sediment is routinely carried out by up to 0.40 wt% for distillate fuel oils containing residual components and to 0.50 wt% in residual fuel oils having a viscosity of 55 cSt (mm^2/s) at 100°C. However, some fuels may exceed the limitations specified in this test method due to various factors, such as the presence of significant quantities of insoluble organic or inorganic material. The method quantifies total sediments as the sum of the insoluble organic and inorganic material that is separated from the bulk of the hydroprocessed oil by filtration through a Whatman GF/A filter medium, and that is also insoluble in a predominantly paraffinic solvent. The segments can also be characterized by different spectroscopic techniques as well as chemical treatments (Figure 5.2). However, there are other methods for determining sediment content for different fractions of petroleum, as listed in Table 5.3. The analysis method is applicable to residual products from virgin crude oil, atmospheric and vacuum residua, thermal, catalytic, and hydrocracking processes providing these products contain 0.05 wt% or greater concentration of asphaltenes, which are responsible for sediment formation.

In general, sediments are separated from the oil sample and filtered at 100°C. After solvent washing and drying, the total sediment on the filter medium is weighed. Finally sediments are Soxhlet extracted using methylene chloride for about 24 to 48 h. Oil and maltene fractions are isolated by asphaltene precipitation with *n*-pentane or *n*-heptane, while sediment and oil extracts are purified by using alumina microcolumns, and saturated fractions are separated by fractionation on silica gel. This analytical procedure has been validated on reference material sediment (SRM 1941a) and a reference material crude oil (SRM 1582) of Atlantic coast after the Erika oil spill in December 1999.

TABLE 5.3
Methods Used for Sediment Analysis for Different Fraction of Petroleum

ASTM	Detail Analysis Method	Designation
D1796-04	Standard Test Method for Water and Sediment in Fuel Oils by the Centrifuge Method (Laboratory Procedure)	Chapter 10.6[a]
D4007-02	Standard Test Method for Water and Sediment in Crude Oil by the Centrifuge Method (Laboratory Procedure)	Chapter 10.3[a]
D473-07	Standard Test Method for Sediment in Crude Oils and Fuel Oils by the Extraction Method	53/00[b]
D4807-05	Standard Test Method for Sediment in Crude Oil by Membrane Filtration	Chapter 10.8[a]
D4870-07a	Standard Test Method for Determination of Total Sediment in Residual Fuels	375/99[b]

[a] Manual of Petroleum Measurement Standards (MPMS)
[b] IP

5.3 POSSIBLE REACTION MECHANISM AND COMPATIBILITIES BETWEEN THE COMPONENTS

Upgrading of heavy crude oils is due to the increasing demand for light crude oils that are on a decline worldwide. Future upgrading refineries must deal with processing heavier and more complex feedstocks for the production of light or transportation fuels (Ancheyta and Rana, 2008). It is a common practice in current refineries that such feedstocks are processed using thermal route, e.g., delayed coking that results in low value fuel. Therefore, in order to maximize fuel quality, as well as light fraction quantity, from heavier fractions of petroleum, treatment in the presence of hydrogen and a catalyst (catalytic hydroprocessing) is a more interesting alternative that has lower affinity to form coke by free radical polymerization reactions. However, the catalyst experiences relatively fast deactivation at elevated temperatures that are necessary in order to increase conversion of feedstocks. The deactivation in heavy oil processing catalysts is due either to the burying of catalytic sites by metal deposition (i.e., slow but progressive in nature) or to the deposition of solid hydrocarbon (i.e., coke, fast at initial time-on-stream (TOS) and due to the nature of acidic sites) on the surface of catalyst. More details about catalyst deactivation due to asphaltenes are given in Chapter 4. Deactivation is believed to occur due to the imbalance between the reactant and product compatibilities of the three principal components of crude oil, as shown in Figure 5.3 and Figure 5.4, where the unstable region corresponds to the semisolid or sediment formation

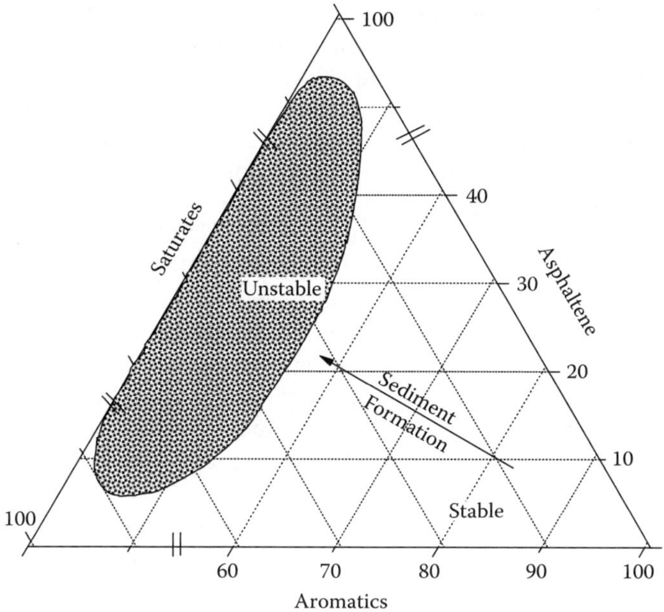

FIGURE 5.3 Representation of petroleum as a three-phase system showing a region of instability. (From Speight, J.G. 1999. *J. Petrol. Sci. Eng.* 22: 3–15. With permission.)

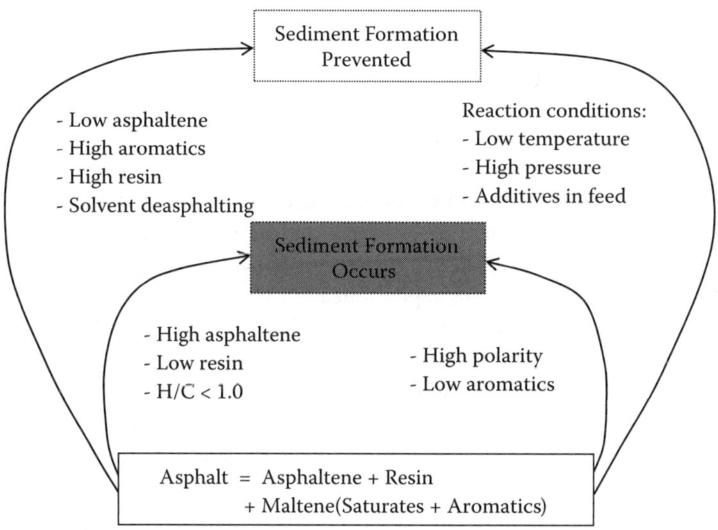

FIGURE 5.4 Possible changes in solubility of asphaltene and presence of different components that leads to sediment formation.

(Speight, 1999; Takatsuka et al., 1989a, 1989b; Stanislaus et al., 1998, 2005; Marafi, 2005a, 2005b; Storm et al., 1994; Strausz et al., 1999).

Asphaltenes play a principal role to formation of sediments, which are present in high concentration and behave as colloidal species. Asphaltene has the ability to form micelles, aggregates, and can flocculate depending on the reaction temperature, pressure, and solvating power of the other hydrocarbons present in the feedstock. On the other hand, the use of high temperatures in hydroprocessing units results in a shorter catalyst life and a dry sludge (condensed carbonaceous material) formation. When residue conversion exceeds certain limits (generally 50%), dry sludge formation dramatically increases due to the incompatibility of polymerized and condensed asphaltenic components in the product. Once it is formed (sludge), it easily deposits on catalyst surface, flash drums, fractionators, heat exchangers, and other downstream equipments (Marafi et al., 2005b; Bartholdy and Andersen, 2000). The maximum conversion attained is then determined by the limit of dry sludge formation rather than by the limit of catalyst activity or other operating and equipment constrains. Analysis of dry sludge formation was performed by Takatsuka et al. (1989a) with the aid of mathematical models. Figure 5.3 shows the ternary composition diagram obtained from the model calculations, showing the dependence of asphaltene solubility on the components of oil products. Line A and line B in the ternary compatibility diagram, shown in Figure 5.5, mark the maximum conversion boundary regulated by asphaltenes content and hydrogenation of product oil, respectively. In addition, stabilization of asphaltene molecules as a dispersion in crude oil has been attributed to a peptizing effect of resins (Cimino et al., 1995; Clarke and Pruden, 1998; Rahimi et al., 2001; Tojima et al., 1998), which is extremely important from a processing point of view because of the severe operational problems encountered in an undesired separation of asphaltenes (Strausz et al., 1999; Marafi et al., 2005a).

Usually, spherical asphaltenes aggregate in petroleum medium and are surrounded by an average resin molecule in the shell, as shown in Figure 5.6a and Figure 5.6b. This is a state of equilibrium existence between components, and the micelles are considered to be peptized (i.e., colloidally dispersed). If, however, the H/C ratio of maltenes is lowered, say by the introduction of a paraffinic diluent, the resins, which are absorbed on the asphaltenes, are desorbed to a certain extent, which results in the asphaltenes not being completely surrounded by resins and they are mutually attracted. This leads to precipitation that appears as sludge. Under the hydroprocessing reaction conditions, asphaltenes are first converted into intermediates called *asphaltene cores*. When their concentration exceeds the solubility limit in the reaction medium, phase separation takes place; if temperature is sufficiently high, this separation can lead to irreversible polymerization and to coke formation (Figure 5.7). It has been proposed that thermal cracking of asphaltene is attributed to the low molecular weight fractions of free radical hydrocarbon, which extract hydrogen from other hydrocarbons to be stabilized or react with other radicals via condensation reaction and form hard

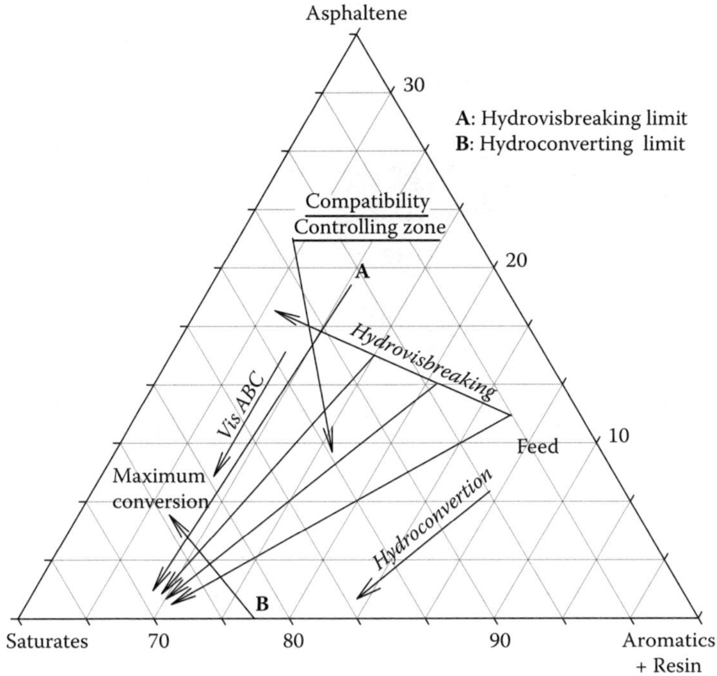

FIGURE 5.5 Compatibility control diagram for dry sludge. (From Takatsuka et al. 1989. *J. Chem. Eng. Jpn.* 22 (3): 298–303. With permission.)

asphaltene (Kawai and Kumata, 1998). It has also been found that the H/C of asphaltenes decreases with increasing the hydrogen partial pressure indicating that hydrocracking and dealkylation are the reactions that occur predominantly. A mechanism has been also proposed by (Kawai and Kumata (1998), presented in Figure 5.7, which intends to establish the way in which sludge is created from asphaltenes. The authors have also pointed out that as a consequence of the temperature, asphaltenes generate free radicals and reduce the molecular weight. Therefore, three different events are plausible to occur:

1. New free radicals extract hydrogen from other molecules.
2. Radicals are stabilized by hydrogen.
3. Radicals can recombine with other radicals to form "hard" asphaltenes.

When a free radical is formed, the next step is to break sulfur linkages by means of free radicals that also attack the aromatic structure of asphaltenes. Moreover, dealkylation reaction also leads to the asphaltenes with smaller size. A role of the smaller asphaltene molecule and the heavy molecule is reported for the peptization of asphaltene in residue hydroprocessing (Tojima et al., 1998). New free

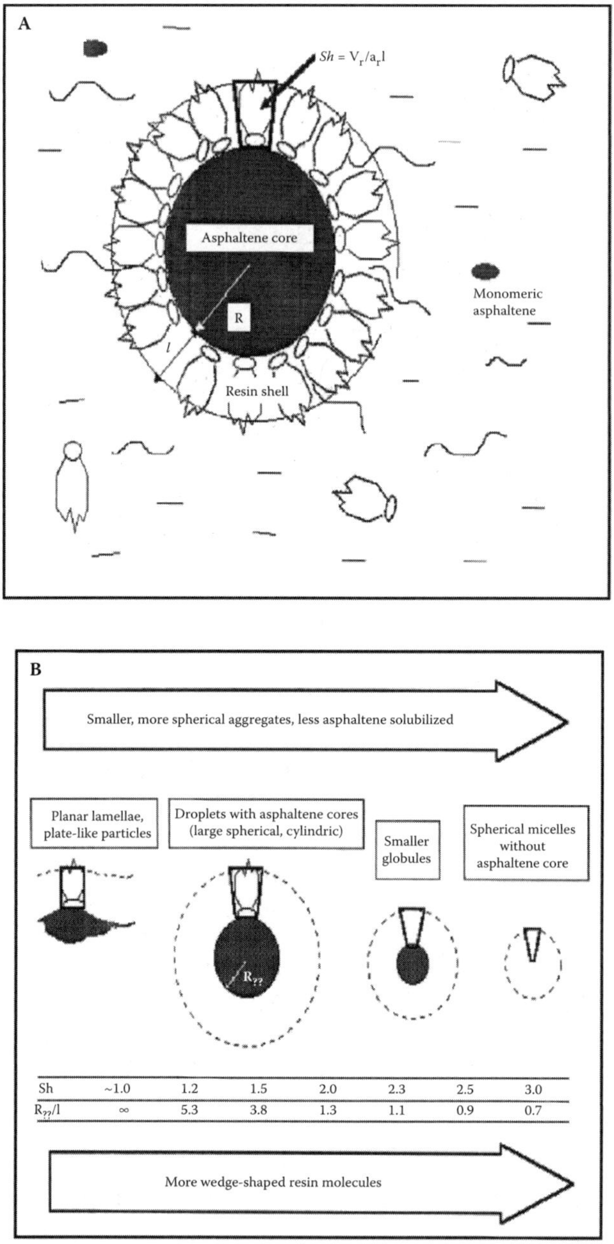

FIGURE 5.6 (a) Spherical asphaltenic aggregate in petroleum medium. Space occupied by an average resin molecule in the shell is indicated by the wage having shape factor (Sh). (b) The effect of the resin molecule shape on size and shape of the aggregate, as predicted by the model. (The fitted asphaltene molar mass in bitumen versus the resin to-asphaltene ratio.) (From Victorov and Smirnova 1999. *Fluid Phase Equilib.* 158–160: 471–480. With permission.)

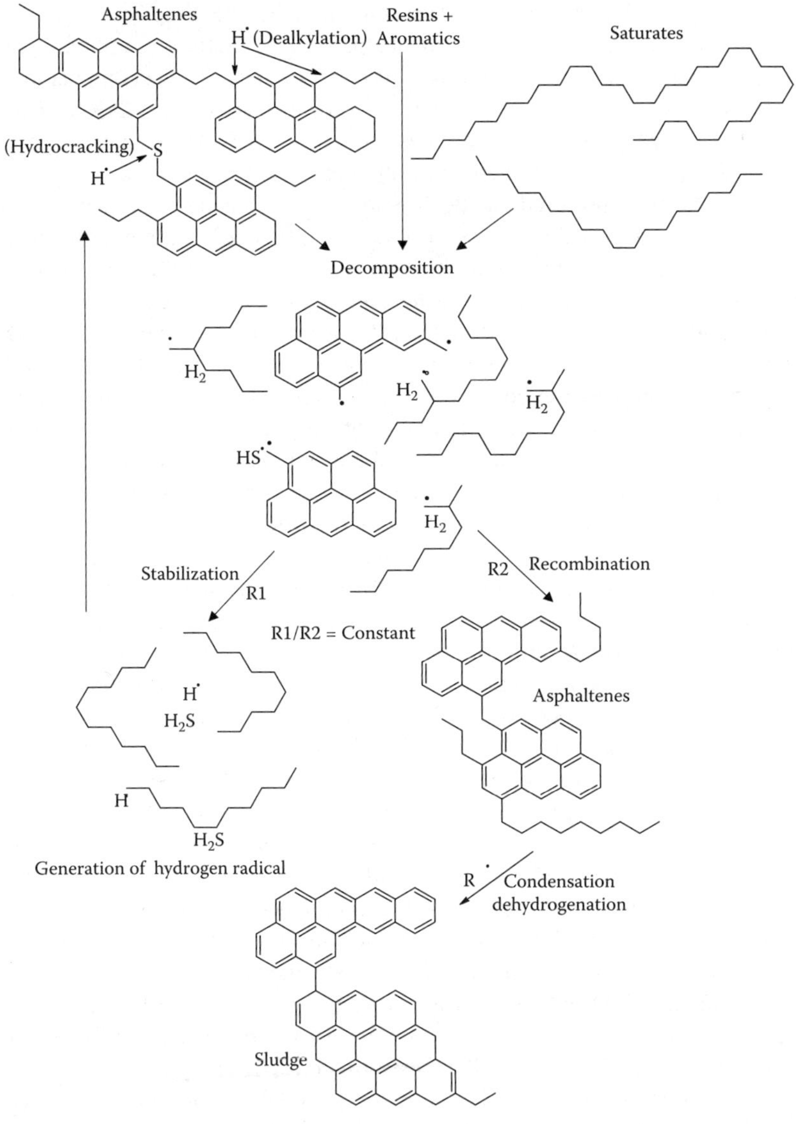

FIGURE 5.7 Possible sludge formation from asphaltenes. (From Kawai, H. and Kumata, F. 1998. *Catalysis Today* 43 (3–4): 281–289. With permission.)

radicals can suffer two reactions: stabilization or recombination. Recombination makes asphaltenes more compact structures and depending on the reaction conditions, dehydrogenation and condensation occur and the final product is highly condensed asphaltenes having minimal alkyl chains. Hard asphaltenes, which are mainly constituted by aromatic rings, turn into less soluble species and they are separated as sediment.

Chemistry has demonstrated to be essential for understanding how variations in composition affect formation of sediments, particularly saturates, aromatic, resin, and asphaltene (SARA) along with their physical properties. The composition can be estimated by using individual analysis of each component (SARA analysis, ASTM D 4124). Figure 5.8 shows a tendency of polarity, molecular weight, and aromaticity of a crude oil. In general, asphaltene and resin in crude oil emulsion stability have been reported in the literature (McLean and Kilpatrick, 1997a, 1997b; Schabron and Speight, 1998; Strassner, 1968), which is often said to be controlled by asphaltenes weight content. However, almost all of the published literature on heavy oil or residue, where resins and asphaltenes content were accounted for the stability of the crude oil and sediment formation, depended on balance between asphaltenes and resins concentrations, which vary in crude oil as well as during the hydroprocessing conditions and conversion. The determination of the particular resin components that interact with asphaltenes is more complicated due to their interaction with asphaltene and has not been widely reported because most of the research has focused on the interaction between the asphaltenes with the traditional bulk resin fraction (McLean and Kilpatrick, 1997a; Wu et al., 1998; Merino-Garcia and Andersen, 2004, Alvarez-Ramirez et al., 2006). Solvent fractionation of asphaltenes may be estimated by using molecular weight distribution by means of steric exclusion chromatography and x-ray diffraction to obtain the Yen parameters of asphaltenes in the solid state in order to reach the "hard core" asphaltene, which is considered to be important in sediment formation (Wandas, 2007; Wandas and Chrapek, 2004). Therefore, crude oil stability is the ability to maintain asphaltenes in a peptized or dissolved state in maltenes (saturates and aromatics) and not undergo flocculation or precipitation. However, compatibility of crude oil corresponds to the property of mixing of two or more oils without precipitation or flocculation of asphaltenes. The resin fraction is comprised of polar molecules often containing nitrogen, sulfur, and oxygen. Thus, the common definition of the resin fraction is that material soluble in light alkanes, but insoluble in liquid propane (Speight; 2006; 1999; Andersen and Speight, 2001). It is then expected that resin fraction overlaps with both the saturate and aromatic fractions.

The different solid fractions of crude oil and of hydrotreated Maya crude physical materials appearance are shown in Figure 5.9 where comparison is shown for virgin asphaltene and resin as well as hydrotreated (HDT) asphaltenes along with sediment obtained from the product of hydrotreated crude oil. It can be observed that asphaltene from virgin crude oil is black, while hydrotreated asphaltene (light brown) and sediment (dark brown) are shiny, and friably solid. However, resin is dark black, shiny, and gummy in nature. The color of the materials may depend on the nature of polar molecules present in the fraction. The dipole moment of different fraction of crude oil is seen in Figure 5.10 using different crudes that have different chemical composition and API gravities (9 and 56°). In Figure 5.10, the different fractions of asphaltene, resin, and oil are diluted in toluene, which showed increasing dielectric constant with its concentration except for high API gravity oil fraction whose dielectric constant decreases with its concentration. There has been a little work carried out on the measurement of asphaltene and resins dipole

Oil component		Method	Physical Properties			Analysis or Separation Methods	
			Non-polar	Weight light	Aromaticity low		
Saturates	MALTENE	S	↓	↓	↓	ASTM D 4124 ASTM D 2007	n-C_7 as solvent in a column packed with alumina n-C_5 as solvent in a column packed with silica/clay
Aromatics	MALTENE	A	↓	↓	↓	ASTM D 4124	Toluene as solvent in a column packed with alumina
Resins	MALTENE	R	↓	↓	↓	ASTM D 4124	Methanol/Toluene (1/1) and trichloroethylene in a column packed with alumina
Asphaltenes		A	Most polar	Weight heavy	Aromaticity high	ASTM D 893 ASTM D 4124	n-C_5 as solvent at 65 ± 5°C, centrifugation for 20 ± 1 min n-C_7 as solvent at boiling point temperature and stirring 1 h

FIGURE 5.8 Composition of crude oil and its physical properties along with the separation or analysis methods used for different hydrocarbons in petroleum fractions.

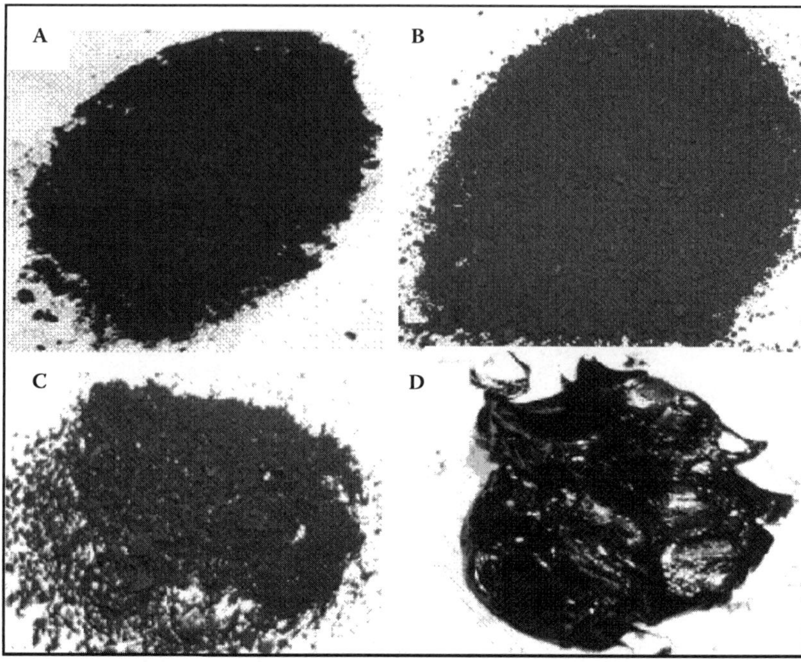

FIGURE 5.9 Asphaltene and resin separated from Maya heavy crude oil (21.3° API): (A) asphaltenes separated from virgin crude, (B) asphaltene separated from HDT products, (C) sediment obtained from HDT products at 400°C, (D) resin separated from virgin crude oil.

moment, despite their polarity. The importance of this measurement is to understand the phenomenon of aggregation of asphaltenes and their interactions with metals and resins, thus, it is vital to know their chemical and electronic structures. This information will allow researchers to design suitable catalysts for their decomposition, as well as to minimize sediment deposition, and to characterize the heavy fraction of crude oil. The polarity of asphaltenes has often been related to the heteroatoms and/or metal present in it (Nalwaya et al., 1999; Kaminski et al., 2000; Swanson, 1942). The resin fractions possess dielectric constant values intermediate between those of the asphaltenes and those of the oily constituents, but they show no measurable dielectric constant dispersion in some other studied regions (Goual and Firoozabadi, 2002; Swanson, 1942). On the other hand, Ese et al. (1998) analyzed adsorptive properties of the asphaltene and resin and concluded that resins are more polar than asphaltene. Since asphaltenes and resins are two adjoining classes of components separated from a continuum of molecules according to their solubilities in a low molecular-weight alkane, the size and chemical composition of some asphaltene molecules may be quite similar to those of some resin molecules. Hence, it appears that polarity of resin and asphaltenes is most likely depending on the origin of crude oil, chemical composition, and method of their fractionation (separation) used. Resins are relatively small molecules and so

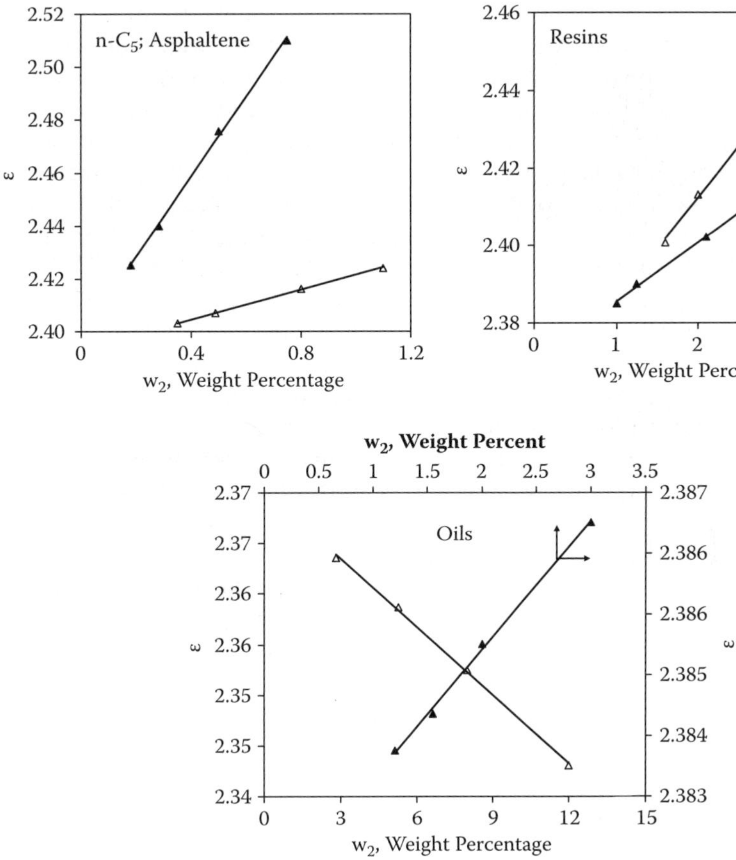

FIGURE 5.10 Dielectric constant (ε_r) and refractive index versus concentration of n-C_5 asphaltene, resins, and oils (eluted in n-C_5) in toluene for two different crudes: (▲) 9° API and (Δ) 56° API. (From Goual, L. and Firoozabadi, A. 2002. *AIChE J.* 48 (11): 2646–2663. Reprinted with permission of John Wiley & Sons, Inc.)

are the values of their molecular weights where their total molar polarization can be computed from the dielectric constant and density data. The dielectric-constant data also can be used to establish the size and character of the asphaltenes and to develop a qualitative explanation of their solubility in various solvents.

5.4 ROLES OF CATALYST AND FEED ON SEDIMENTS FORMATION

Recently the use of heavy and extra heavy crude oils has increased due to depletion of the light petroleum production, thus, processing of such complex feedstocks in an effective way (for instance, by using catalytic processes) is now becoming mandatory. Hydroprocessing is a method where hydrocarbon molecules undergo reconstruction (hydrotreating) as well as break down (hydrocracking) into smaller fractions in the presence of hydrogen (hydrogenation). Heavy or extra heavy oils contain large, complex, high molecular weight constituents, which are relatively easy to change into smaller molecules. Complex hydrocarbon molecules also contain heteroatoms (S, N, O, etc.) that require a catalyst in order to upgrade or refine the bottoms of barrel or crude oils. Unfortunately, during catalytic upgrading of crude oil, several operating problems are encountered that do not favor the use of catalytic processes for such a kind of feedstock. The major problems come for these processes due to catalyst deactivation, which is either by metal deposition or due to the coke-like carbon species (Absi-Halabi et al., 1991; Stanislaus et al., 1996; Kressmann et al., 1998a, 1998b; Ramirez et al., 2007; Rana et al., 2007b). Further carbon deposition can be due to the acid sites or due to the formation of solid material (sediments), which is the leading objective of the chapter. Sediments are often formed in the products during hydroprocessing due to the aforementioned imbalance between hydrocarbon molecules. Sediment materials are usually deposited on the reactor, downstream vessels, and process lines, as well as on the catalyst surface (Marafi, et al., 2005b; Stanislaus et al., 1994, 1998, 2005; Matsushita et al., 2004a; 2004b; Fukuyama et al., 2004; Hauser et al., 2005; Storm et al., 1994). The formation of sediments in the reactor limits residue conversion level and increases pressure drop (ΔP) of the reactor, which forces premature shutdown of the process run. Thus, prevention of sediment formation is highly demanded by refiners. The formation of sediments depends on several parameters, such as severity of the reaction, properties, and composition of feedstock, and level of conversion. The most affecting factors are the reactor temperature and the composition of asphaltene.

As mentioned in earlier sections, it is generally believed that sediment forms due to asphaltene flocculation during processing. Asphaltene become less soluble either due to the conversion of surrounding molecules, such as resins, or decreases in aromatic concentration or both, which generally stabilize asphaltene molecules. On the other hand, it is also expected that the strong acidic function of a catalyst (typically having been the most effective method to convert heavier fractions of crude oil) produces a large amount of light fraction of paraffins, such as propane, which may play a role in precipitating solid asphaltene in the reactor (Lopez-Salinas et al., 2005; Matsushita et al., 2004a), while a very strong catalyst hydrogenation function may turn hydrogenation of aromatic to cyclic compounds that imbalances the reaction product in the reactor as well as downstream reactor vessels. Thus, these two functions of a catalyst have to balance in order to have proper activity of cracking and hydrogenation. Recently a comparison between acidic and nonacidic catalysts based on sediment formation was reported (Figure 5.11). A tendency to form total sediments was significantly suppressed with carbon black composite support in comparison with that of alumina-based catalysts, especially during the high-activity period (150 to 300 h of operation), remaining low and stable throughout the test. Presumably, carbon black in catalysts aids in suppressing Al_2O_3 acidity, which contributes to lower sediment formation. It has also been reported that the active carbon-supported catalyst significantly diminishes sediments formation than alumina-supported catalysts (Figure 5.12). In a hydrocracking process that uses acidic catalysts either in a

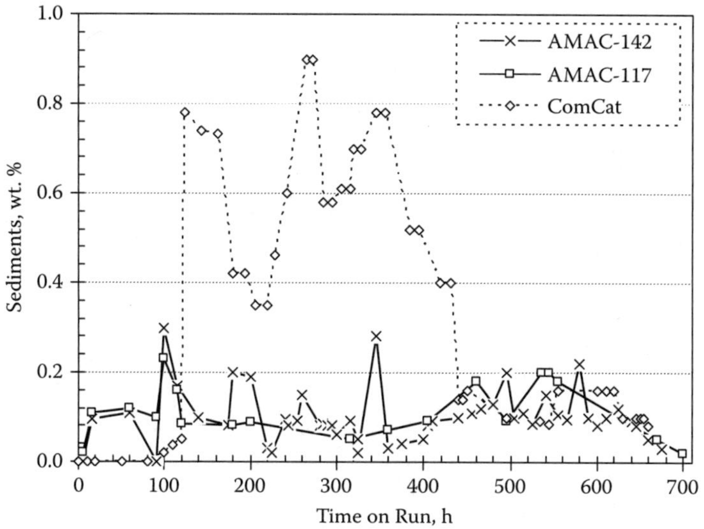

FIGURE 5.11 Evolution of total sediments in the product on carbon black composite supported (AMAC) and alumina (ComCat) supported catalysts. (From Lopez-Salinas et al. 2005. *Catalysis Today* 109: 69–75. With permission.)

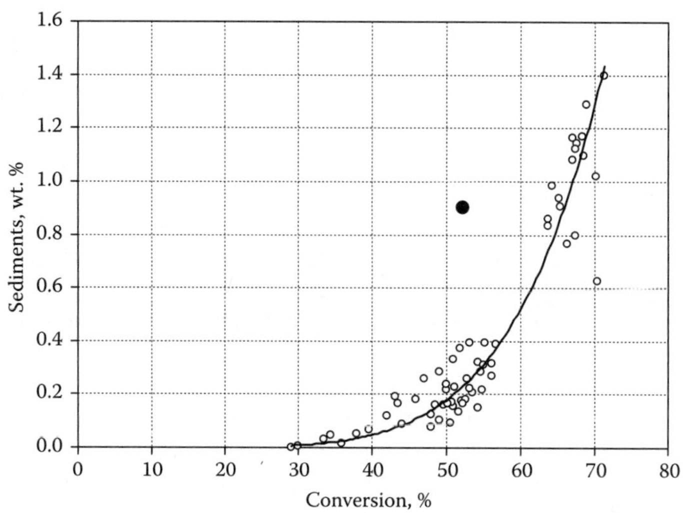

FIGURE 5.12 Effect of catalyst composition: (●) alumina-supported catalyst; (O) active carbon-supported catalyst and conversion on sediment formation. (From Fukuyama et. al. 2004. *Catalysis Today* 98: 207–215. With permission.)

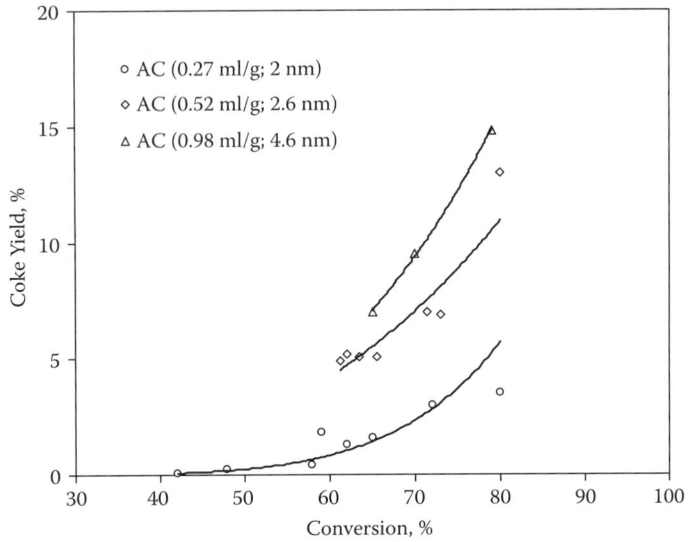

FIGURE 5.13 Relationship between porosity of active carbon (AC) and coke yield for different total pore volume (ml/g) and pore diameter (nm) of the catalysts at different magnitude of conversion (reaction conditions: temp.: 415 to 435°C; pressure: 7 MPa; reaction time: 0.5 to 2 h). (From Fukuyama et. al. 2004. *Catalysis Today* 98: 207–215. With permission.)

fixed-bed or in ebullated-bed, the catalyst deactivation is recognized to be due to coking and deposition of solid material, particularly at high temperature (Rana et al., 2008; Fukuyama, 2006; Fukuyama and Terai, 2007; 2008; Fukuyama et al., 2004). The cracking activity of a conventional alumina-based catalyst is attributed to the Lewis acid sites contributed by the support. It is also well reported that these acid sites suffer from deactivation caused by N-containing compounds or coke formed by polycondensation of heavy hydrocarbon compounds on the catalyst surface (Absi-Halabi et al., 1991; Gray et al., 1992). Deactivation due to support contribution on coke formation using an active carbon (which is neutral or weak base support)-supported catalyst has been also reported (Terai et al., 1999, 2000; Fukuyama et al., 2002) with variation in pore diameter of the catalyst, as shown in Figure 5.13. When increasing the conversion, coke yield also increases due to carbon deposition or to coke-type solid carbon deposition on the surface of the catalyst. Moreover, it is difficult to separate the catalyst deactivation due to the sediment deposition or the coke formation. Usually coke present on the catalyst is classified as soluble coke extracted from spent catalysts, which is rich in alkylated mono- and diaromatics with a low percentage of polyaromatics, whereas the nature of insoluble coke is highly polyaromatic (Sahoo et al., 2003).

The effect of catalyst pore diameter on sediment formation in the product is shown in Figure 5.14. Sediment formation is severely affected by the pore volume. The catalyst having a large number of macropores, particularly in the range of

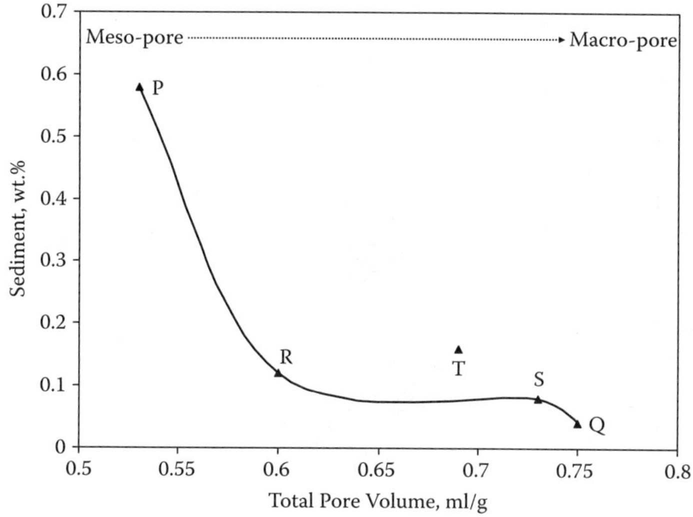

FIGURE 5.14 Influence of catalyst ($NiMo/Al_2O_3$) pore volume on sediment formation during Kuwait vacuum residue (where P, R, T, S, and Q represent the different textural properties of catalysts). (Adapted from Stanislaus et al. 1996. *Stud. Surf. Sci. Catal.* 100: 189–197. With permission.)

80 to 300 nm, produces very low sediments, but at the same time this catalyst has lower activity of asphaltene cracking and residue oil to distillates. These results of sediment formation were explained on the basis of easy diffusion of asphaltene into the larger pores where the balance between asphaltene and resin conversion remains constant, which produces lower sediment formation, while in mesopores the resin and oil diffusion was faster than asphaltene, which increases asphaltene-to-resin ratio, thus the sediment formation (Stanislaus et al., 1996; Gawel et al., 2005). Another explanation for this effect is the higher conversion (>45%) of residue, which affected sediment formation (Figure 5.15). The authors have concluded that the pore diameter plays an important role in the formation of sediments and a probable justification is referred to the rate of conversion of resins. During hydrotreating, resins are converted faster than asphaltenes leaving them unprotected. As reaction progresses, asphaltenes undergo dealkylation and cracking, which turns them less soluble in the media. Asphaltenes are prone to precipitate under these conditions as sediments. Sediment formation may be reduced if catalysts have macropores along with mesopores because a balance between activity to hydrodesulfurization (HDS), hydrodeasphaltenization (HDAsph), and hydrodemetallization (HDM) is necessary to meet (Rana et al., 2004; Ancheyta et al., 2005).

A comparison for two different alumina-supported catalysts (relatively acidic at similar conversion levels) showed higher sediment formation than neutral (carbon)-supported catalysts (Figure 5.12). With increasing temperature or conversion, the pressure drop also increases and strapping problems arise due to the reactor or

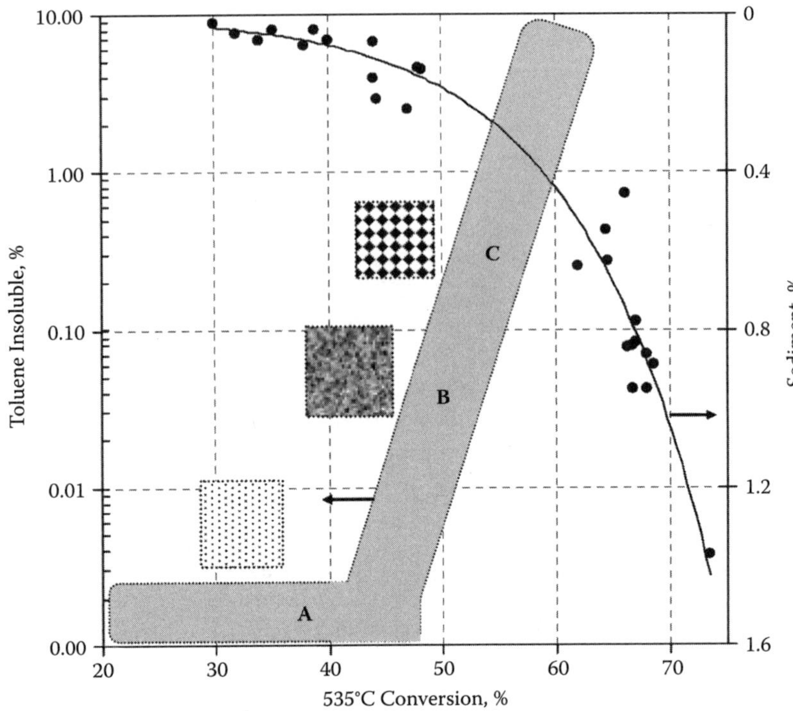

FIGURE 5.15 Effect of conversion on sediment formation (where A, B, and C represent sediment size). (Adapted from Inoue et al. 1998. *Catal. Surv. Jpn.* 2: 87–97. With permission.)

subsequent line plugging, which are due to the precipitation of asphaltene or sediment formation. The above studies were reported in fixed-bed reactors using residue as well as crude oil as feedstock. On the other hand, when using moving or ebullated-bed reactors, sediment formation is also reported, which frequently provokes downstream operational problems, especially by blocking valves, pumps, and hot and cold separators. Sediments are polynuclear aromatics, which result from complex cracking, recombination, and nucleation reactions, such as catacondensed polycyclic and/or pericondensed polycyclic aromatic hydrocarbons along with asphaltenes (Wiehe, 1992; 1993). A study reported by Nowlan and Srinivasan (1996) mentions that it is impossible to operate H-Oil plants at high conversion levels because sediment yields rise, particularly when using heavy residue from heavy crude oils. Thus, once sediment content is higher than 1 wt%, it is expected that the H-Oil plant becomes inoperable. However, a superior catalyst formulation that has considerable fundamental problems of sediment formation can be designed using balance components (support composition, acid-base properties, as well as hydrogenation function of the catalyst), which accounts for these subsequent steps: (1) large pore diameter (macropores) of the catalyst in order to increase

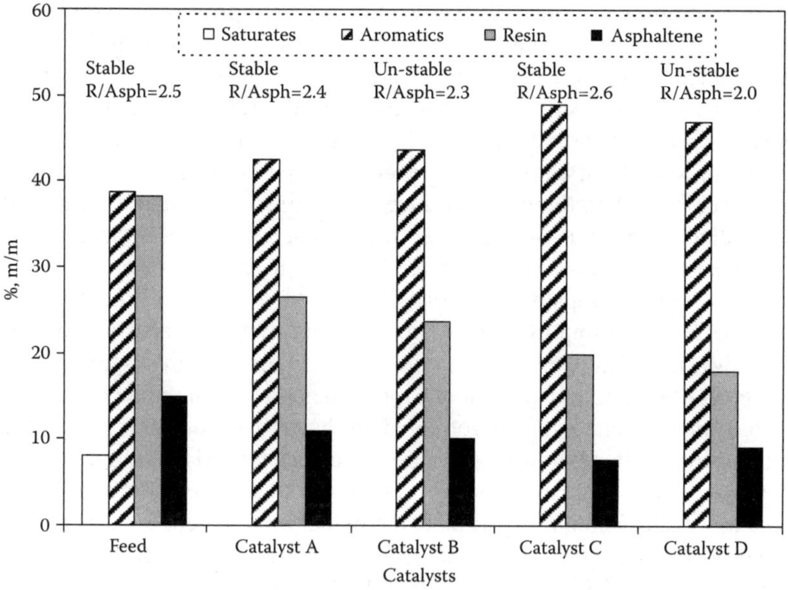

FIGURE 5.16 SARA distribution of H-Oil effluent (+360°C) on different catalysts [LC separation on silica and alumina column using step elusion: heptane, heptane-toluene (2:1, v/v), toluene-dichloromethane-methanol (1:1:1, v/v/v)]. (From Robert et al. 2003. *Petrol. Sci. Tech.* 21: 615–627. With permission.)

asphaltene diffusion and transformation (Sherwood, 1998), and (2) catalytic active sites (hydrogenation and hydrocracking) or support adsorption sites (acid sites) that are less favorable to sediment formation (Cornelius et al., 1955).

Figure 5.16 presents the results of SARA analysis of the H-Oil process obtained at the same level of residue conversion (Robert et al., 2003). Usually, the principal components of reaction products are oil (aliphatic and aromatic), resin, and asphaltene that are in a stable mixture. In that manner, it does not have the tendency to produce asphaltenic sludge; in other words, the mixture is compatible. Thus, incompatibility is the tendency of a fuel to produce a deposit of sludge and that is due to the imbalance between these components occurring after the reaction, e.g., in Figure 5.16, catalysts B and D showed unstable behavior because the ratio of resin and asphaltene was relatively lower than catalysts A and C.

5.5 EFFECT OF HYDROPROCESSING REACTION CONDITIONS ON SEDIMENT FORMATION

This section describes the effect of operating conditions and composition of feedstock on sediment formation during hydroprocessing. It has to be highlighted that there is very limited information about sediment formation particularly during catalytic upgrading of crude oils. Only a few research groups are working on the upgrading of crude oils, among them hardly four or five research groups (Kuwait, Japan, France, Canada, and Mexico) have taken into account sediment formation, despite the fact that it is directly affecting the refinery and deactivation of hydroprocessing catalysts. It has been found that operating conditions are crucial in sediment formation; among them reaction temperature is one of the most relevant variables. When the heavy residue containing large amount of asphaltene was supplied at the start-of-run (SOR) in the reactor, the permanent adsorption of carbonaceous deposit on the active site leads to catalyst deactivation. Thus, in order to acquire supplementary insights about the sediment formation during hydroprocessing, an important role is carried out by asphaltenes and its conversion, as shown in Figure 5.17a and Figure 5.17b. In these figures, sediment formation is proportional to the amount of asphaltenes present in the feedstock as well as asphaltenes present in the hydrotreated products (Wandas, 2007). These operational parameters (temperature, pressure, space velocity, H_2/oil ratio) have shown a direct effect on the conversion of asphaltenes that force a formation of sediments along with feed quality. Hence, the extent of sediment formation depends on the conditions and the asphaltene content present in the feed.

Apart from this, Storm et al. (1997; 1998) stated that the amount of sediments produced per unit weight of vacuum residue feedstock at bench scale was correlated to the degree of condensed polynuclear aromaticity, the average number of alkyl groups, the ratio of heptane insolubles to pentane insoluble/heptane solubles, and the H/C ratio of this fraction. Thus, not only the incompatibility between asphaltenes and maltenes is responsible for sediment formation, but also it is probable that, during hydroprocessing conditions, particularly at higher temperatures, more asphaltenes are created due to the conversion of resins, which can precipitate because the solubility limit is surpassed. It is now an approximation that the sludge formation occurs due to the asphaltenes flocculate, which become less soluble either due to the surrounding hydrocarbon molecules, or they become less well solubilized by the surrounding medium, or they become both due to reactions that occur at the temperatures of processing. This is readily observed that the amount of sludge depends on both processing severity and the origin of the crude oil (residua), as reported in subsequent sections.

5.5.1 Stages of Sediment Formation during Hydrotreating

During the operation of hydrotreating reactors at a high level of conversion (>50%), the main problem is sediment formation, which leads to deposits on the reactor internal parts and downstream vessels (heat exchangers, separators, fractionating towers), and transfer lines. Sedimentation also causes reactor operability problems, catalyst deactivation, increase of catalyst consumption, and decrease in residual fuel stability, which eventually cause shutdown of commercial units (Morel et al., 1997). For this reason refiners commonly limit sediments formation at values of 0.8 to 1.0 wt% to warranty continuous operation of commercial hydrocracking units.

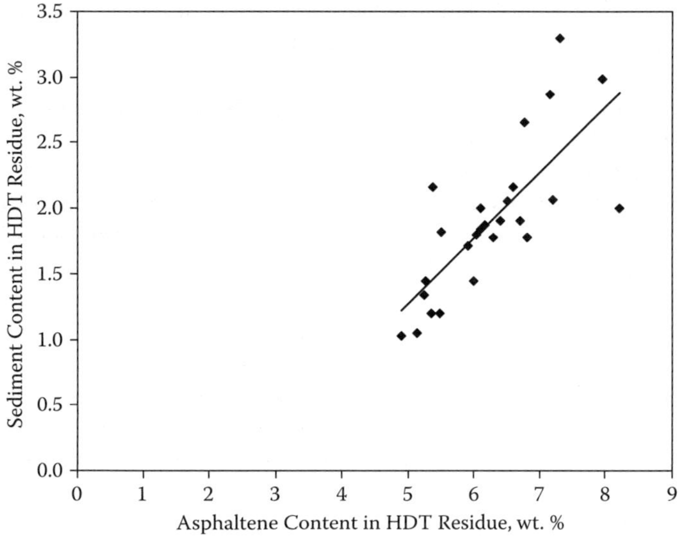

FIGURE 5.17A Correlation between asphaltene content of raw residue and sediment content of desulfurized residue. (From Wandas 2007. *Petrol. Sci. Tech.* 25: 153–168. With permission.)

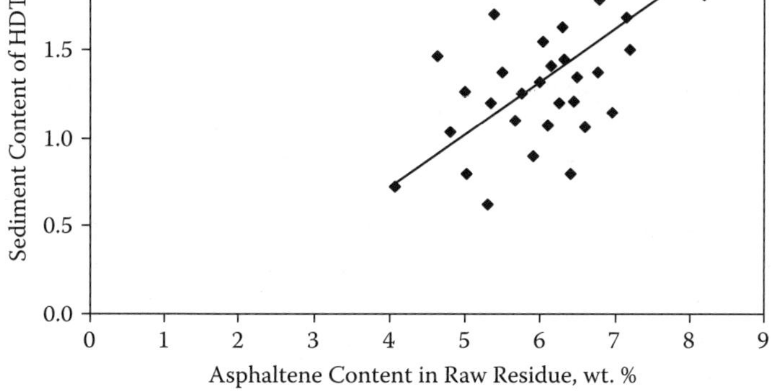

FIGURE 5.17B Correlaon between asphaltene and sediment content of desulfurized residue. (From Wandas 2007. *Petrol. Sci. Tech.* 25: 153–168. With permission.)

Solid formation is attributed in the literature to occur for the following reasons (Storm et al., 1997; Inoue et al., 1998; Speight, 2001):

1. As a result of changes in the relative amounts of the lower-boiling hydrocarbons and higher molecular weight polar species.
2. At the high temperature in the reactor, the liquids split into a light aliphatic phase and a heavy aromatic phase. This separation of the aromatic material is supposed to be the initial step of solid deposition.
3. Reductions of either solubility of asphaltenes in the nonasphaltenes fraction or of the ability of the nonasphaltenic fraction to solubilize the asphaltenes both due to their chemical transformation during hydroprocessing of heavy feeds. More specifically, hydrocracking of the resin fraction (solvent) for large colloidal asphaltenes proceeds faster than the conversion of asphaltenes. The solvent becomes less effective under the severe conditions, causing the precipitation of asphaltenes. This is further enhanced by dealkylation of the asphaltenes, which also decreases solubility.

It has been shown that at temperatures lower than 420°C, only dealkylation of side alkyl chains is observed and, at higher temperatures, hydrocracking of the asphaltene molecule is prominent. Thus, asphaltenes composition and structural parameters change significantly depending on hydroprocessing reaction conditions (Ancheyta et al., 2003), and consequently sediments formation is also influenced by the severity of reaction. For instance, during hydrocracking of a vacuum residue of Arabian light oil (bp >550°C) with a commercial CoMo/Al_2O_3 catalyst in a two-staged fixed-bed microreactor, reaction temperatures of 395, 405, and 418°C are considered the conditions for no formation, beginning of formation, and some formation of sediments, respectively (Mochida et al., 1989). Of course, these temperature values are highly dependent on the type of catalyst and feed. What is very clear is that there is an intimate dependency between the temperature where hydrocracking of asphaltenes is severe and sediments formation.

The limit of 50% conversion to avoid unmanageable sediment formation has been clearly demonstrated with experimental data of sediments and toluene insoluble versus conversion at typical hydroprocessing conditions, as shown in Figure 5.12 and Figure 5.18 (Inoue et al., 1998; Fukuyama et al., 2004). Characterization of sediments collected from different zones of an H-Oil plant has shown that all deposits contain appreciable amounts of metals either coming from the catalyst (Ni, Mo, Al) or from the feed (V, Ni, Fe). Other metals (Mn, Cu, W, and Fe partially) are also present, which are corrosion products (Table 5.4). It was also found that the farther from the reactor, the lower the metal content in the deposit. The chemical composition of sediments also reveals that its content of insoluble organic material increases in the direction from the reactor toward the vacuum column, indicating that the insoluble organics are generated by the

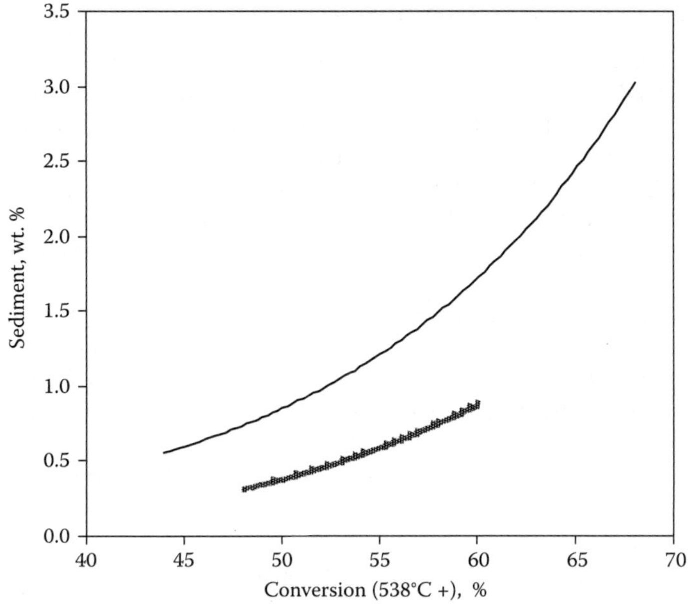

FIGURE 5.18 Effect of feedstock and conversion on sediment formation. Dotted line: VR from light crude oil and asphaltene = 14.2 wt%; solid line: VR from heavy crude oil and asphaltene = 26.3 wt%. (From Ancheyta et al., unpublished results. With permission.)

asphaltenes precipitated from the hydrotreated feed. The inorganics content in the sediments decreases toward the end of the installation, since the majority of the inorganic components comes from the catalyst and, therefore, settles in the installation zone closest to the reactor (Gawel and Baginska, 2004).

TABLE 5.4
Metal Content and Chemical Composition of Sediments from an H-Oil Plant

	Metal Content, ppm			Composition, wt%	
Deposit	**Ni**	**V**	**Fe**	**Insoluble Organics**	**Inorganics**
S	7,279	9,616	16,528	28.02	31.98
D1	3,175	7,294	6,374	35.89	27.39
D2	995	5610	3,905	58.10	6.92

Note: S = sludge precipitated from a sample taken from the filter between the H-Oil reactor and the atmospheric column, D1 = solid deposit collected from the filter after the atmospheric column, D2 = solid deposit collected from the filter after the vacuum column.

5.5.2 Catalytic Reaction Temperature

By increasing hydroprocessing reactor temperature, a decrease in asphaltene content is observed, but at the same time the instability and the tendency toward sludge formation increase. It is well known that in hydrotreating several chemical reactions take place and, hence, affect the product quality and stability. Thus, using a sulfided catalyst we have two different functions corresponding to metallic function (hydrogenation) dominated and cracking (support or sulfhydryl) dominated reactions. However, the conversion of asphaltenes depends more on the thermal stability of its structure, which usually depends on the origin of crude oil. An increase in temperature showed higher sediment formation particularly after 390°C reaction temperature (Mochida, 1990a). Therefore, a closer examination of temperature effect indicated that high molecular weight species were converted into a low molecular weight asphaltenic structure, which is probably more aromatic since the H/C ratio decreases with increasing temperature (Ancheyta et al., 2003).

It has been shown that at temperatures lower than 400°C, only dealkylation of side alkyl chains is observed, and at higher temperatures, hydrocracking of the asphaltene molecule is prominent. Therefore, the reaction may be more controlled by the temperature. Nevertheless, asphaltene composition and structural parameters change significantly depending on hydroprocessing reaction conditions (Ancheyta et al., 2007b) and, consequently, sediments formation is also influenced by the severity of reaction temperature. Mochida et al. (1989) studied a $CoMo/Al_2O_3$ catalyst for hydrocracking of a vacuum residue of Arabian light oil using a two-stage fixed-bed microreactor. Sediment formation was not observed for reaction temperatures of 395, 405, and 418°C, while with further increase in temperature sediment formation increases correspondingly (Mochida et al., 1990a, 1990b). Of course, these temperature values are highly dependent on the type of catalyst and feedstock. Thus, it is obvious that there is an intimate dependency between the temperature and sediments formation particularly when hydrocracking of asphaltenes is severe. Therefore, temperature effect is generally reported in the form of conversion. Several experimental data of sediments formation against conversion using typical hydroprocessing conditions are seen in Figure 5.18 (also see Figure 5.12 and Figure 5.15).

5.5.3 Thermal or Thermal Oxidation of Asphaltenes

So far, we have identified that asphaltene plays an important role in the sediment formation during hydroprocessing. Considering this molecule important and most refractory in the crude oil and residue, it has been reported that asphaltene conversion is lower on catalytic sites, but more likely depends on the temperature (Rana et al., 2007a, 2007b). Recently, Tanaka et al. (2003) reported that asphaltenes vary their molecular weight and size along with increase in temperature, which is shown in Figure 5.19. Thus, it is really a subject of awareness that asphaltene conversion reported in the literature follows catalytic conversion or thermal cracking mechanisms. Apart from temperature effect, origin of crude oil contributes to its

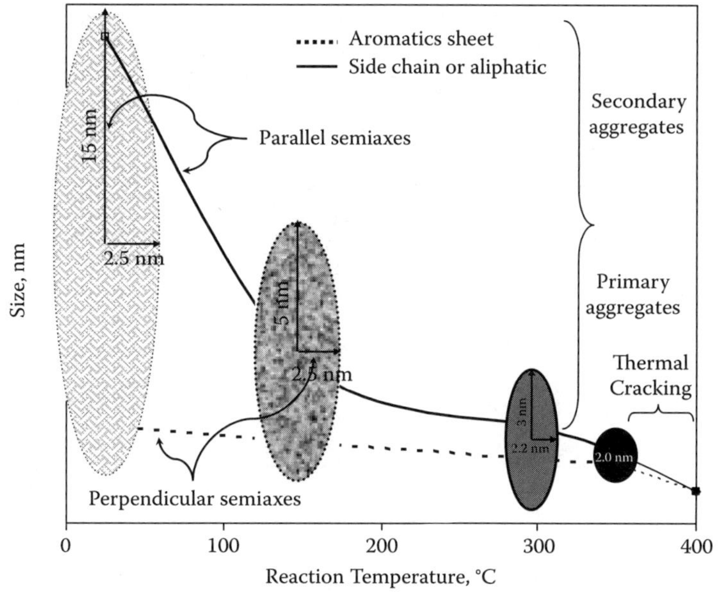

FIGURE 5.19 Temperature dependency of the shape and size of asphaltene extracted from Maya crude oil. (Adapted from Tanaka et al. 2003. *Energy Fuels* 17: 127–134. With permission.)

fractal stability, e.g., Maya asphaltene is more refractory in nature than Iranian light oil and Arabian heavy oil (Khafji) (Tanaka et al., 2003). The decaline precipitated Maya asphaltene showed a fractal network even at 350°C. Hence, thermal cracking starts at around these temperatures, which further indicates the towering coke-making tendency of asphaltene. Considering this complexity of asphaltene at hydroprocessing conditions, the size of asphaltene is about 2 nm in radius. In order to see the decomposition of asphaltene or thermal degradation of asphaltene and resin, some results are reported in Figure 5.20, which indicated that the complete decomposition of asphaltene is impossible particularly in an inert atmosphere. The thermogravimetric (TG) heating was conducted in stationary inert medium and helium flow at 4°C/min rate.

The first sign of molecular change in asphaltene can be seen at around 350°C and severe destruction of molecules appears on the TG curve at around 430°C; as a result total loss (25 to 900°C) was obtained about 53 wt%. Contrary to the asphaltene, the resin decomposition started at a lower temperature, such as 240°C, 410°C, and finally molecular destruction appears on the TG curve at about 490°C, which is relatively higher for resin; however, total loss (25 to 900°C) of resin was at about 91 wt%. The wide difference in the weight loss for resin and asphaltene further confirms the refractory nature of asphaltene, which is relatively numb to the temperature. On the other hand, sediment extracted from Maya crude oil hydrotreated solid product, TG analysis indicated that the low temperature loss

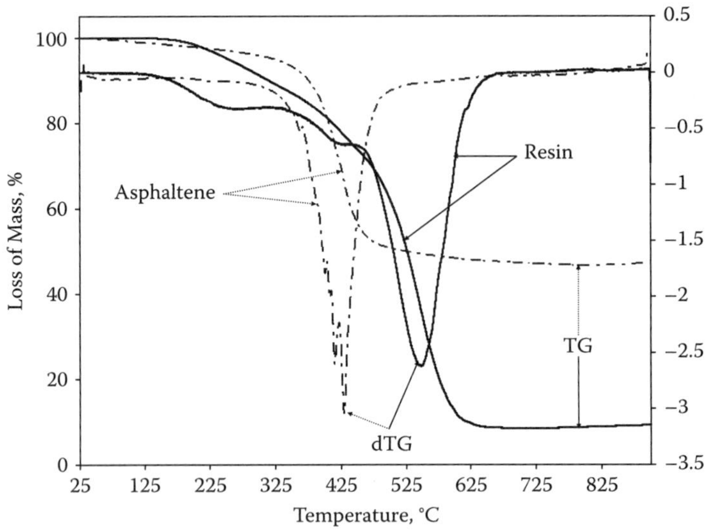

FIGURE 5.20 Results of thermogravimetric analysis (TGA) of asphaltene (----) and resin (——) extracted from heavy crude oil (13° API). (From Rana, unpublished results.)

at 100 and 260°C is due to the side chain cracking while molecular destruction appears at about 415°C, which is almost similar for the asphaltene, as shown in Figure 5.21. Thus, it can be concluded that sediments are the modified structure of the asphaltene, which occurs due to the incompatibility of the different fractions of crude oil along with polar asphaltene molecules. However, when asphaltenes are heated in a stationary oxygen atmosphere or air flow, it is possible to burn 100% of asphaltene (Figure 5.22). The role of air is to start the oxidation reaction; therefore, the decomposition rate is very fast at a higher temperature (ca. 400°C). Similar results were reported by Savelev et al. (2008) using inert and oxygen atmospheric thermal transformations of virgin Argentinean asphaltenes from Rafaelita and Toribia fields.

5.5.4 Pressure

The effect of the presence of hydrogen donor diluents on sediment formation is seen in Figure 5.23, which shows significant difference in sediments reduction with increasing hydrogen partial pressure. Nongbri et al. (1996) indicated that increasing conversion appears in higher amount of sediment formation. Sediment formation results confirm that the process performances can be substantially improved by adding hydrogen or hydrogen donor solvents (Ovalles et al., 2001; Kubo et al., 1994, 1996). Thus, selection of solvent or a proper hydrogen donor is utilized to improve product quality and increase the H/C ratio, which inhibits coke formation over the catalyst as well as sediment formation in the products (Speight, 2004).

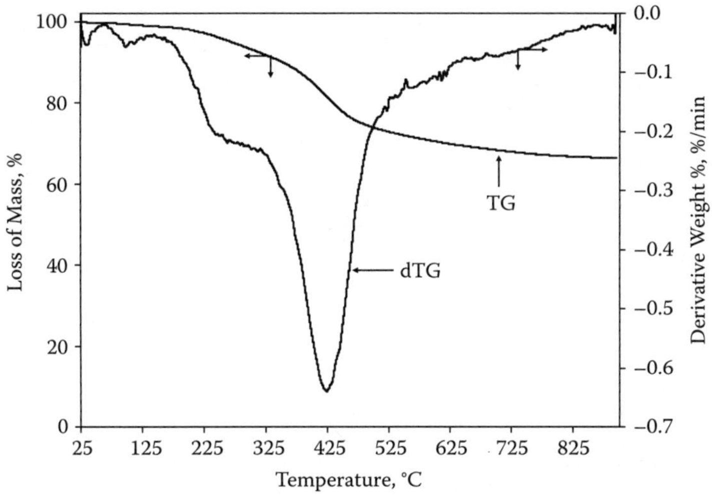

FIGURE 5.21 Results of thermogravimetric analysis (TGA) of sediment extracted from Maya hydrotreated products at 420°C. (From Rana, unpublished results.)

In addition, when increasing hydrogen partial pressure other conversions, such as HDS, HDM, and hydrodenitrogenation (HDN), also enhance without excessive deleterious effects on the operation of the hydrocracking reactor. It has also been recognized that by increasing the efficient use of hydrogen, existing equipment

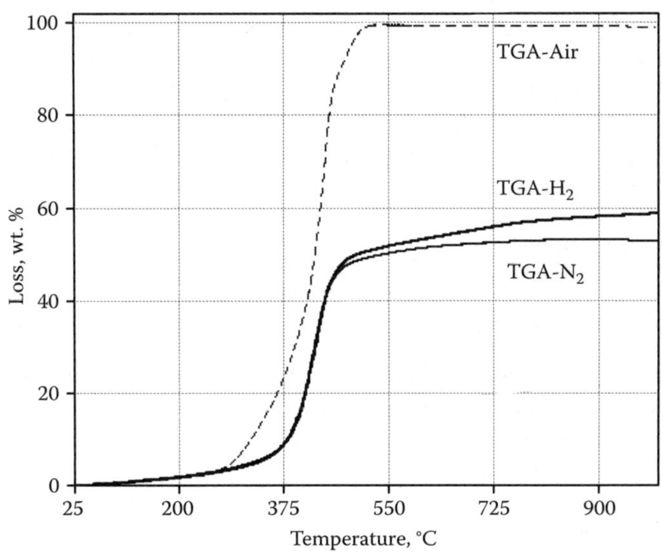

FIGURE 5.22 Effect of thermal liquefaction of asphaltene in different atmospheres. (From Rana, unpublished results.)

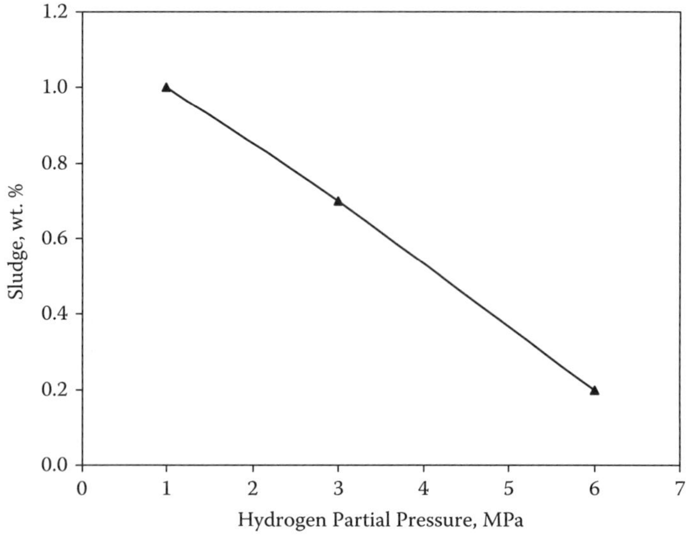

FIGURE 5.23 Effect of hydrogen partial pressure on sludge formation. (From Werner et al. 1998. *Fluid Phase Equilibria* 319–341.)

could be employed to increase the throughput of the feedstock and also increase the yield of liquid hydrocarbons (Scherzer and Gruia, 1996). Contrary to these obvious advantages, an increased hydrogen partial pressure of the process turns into the higher cost of the process. Moreover, higher hydrogen partial pressure results in an increase in the catalyst life cycle. The catalyst life cycle is determined for any given unit by the start-of-run and end-of-run temperatures and the deactivation rate of the catalyst (Ancheyta, 2007a). Start-of-run (SOR) temperature is a function of the catalyst activity, unit hydrogen pressure, and the type of feed being processed. Furthermore, catalyst life cycle is a function of catalyst stability, which is measured in terms of deactivation rate. End-of-run (EOR) temperature is determined by the process economics, usually product yield and hydrogen consumption.

5.5.5 Effect of Feedstock or Diluents

In order to minimize sediment formation or generate compatibility between the products, some aromatics-rich gas oil streams can be added into the crude oil or residue feedstock, such as HCGO and LCO (heavy-cycle gas oil and light-cycle oil) produced in fluid catalytic cracking (FCC) units. These streams have been used as diluents to the vacuum residue fed to the H-Oil process instead of the heavy-vacuum gas oil (HVGO) produced in the same plant, and recycled to the inlet of the H-Oil reactor. No appreciable improvement on sulfur, metals, nitrogen, and other impurities removals was reported. However, sediments formation is substantially reduced when HCGO and LCO are utilized as diluents instead of

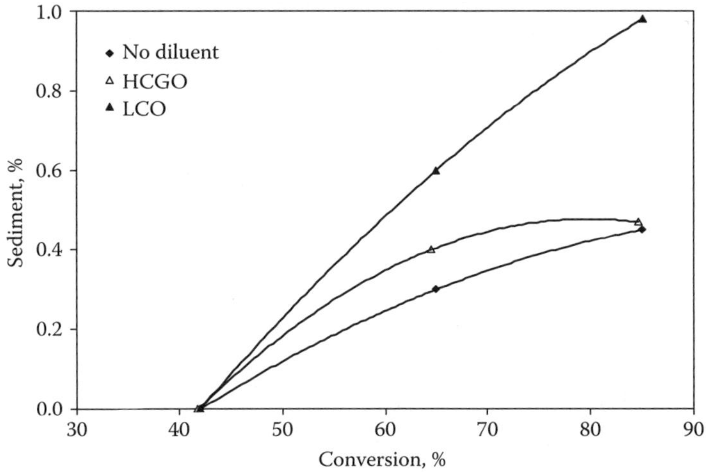

FIGURE 5.24 Effect of diluents in sediment formation at various conversion levels. (From Marafi et al., 2005b. *Petroleum Science and Technology* 23 (7–8): 899–908. With permission.)

HVGO, as can be seen in Figure 5.24. This reduction in sediment formation was attributed to (Marafi et al., 2005b):

1. Reduction in liquid viscosity and improvements in flow dynamics.
2. Diluents may associate the asphaltene aggregates and reduce their size, enhancing their diffusion into the pore and increasing the reaction rate.
3. High aromatic content (>70%) as diluents may help to disperse the scarcely soluble asphaltenes and improve the compatibility between different oil fractions and control the onset of asphaltenes precipitation.

Due to the complexity of crude oil feedstocks (mixture of several thousand molecules), it is not possible to give a full description of these components, therefore, indirect methods are used to obtain an indication about how a specific process parameter affects the product properties. More specifically, hydrocracking of the resin fraction (solvent) for large colloidal asphaltenes proceeds faster than the conversion of asphaltenes. The solvent becomes less effective under the severe conditions, causing the precipitation of asphaltenes. This is further enhanced by dealkylation of the asphaltenes, which also decreases solubility (Ancheyta, 2007a). In order to choose an ideal solvent, Wiehe (1995) developed the two-dimensional solubility parameter approach for solvent selection.

5.5.6 Role of Heteroatoms and Their Effect on Sediment Formation

Compatibility of different components present in petroleum is not the only factor to form sediments, but also the presence of heteroatoms plays an important role. The effect of nitrogen and oxygen compounds on sediment formation was

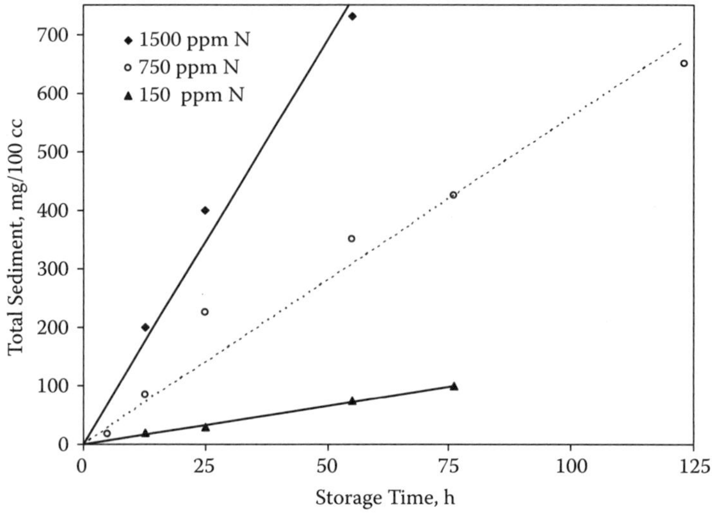

FIGURE 5.25 Effect of nitrogen content on sediment formation versus storage time for 2,5-dimethyl-pyrrole (DMP) in diesel fuel turning dark in storage at 43.3°C. (From Frankenfeld et al. 1983. *Ind. Eng. Chem. Prod. Res. Dev.* 22: 608–614. With permission.)

determined by using pure nitrogen compound (2,5-dimethyl-pyrrol) as model petroleum-derived fuel systems, as shown in Figure 5.25. Even added water showed an accelerating effect on sediment formation if its amount was higher than 100 ppm (Frankenfeld et al., 1983). It is shown that by increasing nitrogen content, a higher amount of sediment is reported, which could be due to the large amount of electrons density (polarity) that appears to participate in condensation reaction during or after the reaction in hydrotreated products. The instability during storage of hydrotreated products along with time also affect sediment formation, as shown in Figure 5.26. In order to confirm the instability, products can be analyzed by the Bromine number, which shows variation with respect to the time of storage (Figure 5.27). The variation in the Bromine number is a characteristic of the presence of unsaturated hydrocarbon molecules in the products. Thus, the rate of sediment formation is dependent on the presence of nitrogen compounds, nature of the diluent employed, and storage time. Detailed studies of the effects of reaction conditions were carried out in presence of air, increased temperature, and dissolved oxygen, and it was found that all strongly accelerated the sediments formed during reaction, while moisture had a variable effect (Frankenfeld et al., 1983). A possible reaction mechanism for such a kind of asphaltene molecule is proposed, with an ipso substitution of aromatic carbon bonded with phenolic groups (Wandas, 2007; Wandas and Chrapek, 2004; McMillen et al., 1987).

Crude oil (or residue) has numerous kinds of nitrogen and oxygen atoms in various forms, which are very large and the fuels produced are indisputably significantly different in chemical composition. Syncrudes from both coal and

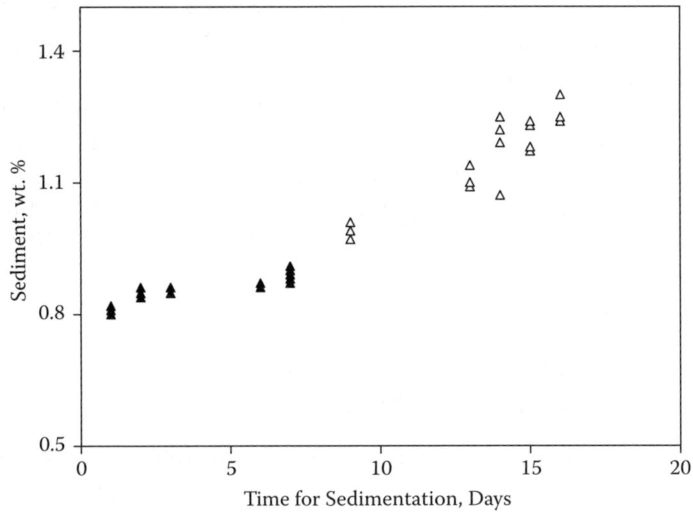

FIGURE 5.26 Effect of storage time on sediment formation for a hydrotreated vacuum residue product of Maya crude oil. (From Gomez et al. 2003. *Rev. Soc. Quim. Mex.* 47 (3): 260–266. With permission.)

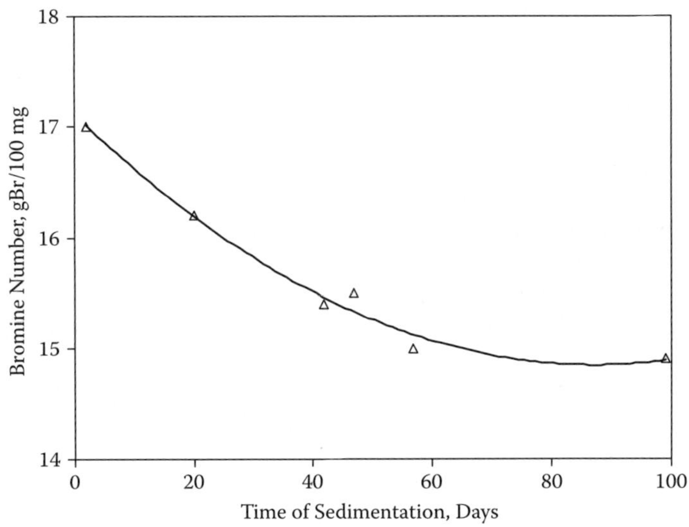

FIGURE 5.27 Variation in Bromine number as function of storage time. (From Gomez et al. 2003. *Rev. Soc. Quim. Mex.* 47 (3): 260–266. With permission.)

shale are much higher in nitrogen content than crude oil (Hauch, 1975). Catalytic denitrogenation, although feasible, is difficult (Satterfield et al., 1975) and it is unlikely that nitrogen levels of petroleum can be reduced economically. As a result, finished fuels can be expected to contain significantly higher levels of nitrogen compounds than their petroleum-derived counterparts. Therefore, nitrogen and oxygen organic compounds are known to be deleterious to the stability of hydrocarbon fuels by promoting the formation of highly obstinate sediment or sludge under storage conditions and during thermal stress (Nixon, 1962; Thompson, 1951).

5.6 PRESUMPTIONS AND REMEDIES FOR SEDIMENT FORMATION

Subjects that include environmental engineering or petroleum refining processes are relevant to review their remedies in order to protect catalyst deactivation as well as valuable fuel from sediment formation over the catalytic sites as well as in products, respectively. The objective of this chapter was to analyze the literature regarding the investigation of the parameters affecting the formation of sediment during hydrotreating, particularly sediments caused by asphaltene. The parameters studied include crude oil components and their compatibility, effect of solvent, effect of reaction conditions, and effect of temperature on sediment formation. Nevertheless, there is little preference that can control the sedimentation during and after the reaction as described below.

Early actions should be used to:

a. Prevent high conversion of bottoms of barrel or operate reaction at lower temperature.
b. Reduce the quantity of asphaltene before hydroprocessing (solvent deasphalting).
c. Add favorable solvent or additives to the feedstock.
d. Select a catalyst that has low or optimum cracking activity and large pore diameter.

Long-term remedial actions should be used to:

a. Attain those objectives listed above that were not accomplished as early actions.
b, Minimize further release of contaminants, such as N, O, and S, surrounding crude oil (heteroatoms responsible for increasing polarity of asphaltene).
c. To find out the fuel market that does not require long storage of processed fuel.
d. Add antifoulant solvent in the HDT products and storage tanks, restore as much of the asphaltene and polar heteroatoms as possible to clean up the storage.

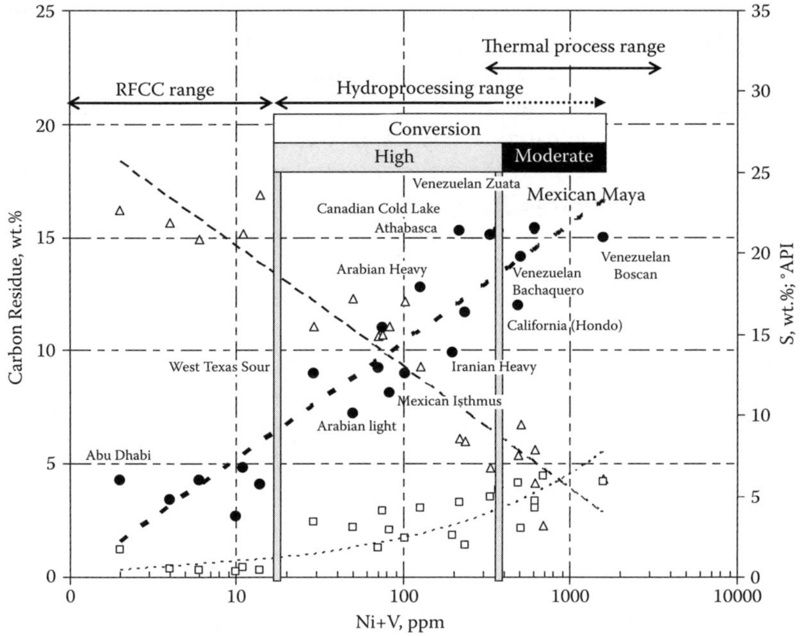

FIGURE 5.28 Choices of process for residue hydroconversion as function of 343°C+ residue properties: (●) carbon residue; (□) sulfur; (Δ) API gravity. (From Rana et al. 2007b. *Fuel* 86: 1216–1231. With permission.)

The selection of process depends on the quality of feedstock, which is usually dependent on several parameters, such as metals (Ni + V), sulfur, and Conradson carbon contents as well as API gravity of the feedstock. Each of these parameters has an independent effect on the process selection, e.g., low metal (Ni + V) content feedstock (<20 wppm) can be effectively processed with a zeolite-based catalyst (RFCC process), while a very high amount of metal content feedstock (>1000 wppm) must be treated in thermal processing. The use of thermal processing is preferable because catalytic processes cannot work with such a high amount of metal because catalysts deactivate very fast. However, most of feedstocks have lower amount of metals, which can be covered by catalytic processes mainly by hydroprocessing using either fixed-bed (working at moderate reaction severity) or moving-bed or ebullated-bed reactor technologies. An overall idea about process selection in order to purify or upgrade bottoms of barrel is shown in Figure 5.28. Recently the hydroprocessing application range has been reported to cover feeds with heavier amount of contaminants with lower API gravity, which is mainly due to the use of catalysts with improved stability as well as operating conditions and proper reactor selection to work under moderate conversion regime so that sediment formation is minimized (Ancheyta et al., 2007b). This process is still at demonstration scale, but the results are very promising. The major problems in fixed-bed residue processing are the coke and metal depositions on the catalyst as has been stated before. Coke deposit is almost impossible to avoid and occurs during the first hours on-stream, while metals are deposited during the whole on-stream time. Hence, to expand the applicability of fixed-bed reactor-based technologies, the use of catalysts with high metal retention capacity together with less coke-like species or sediment deposition is compulsory.

Catalyst deactivation is not only due to the coke or sediment, but also due to the metals. Therefore, the role of asphaltenes should be clearly established, otherwise hydroprocessing becomes less attractive than the carbon rejection methods. Asphaltenes are then the target molecule that deactivates the hydroprocessing catalyst. In this regard, solvent deasphalting could be an important process. A proposed process scheme based on these considerations is shown in Figure 5.29. The integrated process indicates that once asphaltenes are separated from the residue, the deasphalted oil (DAO) is easier to process in hydroprocessing units. The heavy asphalt or pitch can be converted through a thermal process using gasification to complete the conversion, which produces synthesis gas as a major product. Further depending on the need of fuel oil, the synthesis gas may be converted into kerosene or gas oil by using Fischer Tropsch synthesis and isomerization processes (Courthy et al., 1999). Apart from the suggested process, there are very limited possibilities to avoid sediment formation, such as

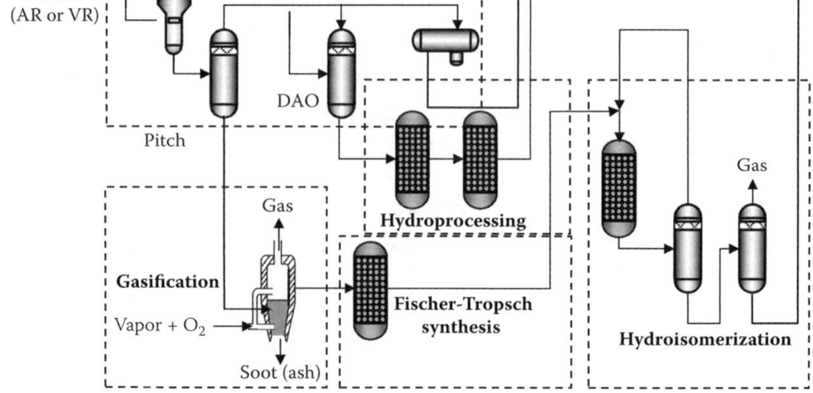

FIGURE 5.29 Integrated solvent deasphalting (SDA), thermal, and hydroprocessing processes. (From Rana et al. 2007b. *Fuel* 86: 1216–1231. With permission.)

a moderate hydrotreating process recently developed by the Mexican Institute of Petroleum (Ancheyta et al., 2007b). For catalytic processes, in order to maximize both performance and cycle length, catalysts must be macroporous to minimize the effect of poisoning of active phases, pore plugging by metal sulfide, and coke deposition with time-on-stream. In any case, using a guard-bed reactor system is highly desired, since it retains multifaceted organometallic and obstinate particulates. Thus, one suitable remedy for preventing sediments formation is to use solvent deasphalting (SDA), since it removes asphaltenes from the oil in which most of the metal, sulfur, and nitrogen are concentrated. The SDA process treats bottoms of barrel at a relatively low cost. If SDA is not feasible, direct hydroprocessing operating at low conversion may be more appropriate.

5.7 CONCLUDING REMARKS

It has been shown that instability of crude oil or residue hydrotreated products has a negative effect in the reactor or in a product as well as in catalyst life. The effect of the crude oil or residue components and their compatibility (physical properties) are the principal cause for sediment formation. The amount of such components depends on crude oil origin before reaction, and on the type of catalyst and reaction conditions after reaction. The resins-to-asphaltenes ratio is the second most important parameter that needs to be highly maintained before and after the hydroprocessing reaction. The loss of side chains (paraffin) always accompanies thermal cracking, and dehydrogenation and condensation reactions are favored by hydrogen deficient conditions, which provoke the recombination or condensation of coke-like species. Even though it is hard to distinguish between the segment formations or the coke species deposited on the catalyst, the nature of sediment is more likely to be the asphaltene molecule.

A catalyst with controlled acid sites or neutral and with larger pore diameter has low sediment formation. An increase in acidic properties will enhance selectivity toward light distillates, which will impact positively on the process global profitability. The catalyst performance can be modified by chemical composition of supports, such as porosity and nature of catalyst (acid-base sites), which improve cracking functionality of the catalyst. Therefore, with respect to catalyst development, additional research is needed to design a catalyst with desirable hydrocracking (HCR) activity (bifunctional), and at the same time be resistant to deactivation. Moreover, any change or improvements in the catalyst need to be under the feasibility of refinery and methodology applicable to the industrial scale. The detrimental effects of coke are a reduction of support porosity, leading to diffusional limitations, and finally blocking the access to active sites. It was formerly believed that coke formation was a polymerization reaction where upon the chemical precursors to coke immediately formed macromolecules when subject to the processing temperatures; the same occurs with sediment formation. Coke formation is a complex process involving both chemical reactions and thermodynamic behavior for reactions that contribute to cracking of side chains from aromatic groups, dehydrogenation of naphthenes to form aromatics, condensation of aliphatic structures to form aromatics, condensation of aromatics to form higher fused-ring aromatics, and dimerization or oligomerization reactions.

REFERENCES

Absi-Halabi, M., Stanislaus, A., and Trimm, D.L. 1991. Coke formation on catalysts during the hydroprocessing of heavy oils. *Appl. Catal.* 72: 193–215.

Alvarez-Ramirez, F., Ramirez-Jaramillo, E., and Ruiz-Molales, Y. 2006. Calculation of the interaction potential curve between asphaltene-asphaltene, asphaltene-resin, and resin-resin systems using density functional theory. *Energy Fuels* 20: 195–204.

Ancheyta, J., Centeno, G., Trejo, F., and Marroquin, G. 2003. Changes in asphaltene properties during hydrotreating of heavy crudes. *Energy Fuels* 17: 1233–1238.

Ancheyta J., Rana M.S., and Furimsky E. 2005. Hydroprocessing of heavy petroleum feeds: Tutorial. *Catalysis Today* 109: 3–15.

Ancheyta, J. 2007a. Reactors for hydroprocessing, in *Hydroprocessing of Heavy Oil and Residua,* Ancheyta, J. and Speight, J.G., Eds., Taylor & Francis: New York, Chap. 5.

Ancheyta, J., Betancourt, G., Marroquin, G., Centeno, G., Muñoz, J.A.D., and Alonso, F. 2007b. Process for the catalytic hydrotreatment of heavy hydrocarbons of petroleum, U.S. patent 2007187294 (Appl. number: U.S.20030563577 20030709).

Ancheyta, J. and Rana, M.S. 2008. Future technology in heavy oil processing, in *Petroleum Engineering-Downstream*, Pedro de Alcantara Pessoa Filho, Ed., Energy Sciences, Engineering and Technology Resources, in *Encyclopedia of Life Support Systems (EOLSS).* Developed under the auspices of the UNESCO, EOLSS Publishers: Oxford, UK. [http://www.edss.net]

Andersen, S.I. and Speight, J.G. 2001. Petroleum resins: Separation, character, and role in petroleum. *Petrol. Sci. Tech.* 19: 1–34.

Bartholdy, J. and Andersen, S.I. 2000. Change in asphaltene stability during hydrotreating. *Energy Fuel* 14: 52–55.

Boussingault, J. 1837, *Annales de Chemie et de Physique.* 64: 141.

Cimino, R., Correra, S., DelBianco, A., and Lockhart, T.P. 1995. Solubility and phase behavior of asphattenes in hydrocarbon media. In *Asphaltenes: Fundamentals and Applications*, Sheu, E.Y. and Mullins, O.C., Eds., Plenum Press: New York, Chap. III.

Clarke, P.F. and Pruden, B.B. 1998. Asphaltene precipitation from Cold Lake and Athabasca bitumens. *Petrol. Sci. Tech.* 16: 287–305.

Cornelius, E.B., Milliken, T.H., Mills, G.A., and Oblad, A.G. 1955. Surface strain in oxide catalysts-alumina. *J. Phys. Chem.* 59: 809–813.

Courty, Ph., Chaumette, P., and Raimbault, C. 1999. Shynthetic or reformulated fuels: A challenge for catalysis. *Oil Gas Sci. Tech.– Rev. IFP* 54 (3): 357–363.

Ese, M.-H., Yang, X., and Sjoblom, J. 1998. Film forming properties of asphaltenes and resins. A comparative Langmuir-Blodgett study of crude oils from North Sea, European Continent, and Venezuela, *Colloid Polym. Sci.*, 276: 800.

EIA (Energy Information Administration), http://www.eia.doe.gov/oil_gas/petroleum/info_glance/petroleum.html.

Frankenfeld, J.W., Taylor, W.F., and Brlnkman, D.W., 1983. Storage stability of synfuels from oil shale. 1. General features of sediment formation in model fuel systems. *Ind. Eng. Chem. Prod. Res. Dev.* 22: 608–614.

Fukuyama, H. 2006. Upgrading technologies for extra-heavy oil and unconventional crude oil. *J. Jpn. Inst. Energy* 85 (4): 277–285.

Fukuyama, H. and Terai, S. 2007. An active carbon catalyst prevents coke formation from asphaltene during the hydrocracking of vacuum residue. *Petrol. Sci. Tech.* 25: 231–240.

Fukuyama, H. and Terai, S. 2008. Preparing and characterizing the active carbon produced by steam and carbon dioxide as a heavy oil hydrocracking catalyst support. *Catalysis Today* 130: 382–388.

Fukuyama, H., Terai, S., Cano, J.L., and Ancheyta, J. 2002. Development of carbon catalyst for heavy oil hydrocracking. Paper presented at the NPRA Annual Meeting, San Antonio, Texas, March 12–19, paper AM-02-14.

Fukuyama, H., Terai, S. Uchida, M., Cano, J.L., and Ancheyta, J. 2004. Active carbon catalyst for heavy oil upgrading. *Catalysis Today* 98: 207–215.

Gawel, I. and Baginska, K. 2004. Characterization of deposits collected from residue hydrotreatment installation. *Prep. Am. Chem Soc. Div. Petrol Chem.* 49: 265–267.

Gawel, I., Bociarska, D., and Biskupski, P. 2005. Effect of asphaltenes on hydroprocessing of heavy oils and residua, *Appl. Catal. A: Gen.* 295: 89–94.

Gomez, M.T., Marroquin, G., Ancheyta, J., Soto, M.I., and Centeno. G. 2003. Inestabilidad de muestras obtenidas durante la hidrodesintegración de residuo del petroleo. *Rev. Soc. Quim. Mex.* 47 (3): 260–266.

Goual, L. and Firoozabadi, A. 2002. Measuring asphaltenes and resins, and dipole moment in petroleum fluids. *AIChE J.* 48 (11): 2646–2663.

Gray, M.R., Khorasheh, F., Wanke, S.E., Achida, U., Krzywicki, A., Sanford, E.C., and Ternan, O.K.Y.M. 1992. Role of catalyst in hydrocracking of residues from Alberta bitumens. *Energy Fuels* 6: 478–485.

Hauck, R.D. 1975. Genesis and stability in peat and coal, *Am. Chem. Soc. Div. Fuel Chem. Prepr.* 20 (2): 85–93.

Hauser, A., Stanislaus, A., Maraf, A., and Al-Adwani, A. 2005. Initial coke deposition on hydrotreating catalysts. Part II. Structural elucidation of initial coke on hydrodemetallation catalysts. *Fuel* 84 (2–3): 259–269.

Inoue, S., Asaoka, S., and Nakamura, M. 1998. Recent trends of industrial catalyst for resid hydroprocessing in Japan. *Catal. Surv. Jpn.* 2: 87–97.

Kaminski, T.J., Fogler, H.S., Wolf, N., and Mairal, A. 1999. Classification of asphaltenes via fractionation and the effect of heteroatom content on dissolution kinetics, *Energy Fuels* 14 (1): 25–30.

Kawai, H. and Kumata, F. 1998. Free radical behavior in thermal cracking reaction using petroleum heavy oil and model compounds. *Catalysis Today* 43 (3–4): 281–289.

Kressmann, S., Morel, F., Harlé, V., and Kasztelan, S. 1998a. Recent developments in fixed-bed catalytic residue upgrading. *Catalysis Today* 43: 203–215.

Kressmann, S., Colyar, J.J., Peer, E., Billon, A., and Morel, F. 1998b H-oil process based heavy crudes refining schemes, Paper presented at the Proceedings of 7th Unitar Conference on Heavy Crude and Tar Sands, Beijing, China, October 27–30, 857–866.

Kubo, J., Higashi, H., Ohmoto, Y., and Arao, H. 1994. The additive effects of hydrogen-donating hydrocarbons derived from petroleum in hydrotreating of heavy oils, Prepr. *ACS* 39 (3): 416–421.

Kubo, J., Higashi, H., Ohmoto, Y., and Arao, H., 1996. Heavy oil hydroprocessing with the addition of hydrogen-donating hydrocarbons derived from petroleum. *Energy Fuels* 10: 474–481.

LePage, L.F., Chatila, S.G., and Davidson, M. 1992. *Resid and Heavy Oil Processing*, Technip: Paris.

Lopez-Salinas, E., Espinosa, J.G., Hernandez-Cortez, J.G., Sanchez-Valente, J., and Nagira, J. 2005. Long-term evaluation of NiMo/alumina–carbon black composite catalysts in hydroconversion of Mexican 538 8°C+vacuum residue. *Catalysis Today* 109: 69–75.

Mansoori, G.A. 1996. Asphaltene, resin, and wax deposition from petroleum fluids. *Arab J. Sci. Eng.* 21 (48): 707–723.

Marafi, A., Stainslaus, A., Hauser, A., and Matsushita, K. 2005a. An investigation of the deactivation behavior of industrial Mo/Al_2O_3 and $Ni\text{-}Mo/Al_2O_3$ catalysts in hydrotreating Kuwait atmospheric residue. *Petrol. Sci. Tech.* 23 (3/4): 385–408.

Marafi, M., Al-Barood, A., and Stanislaus, A. 2005b. Effect of diluents in controlling sediment formation during catalytic hydrocracking of Kuwait vacuum residue. *Petrol. Sci. Tech.* 23 (7/8): 899–908.

Matsushita, K., Koide, R., Hauser, A., Marafi, A., and Stanislaus, A. 2004a. Initial coke deposition on hydrotreating catalysts. Part 1. Changes in coke properties as a function of time on stream. *Fuel* 83 (7/8): 1031–1038.

Matsushita, K., Marafi, A., Stanislaus, A., and Hauser, A. 2004b. Relation between relative solubility of asphaltenes in the product oil and coke deposition in residue hydroprocessing. *Fuel* 83 (11/12) 1669–1674.

McLean, J.D. and Kilpatrick, P.K. 1997a. Effects of asphaltene solvency on stability of water-in-crude-oil emulsions. *J. Colloid Interface Sci.* 189: 242–253.

McLean, J.D. and Kilpatrick, P.K. 1997b. Effects of asphaltene aggregation in model heptane-toluene mixtures on stability of water-in-oil emulsions. *J. Colloid Interface Sci* 196 (23–24): 23–34.

McMillen, D.F., Malhotra, R., Chang, S.-J., Ogier, W.C., Nigenda, S.E., and Fleming, R.H. 1987. Mechanism of hydrogen transfer and bond sciss of strongly bonded coal structures in donar-solvent system. *Fuel* 66: 1611–1620.

Merino-Garcia, D. and Andersen, S.I. 2004. Thermodynamic characterization of asphaltene-resin interaction by microcalorimetry. *Langmuir* 20: 4559–4565.

Mochida, I., Zhao, X.Z., Sakanishi, K., Yamamoto, S., Takashima, H., and Uemura, S. 1989. Structure and properties of sludges produced in the catalytic hydrocracking of vacuum residue. *Ind. Eng. Chem. Res.* 28: 418–421.

Mochida, I., Zhao, X.Z., and Sakanishi, K. 1990a. Catalvtic two-stage hydrocracking of Arabian vacuum residue at a high converiion level without sludge formation. *Ind. Eng. Chem. Res.* 29: 334–337.

Mochida, I., Zhao, X.Z., and Sakanishi, K. 1990b. Suppression of sludge formation by two-stage hydrocracking of vacuum residue at high conversion. *Ind. Eng. Chem. Res,* 29 (12): 2324–2327.

Morel, F., Kressmann, S., Harlé, V., and Kasztelan, S. Process and catalysts for hydrocracking of heavy oil and residues. *Stud. Surf. Sci. Catal.* 106, Elsevier: The Netherlands, 1–16.

Nalwaya, V., Tangtayakom, V., Piumsomboon, P., and Fogler, S. 1999. Studies on asphaltenes through analysis of polar fractions, *Ind. Eng. Chem. Res.* 38: 964.

Nixon, A.C. 1962. Autoxidation and antioxldants of petroleum, in *Autoxidation and Antioxidants*, Vol. 11, Lundberg, W.O., Ed., Interscience: New York, Chap. 17.

Nongbri, G., Nelson, G.V., Self, D.E., and Clausen, G.A. 1996. Ebullated bed process, U.S. patent No. 5,494,570, Feb. 27.

Nowlan, V.J., and Srinivasan, N.S. 1996. Control of coke formation from hydrocracked athabasca bitumen. *Petrol. Sci. Tech.* 14: 41–54.

Ovalles, C., Martinis, J., Pérez-Pérez, A., Cotte, E., Castellanos, L., and Rodríguez, H., 2001. Physical and Numerical Simulation of an Extra-Heavy Crude Oil Downhole Upgrading Process Using Hydrogen Donors under Cyclic Steam Injection Conditions. PDVSA-INTEVEP, SPE 69561

Rahimi, P.M., Gentzis, T., Taylor, E., Carson, D., Nowlan, V., and Cotte, E. 2001. The impact of cut point on the processability of Athabasca bitumen. *Fuels* 80: 1147–1154.

Rahimi, P.M., Teclemariam, A., Patmore, D., De Bruijn, T., Wiehe, I., and Schabron, J. 2004. The effectiveness of deasphalted oil in the stability of visbroken Athabasca bitumen. *ACS Div. Petrol. Chem. Prep.* 49 (2): 147–149.

Ramírez J., Rana, M.S., and Ancheyta, J. 2007. Characteristics of heavy oil hydroprocessing catalysts, in *Hydroprocessing of Heavy Oil and Residua*, Ancheyta, J. and Speight, J.G., Eds., Taylor & Francis: New York, Chap. 6.

Rana, M.S., Ancheyta J., Rayo, P., and Maity, S.K. 2004. Effect of alumina preparation on hydrodemetallization and hydrodesulfurization of Maya crude. *Catalysis Today* 98: 151–160.

Rana, M.S., Maity, S.K., and Ancheyta, J. 2007a. Maya heavy crude oil hydroprocessing catalysts, in *Hydroprocessing of Heavy Oil and Residua.* Ancheyta, J. and Speight, J.G., Eds., Taylor & Francis: New York, Chap. 7.

Rana, M.S., Samano V., Ancheyta, J., and Diaz, J.A.I. 2007b. A review of recent advances on process technologies for upgrading of heavy oils and residua. *Fuel* 86: 1216–1231.

Rana, M.S., Ancheyta, J., Maity, S.K., and Rayo, P. 2008. Maya heavy crude oil hydroprocessing: A zeolite based CoMo catalysts and its spent catalyst characterization. *Catalysis Today* 130: 411–420.

Robert, E.C., Merdrignac, I., Rebours, B., Harlé, V., Kressmann, S., and Colyar, J. 2003. Contribution of analytical tools for the understanding of sediment formation: Application to H-oil® process. *Petrol. Sci. Tech.* 21: 615–627.

Rodolfo, B., Martini, S., Marzin, R., Lopez, J.G., Golding, J.V.R., and Krasuk, J.H. 1988. Process for solid separation from hydroprocessing liquid product. U.S. patent 4732664, March 22.

Sahoo, S.K., Rao, V.C., Rajeshwer, D., Krishnamurthy, K.R., and Singh, I.D. 2003. Structural characterization of coke deposits on industrial spent paraffin dehydrogenation catalysts. *Appl. Catal. A: Gen.* 244: 311–321

Satterfield, C.N., Modell, M., and Mayer, J.F. 1975. Interactions between catalytic hydrodesulfurization (HDS) of thiophene and hydrodenitrogenation (HDN) of pyridine. *AIChe J.* 21: 1100.

Savelev, V., Golovko, A., Gorbunova, L., Kamyanova, V., and Galvalizi, C. 2008. High-sufurous Argentinian asphaltites and their thermal liqifaction products. *Oil Gas Sci. Tech.-Rev. IFP* 63: 57–67.

Scherzer, J. and Gruia, A.J. 1996. *Hydrocracking Science and Technology*. Marcel Dekker: New York.

Schabron, J.F. and Speight, J.G. 1998. The solubility and three-dimensional structure of asphaltenes. *Petrol. Sci. Tech.*. 16: 361–375.

Sherwood, Jr., D.E. 1998. U.S. Patent 5,827,421 and references therein.

Speight, J.G. 1999. The chemical and physical structure of petroleum: effect on recovery operations. *J. Petrol. Sci. Eng*. 22: 3–15.

Speight, J.G. 2001. *Handbook of Petroleum Analysis*. John Wiley & Sons: New York, 223–261.

Speight J.G. 2004. New approaches to hydroprocessing, *Catalysis Today* 98: 55–60.

Speight J.G. 2006. *The Chemistry and Technology of Petroleum*, 4th ed., CRC Taylor & Francis Group, Boca Raton, FL.

Stanislaus, A., Absi-Halabi, M., and Khan, Z. 1994. Fate of asphaltenes during hydroprocessing of heavy petroleum residues. In *Catalytic Hydroprocessing of Petroleum and Distillates*, Oballa, M.C. and Shih, S.S Eds., CRC Press, U.S.A., 159–173.

Stanislaus, A., Absi-Halabi, M., and Khan, Z. 1996. Influence of catalyst pore size on asphaltenes conversion and coke-like sediment formation during catalytic hydrocracking of Kuwait vacuum residues. Catalysts in Petroleum Refining and Petrochemical Industries. *Stud. Surf. Sci. Catal.* 100: 189–197.

Stanislaus, A., Absi-Halabi, M., and Qabazard, H. 1998. Studies on the factors influencing rapid catalyst deactivation and coke-like sediments formation during hydroconversion of heavy petroleum residues. Paper presented at the 15th World Petroleum Congress in Beijing, China, October 12–16. 1997. Proceedings V2 740-41.

Stanislaus, A., Hauser, A., and Marafi, M. 2005. Investigation of the mechanism of sediment formation in residual oil hydrocracking process through characterization of sediment deposits. *Catalysis Today* 109 (1/4): 167–177

Storm, D.A., DeCanio, S.J., and Sheu, E.Y. 1994. Sludge formation during heavy oil conversion. In *Asphaltene Particles in Fossil Fuel Exploration, Recovery, Refining, and Production Processes*, Sharma, M.K. and Yen, T.Y., Eds., Plenum Press: New York, 81–90.

Storm, D.A., Decanio, S.J., Edwards, J.C., and Sheu, E.Y. 1997. Sediment formation during heavy oil upgrading. *Petrol. Sci. Tech.* 15 (1–2): 77–102.

Storm, D.A., Barresi, R.J., Sheu, E.Y., Bhattacharya, A.K., and DeRosa, T.F. 1998. Microphase behavior of asphaltic micelles during catalytic and thermal upgrading. *Energy Fuels* 12: 120–128.

Strassner, J.E.J. 1968. Effect of pH on interfacial films and stability of crude oil-water emulsions. *J. Petrol. Tech.* 20: 303–312.

Strausz, O.P., Mojelsky, T.W., Payzant, J.D., Olah, G.A., and Prakash, G.K.S. 1999. Upgrading of Alberta's Heavy Oils by Superacid-Catalyzed Hydrocracking. *Energy Fuels* 13 (3): 558–569.

Swanson, J.M. 1942. A contribution to the physical chemistry of the asphalts. *J. Phys. Chem.* 46: 141.

Takatsuka, T., Wada, Y., Hirohama, S., and Fukui, Y. 1989a. A prediction model for dry sludge formation in residue hydroconversion. *J. Chem. Eng. Jpn.* 22 (3): 298–303.

Takatsuka, T., Kajiyama, R., Hashimoto, H., Matauo, I., and Miwa, S. 1989b. A practical model of thermal cracking of residual oil. *J. Chem. Eng. Jpn.* 22: 304–310.

Tanaka, R., Hunt, J.E., Winans, R.E., Thiyagarajan, P., Sato, S., and Takanohashi, T. 2003. Aggregates structure analysis of petroleum asphaltenes with small-angle neutron scattering. *Energy Fuels* 17: 127–134.

Terai, S., Fukuyama, H., Sawamoto, S., Ootsuka, K., and Fujimoto, K. 1999. New upgrading process for heavy oil using iron/active carbon mixture catalyst *Prepr. Pap. Am. Chem. Soc., Div. Fuel Chem.* 44 (1): 115–118.

Terai, S., Fukuyama, H., Uehara, K., Fujimoto, K. 2000. Hydrocracking of heavy oil using iron-active carbon catalyst. Importance of mesopore structure of active carbon. *Sekiyu Gakkaishi* 43 (1): 17–24.

Thompson, R., Chenicek, J., Druge, L., and Symon, T. 1951. Stability of fuel oils in storage effect of some nitrogen compounds. *Ind. Eng. Chem.* 43: 935–939.

Tojima, M., Suhara, S., Imamura, M., and Furuta, A. 1998. Effect of heavy asphaltene on stability of residual oil. *Catalysis Today* 43: 347–351.

Victorov, A.I. and Smirnova, N.A. 1999. Description of asphaltene polydispersity and precipitation by means of thermodynamic model of self-assembly. *Fluid Phase Equilibria* 158–160: 471–480.

Wandas, R. and Chrapek T. 2004. Hydrotreating of middle distillates from destructive petroleum processing over high-activity catalysts to reduce nitrogen and improve the quality. *Fuel Proc. Tech.* 85: 1333–1343.

Wandas, R. 2007. Structural characterization of asphaltenes from raw and desulfurized vacuum residue and correlation between asphaltene content and the tendency of sediment formation in H-oil heavy products. *Petrol. Sci. Tech.* 25: 153–168

Wiehe, I.A. 1992. A solvent-resid phase diagram for tracking resid conversion. *Ind. Eng. Chem. Res.* 31 (2): 530–536.

Wiehe, I.A. 1993. A phase-separation kinetic model for coke formation. *Ind. Eng. Chem. Res*. 32 (11): 2447–2454.

Wiehe, I.A. 1995. Polygon mapping with two-dimensional solubility parameters. *Ind. Eng. Chem. Res*. 34 (2): 661–673.

Wu, J., Prausnitz, J.M., and Firoozabadi, A. 1998. Fluid mechanics and transport phenomena molecular-thermodynamic framework for asphaltene-oil equilibria. *AIChE J*. 44 (5): 1188–1199.

Wu, D., Lu, Y., Kong, H., Ye, C., and Jin, X. 2008. Synthesis of zeolite from thermally treated sediment. *Ind. Eng. Chem. Res*. 47 (2): 295–302.

6 Hydrocracking and Kinetics of Asphaltenes

6.1 INTRODUCTION

Hydrocracking is a process which converts high boiling hydrocarbon molecules into low boiling products by simultaneous or sequential hydrogenation (HYD) of C–C bonds breaking. It involves a catalyst and a reactor operated at high hydrogen pressure and moderate temperature. The catalysts used for hydrocracking should have dual functionality, i.e. cracking and hydrogenation functions. Scission of asphaltene alkyl chains generates light petroleum fractions and gases such as CO, CO_2, and H_2S, as a product of thioeter and carboxylic acid decomposition even at low and moderate conditions, is probably the initial stage in asphaltene hydrocracking. Di Carlo and Janis (1992) have reported the following general behavior during hydrocracking of asphaltenes:

1. Dealkylation of alkyl chains attached to naphthenic and/or aromatic rings predominantly contributes to the formation of gas and gasoline fractions.
2. At higher temperatures, cleavage of C–C bonds is carried out producing gases from C_1 to C_4, so that asphaltenes are decomposed also into resins and oils.
3. Dealkylation and cleavage of alkyl side chains occur primordially as a consequence of asphaltenes conversion.
4. The higher the severity of the hydrocracking, the higher the amount of saturates and aromatics due to rupture of aliphatic constituents present in asphaltenes and resins.
5. Saturates and aromatics increase as a result of rupture of polynuclear aromatics and polar resins.
6. Being resins intermediate products, they exhibit a maximum concentration and then reduce their content of time increases.

Takeuchi et al. (1983) proposed that nickel and vanadium are present in asphaltene micelles bonding to constitute large molecules, so that by removing these metals the association of asphaltenes is destroyed. Additional conversion causes rupture of weak bonds reducing significantly the molecular weight. A probable mechanism by which asphaltenes micelles are destroyed is by elimination of vanadium, if asphaltenes contain a large amount of this metal, as pointed out Asaoka et al. (1983) based on the work of Dickie and Yen (1967). On the other hand, depolymerization will occur if weak bonds, such as sulfur linkages, are broken. This rupture is the main cause of molecular weight reduction during cracking of asphaltenes. The stacked part suffers major changes because this portion of asphaltenes is irregularly accommodated. Structural changes involve changes of vanadium-to-carbon

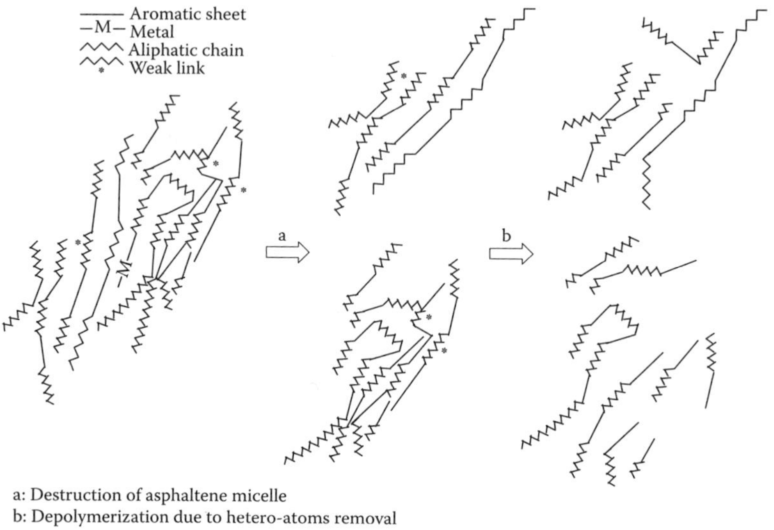

FIGURE 6.1 Proposed mechanism of asphaltenes cracking. (From Asaoka et al. 1983. *Ind. Eng. Chem. Proc. Des. Dev.* 22 (2): 242–248. With permission.)

(V/C) and sulfur-to-carbon (S/C) atomic ratio with increasing the conversion of asphaltenes so that the asphaltenes cracking is not independent of the removal of vanadium and sulfur. For Athabasca bitumen and Boscan crude, the hydrogen-to-carbon (H/C) atomic ratio changed in the range of 1.0 to 1.2, while in the case of Khafji vacuum residue, more pronounced changes were observed (the H/C ratio diminished to 0.8). More aromatic asphaltenes were obtained at higher conversion. Nitrogen and nickel changed only slightly indicating that the roles of these heteroatoms play in the asphaltenes cracking is less important than the role of sulfur and vanadium. This evidence along with the fact that asphaltenes are different-sized molecules makes even a small change in big structures more notorious leading to a reduction of size in the whole molecule. The model proposed by the authors is shown schematically in Figure 6.1.

Deasphalting possesses an important advantage, according to Shiroto et al. (1983) and Takeuchi et al. (1983), who demonstrated that the combination of the ABC process (asphaltenes bottom cracking), which included a proprietary catalyst by Chiyoda Chemical Engineering, along with solvent deasphalting (SDA), allows a high yield of deasphalted oil (DAO) without asphaltenes. The combination of both processes produced a product with only traces of metals and a low sulfur content.

6.2 EFFECT OF REACTION CONDITIONS ON ASPHALTENE HYDROCRACKING

Asphaltenes undergo different chemical transformations depending on the reaction severity conditions to which they are exposed. It has been reported that temperature is one of the most important variables affecting asphaltene structure.

The presence and type of catalyst influences the paths that the reaction will take, whose conversion, selectivity, and yield are highly dependent on the reaction severity, i.e., under certain conditions more coke and gases could be formed due to high temperature. In this section, the influence of temperature, solvent, hydrogen pressure, catalyst, and space velocity are discussed.

6.2.1 Influence of Temperature

Thermal reaction is by far one of the most studied reactions of asphaltenes and several reports regarding it have appeared in technical and scientific texts. According to Hayashitani et al. (1978), during thermal cracking of asphaltenes in the presence of hydrogen, the reactions present are:

- Scission of C-C and C-H bonds
- Breaking of bonds involving metals and heteroatoms
- Aromatization
- Alkylation
- Condensation
- Hydrogenation–dehydrogenation

The complexity of asphaltenes makes the analysis of products very difficult and masks the intrinsic kinetics along with the probable mechanism by which asphaltenes react. For this reason, some authors employ model compounds for studying the possible changes experienced by asphaltenes. Savage and Klein (1987) have used thermal decomposition of *n*-pentadecylbenzene to study the scission of long paraffinic chains attached to aromatic cores. They concluded that free radicals are responsible for the extent of reaction by means of β-scission, and hydrogen abstraction at γ-carbon is an important factor that leads to a thermochemically favorable β-scission. An analogous mechanism is to be expected in asphaltenes thermolysis because of their alkyl chains, which can experience an attack of free radicals. A similar conclusion was obtained with tridecylcyclohexane and 2-ethyltetralin as model compounds (Savage and Klein, 1988). In that case, it was also concluded that the presence of free radicals was imminent and the dominant reaction was dealkylation near or at the aromatic ring. The opening ring is less significant.

Reactivity of asphaltenes is dependent on their structure, which is mainly constituted by condensed aromatic rings in the core with alkyl and naphthenic substituents attached to the core. Scission of peripheral chains is the first step in asphaltenes conversion and, at low temperatures, the major gases produced are H_2S and CO_2, which are the result of fragmentation of thioethers and carboxylic acids. When increasing temperature, breakage of C–C is enhanced, generating hydrocarbon gases. In this way, coke is actually the asphaltenic core stripped of its peripheral substituents (Savage and Klein, 1985). These authors also reported that lighter paraffins were formed from secondary reactions of higher aliphatics and not only by primary decomposition of asphaltenes. Secondary reactions involve cracking of naphthenics and long-chained alkanes. It can be observed that there is an induction period in coke formation around 15 min and coke yield is almost constant after 45 min from the time of reaction. This also yields maltenes. The probable pathway by which asphaltenes are thermally decomposed involves the scission of alkyl chains that generate gases from broken chains, which leaves the aromatic core intact. In addition, maltenes are also formed. If thermal decomposition continues, then smaller maltenes are formed and more gases are released as a consequence of this breaking of alkyl chains.

A similar reaction pathway is reported by Schucker and Keweshan (1980) whose model takes into account secondary reactions of maltenes to produce lower molecular weight maltenes plus gases. When catalytic hydroprocessing of heavy crudes is carried out, asphaltenes that are present in crude oil suffer both the influence of temperature as well as the catalyst effect. Temperature affects asphaltene structure, and thermal breakage does occur in asphaltenes being less soluble in crude oil. At the same time, the catalyst hydrogenates maltenes making the crude very incompatible with asphaltenes, which can precipitate. If small molecules of broken asphaltenes can readily enter into the pores, they can react; however, if molecules of fragmented asphaltenes cannot reach the pores, coke deposition will occur on the catalyst surface.

Yasar et al. (2001) studied the thermal decomposition of residue and isolated the asphaltenes from the product. They proposed a pathway by which asphaltenes that were dissolved in residue reacted during pyrolysis of Maya, Hondo, and Arabian residua at 400, 425, and 450°C from 20 to 180 min of reaction time. The proposed model for lumped resid/asphaltenes involves a reversible reaction that extracts maltenes from asphaltenes. Coke is directly formed from asphaltenes and gases are produced from maltenes and asphaltenes decomposition. However, the kinetic constant of gas from maltenes is negligible, whereas those from asphaltenes to maltenes and vice versa exhibited high values when processing residue. According to the authors, a residue with high asphaltene content is more reactive than a residue with low asphaltene content. More abundant gases were C_1–C_5 paraffins, C_2–C_5 olefins, isoparaffins, H_2S, and CO_2. At 450°C, asphaltenes formed coke at a high amount, whereas, at 400 and 425°C, asphaltenes are selectively decomposed to maltenes. If the feedstock is a residue rich in asphaltenes, maltenes also generate asphaltenes likely due to polymerization reactions, which could occur at higher temperatures. Thus, when isolated asphaltenes are decomposed by pyrolysis, the most prominent reactions were oriented to the production of maltenes and coke.

Some studies related to thermal cracking of asphaltenes have been carried out by Martínez et al. (1997) in a batch reactor. The main objective of this study was to draw a route for which asphaltenes are decomposed into lighter products by using a three-lump model to fit experimental data in which asphaltenes decompose in a parallel reaction as shown below:

$$\begin{array}{c} \text{Coke} \leftarrow \text{Asphaltenes} \rightarrow \text{Gases} \\ \downarrow \\ \text{Oils} \end{array} \tag{6.1}$$

In this case, reaction temperatures were 425, 435, 450, and 475°C and reaction times from 5 to 40 min. Experimental data were fitted to a second-order reaction for asphaltenes at 425, 435, and 450°C, but fitting was poor at 475°C. Hence, their postulated model appears not to be valid at high conversion of asphaltenes promoted by temperature higher than 450°C. It was very evident that coke amount increased as temperature and reaction time increased. Coke is always present, but the smallest amount was obtained at shorter reaction times.

Wang and Anthony (2003) reexamined the data obtained by Martínez et al. (1997) and concluded that at lower temperatures the three-lump model fits well with the data in asphaltenes, and they were decomposed into gas, liquid (oil), and solid (coke). A possible reason by which the model did not fit the data at 475°C was because the highest temperature is responsible for secondary cracking of oil. However, an empirical relation is used to explain the coke yield and the asphaltenes conversion that fit well the data at high conversions or long residence time. The following equation gives a good prediction of the thermal cracking of asphaltenes:

$$y_c = k\frac{y_{Ao} - y_A}{y_A} + y_{Co} \tag{6.2}$$

where y_C and y_{C_o} are the weight fractions of coke in the product and in the feed, respectively; k is the rate constant; and y_A and y_{A_o} are the fractions of asphaltenes in the product and in the feed, respectively.

Butz and Oelert (1995) have performed the cracking of asphaltenes at 410, 435 and 460°C in a batch reactor and observed that the rate of asphaltene conversion decreased continuously with increasing the reaction time. The yields of naphtha and coke increased quickly with increasing reaction time. The continuous decomposition of asphaltenes yielded more oil and at higher temperatures, oil produced more naphtha and gases. In this case, the high amount of coke formed can be attributed to the lack of a good hydrogenating catalyst. The reduction in the reactivity of asphaltenes as reaction time increases is due to the cleavage of weak bonds, which mainly contain disulfides, sulfides, ethers, and carbon acids. The remaining asphaltenes increase their aromaticity and diminish their molecular weight. The aromatic cores are more difficult to be reacted in spite of having longer reaction times. Figure 6.2 shows the behavior of the asphaltene cracking at different temperatures. Trejo et al. (2009) also observed a continuous decomposition of asphaltenes as temperature increased from 380 to 410°C as a function of time of reaction to produce more maltenes. Other products were also coke and gas. Figure 6.3 presents the reduction of weight percent of asphaltenes. The higher the temperature and time of reaction, the lower the amount of asphaltenes.

Singh et al. (2004) have reported a thermal cracking model applied to different residues with variable concentration of asphaltenes following a first-order reaction kinetics. Different products of thermal cracking were evaluated and separated as gas, gasoline, light gas oil (LGO), and vacuum gas oil (VGO). Activation energies for different feedstocks were calculated in the range of 102 to 206 kJ/mol.

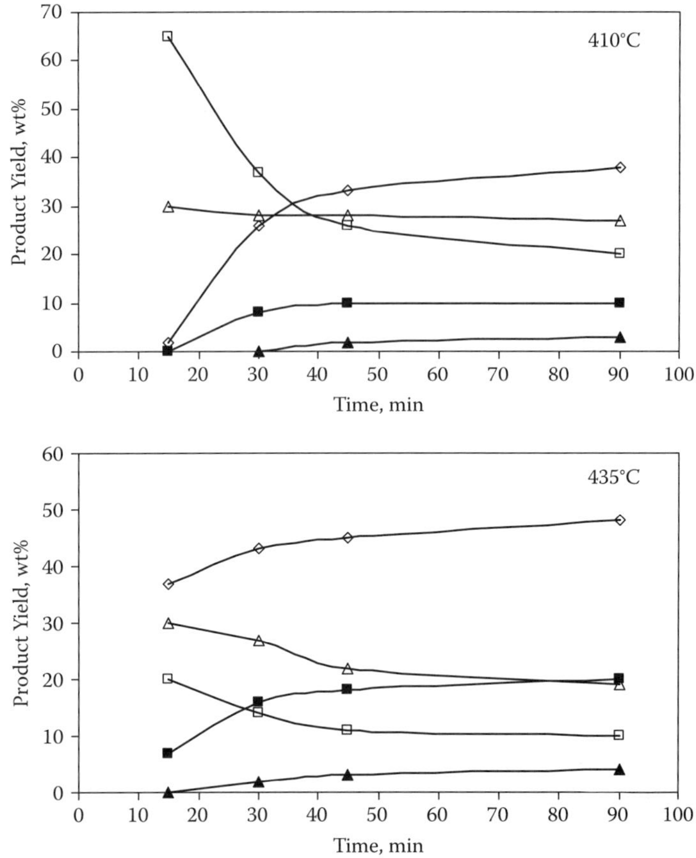

FIGURE 6.2 Asphaltene cracking at 410 and 435°C: (▲) gas, (■) naphtha, (Δ) oil, (□) asphaltenes, (◊) coke. (From Butz, T. and Oelert, H.-H. 1995. *Fuel* 74 (1): 1671–1676. With permission.)

In summary, thermal cracking of residue or asphaltenes has been widely reported in the specialized literature as a common way to explore the possible increase in the quality of petroleum products.

6.2.2 Influence of Solvent

Different authors have carried out experiments on asphaltenes cracking with or without solvent. Such a case is reported by Butz and Oelert (1995) who reported that the addition of solvents increased the decomposition of asphaltenes under mild conditions, but otherwise decreased it. This is due to dilution of asphaltenes in the solvent, which reduces the velocity of the radical cracking reactions so that consecutive reactions of primary radicals generated are decelerated.

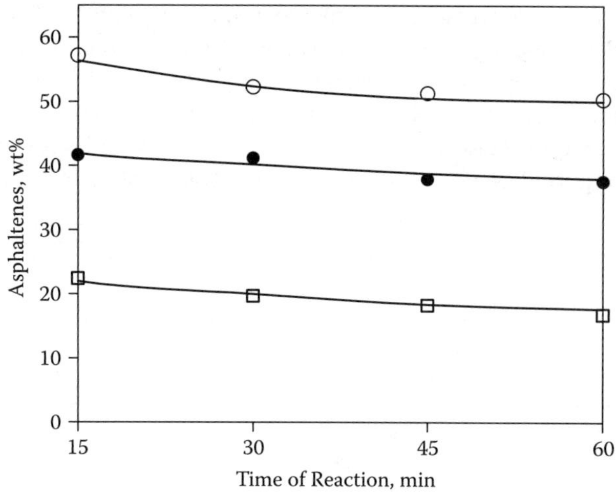

FIGURE 6.3 Variation of asphaltenes as a function of temperature at different reaction times: (○) 380°C, (●) 400°C, (□) 410°C. (From Trejo et al. 2009. Unpublished results. With permission.)

Usui et al. (2004) have carried out hydrocracking reactions at 6.9 MPa of hydrogen and temperatures between 400 and 450°C during 3 h of reaction time in an autoclave reactor with a Ni-Pd/NH_4-Y zeolite catalyst. The feedstock was Brazilian Marlim and Arabian light/medium mixture crudes and an Indonesian natural asphalt-derived asphaltene. The products they obtained after filtration were based on their solubility properties and classified as coke, naphtha, gas oil, and residue. The authors performed two kinds of pretreatments before hydrocracking of asphaltenes with the aim of increasing the asphaltene conversion. In the first case, asphaltenes were dissolved in tetralin (which is a hydrogen donor solvent) without using any catalyst at 420°C during 1 h and at 6.9 MPa of hydrogen pressure. In the second case, asphaltenes were dissolved in benzene with a NiMo/Al_2O_3 catalyst using the same conditions as in the first case to hydrogenate part of the asphaltenes in the presence of the catalyst. Pretreatment of asphaltenes converts at certain extent heavy compounds.

The authors reported three possible reactions during pretreatment of asphaltenes with tetralin:

1. Hydrogen transfer from tetralin to aromatic cores
2. Thermal cleavage of side chains
3. Hydrogen transfer from tetralin to radicals

After pretreatment, the residue sample was hydrocracked in a second stage, where a Ni-Pd/NH_4-Y zeolite catalyst was employed. In this case, the zeolite-supported catalyst has acid sites that are responsible for further cracking leading to a higher conversion with the consequent reduction of the molecular weight of asphaltenes. A good correlation between average molecular weight (measured by size exclusion chromatography [SEC]) and conversion of asphaltenes to lighter fractions was observed and at the same time the reduction of nitrogen content is observed as conversion of asphaltenes is increased (Figure 6.4). In summary, it was reported as advantageous to use the combination of the pretreatment

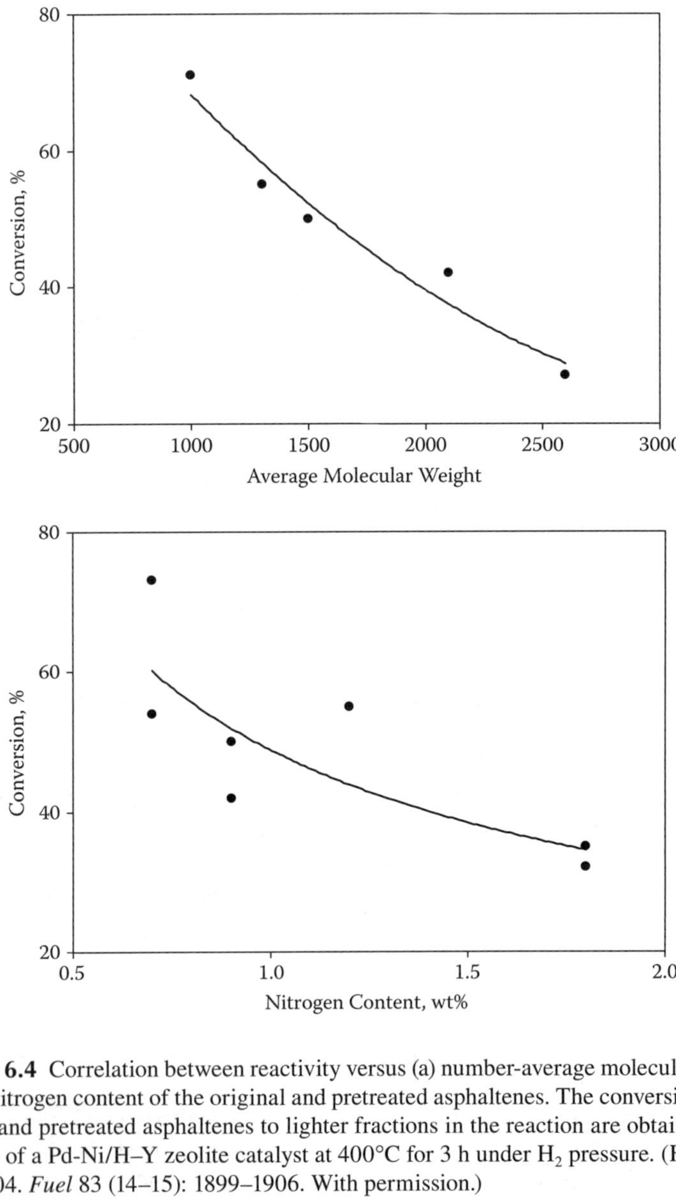

FIGURE 6.4 Correlation between reactivity versus (a) number-average molecular weight, and (b) nitrogen content of the original and pretreated asphaltenes. The conversions of the original and pretreated asphaltenes to lighter fractions in the reaction are obtained in the presence of a Pd-Ni/H–Y zeolite catalyst at 400°C for 3 h under H_2 pressure. (From Usui et al. 2004. *Fuel* 83 (14–15): 1899–1906. With permission.)

of asphaltenes with catalytic hydrocracking using a zeolite-based catalyst because more than 60% of asphaltenes can be converted to lighter fractions.

Other tests, in which hydrogen donor solvents are used in a residue with high content of resins, i.e., Gudao residue, were carried out by Que et al. (1990), who concluded that when hydrogen is added to the feedstock plus a hydrogen-donor solvent, the activation energy during conversion of the residue reduced its value from 216 kJ/mol (51.6 kcal/mol) to 162 kJ/mol (38.7 kcal/mol). The influence of solvent and hydrogen in this case favors the reaction. Resins as well form distillates by cracking reactions and asphaltenes by means of condensation reactions, which can further decompose and generate coke.

Savage et al. (1988) stated that when solvents are added, pyrolysis in toluene reached an asphaltene conversion of ~80%, whereas the same reaction in tetralin as solvent reached only ~35% conversion. However, the selectivity for maltenes is higher in tetralin than in toluene. When hydropyrolysis using toluene as a solvent was carried out, asphaltene conversion was 39.2% and the selectivity to maltenes was almost 80%. These values were obtained at 400°C and 60 min of reaction time. When hydropyrolysis with solvents is carried out coke did not appear, even at 150 min of reaction time, and H_2S, methane, and ethane were the most abundant gases. Propane was also produced during hydropyrolysis with toluene. A relevant fact is that tetralin suppresses coke formation reducing the reaction rate of asphaltenes. Ignasiak and Strausz (1978) worked also with tetralin as a hydrogen-donor solvent and asphaltene conversion was around 50%. The amount of coke at 390°C was insignificant due to the ability of tetralin to add hydrogen during pyrolysis. Tetralin was also used by Al-Samarraie and Steedman (1985) who performed the pyrolysis of petroleum asphaltene at 450°C. They concluded that degradation to mainly lower molecular weight products takes place during pyrolysis; residual asphaltenes have a smaller average size with higher aromaticity and heteroatoms content and show little reaction. The main products are a consequence of β-scission in earlier stages of reaction.

The use of hydrogen donor solvents was also studied by Que et al. (1990) who used tetralin and decalin as solvents. The yields of gases, distillates, oil, and coke increased as conversion increased; however, the presence of hydrogen-donors reduce the formation of gases and coke and can increase the yield of oil and distillates by about 7%. It was confirmed that neither tetralin nor decalin react appreciably during reaction. Solvents that possess aromatic and naphthenic rings (decalin) are very active as hydrogen-donors. Once again, the presence of hydrogen-donor solvents was confirmed as an agent for capping radicals. Radicals are stabilized by subtracting hydrogen atoms from the solvent and the yield of distillates is improved.

6.2.3 Influence of Hydrogen Pressure

It has been stated that scission of C–C is the first step when cracking asphaltenes while using hydrogen or even nitrogen as gas during reaction. Butz and Oelert (1995) have determined that the hydrogen uptake accounts to be 1.5 wt% of the feed into the products. Hydrogen is responsible for capping free radicals generated in alkyl chains in asphaltenes. In this way, hydrogen is consumed and modifies the final composition of the products and also provides a means by which it is possible to control the range of molecular weight and structural-type distribution of liquid products. The presence of hydrogen also inhibits the formation of coke (Oblad et al., 1981). In this regard, Ramakrishnan et al. (1978) have reported activation energies for alkyl radical reactions involving the hydrogen, as shown in:

$$R^{\bullet}(\text{or } Ar^{\bullet}) + H_2 \underset{E_A \approx 38\ \text{kJ mol}^{-1}}{\overset{E_A \approx 58-63\ \text{kJ mol}^{-1}}{\Leftrightarrow}} RH(\text{or } ArH) + H^{\bullet} \tag{6.3}$$

Bunger (1985) has reported that a reversible reaction is carried out at a high rate at 410°C for alkyl hydrogen and around 425°C for aromatic hydrogen. On the contrary, the forward reaction occurs at significant rates at temperatures above 435°C. When hydrogen is present during asphaltene conversion, it is considered that hydrogen is used to convert sulfur compounds into hydrogen sulfide and at the same time it hydrogenates the fragment where sulfur was removed (Takeuchi et al., 1983).

When asphaltenes are converted without the presence of hydrogen (pyrolysis reaction), low conversion to maltene is expected. Hence, when hydrogen is added to the reaction, more conversion to maltenes is obtained. However, when thermal reactions are taking place without using a catalyst, there is generation of hydrogen from asphaltenes, which are the main source of hydrogen in this case. In absence of an efficient hydrogen donor, free radicals formed from asphaltenes subtract hydrogen from the remaining asphaltenes, which enhances the formation of double bounds and eventually the aromatization of naphthenes. Subsequent reactions will turn asphaltenes into more aromatic cores being highly deficient in hydrogen. When a good hydrogen donor is present, such as tetralin, the donor participates in stabilizing free radicals. A similar effect is observed when asphaltenes are dissolved in a good solvent, such as toluene, in presence of hydrogen (hydropyrolysis), the solvent provides an efficient medium to dissolve hydrogen and better contact is achieved. In both cases, hydrogen donors and hydropyrolysis are efficient sources of hydrogen during reaction being that asphaltenes conserve their hydrogen content and retard the coke formation. When solvent is used along with catalysts, conversion of asphaltenes is high and selectivity toward maltenes is equally high. The function of catalysts is to hydrogenate, thus making more efficient the use of hydrogen dissolved in the medium, and then remaining asphaltenes can go inside the pores and react, converting into maltenes. In addition, it has been established that thermal decomposition of asphaltenes occurs prior to catalytic reactions during heavy crude hydrotreating (Savage and Klein, 1989).

The cracking of asphaltenes under hydrogen atmosphere yields a higher amount of oil due to the capping of free radicals impeding the recombination of radicals to form coke. The presence of hydrogen has been reported to increase the yield of oils that have shorter alkyl chains, a higher amount of unsubstituted aromatic carbon atoms, and a smaller proportion of substituted and condensed aromatic carbon atoms (Butz and Oelert, 1995).

Comparison of thermal cracking of asphaltenes with and without hydrogen was carried out by Soodhoo and Phillips (1988a; 1988b), who found that asphaltene conversion is around 99% after 30 min in thermal cracking without hydrogen. However, in thermal hydrocracking, conversion diminishes more slowly reaching a steady conversion after 6 h of 85%. At the same time, coke is produced at higher velocity in thermal cracking yielding conversions of 55 wt% after 30 min and 15 wt% for thermal hydrocracking after 6 h. As can be seen, the increase of temperature is the main reason by which asphaltenes are converted into coke. The presence of hydrogen in thermal cracking reduces significantly the formation of coke during the reaction reducing the occurrence of condensation and polymerization reactions. Resins are thought to be hydrocracked as well giving more aromatic compounds. When hydrogenation is taking place on the less complex aromatic systems, naphthenic compounds are obtained. According to the authors, coke is produced directly from asphaltenes decomposition and not through a series of reactions.

6.2.4 Influence of Catalysts

When asphaltenes are catalytically hydrotreated, the catalyst has a significant influence in the way by which asphaltenes are converted. During catalytic hydrotreating, thermal reactions cannot be discarded. A treatment taking into account thermal and catalytic reactions is suggested by Gollakota et al. (1985). The data treatment was applied to coal liquefaction; however, the application of this concept is not limited to this reaction and it can be extended to petroleum asphaltenes. The following set of equations was proposed:

Thermal reactions:

$$\text{Carbon} + \alpha\ \text{Oxygen} \rightarrow \text{Pre-asphaltenes} \rightarrow \text{Asphaltenes} \rightarrow \text{Oil} \qquad (6.4)$$

$$\text{Carbon} \rightarrow \text{Gas} \qquad (6.5)$$

Catalytic reactions:

$$\text{Preasphaltenes} \rightarrow \text{Asphaltenes} \rightarrow \text{Oils} \qquad (6.6)$$

According to this reaction mechanism, the reaction of preasphaltenes (benzene insolubles and solubles in methylene chloride/methanol mixture) takes place in catalytic sites as well as thermally. However, when a catalyst is present, diffusion limitations will occur, mainly configurationally or restricted diffusion into the catalyst pores.

Catalytic reactions involve the effectiveness factor. Intraparticle diffusion involves the transport of large-derived asphaltenes molecules where the molecular diameter is at least 1/10 of the pore diameter. Catalyst activity diminishes with increasing molecular size of molecules. If asphaltene molecules are in the range of 2 to 15 nm, diffusion limitations are present when a mesoporous catalyst is used, especially between 2 and 20 nm of pore diameter. Köseoglu and Phillips (1988a) stated that conversion of asphaltenes increases with catalyst pore diameter of 20 ± 10 nm, but decreases beyond 20 nm because, as catalyst pore size increases, specific surface area decreases as well as the number of active sites. The authors concluded that hydrocracking reactions are not influenced by mass transfer of reactants when the size of catalyst particles was smaller than 0.3 mm. Two catalysts were used—$CoMo/Al_2O_3$ and $CoMo/TiO_2$-SiO_2—having an average pore size of 9 and 10 nm, respectively, which was considered to be large enough for diffusion of asphaltenes and resins inside the pores.

Studies dealing with thermal and catalytic reactions have been compared by Zhang et al. (2001). They stated that the presence of a catalyst increases the rate of conversion, but does not influence significantly on the reaction pathway except for producing slightly more preasphaltenes. Kobayashi et al. (1987a) carried out catalytic hydrotreating of heavy oils by using different commercial catalysts. Asphaltenes conversion was better carried out over a catalyst having the highest average pore diameter (APD) and pore volume (PV). This is expected since asphaltenes are big molecules that need a catalyst with bigger pores to diffuse inside of its pores. Effectiveness factor was calculated as a function of catalyst pellet size and it increases almost up to the unity as particle size of catalyst is diminished. It is then necessary to reduce the internal and external diffusion limitations when a catalyst is present in order to obtain suitable conversion values to be further used for intrinsic kinetic studies. Trejo and Ancheyta (2005) employed a commercial catalyst that was crushed up in different particle sizes to obtain constant asphaltene conversion at different reaction conditions, as shown in Figure 6.5.

It can be stated that with large pore catalysts, a reaction that is more strongly influenced by diffusion gradients proceeds relatively more easily than that in

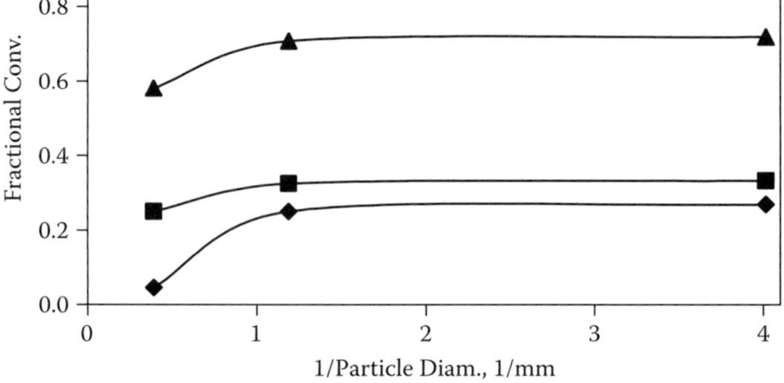

FIGURE 6.5 Evaluation of internal diffusion limitations for asphaltenes conversion: (◆) 360°C, LHSV = 1.0; (■) 360°C, LHSV = 0.33; (▲) 420°C, LHSV = 1.0; (●) 420°C, LHSV = 0.33. (Trejo, F. and Ancheyta, J. 2005. *Catalysis Today* 109 (1–4): 99–103. With permission.)

which contribution of diffusion limitations is bigger. Kobayashi et al. (1987b) found that the optimal pore diameter for asphaltenes removal is around 15 nm at different reaction temperatures. The higher the temperature, the higher the asphaltenes conversion because of the contribution of catalytic sites and thermal reactions. Qader and Hill (1969a) studied the influence of W-Ni bifunctional catalysts supported on Al_2O_3-SiO_2 and found that cracking activity was increased compared with conventional SiO_2-Al_2O_3-and Al_2O_3-supported catalysts. Cracking activity oriented the reaction toward higher yields of gasoline and gas. The presence of a hydrogenating component on bifunctional catalysts also increases the cracking activity and at the same time coke formation is diminished keeping high catalytic activity. The authors have proposed that catalytic hydrocracking of heavy hydrocarbons is a bimolecular reaction that occurs on the catalyst surface with hydrogen and hydrocarbons adsorbed onto adjacent surface sites.

6.2.5 Influence of Space Velocity

Results obtained at different space velocity in a fixed bed reactor have been reported by Trejo and Ancheyta (2005). It was stated that asphaltenes content decreased as space velocity increased. A combination of other reaction conditions gives fewer amounts of asphaltenes if severity of reaction is high. In general, the observed behavior is: the higher the temperature and pressure and the lower the space-velocity, the lower the asphaltene concentration. The most significant effect is observed at the highest reaction conditions 420°C and 100 kg/cm^2, because, at these reaction conditions, asphaltene content is the lowest. It is important to point out that changes in conversion are not proportional to every change in temperature, i.e., if temperature increases from 380 to 400°C, asphaltene conversion increases at different rates than with other increments of temperature, which indicates that asphaltene hydrocracking is higher at an elevated temperature. At low temperatures, only dealkylation of side alkyl chains is observed, and at higher temperatures, asphaltene molecule hydrocracking is prominent. Figure 6.6 shows the influence of space velocity and temperature on asphaltenes conversion.

6.3 REACTION PATHWAYS OF ASPHALTENE HYDROCRACKING

It can be affirmed that asphaltenes react on two different pathways: one of them is by using a catalyst and the other is without a catalyst. When no catalyst is utilized, only thermal reactions take place; however, when a catalyst is employed, not only thermal reactions occur, but also the catalyst has an effect and converts reactants into more valuable products. Both reactions mechanisms are discussed in this section.

6.3.1 Noncatalytic Hydrocracking of Asphaltenes

Nowadays, it is widely accepted that thermal cracking reactions follow first-order kinetics, but this fact was pointed out in the 1950s (Steacie and Bywater, 1955).

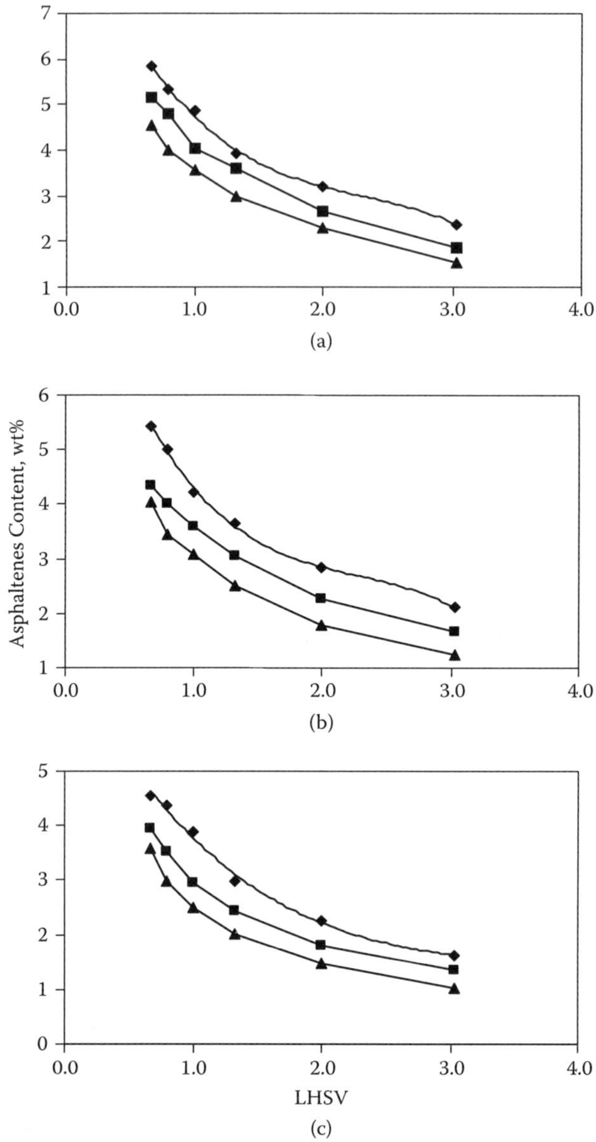

FIGURE 6.6 Asphaltenes content as function of space velocity at: (a) 70 kg/cm², (b) 85 kg/cm², (c) 100 kg/cm² and different temperatures: (◆) 380°C, (■) 400°C, (▲) 420°C. (Trejo, F. and Ancheyta, J. 2005. *Catalysis Today* 109 (1–4): 99–103. With permission.)

This means that complex radical mechanisms might sometimes lead to simple overall kinetics. When hydrocracking reactions without catalyst are present, there are three phases coexisting simultaneously, i.e., solid (coke), liquid (bitumen and/or liquid products), and gas (H_2, C_2-C_4, H_2S, CO and CO_2). Because coke is assumed to be the final product, there are not fluid–solid mass transfer limitations and hydrogen is considered to be in excess (Köseoglu and Phillips, 1988b). Also, since gas–liquid equilibrium is reached very fast, it is supposed that hydrogen concentration is constant in the liquid phase. Qader et al. (1968) and Qader and Hill (1969a; 1969b; 1969c) have stated that the rate of hydrogen transfer across the gas–liquid interphase is negligible, so that reactions in hydrocracking bitumen are assumed to be kinetically controlled rather than diffusion controlled.

In many cases, using lumping different products into groups is a good approach to model a reaction. Lumping techniques have been used for several purposes in petroleum refinery processes. For example, Jacob et al. (1976) made use of lumping to propose a kinetic model for fluid catalytic cracking. Thermal reactions involving either asphaltenes or bitumen can be modeled by using lumps as well. Different models have been proposed to explain the asphaltene conversion during noncatalytic hydrocracking. In some cases, bitumen was the feedstock, while in others, asphaltenes were the feedstock. Earlier studies about decomposition of asphaltenes to give different products were conducted with petroleum and coal liquid derived products. Hayashitani et al. (1977) studied the thermal cracking of Athabasca bitumen in a batch reactor under inert atmosphere where temperature ranged from 300 to 450°C. The main products were classified as coke, asphaltenes, heavy oils, middle oils, light oils, and gases, and a simple pseudoreaction mechanism that correlates the experimental data was obtained. Asphaltenes form heavy oils that in turn form distillates, both reactions being reversible, while coke is directly formed from asphaltenes.

In other studies Hayashitani et al. (1978) worked with Athabasca bitumen and fitted experimental data to first-order reaction proposing that heavy oils also form coke along with asphaltenes in the range of 300 to 450°C.

Yoshiki and Phillips (1985) also studied the thermal decomposition of Athabasca bitumen and observed a good fitting of experimental data to first-order kinetics. Schucker (1983) stated that thermal decomposition reactions of petroleum residua are complex and they cannot be described by a single activation energy model. It was found that activation energies increase with conversion. In the case of asphaltenes, there was an increase of activation energy of 160 kJ/mol from 20 to 90% of volatilization, indicating probably a drastic change in mechanism at the final stages of volatilization, which reflects changes in the strengths of bonds. Weaker bonds are broken at initial stages of pyrolysis, while stronger bonds endure until final stages. In another study, Schucker and Keweshan (1980) stated that noncatalytic hydrocracking of asphaltenes from Cold Lake bitumen followed first-order kinetics with temperatures ranging from 330 to 400°C in a batch reactor where asphaltenes reacted to produce maltenes and unreacted asphaltenes. According to the authors, Cold Lake asphaltenes are hydrogen-deficient core structures with alkyl chains attached to the core and highly substituted aromatic groups. Sulfur is relatively evenly distributed in the core and peripheral groups and nitrogen is concentrated predominantly in the core.

Martínez et al. (1997) carried out the thermal decomposition of asphaltenes without a catalyst and proposed that asphaltenes react to produce oil plus gas and coke that are a function of residence time and where asphaltenes give oil, gas, and coke directly by a parallel reaction. Kinetics in this case are well-fitted to a second-order reaction.

Phillips et al. (1985) studied the thermal cracking of Athabasca bitumen at 360, 400, and 420°C. Asphaltenes formed coke and gas along with heavy oils, which are decomposed to give distillates and these forms of light oils. Heavy oils, distillates, and middle distillates are reversible reactions, all of them following a first-order kinetics. In fact, this pathway is very similar to that proposed by Hayashitani et al. (1978).

Some models involve resins as coke precursors as well. Resins can be classified as soft and hard depending on their molecular weight. Regarding this, Banerjee et al. (1986) have proposed that different fractions of Athabasca bitumen can be separated as asphaltenes, soft resins, hard resins, aromatics, and saturates. Every fraction, except gases, is capable of producing coke in different amounts in the range of 395 to 510°C as a function of time following first-order kinetics as well. Coke formation is a rapid reaction from large aromatics formed as intermediates. Noncatalytic conversion of maltenes also has been reported to be involved in conversion of petroleum residues as proposed by Blažek and Šebor (1993). Conversion of resins can be achieved in two ways: (1) hydrocracking of resins within their natural form (mixed with asphaltenes and oil fractions), or (2) hydrocracking of pure or separated resins. In the second case, changes in resins can be directly evaluated because asphaltenes are not originally present. A kinetic model in which asphaltenes are direct precursors of coke (carbenes and carboids) is shown below:

$$\text{Oils} \underset{k8}{\overset{k7}{\leftrightarrow}} \text{Resins} \underset{k6}{\overset{k5}{\leftrightarrow}} \text{Asphaltenes} \underset{k2}{\overset{k1}{\leftrightarrow}} \text{Carbenes} \underset{k4}{\overset{k3}{\leftrightarrow}} \text{Carboids} \tag{6.7}$$

In this manner, during noncatalytic conversion of maltenes in the presence of hydrogen, asphaltenes and carbenes are formed as well as gaseous and low-boiling point liquids and nonreacted maltenes. In this model, carbenes are likely formed by dehydrogenation and dealkylation of highly polar asphaltenes and cracking of hetero-rings in asphaltenes to give a final product that is rich in oxygen, nitrogen, and metals. Asphaltenes are the result of condensation of at least two molecules of aromatic maltenes. In a separate study, the authors also performed this reaction in the presence of a $CoMo/Al_2O_3$ catalyst obtaining asphaltenes and unreacted maltenes with different properties due to the presence of the catalyst, which modified the product distribution (Blažek and Šebor, 1994).

Soodhoo and Phillips (1988b) have proposed a model based on four lumps in which asphaltenes are precursor of coke, gases, and maltenes. However, maltenes also generate asphaltenes in a reversible reaction, but at higher temperatures, asphaltenes are mainly decomposed to maltenes and the reversible reaction is slowly carried out. Some discrepancies are obtained at the highest temperature (425°C) because higher conversion of asphaltenes is performed making secondary

TABLE 6.1
Kinetic Parameters Obtained by Köseoglu and Philips (1988b)

Temperature, °C	Kinetic Constants					
	k_1	k_2	k_3	k_4	k_5	k_6
375	0.0225	1.1155	0.4250	0.1132	0.1695	0.0862
400	0.1124	4.3541	1.1734	0.2205	0.3263	0.2468
420	0.1606	3.5920	1.3052	0.3114	0.9017	0.7183
Activation energy, kJ/mol (kcal/mol)	168 ± 17 (40 ± 4)	103 ± 7 (25 ± 2)	96 ± 9 (23 ± 2)	85 ± 8 (20 ± 2)	136 ± 12 (32 ± 3)	175 ± 12 (42 ± 3)

Source: Köseoglu, R.O. and Phillips, C.R. 1988b. *Fuel* 67 (7): 906–915. With permission.

reactions (i.e., series and parallel) arise. Henderson and Weber (1985) have postulated this statement as well. The suggested pathway proposes that conversion of asphaltenes is a consequence of breaking heteroatomic linkages, alkyl and naphthenic chains, and subsequent breaking of C–C to produce oils (saturates and aromatics) and more gases. Köseoglu and Phillips (1988b) proposed a kinetic model that involves decomposition of asphaltenes to coke and resins. However, resins can react reversibly to form asphaltenes again. Further resins decomposition gives aromatics, saturates, and gases. A good fit was obtained with:

$$\text{Coke} \xleftarrow{k1} \text{Asphaltenes} \underset{k3}{\overset{k2}{\leftrightarrow}} \text{Resins} \xrightarrow{k4} \text{Aromatics} \xrightarrow{k5} \text{Saturates} \xrightarrow{k6} \text{Gases} \tag{6.8}$$

According to this model, asphaltenes form resins and vice-versa almost with the same activation energy, but decomposition of asphaltenes toward resins is faster. Resins give aromatics with similar value of activation energy as well (Table 6.1).

In most of the cases, all the above studies show that asphaltenes directly form oil (resins, aromatics, olefins, and saturates), gas, and coke. However, coke can be formed by secondary reactions that are a consequence of cracking of oil. Decomposition of asphaltenes and resins only by thermal effects involves both chemical reactions and thermodynamic behavior. The main reactions that contribute to decomposition of heavy fractions are cracking of alkyl chains from aromatic groups, dehydrogenation of naphthenes, condensation of aliphatics and posterior dehydrogenation, and condensation of aromatics to form higher fused-rings aromatics. Finally, it is to be noted that dehydrogenation and condensation reactions are favored when deficient hydrogen atmospheres are used (Marafi et al., 2008).

6.3.2 Catalytic Hydrocracking of Asphaltenes

A summary of the likely steps that occur in the asphaltene conversion in thermal processes, or in the presence of hydrogen and/or the presence of catalyst, are depicted in Figure 6.7, as proposed by Savage et al. (1988). According to the

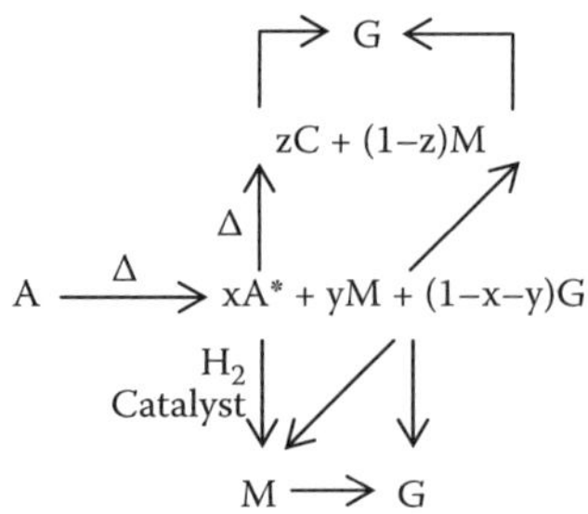

FIGURE 6.7 Asphaltene thermal and catalytic reaction pathways. (From Savage et al. 1988. *Energy Fuels* 2 (5): 619–628. With permission.)

authors, the first step in asphaltenes conversion is the cleavage of peripheral alkyl chains in the original asphaltenes, which called are A. This initial decomposition originates maltenes (M) and gases (G) and reacted asphaltenes (A*). Asphaltenes after reaction possess lower H/C, S/C, and increased nitrogen-to-carbon (N/C) atomic ratios. Two pathways are possible:

1. If thermal decomposition is carrying out, asphaltenes that underwent reaction (A*) produce coke and maltenes. Successive thermal dealkylation of coke and maltenes can produce gases (G).
2. If a hydrogen gas or hydrogen donor is present, dehydrogenation of naphthenes is limited to a low extent by which the H/C atomic ratio is increased in asphaltenes. Additionally, if a catalyst is present, further hydrogenation of asphaltenes (A*) is achieved with a consequent diminution of molecular weight giving maltenes and lighter hydrocarbons. If reactions occur on acid sites of a bifunctional catalyst, then surface reaction controls the overall reaction rates (Weisz and Prater, 1957; Keulemans and Voge, 1959).

A slightly modified pathway from the Savage et al. (1988) model was applied by Adarme et al. (1990). In this model, decomposition of asphaltenes was suggested to begin with dealkylation of side chains in the first hours of reaction at a relatively low temperature. Asphaltenes are converted into intermediate asphaltenes (A*), maltenes (M), and gases (G). At the industrial level, dealkylation may occur in the first half of the reactor when this is a trickle-bed reactor. The intermediate asphaltenes (A*) can undergo further thermal decomposition, which results in the formation of maltenes. However, if a catalyst is present, diffusion of intermediate asphaltenes into the pores can occur and partial demetallization takes place to give more condensed and aromatic asphaltenes (A**). Without using any catalyst, decomposition from (A*) to (A**) is gradual and no demetallization is carried out because of the lack of catalyst. Additional thermal dehydrogenation of (A**) produces coke. Coke is highly formed if higher temperatures (>420°C) are used. Figure 6.8 shows the mechanism postulated by Adarme et al. (1990).

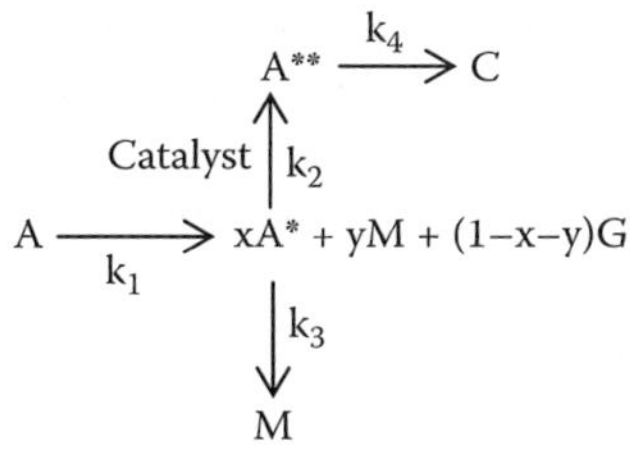

FIGURE 6.8 Mechanism for asphaltene decomposition. (From Adarme. 1989. Effects of asphaltenes on the performance of HDS catalysts used in graded beds. Doctoral thesis University of Oklahoma.)

Savage et al. (1988) tested two $CoMo/Al_2O_3$ catalysts with different supports to compare the effect of pore size. One catalyst had an average pore diameter of 13.2 nm and the other one 19.4 nm. It was observed that the yield of maltenes was higher with the catalyst with smaller average pore diameter (59% after 2 h), whereas with the bigger average pore diameter catalyst the yield to maltenes was 60% at the end of the reaction. It is worthy to mention that the pore structure is important to improve the yield of maltenes when a catalyst is used.

Different studies with using catalysts have been carried out and reaction pathways also have been proposed. In this respect, Papayannakos (1986) found that the hydrodesulfurization (HDS) of an asphaltenic residual fraction fit to a third order by using a model based on the summation of two parallel desulfurization rates of the asphaltenic and nonasphaltenic fractions. This indicated that different sulfur compounds are present in asphaltenes, which can exhibit a wide range of reactivity. On the other hand, nonasphaltenic and deasphalted crude both fit the second-order kinetics. An important conclusion obtained by Papayannakos was that the reaction rate of deasphalted crude was the same as that of the sulfur removal rate in the nonasphaltenic fraction in the presence of asphaltenes, indicating that the asphaltene fraction does not compete for active catalytic sites with the nonasphaltenic fraction, and the asphaltene fraction suffers cracking during HDS, which turns them into lighter compounds.

According to Papayannakos's (1986) mechanism, asphaltenes diminish their size and can be desulfurized or not desulfurized. In addition, desulfurized molecules without asphaltenic molecules are obtained as well and reduction of molecular weight is a consequence of sulfur removal, which is linking two or more aromatic cores. This means that asphaltenes are depolymerized during the cracking of asphaltenes, as reported previously by Asaoka et al. (1983). Activation energy in the HDS of deasphalted crude and nonasphaltenic fraction was 19 kcal/mol and 53 kcal/mol for HDS of asphaltenic fraction. It is well known that there is a big difference in reactivity between asphaltenic and nonasphaltenic feedstocks. Papayannakos and Marangozis (1984) observed during the HDS of a residue feedstock that saturated hydrocarbons content increased with HDS conversion up to 400°C. Aromatics content decreased at low temperatures, remaining constants at higher temperatures indicating that aromatics are converted to saturated compounds. Polar aromatics are created from asphaltenes and at low temperatures their content is increased, remaining almost constant at higher temperature due to their conversion to aromatics or saturated compounds.

Benito et al. (1997) performed the hydrotreating of synthetic oil in the presence of a catalyst that favored the production of gases and oils as main products. Coke amount was kept as a constant due to the presence of hydrogen. The catalyst diminished greatly the activation energy during reaction compared with thermal reaction (5.3 kcal/mol in catalytic hydrotreating against 26.3 kcal/mol in thermal cracking). Oils were produced in higher amounts due to the catalytic activity and coke amount was always constant by which it can be excluded from the following kinetic model:

$$\text{Oil} \xleftarrow{k1} \text{Asphaltenes} \xrightarrow{k2} \text{Gases} \tag{6.9}$$

Köseoglu and Phillips (1988c) proposed a kinetic model, which involved asphaltenes, maltenes, and gases where the heaviest molecular weight components produce lighter products, as shown below:

$$\text{Asphaltenes} \underset{k2}{\overset{k1}{\leftrightarrow}} \text{Maltenes} \xrightarrow{k3} \text{Gases} \tag{6.10}$$

Asphaltenes content diminished continuously with higher temperature as the reaction was carried out. Asphaltenes and resins were mainly consumed during initial heating especially at high temperatures.

Trejo et al. (2009) studied the catalytic hydrocracking of asphaltenes from Maya crude and found that the pathway can be represented by the following model:

$$\begin{array}{c} \text{Coke} \xleftarrow{k_1} \text{Asphaltenes} \xrightarrow{k_2} \text{Maltenes} \\ \downarrow \\ \text{Gases} \end{array} \tag{6.11}$$

This reaction scheme is similar to that proposed by Benito et al. (1997), but in this case coke increased as the temperature also increased. The scheme is also similar to that proposed by Soodhoo and Phillips (1988b), but the reaction from maltenes to asphaltenes is not present in this case, which indicates that asphaltenes are rapidly decomposed without any opportunity for maltenes to react between them, condense, and form asphaltenes again. The set of differential equations to be solved proposed by Trejo et al. (2009) is:

$$\frac{dy_1}{dt} = -(k_1 + k_2 + k_3)y_1 \tag{6.12}$$

$$\frac{dy_2}{dt} = k_1 y_1 \tag{6.13}$$

$$\frac{dy_3}{dt} = k_2 y_1 \tag{6.14}$$

$$\frac{dy_4}{dt} = k_3 y_1 \tag{6.15}$$

where y_1 corresponds to asphaltenes, y_2 is coke, y_3 is maltenes, and y_4 is gases. The authors used *n*-hexadecane for maintaining asphaltenes dispersed inside the batch reactor. An alumina-supported NiMo commercial catalyst sulfided *ex situ* was employed. The ratios of hexadecane-to-asphaltenes and asphaltenes-to-catalyst were kept constant at 40 (wt/wt) and 5 (wt/wt), respectively. The reaction began when temperatures of 380, 400, and 420°C at 2 MPa were reached. Reaction time was within 0 and 1 h. During heating of the reactor up to the desired reaction temperature, some degree of asphaltenes conversion was achieved and said conversion was considered in the kinetic model.

In all cases, it was observed that the amount of asphaltenes diminished as time of reaction increased at different temperatures (Figure 6.9). At the same time, coke and gases yields are increased. Coke reaches its higher yield at the highest temperature because of faster decomposition of asphaltenes whereas the amount of maltenes increased slightly. Not only higher temperature increases the amount of coke, but also the influence of solvent plays a key role. In this case, *n*-hexadecane is a poor hydrogen-donor solvent and favors condensation reactions. Gases only increased steadily with increasing temperature and time of reaction. Gas chromatography (GC) analysis indicated the presence of unreacted H_2 and H_2S, which are originated from thermally labile bonds. The origin of H_2S in the reactor is due to the conversion of sulfur molecules present in asphaltenes.

The experimental data were well fitted to the proposed model and Equation (6.12) through Equation (6.15) were used to determine the kinetic parameter values, which are summarized in Table 6.2. It is observed that maltenes and gases are formed almost at the same reaction rate. Nevertheless, coke amount was almost double compared with maltenes and gases. The high value of activation energy in the formation of coke is an indication that reaction proceeded by free radical originated by cleavage of alkyl chains. The remaining asphaltenes are difficult to hydrogenate and the preferred pathway that they follow is the condensation into larger molecules originating the formation of coke. For improving the hydrogenation of asphaltenes, it is desirable to use a hydrogen donor (i.e., tetralin or 1-methylnaphthalene). The presence of a good solvent along with higher hydrogen pressure and a catalyst would be capable of converting asphaltenes into maltenes with minimum amount of coke and gases. The predicted values of each fraction and experimental values are depicted in Figure 6.10. It can be seen that there is a good agreement between both values. In addition, the correlation coefficient (r) was 0.989 and the residual values were randomly distributed. Validation of a kinetic model was carried out by comparing data that had been reported in the literature. Good reproducibility of data was observed when applying the model to those data obtained by Soodhoo and Phillips (1988b), as seen in Figure 6.11.

6.3.3 Role of Catalytic Sites on Asphaltenes Hydrocracking

When a catalyst is used during cracking of heavy hydrocarbons, catalytic sites play an important role and conduct the reaction for different pathways. Acid sites are responsible for adsorption of basic molecules which contain nitrogen.

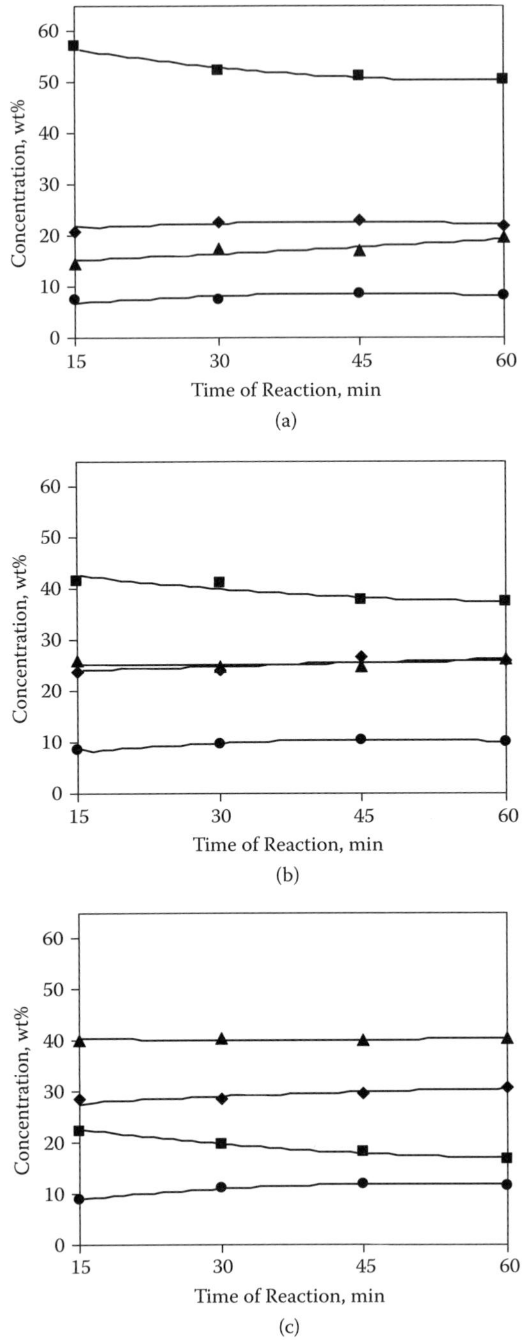

FIGURE 6.9 Distribution of products at a) 380°C, b) 400°C; c) 410°C (■) asphaltenes (◆) maltenes; (▲) coke; (●) gases (Trejo et al., 2009, unpublished results.).

TABLE 6.2
Kinetic Constants Obtained during Catalytic Hydrotreating of Asphaltenes from Maya Crude

Parameter	Kinetic Constants		
Temperature, °C	k_1	k_2	k_3
380	0.0018	0.0024	0.0008
400	0.0032	0.0031	0.0011
410	0.0069	0.0049	0.0017
Preexponential factor, min^{-1}	5.86×10^9	8.15×10^3	8.17×10^3
Activation energy, kcal/mol	37.5	19.6	21.0

Source: Trejo et al. 2009. Unpublished results.

Nitrogen is present in asphaltenes forming part of aromatic rings. Qader and Hill (1969b) found that chemical reactions and not physical processes control the reaction rate during hydrocracking of low temperature tar. Product distribution during hydrocracking of tar proceeds through a mechanism involving simultaneous and consecutive cracking, hydrogenation, and isomerization reactions on the surface of dual-functional catalyst leading to the formation of various products.

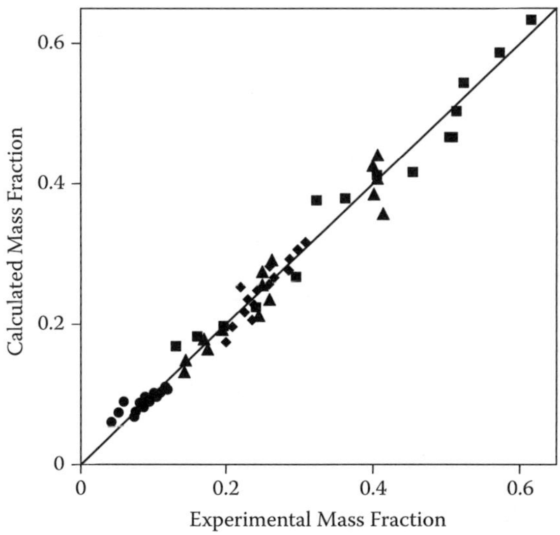

FIGURE 6.10 Experimental and calculated mass fractions obtained with the kinetic pathway. (■) asphaltenes, (◆) maltenes, (▲) coke, (●) gases. (From Trejo et al. 2009. Unpublished results.)

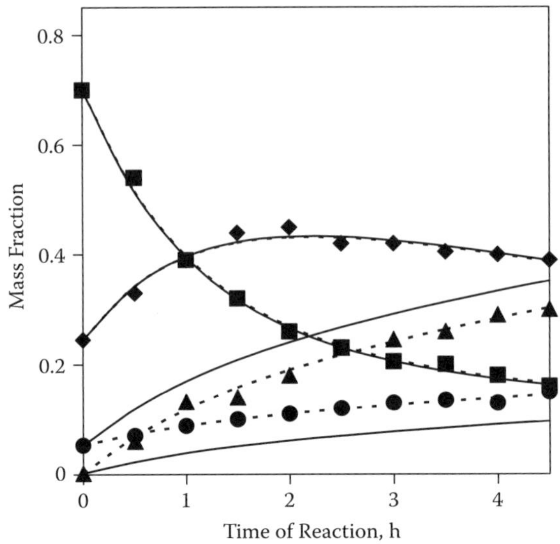

FIGURE 6.11 Validation of the kinetic model at 400°C by using results obtained by Soodhoo and Phillips (—) and those determined by Noyola et al. (- - -). (■) asphaltenes, (◆) maltenes, (▲) coke, (●) gases. (From Trejo et al. 2009. Unpublished results.)

A dual-functional catalyst contains two types of active sites: The acid sites of supports which promote cracking (isomerization) reactions and the metallic sites of coblate-molybdate (Co-Mo) promote hydrogenation function of the reactions. The authors have suggested that hydrocracking of hydrocarbons produces predominantly C_3 and C_4 gases and proceeds essentially through a carbonium ion mechanism, which is characteristic of catalytic cracking, whereas production of mainly C_1 and C_2 gases proceeds through a mechanism by free radicals, which is characteristic of thermal cracking. Cracking of hydrocarbon molecules proceeds by a carbonium ion mechanism wherein they either lose a hydride ion or add a proton to form a carbonium ion and heteromolecules.

Formation of carbonium ions is followed by splitting of C–C bonds to form products. These ions can react with the original molecule or with other ions. In addition, reactions continue in different ways, for example, through isomerization of olefins and hydrogen addition.

C, S, N, and O are removed and the main surface reactions that tar or bitumen suffer during hydrocracking are C–C, C–S, C–N, and C–O scission, along with isomerization and hydrogenation reactions. When hydrogen is in excess, olefins reaction is very fast.

Leyva et al. (2007) have suggested that hydrocracking of heavy hydrocarbons (especially asphaltenes) is enhanced by carbonium ions, which react with the original molecule or with other carbonium ions to stabilize. Additional reactions

$$\text{Asph-R} + H^+(\text{acidic site}) \longrightarrow [\text{Asph-}R^+(\text{carbonium ion}) + H^-]$$

$$\downarrow$$

$$\underset{\text{Aromatic (asphaltene) fraction}}{\text{Asph}} + \underset{\text{alkyl-aliphatic chain}}{R_1H} \qquad \text{(i)}$$

$$\text{Asph} + H^+(\text{acidic site}) \xrightarrow[\text{Pressure } H_2]{} \text{MAL} + R_2H \qquad \text{(ii)}$$

$$\text{MAL} \xrightarrow[\text{Cracking, isomerization}]{} \text{AN} + R_3H \qquad \text{(iii)}$$

$$\downarrow$$

$$[\text{AN}^+ + H^-]$$

$$\downarrow \text{Aromatization}$$

$$\underset{\text{Smaller hydrocarbon}}{AR_1 + R_4H} \qquad \text{(iv)}$$

$$H^- + H^+ \longrightarrow H_2 \qquad \text{(v)}$$

$$H^- + \text{Asph-}R^+ \longrightarrow \text{Asph-RH} \qquad \text{(vi)}$$

FIGURE 6.12 Proposed reaction mechanism of heavy oil hydrocracking involving asphaltenes. Asph: asphaltene or fused aromatic ring, R: alkyl chain, AN: hydroaromatic (naphthenic ring), and MAL: maltene. (From Leyva et al. 2007. *Ind. Eng. Chem. Res.* 46 (23): 7448–7466. With permission.)

during hydrocracking are isomerization and hydrogenation. A mechanism that explains this is depicted in Figure 6.12, where the initial step corresponds to the adsorption of heavy molecules (i.e., asphaltenes) over acid sites on the catalyst, which generates a carbonium ion and a saturate alkane. The cracking of asphaltenes can generate small fragments that can be combined with hydrogen up until they are neutralized and form maltenes and alkanes. After that, maltenes can undergo cracking or even isomerization to give naphthenic rings that can be exposed to dehydrogenation at high reaction temperatures and form aromatic rings. Another option is that carbonium ions can be capped with hydrogen that has been dissociated on the catalytic sites.

The influence of acid sites on asphaltenes conversion has been reported by Breysse et al. (2002). Acid sites can donate a proton that is required for hydrocracking and isomerization reactions and, at temperatures higher than 400°C, the –SH species present on acid sites are good proton donating, and zeolite- and silica-alumina-supported catalysts are rich in acid sites. Olefins and aromatics can readily accept a proton and form a carbocation, which can be formed also by donating a hydride H^- from hydrocarbons to a Lewis site on the catalyst site, according to the following mechanism proposed by Furimsky (2007):

$$\text{RH} + \text{Brønsted site} \rightarrow \text{RH}_2^+ + \text{neutral site}$$

$$\text{RH} + \text{Lewis site} \rightarrow R^+ + \text{Lewis site} - H \qquad (6.16)$$

In the asphaltene structure, carbocations are prone to be formed, especially on aliphatic carbons linked to aromatic carbons. However, further rearrangement of carbocations can occur with skeletal modifications can suffer. It is probable but not confirmed that carbocations can interact with porphyrins in feedstocks rich in metals. This interaction would neutralize carbocations and intramolecular rearrangements of the electron acceptors would change.

6.4 KINETIC MODELING OF ASPHALTENES HYDROCRACKING

The way in which asphaltenes react is very important during hydrocracking. When the reaction rate is known, it is possible to make improvements in the process or even design better catalysts, which are able to process these complex molecules. The knowledge of kinetics as well as the influence of catalyst properties on diffusion limitations, along with other concepts (mass and heat transfer, thermodynamics, transport phenomena, etc.) permits the design of reactors not only for asphaltene hydrocracking but also for any catalytic reaction in which heavy petroleum fractions are involved. It is to be remembered that asphaltenes reduce the reaction rate of other reactions during hydroprocessing, i.e., HDS, since asphaltenes are coke precursors that deactivate catalysts by plugging the catalytic sites. Different approaches have been carried out to represent the way in which asphaltenes react, and kinetic parameters, e.g., reaction order and kinetic constant, can be extracted from such models. Other models give a structural representation of asphaltenes that allows for supposing what bonds will be broken more easily. Based on the fact that asphaltenes are not a unique type of molecule in heavy crudes, but also they are polydispersed in nature, a few models take into account the molecular weight distribution also, which improves the comprehension of the asphaltene reactivity. In this section, different models for asphaltenes hydrocracking reported in the literature are discussed.

6.4.1 Power Law Models

Power law models are by far the most simplified ones to represent the kinetics of asphaltenes hydrocracking. With this approach, different types of reactivity of asphaltenes are not taken into account. Instead, all asphaltene molecules are considered to react at an average rate expressed in the kinetic constant. When the complexity of the reaction system is high, a power law model can be selected to adjust kinetic data. Regarding this, Soodhoo and Phillips (1988b) reported a model that represents the simple decomposition of asphaltenes into products (maltenes, gases, coke). Two steps are involved: (1) cracking of high molecular weight hydrocarbons, and (2) hydrogenation of unsaturated fractions formed during cracking. The proposed model was

$$\text{Asphaltenes} + \text{H}_2 \xrightarrow{k} \text{Products (maltenes, gas, coke)} \qquad (6.17)$$

where k is the global rate constant.

Experiments were carried out at 350, 375, 400, and 425°C and the initial pressure of hydrogen was 7.0 MPa measured at 20°C. The activation energy calculated was 38.43 kcal/mol (160.9 kJ/mol) and the preexponential factor was 2.3×10^{12} h^{-1}. Data were well fitted to first-order kinetics. However, the main consideration made by authors was to exclude the asphaltene conversion higher than 70% because it was assumed that the mechanism could change at higher asphaltene conversion since the overall reaction could proceed through parallel and consecutive secondary reactions (Soodhoo and Phillips 1988b).

Most of the time the reactor employed for obtaining experimental data of hydrotreating of heavy oils is a trickle bed reactor. When a packed bed reactor is used to determine kinetic parameters, it is recommended to take into account the following considerations in order to estimate intrinsic Kinetics as proposed by Korsten and Hoffmann (1996):

- Fluid velocities do not change along the reactor.
- Gas and liquid velocities are constant across the reactor.
- Radial and axial concentration gradients are neglected.
- Reactor is operated in a steady state.
- Reactor is kept at constant pressure and temperature.
- Chemical reactions only take place on the catalyst.

Philippopoulos and Papayannakos (1988) studied the asphaltene decomposition from a Greek atmospheric residue in a trickle-bed reactor packed with a commercial $CoMo/Al_2O_3$ catalyst at 5 MPa of pressure and temperatures ranging from 350° to 430°C. Reaction order was fitted to 2 with activation energy of 27.8 kcal/mol. Callejas and Martínez (2000a; 2000b) reported the kinetics of asphaltenes conversion from a Maya residue in a temperature range of 375° to 415°C and hydrogen pressure of 12.5 MPa, in a continuously stirred tank reactor. Data were well fitted to an order of 0.5 and activation energy was calculated to be 41.5 kcal/mol.

Köseoglu and Phillips (1988c) employed a power law model and used a $CoMo/Al_2O_3$ catalyst. The first model they proposed took into account the excess of hydrogen. Experimental data were well fitted to a first order and activation energy was 26.8 kcal/mol, which is lower than that obtained when noncatalytic hydrocracking is carried out (35.9 kcal/mol). When moderate pressure is used, asphaltenes hydrocracking is more pronounced at temperatures above 400°C; however, behavior of the reaction is dependent on pressure as well. That is why different kinetic regimes have been proposed when developing kinetic models for hydrocracking of heavy oils taking into account pressure and temperature of reaction, as reported by Botchwey et al. (2004) and Sánchez et al. (2005).

When heavy feedstocks are processed and reaction order is higher than unity, a good approach for representing the asphaltenes hydrocracking by means of a power law model is to separate the reaction into two reactions in parallel, as proposed by Kwak et al. (1992). It was suggested that decomposition of heavy hydrocarbons can be fitted to:

$$A\begin{cases} A_1 = \gamma A + H_2 & \xrightarrow{k1} \\ A_2 = (1-\gamma)A + H_2 & \xrightarrow{k2} \end{cases}\Bigg\}\text{Hydrotreated products} \tag{6.18}$$

where A_1 represents the fraction of the heavy hydrocarbon that reacts slowly, whereas A_2 is the less refractory fraction and reacts more quickly. It is assumed that reaction order for heavy hydrocarbon is 1 in each case. According to the power law model, concentration of hydrogen dissolved into the hydrocarbon is required. For calculating this value, Henry's law needs to be applied. Korsten and Hoffmann (1996) have reported the sequence of calculations to be done to calculate the concentration of dissolved gas.

A comparison of a simple power law model and models based on assumptions of Kwak et al. (1992) is given by Trejo et al. (2007) for asphaltenes hydrocracking. One suggested model follows a power law kinetics as shown below:

$$\text{Asphaltenes} + H_2 \rightarrow \text{Products} \tag{6.19}$$

$$-r_A = k\, C_A^{\alpha}\, C_{\text{Hydrogen}}^{\beta} \tag{6.20}$$

where C_A is the asphaltene concentration, $C_{Hydrogen}$ the hydrogen concentration, and α and β are reaction orders for asphaltenes and hydrogen, respectively.

The other model is an adaptation of that represented by Equation (6.18):

$$\text{Asph}\begin{cases} \text{Asph}_1 = \gamma\text{Asph} + H_2 & \xrightarrow{k1} \\ \text{Asph}_2 = (1-\gamma)\text{Asph} + H_2 & \xrightarrow{k2} \end{cases}\Bigg\}\text{Hydrocracked asphaltenes} + \text{Products} \tag{6.21}$$

where γ is the fraction of hard-to-react asphaltenes and $(1 - \gamma)$ is defined as the fraction of easy-to-react asphaltenes. Reaction order for both types of asphaltenes is one according to the assumptions of Kwak et al. (1992). The proposed kinetic model was

$$-r_A = \gamma k_1\, C_A\, C_{\text{Hydrogen}}^{\beta} + (1-\gamma)k_2\, C_A\, C_{\text{Hydrogen}}^{\beta} \tag{6.22}$$

Since γ, $(1 - \gamma)$, k_1, and k_2 are constants, they can be rearranged and grouped to give:

$$-r_A = [\gamma k_1 + (1-\gamma)k_2]C_A\, C_{\text{Hydrogen}}^{\beta} \tag{6.23}$$

$$-r_A = k_o\, C_A\, C_{\text{Hydrogen}}^{\beta} \tag{6.24}$$

As can be seen, Equation (6.24) is again a power law model in which reaction order of asphaltenes is first order. Equation (6.22) allows for determining the values of refractory and nonrefractory fractions of asphaltenes, e.g., the fraction that is more feasible to react. Taking into account the concept of easy and hard-to-react asphaltenes, the second model can be generalized if each fraction of asphaltenes reacts with a different reaction order instead of first-order kinetics. Thus, another power law model was also proposed:

$$-r_A = \gamma k_1 C_A^{\alpha 1} C_{\text{Hydrogen}}^{\beta} + (1-\gamma) k_2 C_A^{\alpha 2} C_{\text{Hydrogen}}^{\beta} \tag{6.25}$$

In all cases, the objective function to determine the best set of kinetic parameters was expressed as the sum of square errors between experimental and calculated concentrations of asphaltenes, which was minimized with the modified Levenberg–Marquardt's algorithm incorporated as subroutine in commercial packages as Matlab® and IMSL 3.0 routine from Visual Fortran®.

For the first model (power law model), the reaction orders for asphaltenes and hydrogen were 2.1 and 1.28, respectively, whereas activation energy was 10.35 kcal/mol. The final kinetic model is shown in Equation (6.26). The average absolute error with this model between experimental and calculated concentrations was 2.79%.

$$-r_A = 5.65 \times 10^6 e^{-\frac{10350}{RT}} C_A^{2.1} C_{\text{Hydrogen}}^{1.28} \tag{6.26}$$

For the second kinetic model, the average absolute error between experimental and calculated concentrations was 2.38%. The final model was

$$-r_A = (0.4) 8.95 \times 10^6 e^{-\frac{10433}{RT}} C_A C_{\text{Hydrogen}}^{1.23} + (0.6) 2.86 \times 10^7 e^{-\frac{9219}{RT}} C_A C_{\text{Hydrogen}}^{1.23} \tag{6.27}$$

According to this model, the fraction of asphaltenes, which is more reactive, is 0.6 and, for the less reactive fraction, it is 0.4. As assumed previously, reaction order of asphaltenes in both fractions is 1. Similar activation energies were calculated for easy and hard-to-react asphaltenes, but a higher preexponential factor is obtained for easy-to-react asphaltenes. This similarity in activation energy values could be a consequence of polydispersity of asphaltenes since some molecules can access inside the pores without a problem depending on their size. It can be observed that both activation energies are quite similar to that obtained by the simple power law model by which the total contribution to the activation energy could be represented as the sum of those from easy- and hard-to-react asphaltenes, along with their respective fractions, as:

$$E_{A\,\text{asphaltenes}} = \gamma E_{A\,\text{hard-to-react}} + (1-\gamma) E_{A\,\text{easy-to-react}} \tag{6.28}$$

$$E_{A\,\text{asphaltenes}} = (0.4)(10{,}433) + (0.6)(9219) = 9{,}705\ \text{cal/mol} \tag{6.29}$$

This value is quite similar to that obtained in Equation (6.26).

The best set of kinetic parameters obtained with the third model is shown in Equation (6.30). For this case, the average absolute error was 2.19%.

$$-r_A = (0.3)3.54\times10^6 e^{-\frac{8650}{RT}} C_A^{0.93} C_{\text{Hydrogen}}^{1.3} + (0.7)5.07\times10^6 e^{-\frac{9940}{RT}} C_A^{2.26} C_{\text{Hydrogen}}^{1.3} \tag{6.30}$$

Values of fractions of asphaltenes are similar in the second and third models. However, according to the results, easy-to-react asphaltenes react with higher reaction order and activation energy. Comparison between experimental and calculated concentrations of asphaltenes with these three kinetic models at different hydrotreating conditions are shown in Figure 6.13. Residuals determined as the difference between experimental and calculated asphaltene concentrations for each model are listed in Figure 6.14. As can be seen in all cases, there is no tendency and the error is randomly distributed, which indicated that kinetic parameters are properly determined.

Physically, easy- and hard-to-react asphaltenes could represent smaller and bigger asphaltene molecules, respectively. Under hydroprocessing conditions, asphaltenes are disintegrated and small fragments could be obtained as severity of reaction conditions increases. If small fragments do not suffer any condensation reaction, then they can go inside the pores and additionally react to produce maltenes. On the contrary, when big fragments of asphaltenes are present, they cannot penetrate easily inside the pores and can be considered as refractory fraction formed by high metals content or sulfur. In this case, diffusion limitations are observed and this could be the reason why activation energies in all cases are relatively small.

6.4.2 Discrete Models

Discrete models describe reactions of chemical components, which are identified in separate ways, mainly by GC. This type of model requires an extensive experimental work and sometimes it is practically impossible to determine chemical structures and reactivity of asphaltenes because they suffer several chemical changes. Boiling point is an important property that permits one to identify with precision different components in heavy hydrocarbons. One successful application of discrete models of asphaltene conversion is based on the Monte Carlo method, which applies differential equations using detailed kinetic and structural data of reactants and products, as reported by Neurock et al. (1994), Trauth et al. (1994), and Campbell and Klein (1997).

Trauth et al. (1994) have reported the construction of resid molecules by a Monte Carlo simulation, assembling between 1,000 and 100,000 resid molecules to have a good statistical representation of the feedstock. Molar concentration distribution allows determining if the molecule is a paraffin, naphthene, aromatic, resin, or

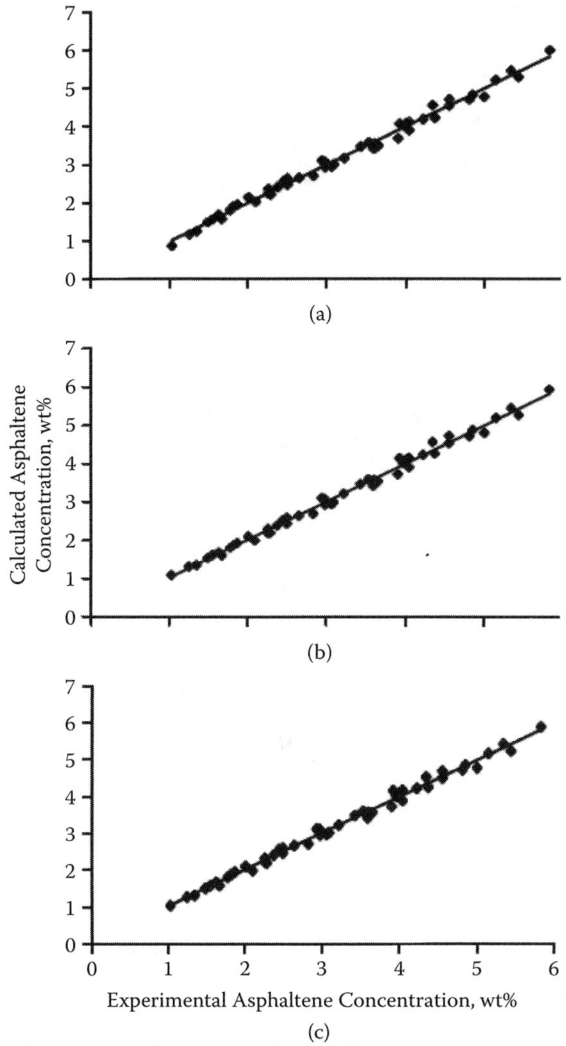

FIGURE 6.13 Comparison between experimental and calculated concentrations of asphaltenes using three kinetic models at different hydrotreating conditions: (a) Model 1, (b) Model 2, and (c) Model 3. (From Trejo et al. 2007. *Petrol. Sci. Tech.* 25 (1–2): 263–275. With permission.)

asphaltene. The optimization sequence starts by taking into account conditional arguments calculated from initial boiling point data and basic structural logic, i.e., an aromatic ring molecule is formed by at least one aromatic ring. If structural characterization is available (H/C atomic ratio, molecular weight, structural parameters by nuclear magnetic resonance [NMR]), it is possible to predict analytical results. Boiling point fractions are determined by normalizing the weight fraction

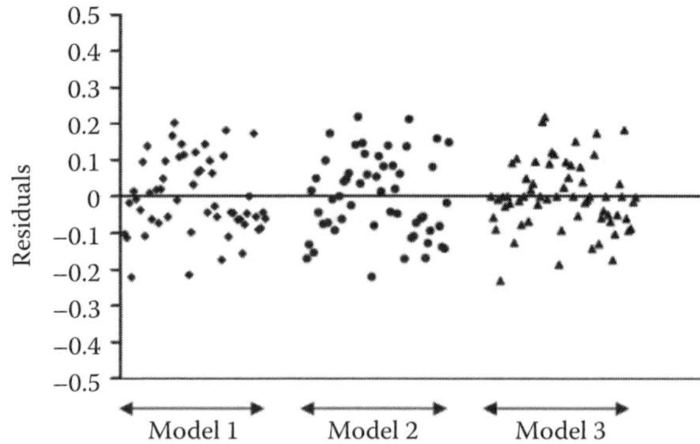

FIGURE 6.14 Residual analysis of three kinetic models for hydrocracking of asphaltenes. (From Trejo et al. 2007. *Petrol. Sci. Tech.* 25 (1–2): 263–275. With permission.)

of molecules present in the feedstock which vaporize within specified temperature ranges. When SARA (saturates, aromatics, resins, asphaltenes) fractionation is used, some considerations need to be done. For example, if the structure is having zero aromatic rings, it is assumed as saturates; one or more aromatic rings with only one polycyclic core is considered as resins or aromatics, and when the polycyclic core is greater than one, it is assumed as asphaltenes. The authors employed a chi-square distribution to approximate structural groups. A comparison between experimental and calculated values obtained by simulation is shown in Table 6.3.

Sheremata et al. (2004) have also used Monte Carlo simulation to represent the asphaltene structure based on analytical tests, such as elemental analysis, ^{1}H and

TABLE 6.3
Comparison of Experimental and Predicted Analytical Results Using Optimized Probability Density Functions

Analytical Test	Experimental Result	Simulation Result
Molecular weight	862	921
H/C atomic ratio	1.47	1.47
Saturates, wt%	13.3	16.1
Aromatics and resins, wt%	63.7	67.5
Asphaltenes, wt%	23.0	16.3
Simulated distillation, wt%		
610–800°F	1	2
800–1000°F	7	12
>1000°F	92	87

Source: Trauth et al. 1994. *Energy Fuels* 8 (3): 576–580. With permission.

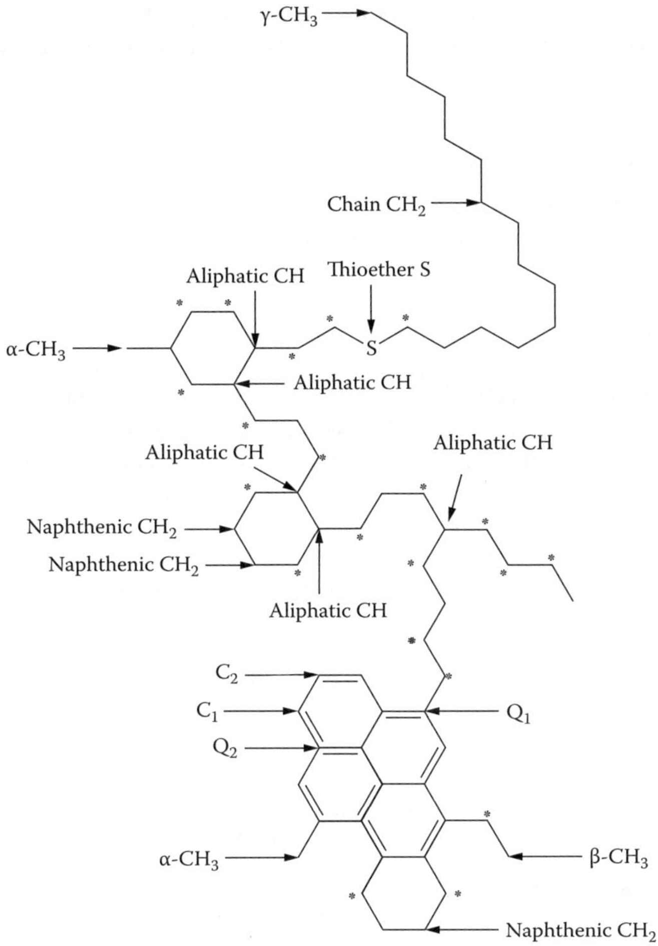

FIGURE 6.15 Various aliphatic and aromatic carbons present in asphaltene representation. Q_1 = alkyl-substituted aromatic quaternary carbon, Q_2 = bridgehead aromatic quaternary carbon, C_1 = aromatic CH beside a Q_2 carbon, C_2 = all aromatic CHs that are not a C_1 carbon. Star structure (*) = other aliphatic carbon. (From Sheremata et al. 2004. *Energy Fuels* 18 (5): 1377–1384. With permission.)

^{13}C NMR, and VPO. Quantitative molecular representations based on analytical data can be used to provide a visual representation of asphaltenes and study their kinetics and reactivity based on these models (Joshi et al., 1999), because these models gather different information analytically obtained. Figure 6.15 shows different representations of types of carbon present in asphaltenes to be simulated. As can be seen, enough structural characterization data are necessary to simulate the complexity of asphaltenes. However, once an asphaltene model is obtained, valuable information is determined and reactivity can be studied because it is relatively easy to deduce which bonds are broken based on the structure modeled by simulation.

6.4.3 Continuous Distribution Kinetics

Continuous distribution kinetics has been applied to kinetic analysis of macromolecular reactions, which include polymer degradation, as reported by several authors (Gloor et al., 1994; Browarzik and Koch, 1996; Kodera and McCoy, 1997; Sezgi et al., 1998).

Kodera et al. (2000) proposed a set of simple chemical reactions to model the overall reactions that occur in asphaltenes, considered as macromolecular species, by using the moments method. The weights and molecular weight distribution of asphaltenes were converted to zero (molar concentration) and first (mass concentration) moments as a function of time. Monitoring of molecular weight distribution is analyzed by SEC with a refractive index (RI) detector. Macromolecular degradation is expressed in one or more reaction schemes to obtain a set of time-differential equations using rates of reaction for each component that is quantitatively analyzed. Because macromolecules are polydisperse and do not have a unique molecular weight, thus it is necessary to employ molecular weight distribution (MWD), which is effectively used to model asphaltenes hydrocracking. However, MWD can be represented by Gamma, Gaussian, and Poisson distribution to approximate the MWD curve of a sample. The mild reaction conditions to have products in liquid phase make that formation of gas and coke to be limited. It is also assumed that molybdenum has enough catalytic activity to hydrogenate and neutralize free radicals when reaction is carried out in a batch reactor; however, hydrogenation of aromatic rings is not present at long extent. With experimental data of asphaltenes, some assumptions need to be done in order to apply the kinetic model, which include three sets of reaction schemes: (1) cleavage of a linkage between fused-ring units (2) alkyl-chains liberation from fused rings, and (3) random-chain scission of aliphatic compounds and their reversible reactions. The possible reaction pathways are:

$$A(x)\underset{K_A}{\overset{k_a}{\leftrightarrow}}\mathrm{A}(x')+\mathrm{A}(x-x') \tag{6.31}$$

$$A(x)\underset{K_B}{\overset{k_b}{\leftrightarrow}}P(x')+A(x-x') \tag{6.32}$$

$$P(x)\underset{K_R}{\overset{k_r}{\leftrightarrow}}P(x')+P(x-x') \tag{6.33}$$

Equation (6.31) shows the cleavage of linkages between fused rings, Equation (6.32) is the alkyl chains cleavage from fused rings, and Equation (6.33) is the bond-scission of aliphatic hydrocarbons. In all cases, recombination of radicals is considered. For all schemes, $A(x)$ represents asphaltenes defined as hexane insoluble components, x is the molecular weight, and $P(x)$ represents aliphatic compounds, such as alkanes and alkenes having a molecular weight x. Balance equations for asphaltenes and aliphatics are given on the basis of conversion and time, i.e., $a(x,t)$ and $p(x,t)$, respectively.

Integro-differential equations are obtained for asphaltenes and aliphatics, which can be simplified by applying the moment operation to obtain the following expressions:

$$\frac{da^{(n)}}{dt} = [(2Z_{n0} - 1)k_a + (Z_{n0} - 1)k_b]a^{(n)} + k_A\left[\sum_{j=0}^{n} \binom{n}{j} a^{(j)}a^{(n-j)} - 2a^{(n)}a^{(0)}\right] + k_B\left[\sum_{j=0}^{n} \binom{n}{j} p^{(j)}a^{(n-j)} - a^{(n)}p^{(0)}\right] \quad (6.34)$$

$$\frac{dp^{(n)}}{dt} = k_B Z_{n0} a^{(n)} - k_B p^{(n)} a^{(0)} + (2Z_{n0} - 1)k_n p^{(n)} + k_R\left[\sum_{j=0}^{n} \binom{n}{j} p^{(j)}p^{(n-j)} - 2p^{(n)}p^{(0)}\right] \quad (6.35)$$

Finally, zero moment ($n = 0$) for asphaltenes and aliphatics are represented as:

$$\frac{da^{(0)}}{dt} = (k_a - k_A a^{(0)})a^{(0)} \quad (6.36)$$

$$\frac{dp^{(0)}}{dt} = k_b a^{(0)} + (k_r - k_B a^{(0)})p^{(0)} \quad (6.37)$$

First moment ($n = 1$) for asphaltenes and aliphatics, respectively, are:

$$\frac{da^{(1)}}{dt} = -\frac{1}{2}k_b a^{(1)} + k_B a^{(0)} p^{(1)} \quad (6.38)$$

$$\frac{dp^{(1)}}{dt} = \frac{1}{2}k_b a^{(1)} - k_B a^{(0)} p^{(1)} \quad (6.39)$$

The obtained moment values are represented in Figure 6.16. At the highest hydrocracking temperature, there is not a well fitting for zero moment, which could be due to the undergoing degradation of paraffins (ignored in the mathematical model). Activation energy for the forward reaction in Equation (6.31) representing the cleavage of linkages between fused rings was 6 ± 3 kcal/mol, whereas activation energy for the forward reaction in Equation (6.32) representing the alkyl chains cleavage from fused rings was 8 ± 1 kcal/mol.

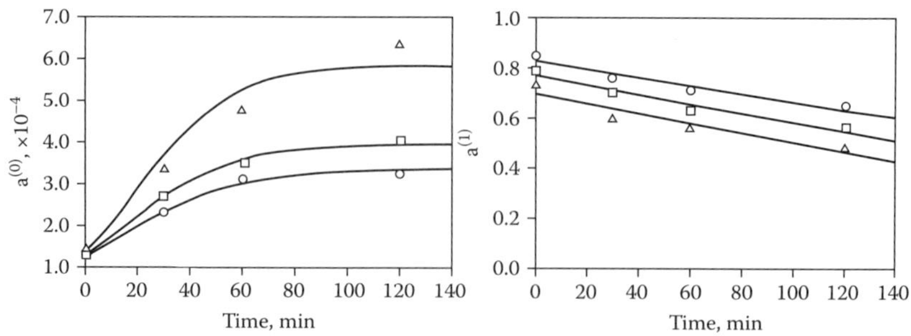

FIGURE 6.16 Graphical representation of zero moment (left) and first moment (right). (O) 380°C, (□) 400°C, (Δ) 420°C. (From Kodera et al. 2000. *Energy Fuels* 14 (2): 219–296. With permission.)

In conclusion, this method is useful to evaluate asphaltene conversion during hydrocracking in terms of changes in molecular weight distribution with the possibility to extend this analysis not only to SEC but also to mass spectrometry that also gives a molecular weight distribution.

6.5 CONCLUDING REMARKS

Increases in yield of light petroleum products by noncatalytic cracking and hydrocracking of petroleum residues represent one of the ways of effectively exploiting this natural resource. The most important reactions taking place during hydrocracking of asphaltenes are

- Breakage of C–C and C–H bonds
- Breakage of bonds with metals and heteroatoms
- Aromatization
- Alkylation
- Condensation
- Hydrogenation–dehydrogenation

Experimental studies indicate that the catalyst and hydrogen pressure mainly influence asphaltene conversion in heavy fraction of petroleum, asphaltenes are more hydrogenated and more maltenes are obtained. However, the catalyst pores could be responsible for diffusion limitations if their size is not adequate for bigger molecules, i.e., asphaltenes and resins. Modifications in reaction mechanism by which asphaltenes transform could be observed at temperatures above 400°C, where reactions change from being hydrogenating-dominated to hydrocracking-dominated.

Selection of processes for proper transformation of asphaltenes present in heavy crude oils and residua includes the use of thermal reactions without hydrogen, hydrothermal reactions under hydrogen pressure, and cracking in presence of catalysts and hydrogen pressure. Different yields of products will be obtained depending on the reaction conditions. In some cases, more coke or gases could be

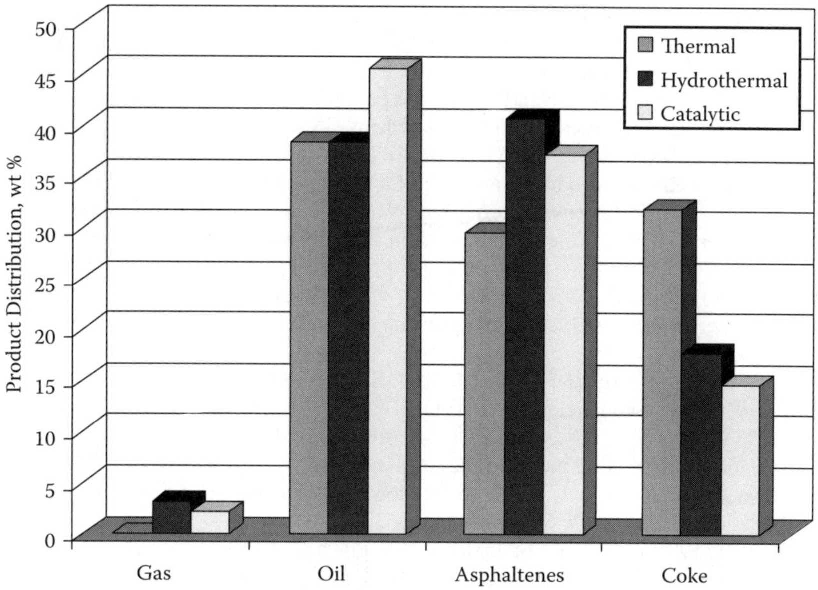

FIGURE 6.17 Comparison of product distribution of different processes involving thermal, hydrothermal, and catalytic cracking of residue. (Adapted from several sources.)

expected to be formed. Catalytic hydrotreating has the advantage of giving higher oil (liquid) yield due to the presence of hydrogenating catalysts. A brief comparison of product yields obtained with these processes is presented in Figure 6.17, in which the differences are clearly established.

Because there are too many reactions involving both cracking and hydrogenation taking place during upgrading processes, it is difficult to give an overall description of them. Not even a unique kinetic model approach can be stated as universal for each reaction condition or feedstock in catalytic or noncatalytic hydrocracking of asphaltenes; however, the purpose of this chapter was to give to the reader enough information about different kinetic models proposed in the literature to represent experimental data. Previous experience and literature reports concerning modeling kinetics, not only for reaction of asphaltenes hydrocracking but also for other reactions, will help us choose the best approach.

REFERENCES

Adarme, R., Sughrue, E.L., Johnson, M.M., Kidd, D.R., Phillips, M.D., and Shaw, J.E. 1990. Demetallization of asphaltenes: Thermal and catalytic effects with small-pore catalysts. *Am. Chem. Soc., Div. Petr. Chem.–Prepr.* 35 (4): 614–618.

Al-Samarraie, M.F. and Steedman, W. 1985. Pyrolysis of petroleum asphaltene in tetralin. *Fuel* 64 (7): 941–943.

Asaoka, S., Nakata, S., Shiroto, Y., and Takeuchi, C. 1983. Asphaltene cracking in catalytic hydrotreating of heavy oils. 2. Study of changes in asphaltenes structure during catalytic hydroprocessing. *Ind. Eng. Chem. Proc. Des. Dev.* 22 (2): 242–248.

Banerjee, D.K., Laidler, K.J., Nandi, B.N., and Patmore, D.J. 1986. Kinetic studies of coke formation in hydrocarbon fractions of heavy crudes. *Fuel* 65 (4): 480–484.

Benito, A.M., Callejas, M.A., and Martínez, M.T. 1997. Kinetics of asphaltene hydroconversion. 2. Catalytic hydrocracking of a coal residue. *Fuel* 76 (10): 907–911.

Blažek, J. and Šebor, G. 1993. Formation of asphaltenes and carbenes during thermal conversion of petroleum maltenes in the presence of hydrogen. *Fuel* 72 (2): 199–204.

Blažek, J. and Šebor, G. 1994. Formation of asphaltenes and carbenes during catalytic conversion of petroleum maltenes in the presence of hydrogen. *Fuel* 73 (5): 695–699.

Botchwey, C., Dalai, A.K., and Adjaye, J. 2004. Kinetics of bitumen-derived gas oil upgrading using a commercial $NiMo/Al_2O_3$ catalyst. *Can. J. Chem. Eng.* 82 (3): 478–487.

Breysse, M., Furimsky, E., Kasztelan, M., Lacroix, M., and Perot, G. 2002. Hydrogen activation by transition metal sílfides. *Catal. Rev.–Sci. Eng.* 44 (4): 651–735.

Browarzik, D. and Koch, A.J. 1996. Application of continuous kinetics to polymer degradation. *J. Macromol. Sci. A* 33 (11): 1633–1641.

Bunger, J.W. 1985. Reactions of hydrogen during hydropyrolysis processing of heavy crudes. *Am. Chem. Soc., Div. Petr. Chem.–Prepr.* 30 (4): 658–663.

Butz, T. and Oelert, H.-H. 1995. Application of petroleum asphaltenes in cracking under hydrogen. *Fuel* 74 (11): 1671–1676.

Callejas, M.A. and Martínez, M.T. 2000a. Hydroprocessing of a Maya residue. I. Intrinsic kinetics of asphaltenes removal reaction. *Energy Fuels* 14 (6): 1304–1308.

Callejas, M.A. and Martínez, M.T. 2000b. Hydroprocessing of a Maya residue. II. Intrinsic kinetics of the asphaltenic heteroatom and metal removal reaction. *Energy Fuels* 14 (6): 1309–1313.

Campbell, D.M. and Klein, M.T. 1997. Construction of a molecular representation of a complex feedstock by Monte Carlo and quadrature methods. *Appl. Catal. A* 160 (1): 41–54.

Di Carlo, S. and Janis, B., 1992. Composition and visbreakability of petroleum residues. *Chem. Eng. Sci.* 47 (9–11): 2695–2700.

Dickie, J.P. and Yen, T.F. 1967. Macrostructures of asphaltic fractions by various instrumental methods. *Anal. Chem.* 39 (14): 1847–1852.

Furimsky, E. 2007. *Catalysts for Upgrading Heavy Petroleum Feeds,* Vol. 169 in *Studies in Surface Science and Catalysis* series, Elsevier, Amsterdam, The Netherlands, pp. 126–132.

Gloor, P.E., Tang, Y., Kostanska, A.E., and Hamielec, A.E. 1994. Chemical modification of polyolefins by free-radical mechanisms—A modeling and experimental study of simultaneous random scission, branching and cross-linking. *Polymer* 35 (5): 1012–1030.

Gollakota, S.V., Guin, J.A., and Curtis, C.W. 1985. Parallel thermal and catalytic kinetics in direct coal liquefaction. *Ind. Eng. Chem. Proc. Des. Dev.* 24 (4): 1148–1154.

Hayashitani, M., Bennion, D.W., Donnelly, J.K., and Moore, R.G. 1977. Thermal cracking of Athabasca bitumen. Proceedings of Canada Venezuela Oil Sands Symposium, Canadian Institute of Mining, Metallurgy and Petroleum, Edmonton, Canada, May 30 to June 3, 1977, CIM Special Volume 17, pp. 233–247.

Hayashitani, M., Bennion, D.W., Donnelly, J.K., and Moore, R.G. 1978. Thermal cracking models for Athabasca oil sand oil. Paper SPE-7549 presented at 53th Annual Technical Conference and Exhibition of Society of Petroleum Engineers, Houston, TX, October 1.

Henderson, J.H. and Weber, L. 1985. Physical upgrading of heavy crude oil by the application of heat. *J. Can. Petrol. Tech.* 24 (4): 206–208.

Ignasiak, T.M. and Strausz, O.P. 1978. Reaction of Athabasca asphaltene with tetralin. *Fuel* 57 (10): 617–621.

Jacob, S., Gross, B., Voltz, S.E., and Weekman, V.W. 1976. Lumping and reaction scheme for catalytic cracking. *AIChE J.* 22 (4): 701–713.

Joshi, P.V., Kumar, A., Mizan, T.I., and Klein, M.T. 1999. Detailed kinetic models in the context of reactor analysis: Linking mechanistic and process chemistry. *Energy Fuels* 13 (6): 1135–1144.

Keulemans, A.I.M. and Voge, H.H. 1959. Reactivities of naphthenes over a platinum reforming catalyst by a gas chromatographic technique. *J. Phys. Chem.* 63 (4): 476–480.

Kobayashi, S., Kushiyama, S., Aizawa, R., Koinuma, Y., Inoue, K., Shimizu, Y., and Egi, K. 1987a. Kinetic study on the hydrotreating of heavy oil. 1. Effect of catalyst pellet size in relation to pore size. *Ind. Eng. Chem. Res.* 26 (11): 2241–2245.

Kobayashi, S., Kushiyama, S., Aizawa, R., Koinuma, Y., Inoue, K., Shimizu, Y., and Egi, K. 1987b. Kinetic study on the hydrotreating of heavy oil. 2. Effect of catalyst pore size. *Ind. Eng. Chem. Res.* 26 (11): 2245–2250.

Kodera, Y. and McCoy, B.J. 1997. Distribution kinetics of radical mechanisms: Reversible polymer decomposition. *AIChE J.* 43 (12): 3205–3214.

Kodera, Y., Kondo, T., Saito, I., Sato, Y., and Ukegawa, K. 2000. Continuos-distribution kinetic analysis for asphaltene hydrocracking. *Energy Fuels* 14 (2): 291–296.

Korsten, H. and Hoffmann, U. 1996. Three-phase reactor model for hydrotreating in pilot trickle-bed reactors. *AIChE J.* 42 (5): 1350–1360.

Köseoglu, R.Ö. and Phillips, C.R. 1988a. Effect of reaction variables on the catalytic hydrocracking of Athabasca bitumen. *Fuel* 67 (9): 1201–1204.

Köseoglu, R.Ö. and Phillips, C.R. 1988b. Kinetic models for the non-catalytic hydrocracking of Athabasca bitumen. *Fuel* 67 (7): 906–915.

Köseoglu, R.Ö. and Phillips, C.R. 1988c. Kinetics and product yield distributions in the $CoO\text{-}MoO_3/Al_2O_3$ catalysed hydrocracking of Athabasca bitumen. *Fuel* 67 (10): 1411–1416.

Kwak, S., Longstaff, D.C., Deo, M.D., and Hanson, F.V. 1992. Hydrotreating process kinetics for bitumen and bitumen-derived liquids. *Proceedings of the Eastern Oil Shale Symposium*, University of Kentucky, Lexington, November 13–15, pp. 208–215.

Leyva, C., Rana, M.S., Trejo, F., and Ancheyta, J. 2007. On the use of acid-base-supported catalysts for hydroprocessing of heavy petroleum. *Ind. Eng. Chem. Res.* 46 (23): 7448–7466.

Marafi, A., Kam, E., and Stanislaus, A. 2008. A kinetic study on non-catalytic reactions in hydroprocessing Boscan crude oils. *Fuel* 87 (10–11): 2131–2140.

Martínez, M.T., Benito, A.M., and Callejas, M.A. 1997. Thermal cracking of coal residues: Kinetics of asphaltene decomposition. *Fuel* 76 (9): 871–877.

Neurock, M., Nigam, A., Trauth, D.M., and Klein, M.T. 1994. Molecular representation of complex hydrocarbon feedstocks through efficient characterization and stochastic algorithms. *Chem. Eng. Sci.* 49 (24): 4153–4177.

Oblad, A.G., Shabtai, J., and Ramakrishnan, R. 1981. Hydropyrolysis process for upgrading heavy oils and solids into light liquid products. U.S. Patent 4,298,457.

Papayannakos, N. 1986. Kinetics of catalytic hydrodesulphurization of a deasphalted oil and of the asphaltenic and non-asphaltenic fractions of a petroleum residue. *Appl. Catal. A* 24 (1–2): 99–107.

Papayannakos, N. and Marangozis, J. 1984. Kinetics of catalytic hydrodesulfurization of a petroleum residue in a batch-recycle trickle bed reactor. *Chem. Eng. Sci.* 39 (6): 1051–1061.

Philippopoulos, C. and Papayannakos, N. 1988. Intraparticle diffusional effects and kinetics of desulfurization reactions and asphaltenes cracking during catalytic hydrotreatment of a residue. *Ind. Eng. Chem. Res.* 27 (3): 415–420.

Phillips, C.R., Haidar, N.I., and Poon, Y.C. 1985. Kinetic models for the thermal cracking of Athabasca bitumen. The effect of the sand matrix. *Fuel* 64 (5): 678–691.

Qader, S.A. and Hill, G.R. 1969a. Catalytic hydrocracking. Hydrocracking of a low temperature coal tar. *Ind. Eng. Chem. Proc. Des. Dev.* 8 (4): 450–455.

Qader, S.A. and Hill, G.R. 1969b. Catalytic hydrocracking. Mechanism of hydrocracking of low temperature coal tar. *Ind. Eng. Chem. Proc. Des. Dev.* 8 (4): 456–461.

Qader, S.A. and Hill, G.R. 1969c. Hydrocracking of petroleum and coal oils. *Ind. Eng. Chem. Proc. Des. Dev.* 8 (4): 462–469.

Qader, S.A., Wiser, W.H., and Hill, G.R. 1968. Kinetics of the hydroremoval of sulfur, oxygen, and nitrogen from a low temperature coal tar. *Ind. Eng. Chem. Proc. Des. Dev.* 7 (3): 390–397.

Que, G., Sun, B., and Liang, W. 1990. Hydrovisbreaking of Gudao vacuum residue in the presence of hydrogen donor. *Am. Chem. Soc., Div. Petr. Chem.–Prepr.* 35 (4): 626–634.

Ramakrishnan, R., Shabtai, J., and Oblad, A.G. 1978. Hydropyrolysis of model compounds. *Am. Chem. Soc., Div. Petr. Chem.–Prepr.* 23 (1): 159–162.

Sánchez, S., Rodríguez, M.A., and Ancheyta, J. 2005. Kinetic model for moderate hydrocracking of heavy oils. *Ind. Eng. Chem. Res.* 44 (25): 9409–9413.

Savage, P.E. and Klein, M.T. 1985. Asphaltene reaction pathways. 1. Thermolysis. *Ind. Eng. Chem. Proc. Des. Dev.* 24 (4): 1169–1174.

Savage, P.E. and Klein, M.T. 1987. Asphaltene reaction pathways. 2. Pyrolysis of n-pentadecylbenzene. *Ind. Eng. Chem. Res.* 26 (3): 488–494.

Savage, P.E. and Klein, M.T. 1988. Asphaltene reaction pathways. 4. Pyrolysis of tridecylcyclohexane and 2-ethyltetralin. *Ind. Eng. Chem. Res.* 27 (8): 1348–1356.

Savage, P.E. and Klein, M.T. 1989. Asphaltene reaction pathways. 5. Chemical and mathematical modeling. *Chem. Eng. Sci.* 44 (2): 393–404.

Savage, P.E., Klein, M.T., and Kukes, S.G. 1988. Asphaltene reaction pathways. 3. Effect of reaction environment. *Energy Fuels* 2 (5): 619–628.

Schucker, R.C. 1983. Thermogravimetric determination of the coking kinetics of Arab Heavy vacuum residuum. *Ind. Eng. Chem. Proc. Des. Dev.* 22 (4): 615–619.

Schucker, R.C. and Keweshan, C.F. 1980. Reactivity of Cold Lake asphaltenes. *Am. Chem. Soc., Div. Fuel Chem.* 25 (3): 155–165.

Sezgi, N.A., Cha, W.S., Smith, J.M., and McCoy, B.J. 1998. Polyethylene pyrolysis: Theory and experiments for molecular-weight-distribution kinetics. *Ind. Eng. Chem. Res.* 37 (7): 2582–2591.

Sheremata, J.M., Gray, M.R., Dettman, H.D., and McCaffrey, W.C. 2004. Quantitative molecular representation and sequential optimization of Athabasca asphaltenes. *Energy Fuels* 18 (5): 1377–1384.

Shiroto, Y., Nakata, S., Fukui, Y., and Takeuchi, C. 1983. Asphaltene cracking in catalytic hydrotreating of heavy oils. 3. Characterization of products from catalytic hydroprocessing of Khafji vacuum residue. *Ind. Eng. Chem. Proc. Des. Dev.* 22 (2): 248–257.

Singh, J., Kumar, M.M., Saxena, A.K., and Kumar, S. 2004. Studies on thermal cracking behavior of residual feedstocks in a batch reactor. *Chem. Eng. Sci.* 59 (21): 4505–4515.

Soodhoo, K. and Phillips, C.R. 1988a Non-catalytic hydrocracking of asphaltenes. 1. Product distributions. *Fuel* 67 (3): 361–374.

Soodhoo, K. and Phillips, C.R. 1988b. Non-catalytic hydrocracking of asphaltenes. 2. Reaction kinetics. *Fuel* 67 (4): 521–529.

Steacie, E.W.R. and Bywater, S. 1955. Mechanism for thermal decomposition of hydrocarbons, in *The Chemistry of Petroleum Hydrocarbons*, Vol. II, Brooks, B.T., Cecil, E.B., Stewart, S.K. Jr., and Schmerling, L., Eds., Reinhold Publishing: New York, chap. 22.

Takeuchi, C., Fukui, Y., Nakamura, M., and Shiroto, Y. 1983. Asphaltene cracking in catalytic hydrotreating of heavy oils. 1. Processing of heavy oils by catalytic hydroprocessing and solvent deasphalting. *Ind. Eng. Chem. Proc. Des. Dev.* 22 (2): 236–242.

Trauth, D.M., Stark, S.M., Petti, T.F., Neurock, M., and Klein, M.T. 1994. Representation of the molecular structure of petroleum resid through characterization and Monte Carlo modeling: Resid upgrading. *Energy Fuels* 8 (3): 576–580.

Trejo, F. and Ancheyta, J. 2005. Kinetics of asphaltenes conversion during hydrotreating of Maya crude. *Catalysis Today* 109 (1–4): 99–103.

Trejo, F., Ancheyta, J., Sánchez, S., and Rodríguez, M.A. 2007. Comparison of different power-law kinetic models for hydrocracking of asphaltenes. *Petrol. Sci. Tech.* 25 (1–2): 263–275.

Trejo, F., Ancheyta, J., Rana, M.S., and Noyola, A. 2009. Kinetics of dipholtene cracking in a batch reactor. Unpublished results.

Usui, K., Kidena, K., Murata, S., Nombra, M., and Trisunaryanti, W. 2004. Catalytic hydrocracking of petroleum-derived asphaltenes by transition metal-loaded zeolite catalysts. *Fuel* 83 (14–15): 1899–1906.

Wang, J. and Anthony, E.J. 2003. A study of thermal-cracking behavior of asphaltenes. *Chem. Eng. Sci.* 58 (1): 157–162.

Weisz, P.B. and Prater, C.D. 1957. *Advances in Catalysis*, Vol. 9. Academic Press: New York, 575.

Yasar, M., Trauth, D.M., and Klein, M.T. 2001. Asphaltene and resid pyrolysis. 2. The effect of reaction environment on pathways and selectivities. *Energy Fuels* 15 (3): 504–509.

Yoshiki, K.S. and Phillips, C.R. 1985. Kinetics of the thermo-oxidative and thermal cracking reactions of Athabasca bitumen. *Fuel* 64 (11): 1591–1598.

Zhang, C., Lee, C.W., Keogh, R.A., Demirel, B., and Davis, B.H. 2001. Thermal and catalytic conversion of asphaltenes. *Fuel* 80 (8): 1131–1146.

7 Fractionation of Heavy Crudes and Asphaltenes

7.1 INTRODUCTION

In today's world, composition of petroleum is an important topic in the study of its physical and chemical changes during processing. In this sense, separation by fractionation is a useful tool for understanding more deeply its composition. Asphaltenes are the important and complex component of petroleum, particularly for heavy, extra-heavy crude oil and its residues. It is generally thought that asphaltenes are formed by similar molecules varying mainly in molecular weight, as reported by Yen et al. (1984) and Pelet et al. (1986). However, there is evidence that chemical properties and average molecular structure may vary considerably with molecular size, such as distribution of metals in asphaltenes (Hall and Herron, 1979). In this regard, different fractionation techniques have been reported in the literature, e.g., adsorption and ion exchange chromatography (McKay et al., 1978), differential precipitation (Yang and Eser, 1999), membrane fractionation (Acevedo et al., 1997), and size exclusion chromatography (SEC) (Semple et al., 1990), all reported in Gutiérrez et al. (2001). Liao and Geng (2002) stated that a significant quantity of hydrocarbons (including alkanes) is occluded in the skeleton of asphaltene molecules. The hydrocarbons are probably traces of the "original oil" that had been retained within the asphaltene matrix and protected from the secondary alteration processes, which occurred subsequently in the oil reservoirs. The authors oxidized asphaltenes without releasing significant amounts of occluded hydrocarbons. These hydrocarbons can be desorbed from asphaltenes by washing with solvents.

Wiehe (1992) has reported that a useful way to measure the distribution of properties in residue and their products is by dividing them into pseudocomponents. Each pseudocomponent occupies a unique area and it can be represented in a solvent-resid phase diagram, which is a plot of molecular weight versus hydrogen content. In this manner, one pseudocomponent is differentiated from others in the function of its molecular weight and its hydrogen content. A schematic representation of the solvent-resid phase diagram is shown in Figure 7.1. The dashed line covers the unconverted resid. Saturates and aromatics occupy almost the same area under the curve, whereas resins and asphaltenes, after conversion, displacc to lower hydrogen content having equal or lower molecular weight. The area occupied by asphaltenes is higher than that of resins, aromatics, or saturates. The area of coke could be even higher than that of asphaltenes, but the limitation in this case was the molecular weight determination for coke, which was only partially soluble in *o*-dichlorobenzene used as solvent for determining the molecular weight by vapor pressure osmometry (VPO). From the diagram, it can

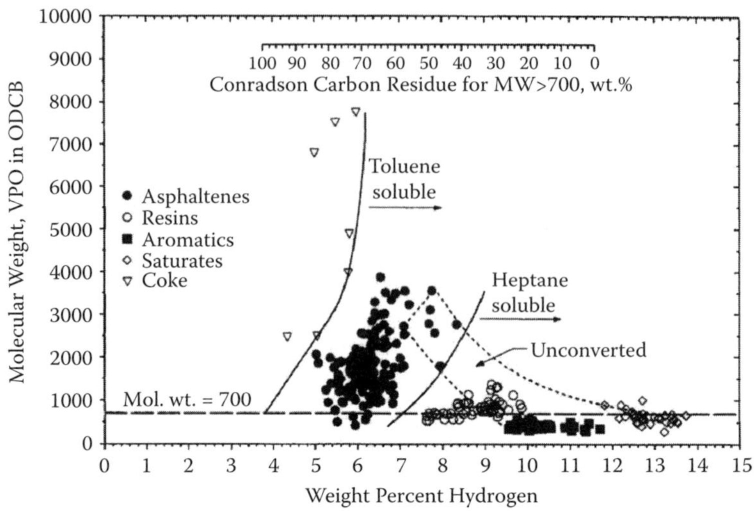

FIGURE 7.1 Solvent-resid phase diagram. (From Wiehe, I.A. 1992. *Ind. Eng. Chem. Res.* 31 (2): 530–536. With permission.)

be observed that resins are formed from aromatics by diminishing the hydrogen content or by increasing the molecular weight by means of condensation. Asphaltenes could be the product of resins, which increase their molecular weight or decrease their hydrogen content.

7.2 DISTILLATION AND SEPARATION OF CRUDE OIL

In this first section, distillation and separation of crude oil is described by different methods, such as saturates, aromatics, resins, and asphaltenes (SARA) fractionation, supercritical fluid extraction and simulated distillation.

7.2.1 SARA Fractionation of Crude Oil

One of the most common and widely known methods for separation of crude oil into four fractions, i.e., saturates, aromatics, resins, and asphaltenes (SARA), was proposed earlier by Jewell et al. (1972), who performed a separation scheme to remove polar nonhydrocarbon compounds as acid and basic fractions with anion and cation exchange resins, respectively. In addition, neutral nitrogen fractions were also separated by coordination complex formation with ferric chloride supported on Attapulgus clay. The remaining nonpolar hydrocarbons were separated into saturates and aromatic fractions by adsorption chromatography with silica gel. On the other hand, Suatoni and Swab (1975) performed a method based on high performance liquid chromatographic (HPLC) to determine analytically the saturates, aromatics, polar compounds, and *n*-hexane insolubles applicable to the whole crude, gasoline, kerosene, middle distillates, fuel oils, residuals, among others. Grizzle and Sablotny (1986) also performed an HPLC method for separations of crude oils, bitumens, and related materials into either group type fractions (saturates, aromatics, and polars) or aromatic ring number fractions (saturates, mono-, di-, tri-, and polyaromatics). From these reports, it can be observed that fractionation of crude to obtain and analyze different components of petroleum or bitumen has been a useful tool in analytical chemistry.

A standard method has been provided by the American Society for Testing and Materials (ASTM) (ASTM D 4124), which describes a procedure to fractionate petroleum asphalts into four fractions. The four fractions (saturates, naphthene-aromatics, polar-aromatics, and asphaltenes) are separated on the basis of their solubility properties. However, the method also can be applied to separate saturates, naphthene-aromatics, and polar-aromatics from vacuum gas oil, lubricating oils, and cycle stocks, among other distillates of petroleum. Asphaltenes must be separated first and maltenes are passed through a glass chromatographic column packed with calcined CG-20 alumina. Addition of solvents or mixtures will give different fractions, i.e., saturates will be separated with *n*-heptane, naphthene-aromatics with toluene and toluene/methanol (50/50), and resins (polar-aromatics) with trichloroethylene.

Kharrat et al. (2007) have compared different SARA methodologies and observed that they can generate different results. They concluded that comparison of results from different procedures based on SARA separation could lead to erroneous statements. SARA fractionation is a useful method when it is recovered around 100% of the material, when topping of the sample is performed initially to avoid loss of volatiles during solvent removal to achieve a good mass balance, and when solvents should be used in quantities that do not allow for low recoveries. Topping is the removal of volatiles by distillation, rotary evaporation, or heating under atmospheric pressure or vacuum at conditions in which hydrocarbons are not lost by further heating. When using a chromatographic separation, the effect of solvent and packing material is very important because recovering of saturates is attributed to the nature of solvent, i.e., the interaction of this fraction with the column is minimal because saturates are not polar species. When recovering of samples is far from 100%, normalization generates erroneous results and this issue must be carefully taken into account for a correct interpretation of results.

7.2.2 Supercritical Fluid Extraction and Fractionation of Crude Oil

Conventional vacuum distillation commonly distillates up to temperatures of ~530°C (538°C = 1,000°F is widely used as final boiling point, above which all material is vacuum residue). When using high-vacuum distillation, the temperature can reach up to 700°C. However, nondistillable residue remains after high-vacuum distillation. Fractionation with supercritical solvents allows for obtaining fractions based on the solubility of residua in different solvents. The use of supercritical fluid extraction and fractionation (SFEF) applied to petroleum residue gives yields up to 75 to 90% of narrow cuts (Chung et al., 1997). The study of narrow cuts allows for knowing their properties in detail, and fractionation of heavy fractions is recommended to avoid properties of heavier fractions that are masked by those of lighter fractions. SFEF also has been developed to reach a clean separation of narrowcuts with total yields up to 75 to 90% (Yang and Wang, 1999). The major advantage of this technique is its use for preparing enough amounts of narrow cuts of residue for characterization. This approach allows the distribution of key species to be determined indicating the variation of residue properties as fractions become heavier.

Zhao et al. (2005) fractionated different types of residue by SFEF and stated that saturates are concentrated in the front fractions of residue decreasing as fractions become heavier. Hydrogen-to-carbon (H/C) atomic ratio for narrow cuts decreased corresponding to more aromatic compounds and resins in fractions, which have higher molecular weight. Sulfur content, Conradson carbon residue (CCR), and metals were concentrated in heavier fractions and unevenly distributed. It is observed in Figure 7.2 how sulfur content is increased as the distillation yield also increases for different feedstocks. Ni and V contents also increase as the percentage of distillation is higher, as shown in Figure 7.3. These results show that most of contaminants are present in end-cut fractions, which are rich in aromatics, such as resins and asphaltenes, having a high coke-forming tendency as indicated by their high CCR values. The use of SFEF applied to Chinese atmospheric residua has permitted obtaining correlations in which density (ρ) at

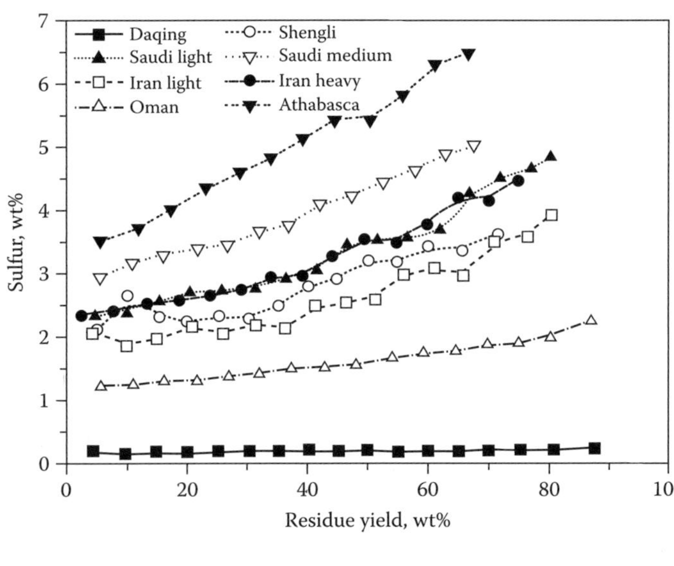

FIGURE 7.2 Distribution of sulfur in different types of residue. (From Zhao et al. 2005. *Fuel* 84 (6): 635–345. With permission.)

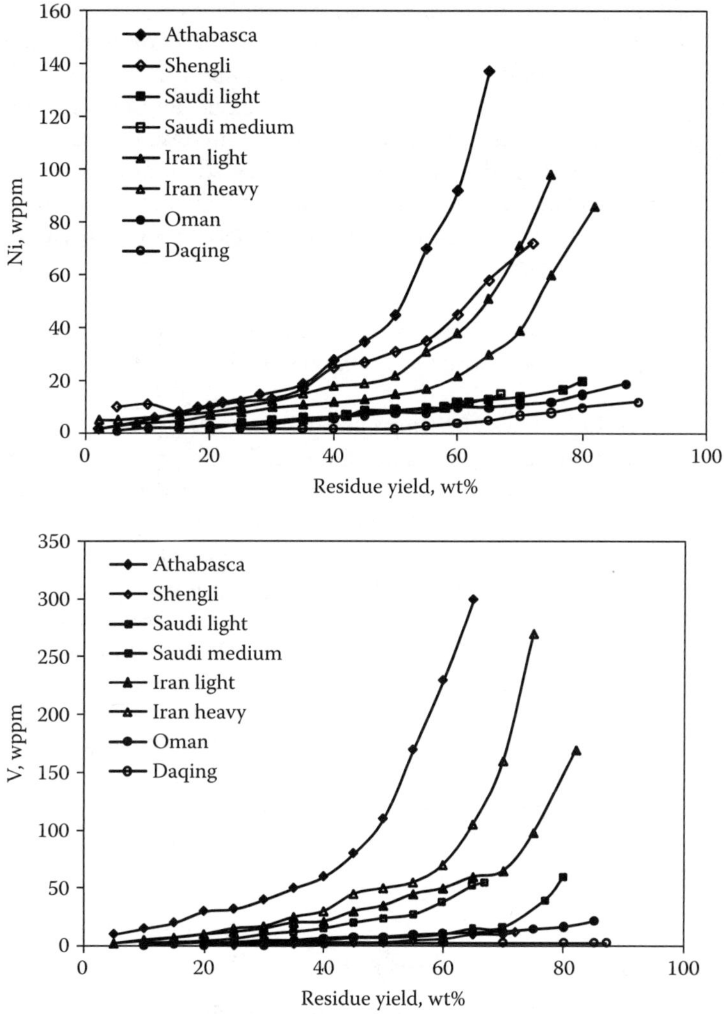

FIGURE 7.3 Distributions of metals in different types of residue. (From Zhao et al. 2005. *Fuel* 84 (6): 635–345. With permission.)

20°C, average molecular weight (M), and refractive index (RI) at 70°C are used to calculated the boiling point (T_b) in Kelvin, as published in Zhao et al. (2005).

$$T_b = 85.66\rho^{0.2081}\ M^{0.3547} \tag{7.1}$$

$$T_b = 70.13\ RI^{0.3067}\ M^{0.3644} \tag{7.2}$$

Temperature can be extrapolated up to 1,000°C with an average error of 2%.

Fractionation along with characterization techniques is a powerful tool for obtaining several properties and parameters of each fraction with some detail. For instance, the use of ^{13}C NMR (nuclear magnetic resonance), ^{1}H NMR, Fourier transform infrared (FTIR), molecular weight, and elemental analysis gave hypothetical structures of three different cuts (first, middle, and end cuts). Initial cuts are characterized in all cases for having small aromatic rings along with naphthenics and alkyl chains linked to the core. On the other hand, middle cuts are characterized for having more aromatic rings and, in some cases, having small alkyl chains. End cuts consist of large aromatic cores and polymerization is frequent in end cuts where the number of basic units is almost two. In addition, highly condensed aromatic cores are present. Yang et al. (2005) stated that the SFEF end cut was the most refractory fraction of residue and had a higher coking propensity.

The SARA technique was used by Chung et al. (1997) for determining saturates, aromatics, resins, and asphaltenes for each fraction recovered from a residue. Results of nine fractions obtained by SFEF revealed that around 60% of the bitumen residue is composed of small molecules having a molecular weight between 500 and 1,500 g/mol. Large molecules constitute the remaining amount of residue with molecular weight from 1,500 to 4,200 g/mol. From this analysis, it can be observed that saturates decrease quickly as the fraction becomes heavier. Aromatics are more or less constant, but its content reduces significantly in the last fractions. The resins concentration increases as resin molecular weight increases as well. The end cut is formed mainly by asphaltenes and a small amount of resins and aromatics. The entire residue was also fractionated by SARA analysis and the results are presented in Figure 7.4. It can be seen that SARA composition for the entire residue compared with those of each fraction does not represent correctly its behavior and composition. The final analysis determined for the entire residue is only "an average result."

7.2.3 Chromatographic Analysis for Crude Oil Separation

Chromatographic analysis was first applied to separation of crude oil in different fractions. Simulated distillation (SIMDIS) is a chromatographic application to allow for determining true boiling point (TBP) curves with a reduced amount of time spent in each analysis. Other advantages of this technique are the small amount of sample needed (0.2 to 0.5 μL) and minimal contamination of samples. SIMDIS is a useful tool characterized by a greater level of confidence because analysis is obtained in only one step without additional correction to account for pressure differences. Espinosa-Peña et al. (2004) have applied SIMDIS based on the ASTM D-5307 method for two Mexican heavy crudes and obtained good agreement between mass and volume percentage curves compared with physical distillation. Other studies involving semipreparative liquid chromatography for separating aromatic rings present in crude oils have been performed by Radke et al. (1984).

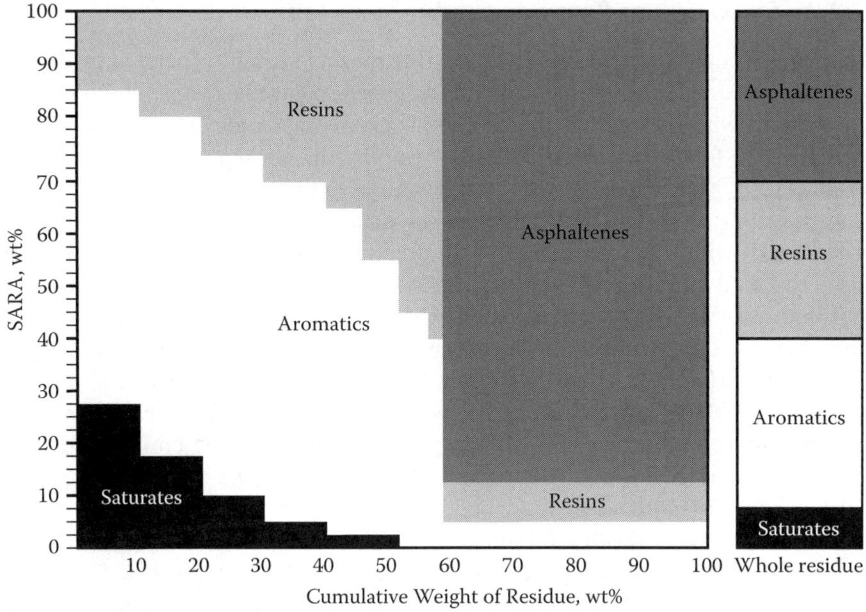

FIGURE 7.4 SARA (saturates, aromatics, resins, and asphaltenes) analysis of Athabasca bitumen residue. (From Chung et al. 1997. *Oil Gas J.* 95 (3): 66–69. With permission.)

Vale et al. (2008) used two crude oil samples to separate asphaltenes from maltenes with *n*-heptane, which were further separated on a silica column and eluted with solvents of increasing polarity to obtain four fractions: F1 (saturated and light aromatics), F2 (polyaromatics), F3 (resins), and F4 (polar compounds). All fractions were characterized by elemental analysis and only fractions F1 and F2 were analyzed by gas chromatography (GC). Aromatics increased their content progressively in fractions F2, F3, F4, and asphaltenes. Electrothermal atomic absorption spectrometry (ET-AAS) was also used for determining Ni and V contents, and distinguishing between volatile and nonvolatile nickel and vanadium compounds present in the crudes. Volatile species of Ni and V were associated to fractions F3 and F4, whereas only thermally stable Ni and V precipitated in part along with asphaltenes. Other analyses involving chromatographic techniques have been done with SEC. Semple et al. (1990) employed this chromatographic technique to fractionate Cold Lake oil and study the molecular size dependence of chemical structural properties. The oil was divided into five fractions and it was observed that metallo-porphyrins were mainly concentrated in the last fraction, which corresponded to the lighter fraction having a molecular weight of around 800 g/mol. The size of aromatic moieties decreased with the molecular weight of each fraction in general, but it increased significantly in the last fraction. The degree of substitution on aromatic carbons decreased with decreasing molecular weight, making aromaticity higher in low molecular weight fractions.

7.2.4 Separation of Porphyrins from Crude Oil

Porphyrins concentrate mostly in heavy fractions and can be separated by different techniques. Vanadium in crude oils is present mainly as vanadyl ion (VO^{+2}) in chelates with porphyrins and nonporphyrins compounds. The use of HPLC has allowed for separating mixtures of porphyrins, as reported by Ysambertt et al. (1995), Xu and Lesage (1992), and Márquez et al. (1989). Molecular weight and characterization at a molecular level of vanadyl-porphyrin and nonporphyrin compounds with HPLC, combined with graphite furnace atomic absorption, has been reported by Fish and Komlenic (1984). Different schemes for separation of porphyrins have been proposed; among them, the SARA fraction is used mainly in heavy fractions without further isolation. Various solvents can be employed to remove different fractions according to their polarity. Márquez et al. (1999) proposed a method in which Boscan crude was first dissolved in methylene chloride and passed through a chromatographic column (2.54 × 75 cm) packed with alumina. Saturated fraction (fraction 1) was extracted with a mixture of hexane/cyclohexane (1:1 vol/vol) giving a soft yellow fraction. Aromatic fraction (fraction 2) was extracted with a mixture of hexane-toluene (7:3 vol/vol), which was yellow-colored. Resins (fraction 3) were extracted with a mixture of CCl_4/CH_3Cl (7:3 vol/vol) giving a dark-colored fraction. Finally, asphaltenes (fraction 4) were separated with a mixture of acetonitrile and methanol (1:1 vol/vol) giving a red wine-colored fraction. Each fraction was further analyzed by UV-vis spectroscopy. Organometallic compounds present in petroleum are characterized by ultraviolet (UV)-vis spectroscopy because metalloporphyrins show weak visible absorption bands at wavelengths above 500 nm, as noted by Gouterman (1961). UV absorption at around 410 nm (Soret band) always is present along with those visible signals indicating electronic transitions in metalloporphyrins (Ferrer and Baran, 1991; Reynolds and Biggs, 1985; Galiasso et al., 1985; Rankel, 1981). Asphaltenes exhibit signals at 578 nm (α-band) and 532 nm (β-band), while resins present signals in the UV region between 450 and 350 nm with a Soret band as well, which suggests that most of vanadium and nickel in resins are nonporphyrinic compounds (Spencer et al., 1982; Fish et al., 1984).

Another way to separate porphyrins is by washing with the Soxhlet process. In this method, heavy crude is dissolved in methylene chloride and used to impregnate neutral alumina, which is further dried with a flow of nitrogen. The resultant powder is submitted to Soxhlet extraction with different solvents to separate the crude into fractions. Solvents used were acetonitrile, which gave a red wine-colored fraction, methanol gave an orange fraction, and *n*-heptane a brown dark fraction. Finally benzene was utilized to extract the residue (Márquez et al., 1999). In this case, the fraction extracted with acetonitrile exhibited stronger UV-vis bands than methanol and *n*-heptane, which indicates more content of metalloporphyrins. The last fraction did not exhibit these porphyrin bands. Ni and V contents increased in the last two fractions, but according to the UV-vis spectra, these metals could be considered as nonporphyrinic compounds. The results of both methods of fractionation reported by the authors are shown in Table 7.1.

TABLE 7.1
Separated Fractions from Boscan Crude Oil by SARA and Soxhlet Extraction Method

Fraction	Original Crude, wt%	V, wppm	Ni, wppm
Crude oil		64	1150
		SARA Method	
Saturates	21		
Aromatics	26		
Resins	36	900	80
Asphaltenes	15	1100	100
		Soxhlet Extraction	
Fraction 1 (acetonitrile-extracted)	7.7	120	2
Fraction 2 (methanol-extracted)	28.9	75	2
Fraction 3 (*n*-heptane-extracted)	48.6	950	26
Fraction 4 (benzene-extracted)	10.8	1060	39

Source: Adapted from Márquez et al. 1999. *Analytica Chim. Acta* 395 (3): 343–349.

Different metal compounds present in heavy crude have been reported by researchers, i.e., Bachaquero crude oil was analyzed by Dunning et al. (1960), finding four types of vanadium: volatile and vanadyl porphyrin, volatile and nonvanadyl, nonvolatile and vanadyl, and nonvolatile and nonvanadyl compounds. Analyses by extended x-ray absorption fine structure (EXAFS) and x-ray absorption near edge structure (XANES) of residue have demonstrated that the coordination number of vanadyl ion in porphyrin and nonporphyrin compounds is very similar having four bonds with N in a square-planar geometry (Goulon et al., 1984). Miller et al. (1999) found that after hydrocracking of residue to about 60% conversion, the coordination number of Ni and V in hydrocracked asphaltenes was almost the same, which is due to its chemical stability.

Nali et al. (1997) proposed a separation procedure for nickel and vanadyl porphyrins from petroleum, which is based on adsorption chromatography using cyanopropyl-bonded silica gel as packing material, followed by solid phase extraction for purification to finally obtain nickel and vanadyl porphyrins. Another procedure to separate and isolate porphyrins is by using a modified chromatographic column with different solvents, as reported by Lee et al. (1995). Separation based on polarity indicated that the least polar fraction (*n*-hexane fraction) did not have porphyrins. Ni porphyrins were extracted with methylene chloride (moderate polar fraction) and 40 wt% of nickel content is present as Ni porphyrins. The most polar fraction was extracted with methanol. Vacuum residua had approximately

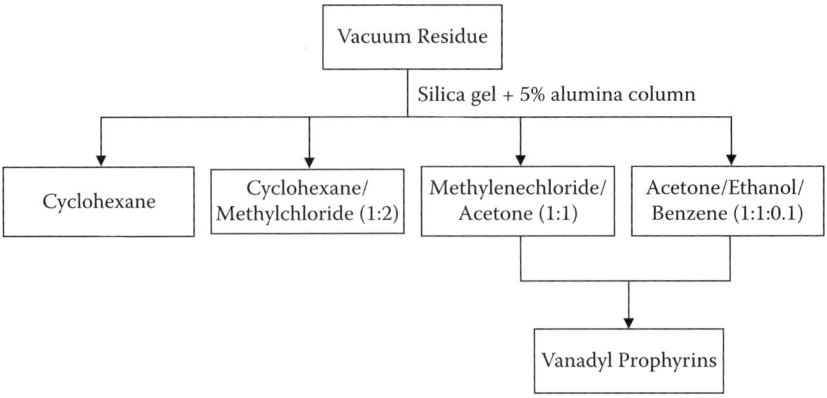

FIGURE 7.5 Separation scheme for the polar components and vanadium distributions of residue. (From Zhu et al. 1998. *Am. Chem. Soc. Div., Petrol. Chem.–Prepr.* 43 (1): 146–140. With permission.)

20 wt% of Ni bound as Ni porphyrins, indicating that vacuum distillation destroys porphyrins.

Some schemes of separation of vanadium compounds have been proposed based on polarity properties of vanadyl complexes. Fractionation of vacuum residue leads to obtaining some fractions rich in metals, as noted by Zhu et al. (1998). Vacuum residue can be dissolved in cyclohexane and passed through a column packed with silica gel plus 5% alumina. Different fractions are obtained by adding cyclohexane, cyclohexane/methylenechloride (1:2 vol/vol), methylenechloride/acetone (1:1 vol/vol), and acetone/ethanol/benzene (1:1:0.1 vol/vol). Vanadyl complexes are obtained in the most polar fractions (Figure 7.5). According to the reported results, it was concluded that 25 to 35% of vanadium is present in resins and the 65 to 75% in asphaltenes within vacuum residue.

Ali et al. (1993) performed two schemes of separation for porphyrins: one of them was based on extraction of porphyrins from crude oil and the second corresponded to the separation of porphyrins from asphaltenes. In the first case, Arabian Heavy crude oil was distillated up to 535°C and the obtained residue (535°C+) underwent an extraction with methanol. Further separation through a silica gel column with different mixtures of solvents was carried out, i.e., benzene/cyclohexane (3:7 vol/vol) gave nickel porphyrins. Benzene, benzene/ethylacetate (1:1 vol/vol), and methanol were also used to elute vanadyl porphyrins. The eluents of these three solvents were mixed together and passed through an alumina column, and vanadyl porphyrins were separated from the column with a mixture of methylene chloride/*n*-hexane (1:1 vol/vol). In addition, the authors separated porphyrins from isolated asphaltenes. In this case, Arabian Heavy crude oil was mixed with *n*-heptane as a solvent to precipitate asphaltenes. Asphaltenes were washed with acetonitrile/benzene in different volumetric ratios and passed through a silica gel column. The solvents used to extract porphyrinic compounds

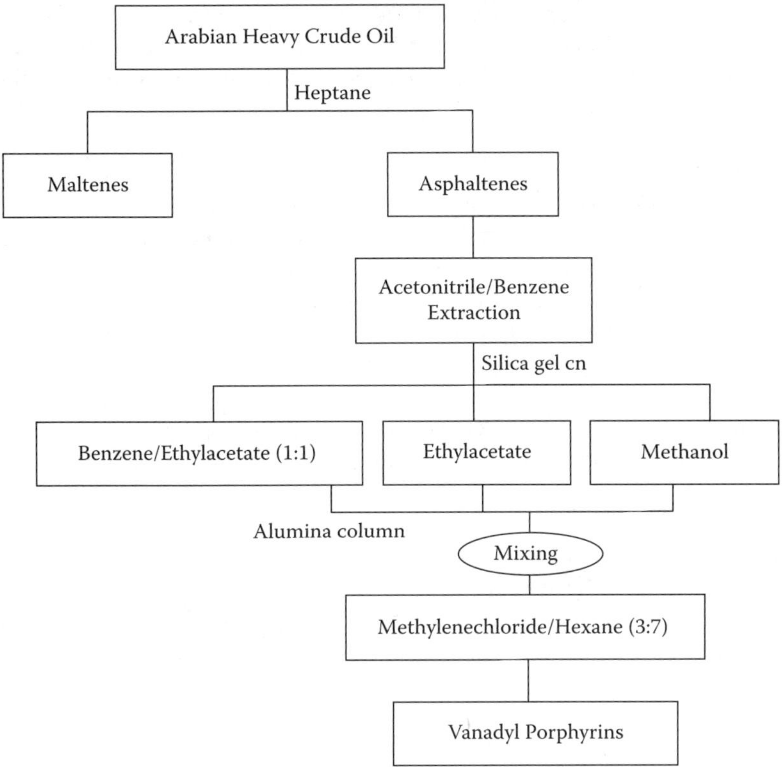

FIGURE 7.6 Separation of porphyrins from Arab Heavy crude asphaltenes. (From Ali et al. 1993. *Energy Fuels* 7 (2): 179–184. With permission.)

were benzene/ethylacetate (1:1 vol/vol), ethylacetate, and methanol, and all eluents were mixed together and passed through an alumina column as well. Finally, vanadyl porphyrins were separated with a mixture of methylene chloride/*n*-hexane (3:7 vol/vol). A scheme of porphyrins separation procedure from asphaltenes is seen in Figure 7.6.

7.2.5 Summary of Standards Methods for Distillation and Fractionation of Petroleum

Several methods for distillation and fractionation of petroleum have been reported in the literature and some of them have been standardized. These methods can be classified as: (1) physical distillation and (2) distillation by chromatographic methods.

7.2.5.1 Physical Distillation

Physical distillation is the most important approach to separate petroleum into different components (lights, middle distillates, and residue). This method has the advantage of determining the true boiling points; however, they require a large amount of sample and are time-consuming. Various standard methods have been implemented to separate different feedstocks by distillation. Scope and limitations of each method is given below.

7.2.5.1.1 ASTM D-2892

Scope—This method is one of the most widely used to determine the true boiling point (TBP) curves. It is applied to distill stabilized crude in a range from room temperature up to 400°C atmospheric equivalent temperature (AET). Fractionation is carried out in a column with a reflux ratio of 5:1 with 15 theoretical plates. The determination of yields can be done by mass or volume and the results are visualized on a plot of temperature versus mass or volume percent distilled.

Limitations—This method cannot be applied to liquefied petroleum gases, very light naphtha, and fractions with initial boiling points above 400°C.

7.2.5.1.2 ASTM D-5236

Scope—This method is based on the ASTM D-2892 and is applied to heavier portions of crudes. The method covers the procedure for distillation of heavy hydrocarbon mixtures with an initial boiling point greater than 150°C, i.e., heavy crude oil, petroleum distillates, and residue or mixtures thereof. The highest AET depends on the nature of the feedstock, but for most of samples, temperature can be around 565°C. In this method, a pot still at low pressure is used. The procedure is used for obtaining distillate fraction of standardized quality in the gas oil, lubricating oil range, and production of standard residue.

Limitations—Although the ASTM D-2892 method can be used for distillation up to 400°C AET and ASTM D-5236 requires an initial boiling point higher than 150°C, distillation curves and fraction qualities obtained by both methods are not comparable. Very heat-sensitive samples could attain lower AET values when applying this method.

7.2.5.1.3 ASTM D-1160

Scope—This method is based on one theoretical plate physical distillation. It covers the determination of the range of boiling points at reduced pressure for petroleum product that can be partially or completely vaporized at a maximum temperature of 400°C. This method can be manually or automatically operated.

Limitations—Vacuum must be properly controlled, otherwise distillation results are not accurate and the 538°C cut point (the typical initial boiling point of vacuum residue) is not achieved. It is not applicable for complete crude oils or light distillates.

7.2.5.1.4 ASTM D-86

Scope—It is applied to atmospheric distillation of petroleum in a batch unit consisting of one theoretical plate. It is applicable to determine quantitatively the boiling range of light and middle distillates, automotive spark-ignition engine fuels, automotive spark-ignition engine fuels having up to 10% of ethanol, fuel oils, diesel fuels, marine fuels, naphtha, kerosene, and grades 1 and 2 burner fuels.

Limitations—This method is not applicable to residual material.

7.2.5.2 Chromatographic Methods

Chromatographic methods have the advantage of being much faster than those based on physical distillation giving similar results. In addition, only a small amount of sample is required and good repeatability can be achieved. They, of course, are simulated values of a real physical distillation. Among these methods are the following:

7.2.5.2.1 ASTM D-3710

Scope—This method corresponds to a simulated distillation for gasoline and gasoline components with either high or low Reid vapor pressure (RVP) and is applicable to petroleum products and fractions having a final boiling point of 260°C or lower.

Limitations—The method has not been validated for gasoline containing oxygenated compounds, such as alcohols or ethers.

7.2.5.2.2 ASTM D-2887

Scope—The method is applicable to petroleum products and fractions having a final boiling point of 538°C or lower at atmospheric pressure, such as jet fuel, kerosene and diesel, light oils, and heavy gas oils from coker and hydrotreating units.

Limitations—This test method is limited to samples having a boiling range greater than 55°C, and vapor pressure sufficiently low to permit sampling at ambient temperature. It is not used for analysis of gasoline samples or gasoline components, which must be analyzed by the ASTM D-3710 method.

7.2.5.2.3 ASTM D-5307

Scope—This method is used in refining industry laboratories as a good alternative to physical distillation. It offers the possibility of calculating the residue content as a mass percentage for samples with nonvolatile material up to 538°C. It is applicable to whole crude samples that can be solubilized in a solvent to permit sampling by means of a microsyringe.

Limitations—It is not applicable to petroleum or petroleum products having low molecular weight components.

7.2.5.2.4 ASTM D-6352

Scope—This method covers the determination of the boiling range distribution of petroleum distillate fractions being applicable to petroleum distillate fractions with an initial boiling point greater than 174°C and a final boiling point of less than 700°C, corresponding to C_{10}-C_{90} samples at atmospheric pressure.

Limitations—It is not applicable for analysis of petroleum or petroleum products containing low molecular weight components, i.e., naphtha, reformates,

TABLE 7.2
Summary of Standard Methods Applied to Simulated Distillation

	ASTM Methods					
	D 3710	D 2887	D 5307	Extended 2887	D 6352	D 7169
Carbon number	C_{20}	C_{44}	C_{44}	C_{60}	C_{90}	C_{120}
Sample type	Gasoline Naphtha	Jet fuel Diesel	Crude oil	Lube oil based stocks	Lube oil based stocks	Residue Crude oil
	ISO, DIN and IP methods					
	ISO 3924 IP 406	DIN 51435-1	DIN 51435-2	IP PM-CF/98	IP 480 Type A	IP 480 Type B/C
Carbon number	C_{44}	C_{60}	C_{120}	C_{120}	C_{120}	C_{120}
Sample type	Jet fuel Diesel	Lube oils	Residues Crude oil Lube oil	Residues Crude oil	Lube oils Heavy distillate	Residues Crude oil

Source: Adapted from several sources.

gasoline, or crude oils. Materials containing heterogeneous components (alcohols, ethers, acids, or esters) or residue cannot be analyzed by this method.

7.2.5.2.5 ASTM D-7169

Scope—This test method extends the applicability of simulated distillation to samples that do not elute completely from the chromatographic system and it is used to determine the boiling point distribution through a temperature of 720°C or even higher (corresponding to C_{100} or higher). This test method is also designed to obtain the boiling point distribution of other incompletely eluting samples, such as atmospheric residues, vacuum residues, etc.

Limitations—It is not applicable for the analysis of materials containing heterogeneous components, such as polyesters and polyolefins.

Table 7.2 presents a summary of the different ASTM methods and their equivalents in ISO, DIN, and IP methods along with the sample type.

7.3 FRACTIONATION OF ASPHALTENES BY DIFFERENT TECHNIQUES

Fractionation allows asphaltenes to be separated into different components according to their solubility properties. In this section, a summary of fractionation of asphaltenes mainly by solvents and chromatographic techniques is discussed. In spite of having several and different models that describe an average structure of asphaltenes reported in the literature, one is advised to be careful when dealing with compounds that have a very wide range of molecular weight, such as asphaltenes. Thus, it is worthwhile to fractionate asphaltenes to have samples more or less homogeneous and characterize them correctly without any interference from higher or smaller molecules (Ali et al., 2006).

7.3.1 Solvent Fractionation of Asphaltenes

Solubility of asphaltenes in different solvents is the basis for fractionation. Solubility is modified as a function of the asphaltene structure. It is important to remember that asphaltenes are a solubility-class compounds defined as heptane insolubles and toluene solubles. These two solvents can precipitate or dissolve asphaltenes depending on which of them has a higher amount when they are present in a mixture of solvents. Gutiérrez et al. (2001) stated that asphaltenes are formed by two types of "components" (colloidal and soluble phase) when *o*-paranitrophenol was used. This solvent has been considered to form charge-transfer complexes with aromatic compounds (Baba et al., 1964). Their results showed that a large percentage of solid asphaltenes is actually insoluble in toluene and the high solubility of the whole sample in this solvent is only due to the surfactant or dispersing effect of the rest of asphaltenes. In this way, the fraction that is insoluble (A_1) in toluene corresponds to the colloidal phase and it is dispersed by the soluble fraction of asphaltenes (A_2).

The colloidal nature of asphaltenes has been also stated by Miller et al. (1998), who separated asphaltene into two fractions. One fraction is colloidal, or not associated, in solution while the other is colloidal, forming polymolecular aggregates. The greatest difference between the two fractions is the degree to which they associate in solution. The extracted asphaltenes are not associated in aromatic solvents and, thus, is noncolloidal. This results in a smaller size and lower molecular weight. In different studies monotonic trends are found when asphaltenes are fractionated using the *n*-heptane–toluene solvent. However, when acetone–toluene was used, the behavior of the resulting subfractions was no longer monotonic in properties, such as molecular size, presumably due to the increased importance of polarity with this solvent system (Groenzin and Mullins, 2000; Buenrostro-González et al., 2001; Buch et al., 2003).

Douda et al. (2004) performed a Soxhlet fractionation and compared the structures of maltene samples with that of the original Maya crude asphaltene. The maltene fractions with different extraction times may reflect the chemical composition of the asphaltenes–resin micelle. They observed that the alkyl chains are longer outside the micelle and shorter inside the micelle, making the solubility of asphaltene molecules more efficient. The asphaltene sample has more significant aromaticity and a higher content of S- or N-heterocyclic compounds and ketones compared with the maltene samples. The aromaticity of the maltene fraction increases significantly for longer Soxhlet extraction time. The polarities of aldehydes, esters, ketones, and carboxylic acids in maltenes decrease inside the micelle, while the polarities of the sulfoxide group and carbonyl group of amides and some sulfur heterocyclic compounds increase.

Tojima et al. (1998) used a toluene–heptane binary solvent system to fractionate *n*-C_7-asphaltene into heavy and light fractions. They found that the lowest soluble fraction in *n*-C_7-asphaltene, defined as heavy asphaltene, consisted of the most highly condensed polynuclear aromatics. This method was applied to precipitate asphaltenes from vacuum residue and considers that the relationship between the amount of precipitated asphaltenes and *n*-heptane concentration in the binary solvent mixture is not linear. Asphaltenes were separated into four fractions with volume ratios of toluene/heptane of 35:65, 25:75, and 18:82.

Trejo et al. (2004) applied the aforementioned method and separated asphaltenes from Maya and Isthmus crudes into three fractions by using Soxhlet extraction. Fractionation consisted of dissolving asphaltenes in toluene completely under reflux conditions, and *n*-heptane is added to the solution. After cooling at room temperature, the fraction including the lowest soluble asphaltene, precipitated out from the solution, is recovered. The remaining solution portion of asphaltene in the toluene/heptane solvent is recovered from the solution by evaporation. For the subsequent separation, the ratio of binary solvent is changed and the same procedure is repeated. Hence, for obtaining three fractions of asphaltenes, i.e. A, B, and C, the volumetric ratios of toluene/heptane of 33.3:66.7 and 66.7:33.3 were employed. Figure 7.7 shows the way in which asphaltenes were fractionated with solvents. Because 15 g of asphaltenes are needed for fractionation and characterization, various precipitations were performed. Then, about 5 g of each of the fractions were recovered. Figure 7.8 shows the amount of precipitated asphaltenes against the volume of *n*-heptane volume. It is seen that the amount of precipitate has indeed a linear relationship with the toluene/*n*-heptane volumetric ratio, which is practically the same for both crude oils where fractions were obtained roughly in the same amount in mass.

Kaminski et al. (2000) proposed a fractionation procedure in which asphaltenes are first dissolved in methylene chloride. Then, enough pentane is added to precipitate the first asphaltene, which is recovered by filtration. This process of adding pentane to the supernatant liquid and collecting the asphaltenes by filtration is repeated to achieve methylene chloride-to-pentane weight ratios of 30:70, 25:75, 20:80, and 10:90. They found that the fraction with the highest polarity (ratio of 30:70) contained more metals and dissolved more slowly and to a lesser extent than the lowest polarity fraction (ratio of 10:90). They also reported that unfractionated asphaltene samples appear to behave as a sum of their fractions.

Andersen et al. (1997) precipitated asphaltenes from Boscan crude by using two procedures: (1) with *n*-heptane as ordinary solvent at room temperature and (2) with mixtures of toluene/*n*-heptane at temperatures of 24, 50, and 80°C. Asphaltene yield is different by the two processes at similar conditions although both increases in temperature or toluene content lead to lower solid yield. Differences in properties were observed in both types of asphaltenes; however, they followed a regular trend in H/C and nitrogen-to-carbon (N/C) atomic ratios observing an increase

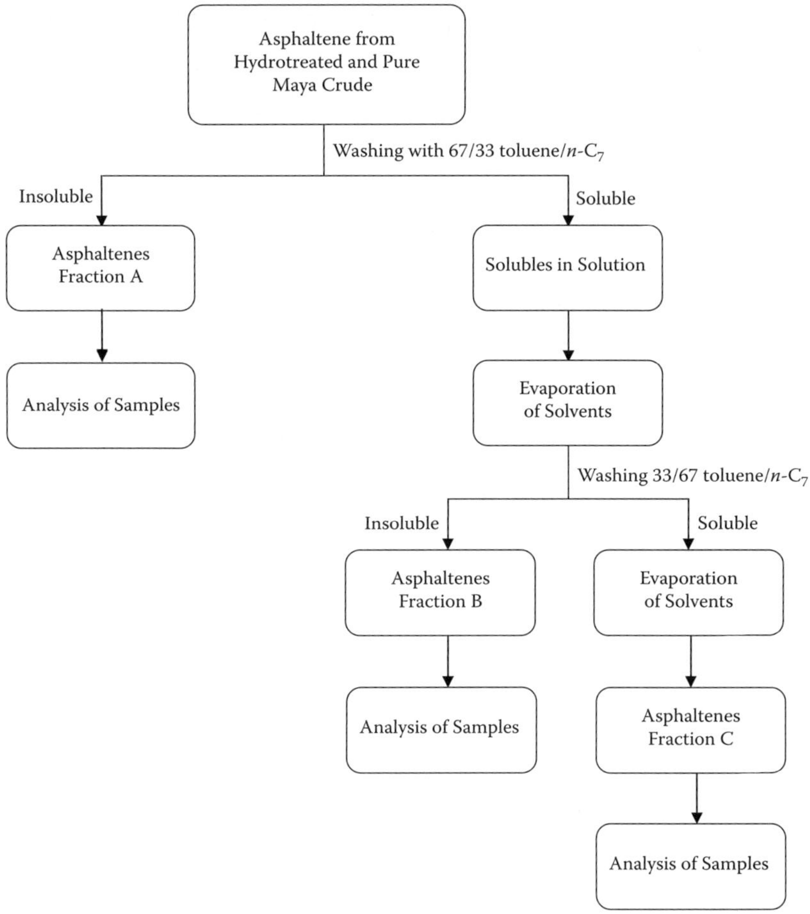

FIGURE 7.7 Scheme of asphaltenes fractionation with using a mixture of solvents (toluene + n-C_7).

in aromaticity and nitrogen content. Porphyrins concentration diminished rapidly with solid yield decrease in both precipitation types.

Robert et al. (2003) fractionated asphaltenes based on their solubility with mixtures of toluene and n-heptane, which gave a different solubility class of asphaltenes containing those that were denominated as "hard-core asphaltenes" coming from an H-Oil® pilot unit, in which four catalysts named as A, B, C, and D were tested. Catalyst D is microporous, whereas the remaining are macroporous in the following order: C > A > B. With a mixture of toluene/heptane of 15/85, almost 80% of insolubles were obtained due to the higher volume of heptane, which tends to keep asphaltenes as insolubles, and no significant difference between insolubles from the feedstock and products was obtained. With

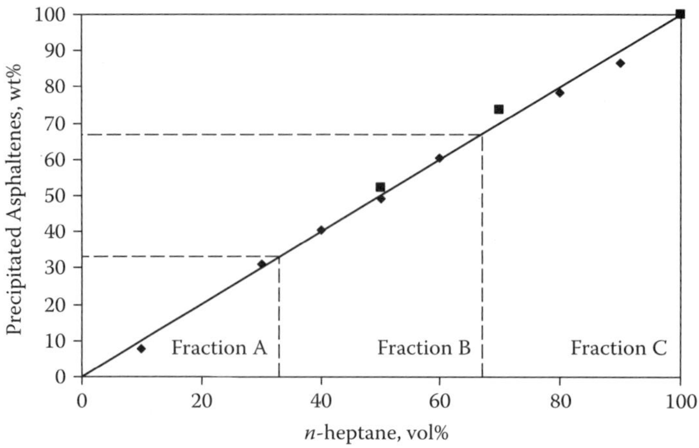

FIGURE 7.8 Amount of precipitated versus *n*-heptane concentration in binary solvent system of toluene and *n*-heptane: (♦) Maya, (■) Isthmus. (From Trejo et al. 2004. *Fuel* 83 (16): 2169–2175. With permission.)

increasing the volume percent of toluene in the mixture (30/70 toluene/heptane), less amount of asphaltenes was kept as insolubles, i.e., 40% of the total was insoluble. Insoluble asphaltenes from the products were between 60 and 70%. For 50/50 toluene/heptane, almost zero content of asphaltenes in the feedstock was obtained; however, the insolubles were between 25 and 40% in products. For 70/30 toluene/heptane, there was not any amount of insoluble asphaltenes from the feedstock, but a little percentage from the products was kept as insoluble asphaltenes (around 10%). Regarding the catalysts, the products from B and D were the most unstable due to their smaller pore size. In the case of catalyst D, microporous in nature, asphaltenes cannot be converted to smaller molecules and remain in high proportion in the products. The amount of insoluble asphaltenes was reduced as toluene increased since they were mainly solubilized by this solvent. This is in agreement with results reported by Trejo et al. (2004).

Spiecker et al. (2003) precipitated asphaltenes from B6 (B6), Hondo (HO), Arab Heavy (AH), and Canadon Seco (CS) crude oils and dissolved them with mixtures of toluene and *n*-heptane, which were called *heptol*. Asphaltenes were solubilized in heptol mixtures having mainly toluene (>52% vol/vol toluene/heptane). However, the addition of *n*-heptane favored the precipitation of asphaltenes. The critical heptol concentration for the initial formation of precipitates ranged from 45 to 52 vol% of toluene depending on the asphaltene type. Figure 7.9 shows a curve in which asphaltenes are solubilized/precipitated at different toluene/*n*-heptane volumetric ratios. It is observed that precipitate is not present when the volume percent of toluene in the heptol mixture is higher than 60%, which indicates that all asphaltenes are solubilized by toluene. Hansen (1969) proposed that the addition of

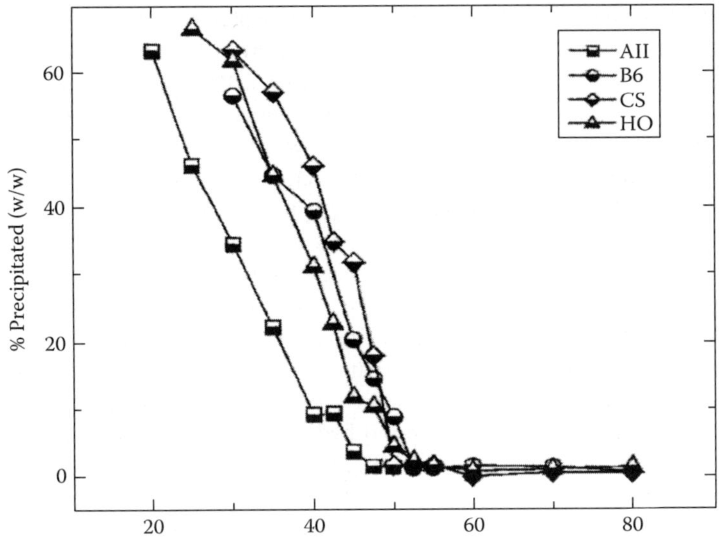

FIGURE 7.9 Solubility profiles of asphaltenes in mixtures of toluene/*n*-heptane from different sources. (From Spiecker et al. 2003. *J. Colloid Interface Sci.* 267 (1): 178–193. With permission.)

n-heptane to toluene reduced the polar components affecting the solubility parameters of toluene, resulting in flocculation and precipitation of asphaltene aggregates. Precipitation of asphaltenes is also controlled by weak dispersion forces (Porte et al., 2003) and π–π bonding interactions between asphaltenes and the mixture of solvents as suggested by refractive index measurements (Buckley, 1996; 1999). Asphaltenes possess high aromaticity and polar heteroatoms able to participate in hydrogen-bonding interactions. Precipitation of asphaltenes when adding *n*-heptane is a consequence of their high percentage of fused aromatic rings and polar heteroatoms, which form large aggregates in solution and precipitate firstly as the solvent quality diminishes below the solubility limit, as stated by Spiecker et al. (2003). These authors concluded that fractionation concentrates the most polar species in the least soluble fraction of asphaltenes. Experimental support for this was obtained by elemental analysis and small angle neutron scattering (SANS). All the less-soluble asphaltenes (more polar) contribute to aggregation in solution and they are probably the main reason for deposition in pipelines and water-in-oil emulsion stabilization. Once asphaltenes were precipitated from different sources, these were fractionated with mixtures of *n*-heptane/toluene with different volume ratio, which allowed for obtaining two types of asphaltenes: a soluble fraction and an insoluble fraction (precipitated).

The amount of precipitated asphaltenes from AH crude when using a mixture of 70/30 vol/vol (n-C_7/toluene) was 35 ± 3% and the rest was soluble. Similar values were obtained for asphaltenes from B6 crude using a mixture of 60/40 vol/vol

(n-C_7/toluene) where the precipitated amount of asphaltenes was 32.4 ± 0.8% and 67.6 ± 0.8% corresponded to soluble asphaltenes. Precipitated asphaltenes from CS and HO crudes were 35.7 and 30.7%, respectively. Precipitated fraction remaining on the filter paper was less soluble in methylene chloride compared with the whole asphaltenes of the same source. Removal of the "soluble" fraction from the whole asphaltenes caused the precipitated fraction to become insoluble at a higher toluene volume fraction than expected, particularly with B6 and HO. In fact, the B6 precipitated fraction was observed to be slightly insoluble in pure toluene. The soluble fraction appears to solvate the precipitated fraction through disruption of strong polar and hydrogen-bonding interactions that drive self-association in the precipitate fraction. It was stated according to elemental analysis that the precipitated fraction (insoluble fraction) was rich in aromatic carbon, whereas the soluble fraction (dissolved asphaltenes) had less content of aromatic carbon. Soluble fraction likely disrupts the strong polar and hydrogen-bonding interactions that promote self-association in precipitated fraction. It was also observed that nitrogen content (present as amines, carbonyl amides, pyrroles, and pyridines) diminished slightly when increasing asphaltene solubility. It is to be remembered that pyridine can participate in strong hydrogen-bonding interactions due to its basicity. On the other hand, sulfur content in soluble and precipitated fractions of asphaltenes was very similar in this case. Sulfur is present in each fraction as thiophene, sulfides, and sulfoxide in small quantities. Metals, such as Fe, Ni, V, and Na, are more concentrated in precipitated fractions. Table 7.3 shows the elemental analysis and metals content in each fraction of different asphaltenes. AH and CS whole asphaltenes contained less nitrogen and their H/C ratio was low compared with B6 and HO whole asphaltenes. The aggregation mechanism for AH and CS asphaltenes was likely dominated by dispersion forces or differences between asphaltene and solvent aromaticity (π–π bonding interactions). In addition, SANS analysis demonstrated that the less soluble (more polar) asphaltenes are responsible for asphaltene aggregation in solution.

Szewczyk et al. (1996) also performed a fractionation procedure of an asphaltene sample with increasing volume of n-heptane, which allowed for separating asphaltenes into two classes: soluble and insoluble fraction. The insoluble fraction contains the aggregates, which flocculate first. Elemental analysis demonstrated that the aggregates that flocculate are more aromatic and have more heteroatoms in their structure and they are formed by monomers with different elemental composition. Current tendencies point to the mechanism suggested by Andersen and Birdie (1991) to describe the behavior of asphaltenes in crude oil according to the following representation:

$$\text{Dissolved monomer} \rightarrow \text{micelle (aggregate)} \rightarrow \text{particulate} \rightarrow \text{precipitate} \quad (7.3)$$

Asphaltene monomers tend to associate to generate intermediate products and the formation of precipitates through continuous association of intermediate products. It has to be kept in mind that at low asphaltene concentrations they behave

TABLE 7.3
Elemental Analysis and Metals Content in Each Fraction of Different Asphaltenes

	H/C			Nitrogen, wt%			Sulfur, wt%			Oxygen, wt%		
Asphaltene	**Soluble**	**Whole**	**Precipitate**	**Soluble**	**Whole**	**Precipitate**	**Soluble**	**Whole**	**Precipitate**	**Soluble**	**Whole**	**Precipitate**
AH	1.17	1.14	1.13	0.92	1.02	1.08	8.06	8.32	7.66	1.92	1.64	2.52
B6	1.30	1.24	1.22	1.81	1.87	1.93	7.25	6.68	6.33	2.67	2.90	2.81
CS	1.12	1.11	1.09	1.32	1.32	1.39	0.52	0.52	0.48	2.11	1.73	2.27
HO	1.30	1.29	1.24	1.95	1.99	2.11	8.42	8.53	8.48	2.51	2.10	2.66
	Iron, ppm			**Nickel, ppm**			**Vanadium, ppm**			**Sodium, ppm**		
AH	14	26	50	84	160	160	350	490	540	31	31	31
B6	8	35	47	350	330	410	1000	1000	1200	27	9300	25000
CS	51	77	150	19	21	28	42	48	48	43	130	180
HO	6.3	16	12	340	360	410	930	950	1100	12	550	1800

AH: Arab Heavy
B6: B6 crude
CS: Canadon Seco
HO: Hondo crude
Source: Spiecker et al. 2003. *J. Colloid Interface Sci.* 267 (1): 178–193. With permission.

as nonassociated molecules (monomers). Small angle x-ray scattering (SAXS) analysis has shown the existence of particles in crude oils diluted with aromatic solvents, as reported by Dwiggins (1965) and Xu et al. (1994).

7.3.2 Fractionation of Asphaltenes by Size Exclusion Chromatography (SEC)

A very common way to fractionate asphaltenes is by SEC. This technique can be used to separate them by molecular weight. Packing material is an important issue since no commercial standard can represent correctly the molecular weight distribution of asphaltenes. The sample is susceptible to suffering aggregation instead of being monomeric species. However, the information obtained by this chromatographic technique is very useful and gives an idea about properties of asphaltenes. Molecular weight distribution obtained by SEC can be strongly influenced by the solvent employed. Different solvents have been used, such as tetrahydrofuran (THF), THF/methanol, THF/triethylamine, benzonitrile, N-methyl-2-pyrrolidinone (NMP), among others. THF has been reported to be an unsuitable eluent for asphaltenes giving a smearing-out of the sample rather than a close approach to size exclusion; NMP has been considered a better choice (Domin et al., 1999; Bartholdy et al., 2001; Li et al., 2004; Al-Muhareb et al., 2007). NMP has the advantage that no compounds have been observed to elute after the permeation limit in coal tars, biomass tars, or petroleum vacuum residues, and the material corresponding to that excluded from the column porosity has been shown to be large, nonpolar molecules and they do not need to be considered as clusters of small polar molecules.

Ali et al. (2006) fractionated asphaltenes separated from Kuwaiti atmospheric and vacuum residue by SEC with THF as the mobile phase. Fractionation was carried out by molecular weight and 11 fractions were obtained. Molecular weight of asphaltenes in most of cases was lower than 2,000 g/mol, which can indicate that asphaltenes are aggregated instead of being monomeric species. Asphaltenes of higher molecular weight from atmospheric and vacuum residue showed similar structural parameters, which suggests that low molecular weight asphaltenes convert to distillates when upgrading is carried out. However, high molecular weight asphaltenes remain almost unaltered as demonstrated by fractionation, which in turn converts these heavier species into coke or sediments during hydroprocessing.

Trejo and Ancheyta (2007) also fractionated asphaltenes into three fractions by SEC using THF as the eluent. Asphaltenes were precipitated from Maya crude previously hydrotreated at different reaction conditions and proposed a possible reaction mechanism based on SEC results. They have elucidated that it is possible to represent the changes of molecular weight (MW) distribution of asphaltenes during hydroprocessing of heavy oils. A comparison of chromatograms was made between nonhydrotreated and hydrotreated asphaltenes. As expected, hydrotreated asphaltenes have a narrower molecular weight distribution due to changes that asphaltenes suffer during hydrotreating. The area under each chromatogram was divided into three sections and molecular weight distribution of asphaltenes from nonhydrotreated Maya crude was taken as reference. Each chromatogram of asphaltenes from hydrotreated crude was overlapped onto the chromatogram of asphaltenes from nonhydrotreated Maya crude and sectioned in the same three intervals. These intervals were selected in order to cover all the area under the chromatogram of asphaltenes from Maya crude. The chosen intervals of molecular weight were: (1) from 0 to 2,350 g/mol, (2) from 2,350 to 5,200 g/mol, and (3) from 5,200 g/mol to the upper limit. In the case of hydrotreated asphaltenes, molecular weight distribution of asphaltenes is not necessarily equal sized as in the case of the chromatogram of asphaltenes from Maya crude taken as reference. The proposed route by which asphaltenes are disintegrated emerges as the breaking of big molecules to give smaller molecules. Hence chromatograms are divided into three parts; one corresponds to lower molecular weight, another to middle molecular weight, and the last one corresponds to high molecular weight, as shown schematically in Figure 7.10.

The percentage of each fraction can be calculated by dividing the area under to curve (A_i) of each fraction (i.e., 1, 2, or 3) by the total area (A_t) according to the following relationship:

$$y_i = \frac{A_i}{A_t} \tag{7.4}$$

where y_1 corresponds to the heavier fraction, y_2 is an intermediate fraction, and y_3 is the lightest fraction. The values of y_i corresponding to asphaltenes from Maya crude were $y_1 = 0.21$, $y_2 = 0.36$, and $y_3 = 0.43$. A proposed route in series-parallel

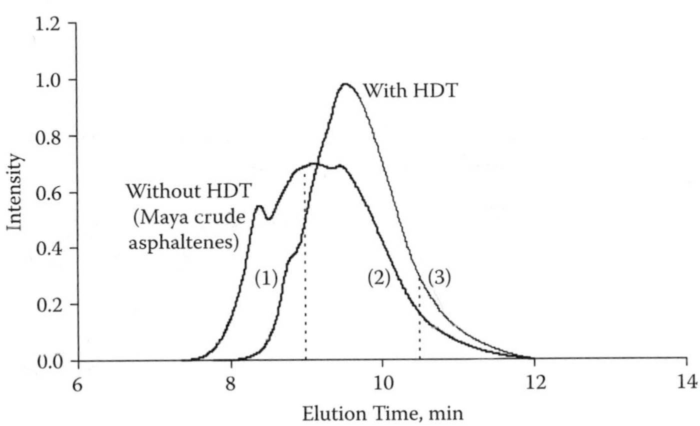

FIGURE 7.10 Example of the way in which distinct molecular weight distributions of asphaltenes were taken from hydrotreated crudes and Maya crude. (From Trejo, F. and Ancheyta, J. 2007. *Ind. Eng. Chem. Res.* 46 (23) 7571–7579. With permission.)

by which larger molecules of asphaltenes decompose to give smaller molecules of asphaltenes is shown below:

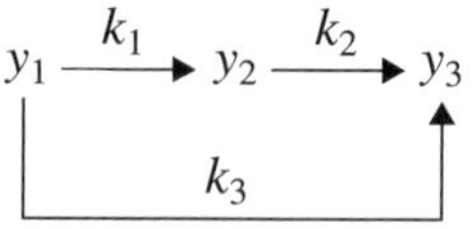

(7.5)

A set of differential equations is generated from the above kinetic pathway and solved simultaneously to find k_1, k_2, and k_3 as shown:

$$\frac{dy_1}{d(1/\text{LHSV})} = -k_1y_1 - k_3y_1$$

$$\frac{dy_2}{d(1/\text{LHSV})} = k_1y_1 - k_2y_2 \qquad (7.6)$$

$$\frac{dy_3}{d(1/\text{LHSV})} = k_3y_1 + k_2y_2$$

The objective function was based on the minimization of the sum of square errors between experimental and calculated fractions as shown in Table 7.4. It is observed that the value of k_3 is zero, indicating that only an in-series mechanism is consistent with experimental data. This means that asphaltenes disintegrate from the highest molecular weight species passing through intermediate molecular weight to the lowest molecular weight asphaltenes. The area under the curve of fraction y_1 was the lowest and that of y_3 the highest in all cases. It can be then established that a rupture in bigger molecules generates more notorious changes in these molecules than the changes of smaller molecules. Fragmentation of big molecules tends to increase the fraction of the smallest molecules (fraction y_3). The lower E_A of hydrocracking of

TABLE 7.4
Kinetic Parameters for Molecular Weight Distribution of Asphaltenes Fractions during Hydrotreating of Maya Crude

	70 kg/cm²			85 kg/cm²			100 kg/cm²		
Temperature, °C	k_1	k_2	k_3	k_1	k_2	k_3	k_1	k_2	k_3
380	0.7117	0.4247	0	0.7888	0.4512	0	0.8776	0.4692	0
400	0.8687	0.4677	0	0.9110	0.5372	0	0.9128	0.6049	0
420	0.9150	0.6320	0	1.0654	0.6475	0	1.1858	0.6650	0
Activation Energies									
E_{A1}, cal/mol		5682			6755			6717	
E_{A2}, cal/mol		8891			8118			7876	

Source: Trejo, F. and Ancheyta, J. 2007. *Ind. Eng. Chem. Res.* 46 (23): 7571–7579. With permission.

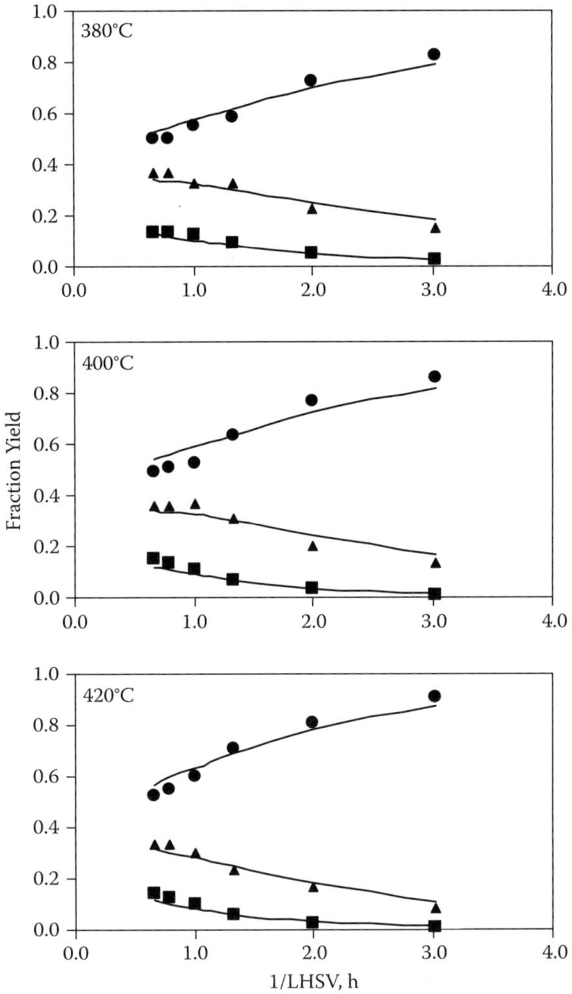

FIGURE 7.11 Molecular weight of asphaltenes from the hydrotreated crude at 70 kg/cm^2, at different temperatures and variable space-velocity: (■) Fraction y1, (▲) Fraction y2, (●) Fraction y3, (symbols) experimental values, (—) model. (From Trejo, F. and Ancheyta, J. 2007. *Ind. Eng. Chem. Res.* 46 (23) 7571–7579. With permission.)

fraction y_1 compared with that of y_2, indicates higher diffusion limitations of the former due to its more complex nature.

Figure 7.11 demonstrates that a good correspondence between experimental and calculated y_i values was obtained. In this way, the molecular weight of each fraction can be calculated despite the problems observed in SEC, such as possible extra-retention of the smaller molecules into the porous material tailing the chromatogram toward larger retention times. Information obtained by SEC could be useful to estimate the molecular weight of asphaltenes and know the changes that they undergo in their molecular weight distribution during hydroprocessing.

All y_i values can be recalculated by solving the set of equations in Equation (7.6) with parameter values reported in Table 7.4. These y_i values along with the total area under the curve (*At*) were used to recalculate the area under the curve of each fraction (*Ai*), according to Equation (7.4), i.e., multiplying y_i by A_t. A good fit of values is shown in Figure 7.12.

Along with the knowledge obtained with this phenomenon, the information generated could be useful in designing catalysts that take into account changes in

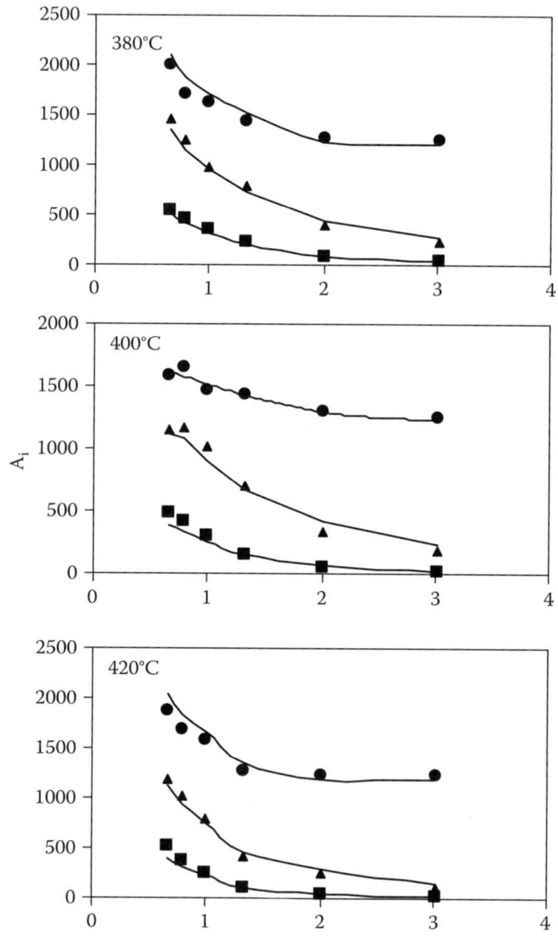

FIGURE 7.12 Molecular weight of asphaltenes from the hydrotreated crude at 70 kg/cm², at different temperatures and variable space-velocity: (■) Fraction y_1, (▲) Fraction y_2, (●) Fraction y_3, (symbols) experimental values, (—) model. (From Trejo, F. and Ancheyta, J. 2007. *Ind. Eng. Chem. Res.* 46 (23) 7571–7579. With permission.)

molecular weight, which is directly related to their size to avoid diffusion restrictions during hydroprocessing of heavy feedstock that is rich in asphaltenes.

7.4 STRUCTURE AND COMPOSITION OF CRUDE AND ASPHALTENES AFTER FRACTIONATION

Wood et al. (2008) fractionated diverse samples that included bitumen, heavy crude, and conventional crude, which were analyzed to determine the nitrogen and sulfur contents in each fraction (Table 7.5). It can be observed that nitrogen and sulfur content of saturates fraction is relatively low in each case. In all cases, nitrogen content increases rapidly in resin and asphaltene fractions. Resins fraction contains more than 40% of total nitrogen and sulfur of the whole crude, which is even higher than in the case of asphaltene fractions. Estimation of the difference with oxygen indicated increasing oxygen content for the heavier fractions, and the estimated oxygen content for each asphaltene fraction was always less than the value for the corresponding resin component. Further fractionation of resins was carried out to analyze the distribution of sulfur and nitrogen contents in each subfraction of resins, which were named as R1, R2, and R3. Each subfraction for different sources showed that only minor differences are present. The R1 subfractions accumulated 20 to 27% of the total nitrogen present in resins, whereas the remaining nitrogen was distributed nearly equally between subfractions R2 (34 to 39%) and R3 (34 to 42%). According to the authors, increasing nitrogen species in the last fractions are eluted during HPLC because of their polarity, which is higher compared with other species of the crude. Sulfur content is higher in fraction R1 and differs from that of nitrogen by which it is thought that sulfur species are less strongly bound to the column than the nitrogen compounds present in fractions R2 and R3. The main constituents of each fraction are pyrrole and thiophenes, being the predominant nitrogen and sulfur species, respectively. In addition, pyrroles are more polar than sulfur compounds and may affect the quality of the final product through the formation of sediments (Mushrush and Speight, 1998).

Zhao et al. (2005) showed that the concentration of thiophenes increases as residue fractions become heavier. About 65 to 80% of the total sulfur in bitumen occurs as thiophenes and the remaining sulfur as sulfide. It was also concluded that hydrocracking removed sulfides more efficiently than thiophenes. However, it was showed that concentrations of sulfides and thiophenes in the end-cut fractions from hydrocracking and coking residues are similar, as a probable consequence of that reaction begin thermally dominated rather than catalytically. A computational approach allowed the authors to draw different structures based on SFEF fractions. Molecules with a single unit represent the light front cut fractions of residue. However, middle cuts that represent around 50 to 60% of the whole residue are formed by at least four aromatic rings and, in some cases, the monomer structure is repeated two times as in Athabasca bitumen. End cuts are formed by several condensed aromatic rings with repetitive monomer structures, as seen in Figure 7.13. End cuts represent more complex structures with highly condensed aromatic rings and more contaminants in their structures.

TABLE 7.5
Nitrogen and Sulfur Analysis by Means of SARA Fractionation from Different Crude Oils

		N (wt% of Each Fraction)				N (wt% of Total N in Crude)			
Crude Type	Identification	Saturates	Aromatics	Resins	Asphaltenes	Saturates	Aromatics	Resins	Asphaltenes
Bitumen	Bit 1	0.15	0.17	0.80	1.21	4.3	7.6	50.2	37.8
	Bit 2	0.29	0.33	0.75	1.21	6.9	12.4	41.5	38.9
Heavy crude	A	0.24	0.33	0.84	1.30	11.8	18.5	45.8	23.9
	B	0.18	0.12	0.57	1.16	10.8	8.9	45.3	35.4
	C	0.23	0.24	0.67	1.19	10.5	14.6	44.2	30.7
	D	0.12	0.16	0.65	1.16	6.5	11.5	47.0	34.9
	E	0.11	0.24	0.70	1.11	5.5	15.0	43.4	35.9
	F	0.19	0.22	0.73	1.16	11.8	11.8	44.9	31.5
	G	0.22	0.26	0.80	1.19	12.8	15.2	44.8	27.2
Conventional	Conventional	0.20	0.23	0.73	1.14	30.9	16.5	41.3	11.4
		S (wt% of Each Fraction)				S (wt% of Total S in Crude)			
Bitumen	Bit 1	0.30	3.79	6.19	7.90	1.1	20.9	47.7	30.3
	Bit 2	0.23	4.40	6.97	8.65	0.7	20.1	46.0	33.2
Heavy crude	A	0.14	3.24	5.10	5.99	1.2	31.5	48.2	19.1
	B	0.43	4.05	7.30	8.60	2.1	25.7	49.8	22.2
	C	0.13	5.59	7.36	8.56	0.6	32.3	46.2	20.9
	D	0.47	5.49	7.75	8.74	2.0	31.7	45.1	21.2
	E	0.14	3.86	5.34	6.10	0.9	31.1	42.6	25.4
	F	0.12	2.69	4.64	5.93	1.2	24.2	47.7	26.9
	G	0.16	3.45	5.15	6.04	1.5	31.6	45.3	21.7
Conventional	Conventional	0.11	2.26	3.82	4.40	3.9	36.9	49.2	9.9

Source: Adapted from Wood et al. 2008. *Oil Gas Sci. Tech.–Rev. IFP* 63 (1): 151–163.

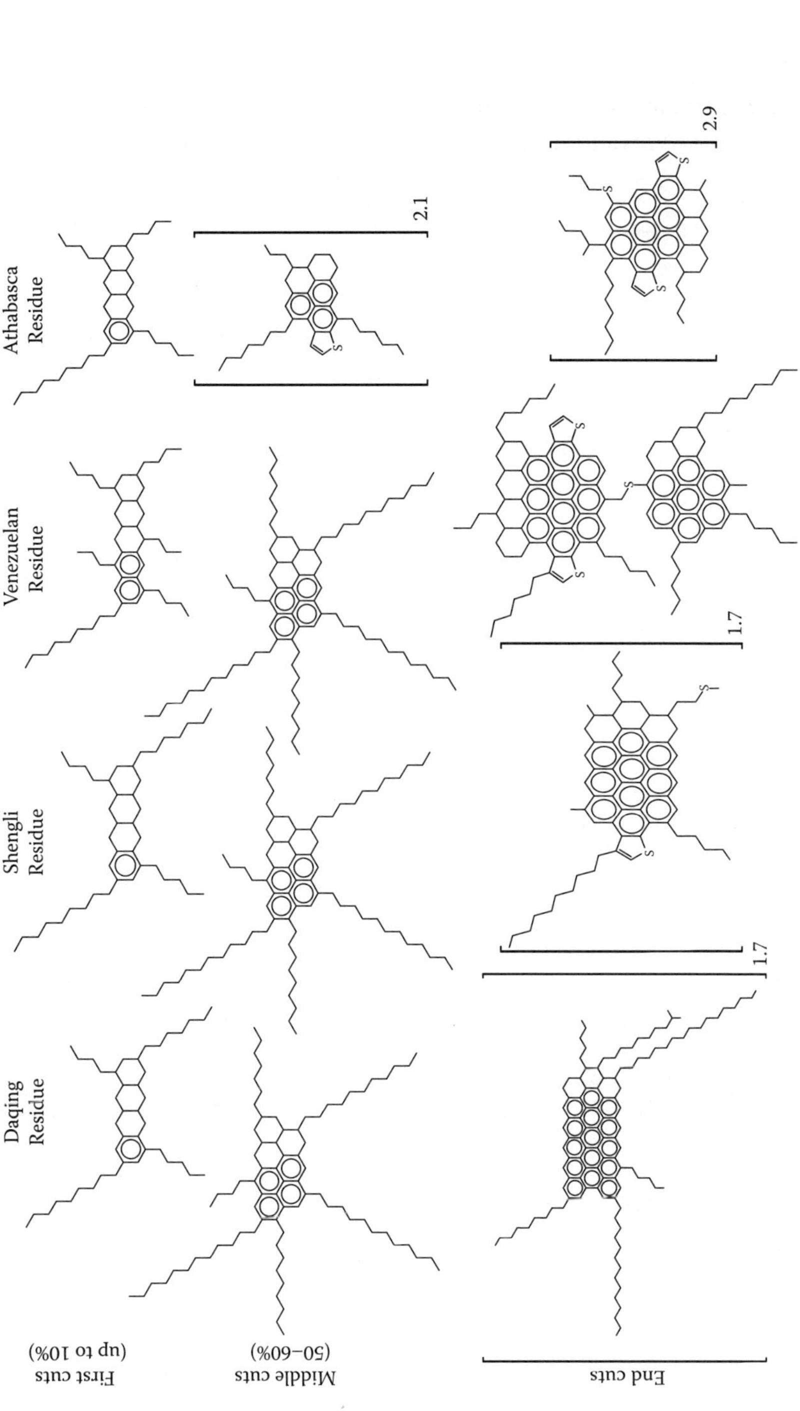

FIGURE 7.13 Average structure of fractions from different residues: (A) Daqing, (B) Shengli, (C) Venezuelan, and (D) Athabasca. (From Zhao et al. 2005. *Fuel* 84 (6): 635–345. With permission.)

TABLE 7.6
Structural Parameters of Asphaltenes by SEC Obtained from Kuwaiti Atmospheric and Vacuum Residue

	Whole asph. AR		**SEC Fraction Number of AR**		
Parameter	**Value**	**2**	**4**	**6**	**8**
MW	3230	5590	5260	4750	2120
Atoms of C	224	420	367	336	149
C_{al}	47%	44%	43%	44%	48%
$C_{a,\ n\text{-alkyl}}$	26%	27%	25%	21%	30%
n	10	10	9	12	9
$C_{al,\ CH3}$	15%	6%	10%	8%	10%
C_{ar}	53%	56%	57%	56%	52%
$C_{ar,\ t}$	11%	8%	6%	7%	8%
$C_{ar,\ q}$	42%	48%	51%	49%	44%
Elemental analysis					
C, wt%	83.29	84.81	83.76	83.94	84.20
H, wt%	7.99	7.94	7.94	8.12	8.24
S, wt%	7.86	6.43	7.43	7.07	6.54
N, wt%	0.86	0.82	0.87	0.87	1.02
Avg. mol. formula	$C_{224}H_{258}S_8N_2$	$C_{423}H_{476}S_{12}N_4$	$C_{367}H_{418}S_{12}N_3$	$C_{332}H_{386}S_{10}N_3$	$C_{149}H_{175}S_4N_2$
XRD parameters					
d_m, Å	3.5		3.6		
L_a, Å	8.9		9.9		
L_c, Å	16.6		13.6		
M	6		5		
	Whole asph. VR		**SEC Fraction Number of VR**		
MW	2670	6254	5360	3220	2120
Atoms of C	191	442	385	233	153
C_{al}	54%	43%	48%	57%	57%
$C_{a,\ n\text{-alkyl}}$	27%	23%	26%	25%	27%
n	11	11	10	12	11
$C_{al,\ CH3}$	8%	8%	8%	8%	10%
C_{ar}	46%	57%	52%	43%	43%
$C_{ar,\ t}$	8%	9%	7%	5%	6%
$C_{ar,\ q}$	39%	48%	45%	38%	37%
Elemental analysis					
C, wt%	85.75	84.75	86.22	86.87	86.52
H, wt%	9.13	8.10	8.23	8.40	8.41
S, wt%	4.17	6.14	4.58	4.00	4.50
N, wt%	0.95	1.01	0.97	0.73	0.51

(Continued)

TABLE 7.6 (CONTINUED)
Structural Parameters of Asphaltenes by SEC Obtained from Kuwaiti Atmospheric and Vacuum Residue

Avg. mol. formula	$C_{191}H_{244}S_3N_2$	$C_{442}H_{507}S_{12}N_5$	$C_{385}H_{441}S_8N_4$	$C_{233}H_{270}S_4N_2$	$C_{157}H_{178}S_3N_1$
XRD parameters					
d_m, Å	3.5		3.6		
L_a, Å	8.4		8.0		
L_c, Å	14.6		11.4		
M	5		4		

Note: MW = Molecular weight, C_{al} = aliphatic carbons, $C_{a,n\text{-}alkyl}$ = alkyl chains attached to aromatic carbons, n = average number of carbons per molecule, $C_{al,CH3}$ = methyl groups attached to aromatic carbons, C_{ar} = total aromatic carbons, $C_{ar,t}$ = total number of tertiary aromatic carbons, $C_{ar,q}$ = total number of quaternary aromatic carbons, d_m = distance between aromatic layers, L_a = size of an aromatic sheet, L_c = diameter of a stack, M = number of aromatic layers.
Source: Ali et al. 2006. *Energy Fuels* 20 (1): 231–238. With permission.

Ali et al. (2006) developed a very complete study of structural representations of asphaltenes when they are fractionated by SEC. Characterization involved elemental analysis, x-ray diffraction (XRD), NMR, among others. It was demonstrated that fractionation by SEC allows asphaltenes to be separated by molecular weight. Asphaltenes were isolated from Kuwaiti atmospheric and vacuum residua and were further fractionated into 11 fractions according to their molecular weight. Only the results of four fractions are presented here. First fractions are heavier in nature and molecular weight shows a drastic change toward last fractions. In general, asphaltenes obtained from vacuum residue are heavier than those from atmospheric residue. The number of carbons is higher in asphaltenes from vacuum residue than atmospheric residue having slightly more carbon per alkyl chains. However, last fractions of asphaltenes from atmospheric residue were more aromatic than those from vacuum residue and quaternary aromatic carbons are also in higher amount, indicating a more condensed structure similar to pericondensed types. The distance between two aromatic layers (d_m), size of an aromatic sheet (L_a), diameter of a stack (L_c), and the number of aromatic layers (M) was only reported for the fraction number four. These parameters were slightly lower for asphaltenes from vacuum residue than those from atmospheric residue indicating a more compact crystalline structure. Sulfur and nitrogen content diminished in general as fractions were lighter and the contents were smaller in asphaltenes from vacuum residue as well. Table 7.6 summarizes the results reported by the authors.

Two studies that referred to characterization of fractionated asphaltenes with and without hydrotreating were carried out by Trejo et al. (2004) and Trejo and Ancheyta (2007). In the first case, asphaltenes from Maya and Isthmus crudes were

TABLE 7.7
Properties of Asphaltene Fractions

		Fraction				Fraction		
Property	**Maya**	**A**	**B**	**C**	**Isthmus**	**A**	**B**	**C**
C, wt%	81.62	81.26	81.47	81.42	83.99	83.43	83.80	83.82
H, wt%	7.26	7.22	7.34	7.77	7.30	7.40	7.47	7.67
O, wt%	1.02	1.23	1.00	0.87	0.79	0.87	0.77	0.67
N, wt%	1.46	1.50	1.46	1.40	1.35	1.47	1.38	1.30
S, wt%	8.46	8.57	8.49	8.36	6.48	6.74	6.49	6.46
H/C ratio	1.067	1.066	1.081	1.145	1.043	1.064	1.069	1.098
Ni, wt%	0.032	0.045	0.038	0.033	0.018	0.020	0.020	0.07
V, wt%	0.151	0.193	0.185	0.145	0.074	0.074	0.071	0.066
f_a	0.52	0.51	0.55	0.53	0.57	0.57	0.59	0.53
n	6.8	10.4	7.6	8.6	5.0	6.5	7.5	8.5

Source: Trejo et al. 2004. *Fuel* 83 (16): 2169–2175. With permission.

separated and further fractionated by using a mixture of solvents, and then characterized by different techniques. It is observed, in general, that, for both crudes, carbon content in the whole asphaltene samples is higher than that found in the fractions, while hydrogen content is lower in the whole asphaltenes, as can be seen in Table 7.7. Oxygen, nitrogen, and sulfur contents in the whole asphaltene samples are among those observed in the three fractions. The least soluble fraction A has higher S, N, O, and metals content than the others and more complex structures are to be expected in this fraction. Fraction A is more aromatic due to its low H/C ratio. Aromaticity factor (f_a) obtained by NMR was almost the same in each fraction and the average length of alkyl chains (*n*) was larger in each fraction than in the case of the whole asphaltenes in both cases. According to aggregate molecular weight obtained by VPO (toluene as solvent), all fractions of asphaltenes from Isthmus crude were lighter than those of the asphaltenes from Maya (Figure 7.14). It can be observed that fraction A is the heaviest, whereas fractions B and C are similar in aggregate molecular weight. It is noted that a big increment of aggregate MW in fraction A of asphaltenes from Maya crude compared with the whole asphaltenes since heavier asphaltenes precipitate first due to their insolubility in the mixture of solvents. It is seen that aggregate MW is higher in fractions than in whole asphaltenes because aggregate MW of a whole sample is only an average value obtained from all fractions. Fractions of asphaltenes from Isthmus crude are more aromatic with fewer average number of carbons per alkyl side chain compared with those of the asphaltenes from Maya crude, especially in fraction A since fractions B and C have almost the same average length of alkyl chains (Figure 7.15).

The composition and properties of asphaltene fractions obtained from hydrotreated Maya crude has been reported by Trejo and Ancheyta (2007). They

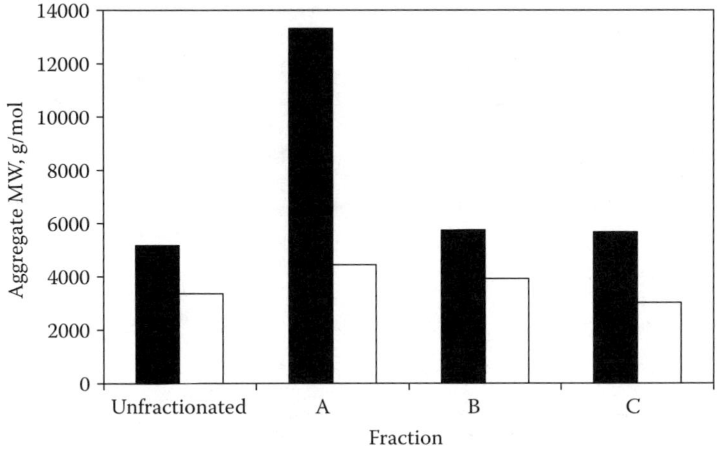

FIGURE 7.14 Aggregate molecular weight obtained by vapor pressure osmometry (VPO) of asphaltenes and their fractions: (■) Maya, (□) Isthmus. (From Trejo et al. 2004. *Fuel* 83 (16): 2169–2175. With permission.)

stated that the changes that asphaltenes suffer after hydrotreating were better observed when fractionating the samples with solvents into three fractions (A, B, and C), where fraction A was the least insoluble and C the most soluble in the mixture of solvents. The authors studied the distribution of metals (V and Ni), elemental analysis, and molecular weight distribution by matrix-assisted laser desorption/ionization (MALDI), and observed that V and Ni tend to concentrate in the most insoluble fraction (fraction A). Figure 7.16 presents the results of

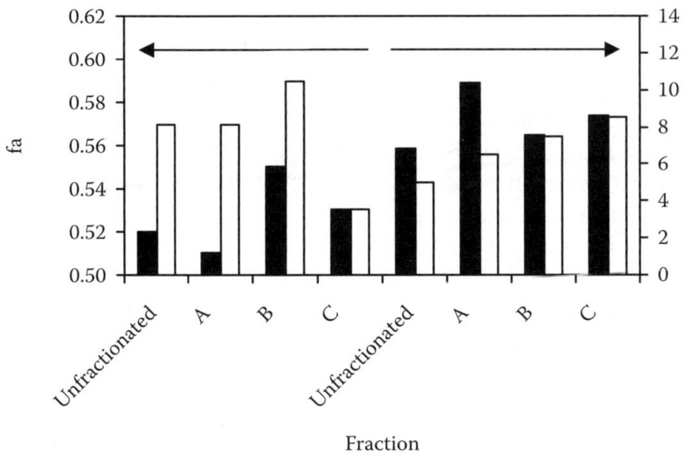

FIGURE 7.15 Aromaticity factor (fa) and average number of carbons per alkyl chain: (■) Maya, (□) Isthmus. (From Trejo et al. 2004. *Fuel* 83 (16): 2169–2175. With permission.)

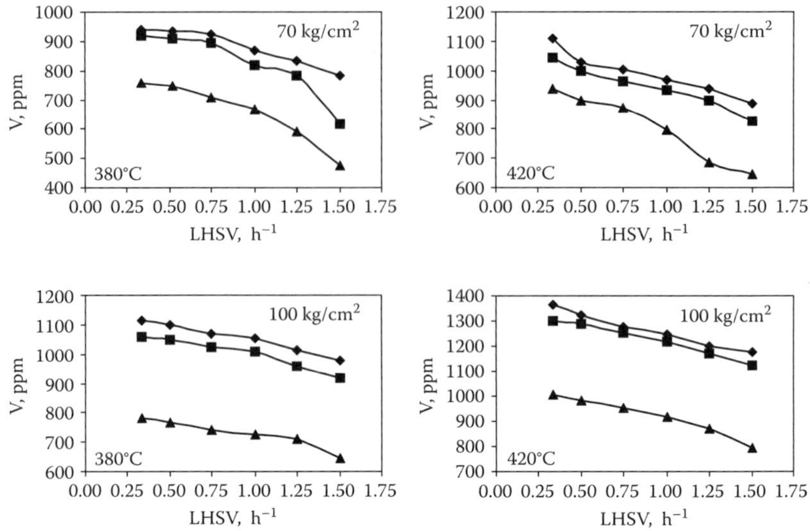

FIGURE 7.16 Vanadium content in different fractions of asphaltenes from hydrotreated Maya crude at different reaction conditions: (◆) Fraction A, (■) Fraction B, (▲) Fraction C. (From Trejo, F. and Ancheyta, J. 2007. *Ind. Eng. Chem. Res.* 46 (23) 7571–7579. With permission.)

vanadium concentration in each fraction, and the content in all fractions increases as pressure and temperature increased. The highest content was obtained as the space velocity is reduced. With fractionation, large molecules of asphaltenes containing high vanadium concentration are separated and grouped in fraction A. Nickel (Figure 7.17) exhibited a similar trend as in the case of vanadium. Both vanadium and nickel contents are higher in asphaltenes from hydrotreated products at higher pressure and temperature and the lowest space velocity. Hydrocracking of asphaltenes produces fragmentation of big molecules forming small fractions with low metals contents, while the most refractory molecules remain almost without changes in which metals are concentrated.

It can be seen that both metals tend to concentrate more or less uniformly in the three fractions and, as the reaction severity increases, metal contents in fractions A and B approach each other, giving the impression that only two fractions are present, (A + B) and C. This is clearly due to differences in changes of the reaction selectivity because at low severity hydrogenation predominates, whereas at high severity hydrocracking is more dominant.

As temperature is increased, the amount of these metals in each fraction is prone to increase as well because asphaltene molecules are hydrocracked in a high degree including the alkyl chains, whereas metals content remains almost constant because they are in the inner section of the structure. This difference in relative amounts of small and large asphaltenes molecules alters their solubility and can contribute in great extent to sediment formation. Figure 7.18 shows the H/C atomic ratio, which

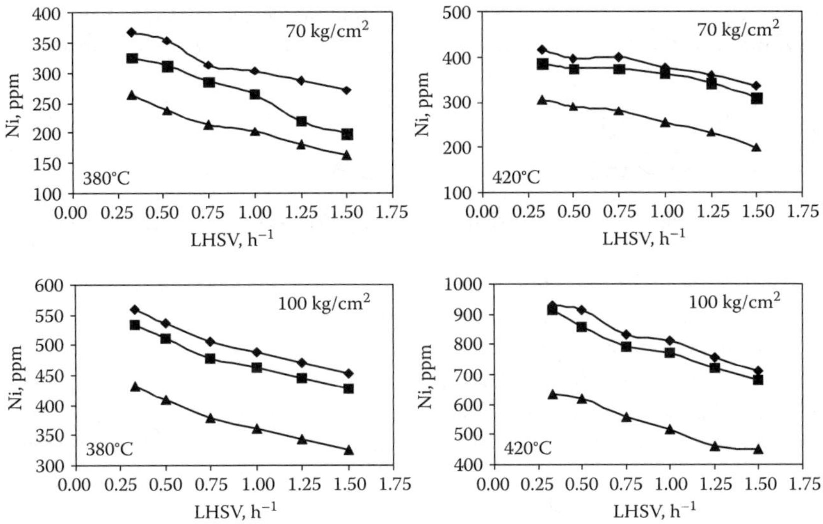

FIGURE 7.17 Nickel content in different fractions of asphaltenes from hydrotreated Maya crude at different reaction conditions: (◆) Fraction A, (■) Fraction B, (▲) Fraction C. (From Trejo, F. and Ancheyta, J. 2007. *Ind. Eng. Chem. Res.* 46 (23) 7571–7579. With permission.)

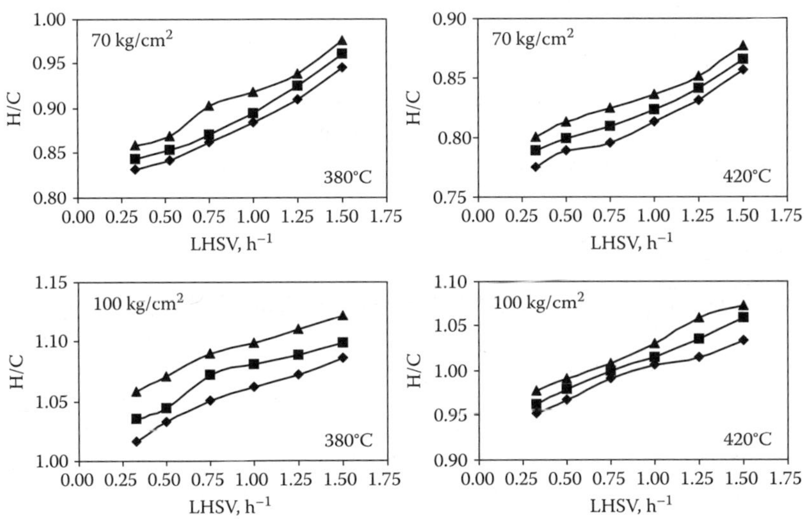

FIGURE 7.18 H/C atomic ratio of different fractions of asphaltenes from hydrotreated Maya crude at different reaction conditions: (◆) Fraction A, (■) Fraction B, (▲) Fraction C. (From Trejo, F. and Ancheyta, J. 2007. *Ind. Eng. Chem. Res.* 46 (23) 7571–7579. With permission.)

was reduced in all fractions as temperature increased and pressure was reduced. Fraction A was the most aromatic fraction as shown by its lower H/C ratio.

7.4.1 Molecular Weight Distribution of Fractionated Asphaltenes by MALDI

Analysis of molecular weight distribution by MALDI was carried out in linear mode without deflection to study the changes that fractions of asphaltenes undergo during hydrocracking. Figure 7.19 corresponds to the asphaltene sample hydrotreated at 400°C, 70 kg/cm^2, and liquid hourly space velocity (LHSV) of 1 h^{-1}. Because of more severe reaction conditions, fraction C shows the sharper molecular weight distribution. Fraction C has the smaller distribution of MW indicating that it is the lightest fraction. It is observed that the three fractions do not present signals below *m/z* 1,000 g/mol indicating that at these conditions (especially moderate temperature, i.e., 400°C) there are no small fragments of asphaltenes having lower MW.

Figure 7.20 shows the asphaltenes hydrotreated at 420°C, 100 kg/cm^2, and LHSV) of 1 h^{-1}. It is to be expected that some asphaltenes of low molecular weight could be evaporated in the ion source vacuum before ionization takes place and, therefore, the ions of mass less than *m/z* (mass-to-charge ratio) 1,000 g/mol may be missing from the spectrum as observed in fraction A. However, fractions B and C show signals at low molecular weights below ~*m/z* 1,000 g/mol. The maximum ion abundance in this case was found at ~*m/z* 1,200 g/mol. Temperature favors more hydrocracking reactions and asphaltenes undergo higher fragmentation. Fractionation evidences that smaller molecules are obtained in fractions B and C. When reducing the space velocity the asphaltene cracking is more remarkable and molecular weight distribution tends to decrease. The maximum ion abundance is around ~*m/z* 1,200 g/mol. The effects of space velocity and temperature are similar. In both cases, the maximum ion abundance is obtained at lower molecular weights.

7.4.2 Microscopic Analysis of Fractionated Asphaltenes

Microscopic analysis was also carried out on asphaltenes from hydrotreated crude at two different reaction conditions: (1) 400°C, 100 kg/cm^2, and LHSV of 1 h^{-1}, and (2) 420°C, 100 kg/cm^2, and LHSV of 1 h^{-1}. Fractionation with solvents was applied to these two samples by using a mixture of solvents of toluene/*n*-heptane of 67/33 vol% and 33/67 vol%. Firstly, ~15 g of whole asphaltenes were washed with the first mixtures of solvents under Soxhlet reflux and insoluble and soluble asphaltenes were obtained after washing. Insoluble asphaltenes correspond to fraction A. Then the solution was distilled to recover soluble asphaltenes and they were washed with the second mixture of solvents. Insoluble asphaltenes are the fraction B and soluble ones are the fraction C. A schematic view of this procedure was illustrated in Figure 7.7. Aliphatic and aromatic carbon contents are responsible for solubility/insolubility of asphaltenes. For example, if alkyl chains are

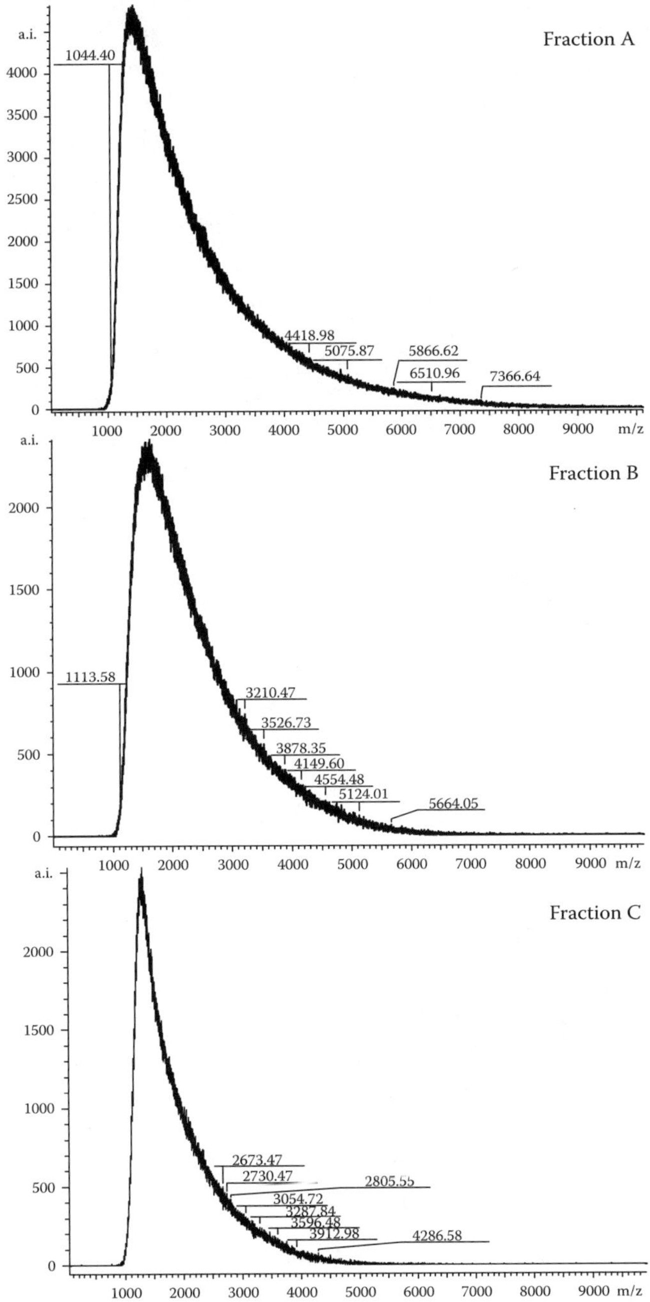

FIGURE 7.19 Molecular weight distribution by MALDI mass spectrometry for fractionated asphaltenes from hydrotreated Maya crude at P = 70 kg/cm², T = 400°C, LHSV = 1.0 h⁻¹. (From Trejo, F. and Ancheyta, J. 2007. *Ind. Eng. Chem. Res.* 46 (23) 7571–7579. With permission.)

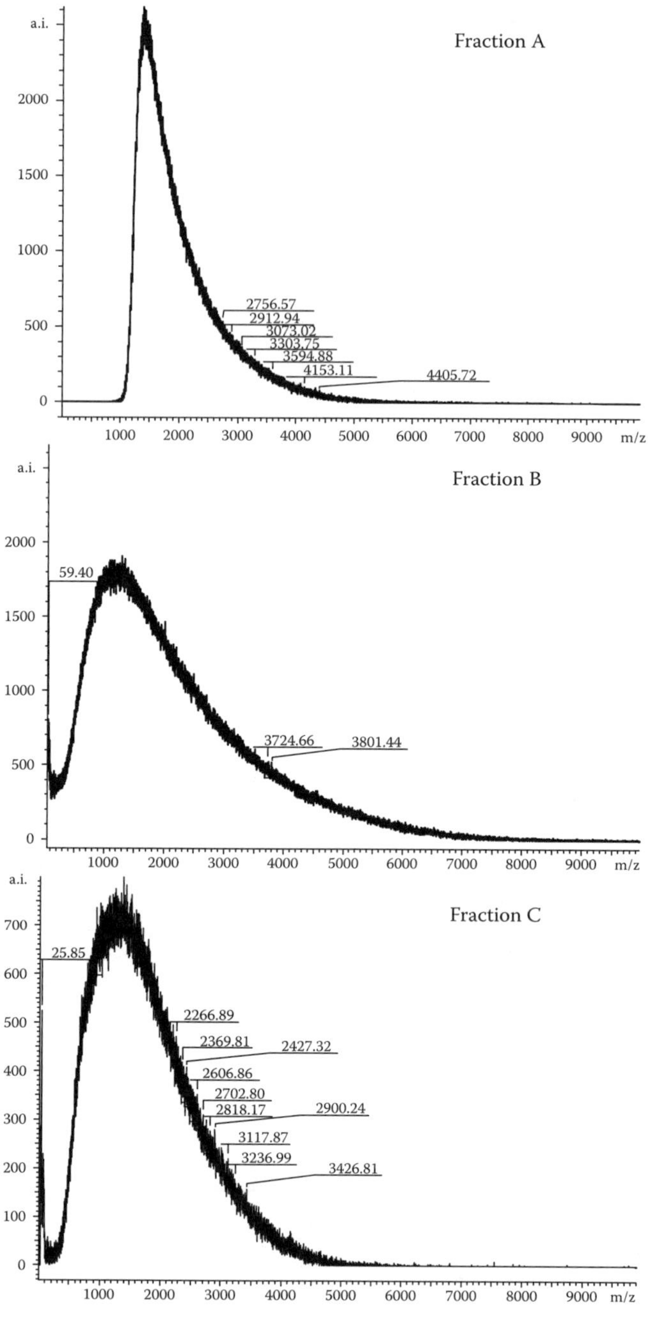

FIGURE 7.20 Molecular weight distribution by MALDI mass spectrometry for fractionated asphaltenes from hydrotreated Maya crude at P = 100 kg/cm^2, T = 420°C, LHSV = 1.0 h^{-1}. (From Trejo, F. and Ancheyta, J. 2007. *Ind. Eng. Chem. Res.* 46 (23) 7571–7579. With permission.)

abundant and their average number of carbons is high, some degree of solubilization could be expected. In fact, in previous studies (Trejo et al., 2004), it has been observed that fraction A possesses the highest average number of carbons in alkyl chains; however, this fraction has also the highest content of aromatic carbon as well as the largest number of aromatic rings per molecule, and for this reason alkyl carbon is not enough for keeping soluble the fraction. Thus, precipitation of fraction A is the result of the insolubility of aromatic cores in the mixture of solvents, which selectively separates bigger molecules having higher molecular weights. It was reported that fraction A is less soluble than Fraction B and C, which are recovered by the distillation process.

7.4.2.1 Scanning Electron Microscope (SEM) Analysis

SEM analysis of asphaltenes showed the large variation in their structure on fractionated products from hydrotreated crude oil as an effect of reaction conditions, nature of crude oil and the fractionation technique applied. SEM analysis applied to hydrotreated asphaltenes at 400°C, 100 kg/cm^2, and LHSV of 1 h^{-1} showed two different pictures, as presented in Figure 7.21. Figure 7.21a shows agglomerate particles of asphaltenes, whereas Figure 7.21b presents a structure with very small pores having some particles deposited with irregular shape and variable size on the surface. Porous structure could be formed during removal of small molecules of adsorbed resins on asphaltenes. It is important to remember that fractions B and C are the lightest ones and can adsorb resins, which also have smaller MW concentrating in the mixture of solvents along with fraction C. Another possible explanation for cavities formation is the tendency of asphaltenes to occlude microparticle precursors of coke that are released from asphaltene surface during washing with solvents because coke precursors are toluene insolubles. This change in morphology is attributed to: (1) temperature at which the crude oil was hydrotreated, and (2) solubility/insolubility of asphaltenes in different mixtures of solvents. It has to be remembered that crude oil is a balance between asphaltenes and resins and, as reaction progresses, resins are transformed at a higher rate than asphaltenes; however, asphaltenes suffer different changes depending on reaction severity. Asphaltenes properties, such as solubility or molecular weight, are a consequence of asphaltene polydispersity, and its conversion not only depends on reaction temperature during hydroprocessing, but also on resins content that influence the degree of asphaltene conversion.

Morphology of hydrotreated asphaltenes at 420°C, 100 kg/cm^2, and LHSV of 1 h^{-1} shows differences between fractions A and B (Figure 7.22). Figure 7.22a corresponds to fraction A and it is seen that solids with irregular shapes are obtained after fractionation and some particles are deposited on the surface. Fraction B (Figure 7.22b) shows a smooth surface having deposited particles on the surface as well. In this case, temperature during hydrotreating could be responsible for disrupting the agglomerate particles showed at lower temperatures (400°C). Changes were produced after hydrotreating led to asphaltenes to modify their chemical nature. Due to the breaking of alkyl chains, asphaltenes are prone to stacking. When precipitating, asphaltenes may stack and form solids with either

FIGURE 7.21 Scanning electron microscope (SEM) analysis of: (A) Fraction A and (B) Fraction B from asphaltenes hydrotreated at 400°C, 100 kg/cm^2, and LHSV of 1 h^{-1}.

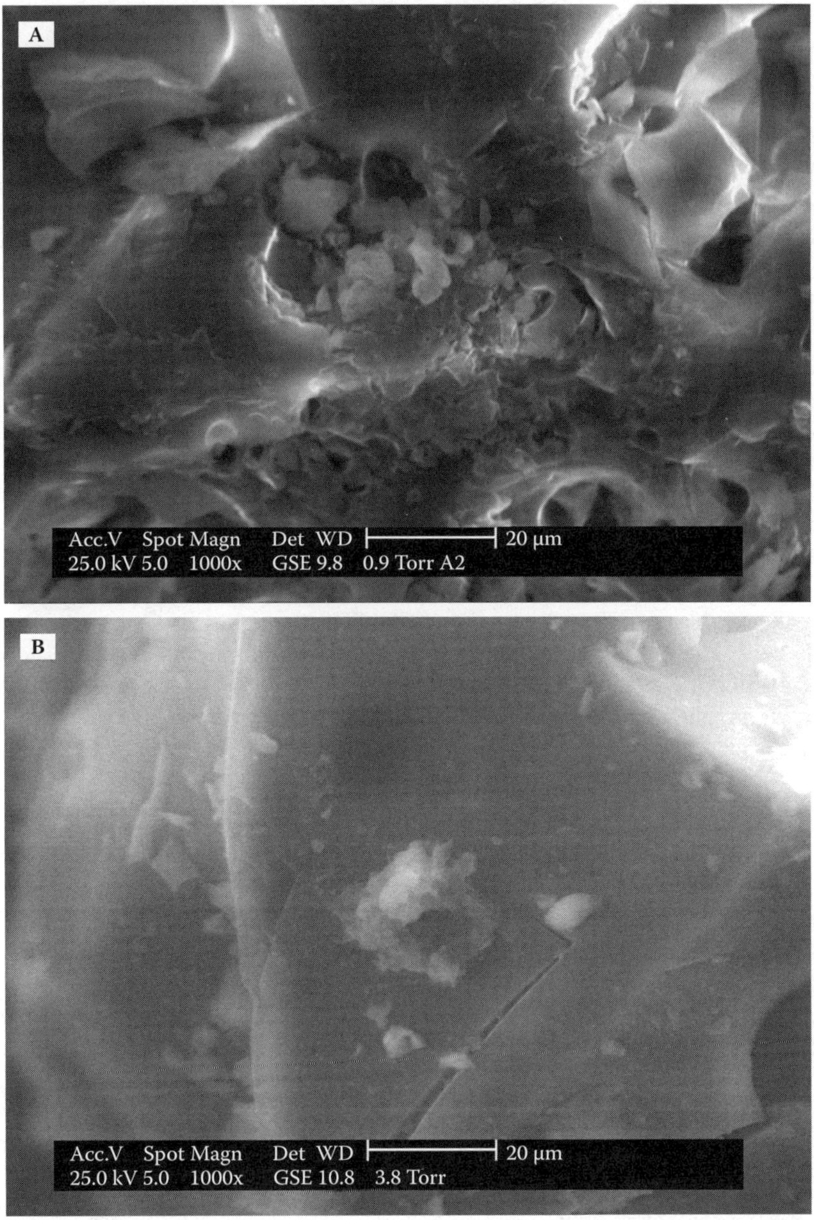

FIGURE 7.22 Scanning electron microscope (SEM) analysis of: (A) Fraction A and (B) Fraction B from asphaltenes hydrotreated at 420°C, 100 kg/cm^2, and LHSV of 1 h^{-1}.

irregular surface or smooth surface. At higher hydrotreating temperatures, porous structures could be less common since resins were converted into lighter components and only remaining asphaltenes are present.

The semiqualitative analysis of composition and elemental mapping of fraction A from hydrotreated crude at 420°C, 100 kg/cm^2, and LHSV of 1 h^{-1} are shown in Figure 7.23. Results of fractions A and B from hydrotreated crude at different conditions are shown in Table 7.8 (Trejo et al., 2009). In most of the cases, it was observed that the higher the temperature, the higher the amount of these deposited metals. A plausible explanation for this is that destruction of alkyl chains of asphaltenes leaves practically intact the core of the micelle where porphyrins are located and the metals amount (Ni, V, Fe) tends to concentrate. An increase of Fe:C and Ni:C ratios is observed as temperature increases comparing the same fractions. A V:C ratio shows more irregular behavior as temperature is raised; however, in some cases this ratio was higher in fractions after hydroprocessing than in nonhydroprocessed asphaltenes. V content was also the highest compared with Fe and Ni, which is expected due to the large amount of V generally present in asphaltenes. It is important to remember that this analysis must be carefully considered because the concentration and mapping of elements depend on the region where asphaltenes are analyzed. Phosphorous has not been commonly reported in asphaltenes; however, it is detected by SEM-EDS (scanning electron microscope-energy dispersive spectroscopy) having in some cases similar amount (wt%) compared with other metals (Fe, Ni, V). The exact place where phosphorous is located inside the asphaltene structure has not been reported as of yet.

7.4.2.2 TEM Analysis

Reaction conditions during hydroprocessing of Maya crude indeed influence asphaltene structure, but it is fractionation with solvents that shows much more clearly the effect on shape and size. Structural rearrangements of asphaltenes after hydroprocessing by using different characterization techniques (e.g., XRD, NMR) have been reported by other authors (Andersen et al., 2005; Tanaka et al., 2004). However, information of structural changes suffered by asphaltenes during hydroprocessing at a microscopic level by transmission electron microscopy (TEM) after being separated by solvent fractionation is lacking.

Figure 7.24a shows some type of rearrangement near the edge having an interlayer distance of around 0.354 nm in fraction A from hydrotreated crude at 400°C, 100 kg/cm^2, and LHSV of 1 h^{-1}, which corresponds to typical separation of aromatic structures of amorphous asphaltenes. However, going deeper inside the sample, another type of rearrangement is observed that agrees to perfectly ordered layers having an interlayer separation around ~0.335 nm, which is similar to the interlayer spacing of graphite-like carbon. Near the edge the amorphous structure is observed, but at the interior of the sample more evident changes were found. In spite of having moderate operating conditions during hydroprocessing, these were strong enough to modify the amorphous structure by alkyl chains cleavage that allowed the asphaltenes to exhibit well-ordered layers because there is not steric

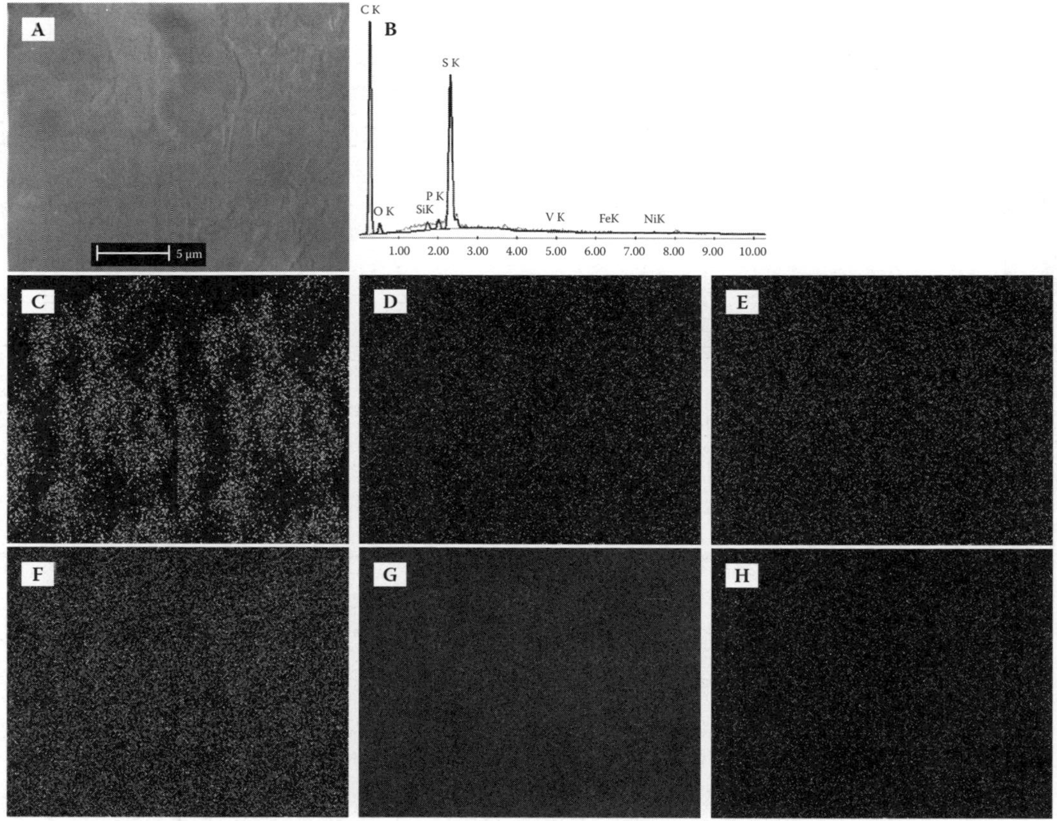

FIGURE 7.23 SEM-EDS (scanning electron microscope-energy dispersive spectroscopy) elemental mapping of hydrotreated asphaltenes (Fraction A) at 420°C, 100 kg/cm^2, and LHSV of 1 h^{-1} by SEM. (A) Original picture, (B) elemental analysis, (C) C*k* mapping, (D) Fe*k* mapping, (E) Ni*k* mapping, (F) P*k* mapping, (G) S*k* mapping, and (H) V*k* mapping.

TABLE 7.8
Elemental Composition of Asphaltene Fractions Obtained by the Scanning Electron Microscope (SEM) (HDT Conditions: Pressure of 100 kg/cm², H_2-to-Oil Ratio of 5,000 ft³/Bbl, and LHSV of 1 h^{-1})

	Elemental Composition of Each Fraction, wt%			
	HDT at 400°C		HDT at 420°C	
Element	Fraction A	Fraction B	Fraction A	Fraction B
C	86.06	86.71	85.07	86.90
O	4.17	4.56	3.39	5.05
Si	0.94	0.34	0.27	0.39
P	0.26	0.43	0.33	0.26
S	7.96	7.23	10.16	6.57
V	0.28	0.15	0.40	0.23
Ni	0.21	0.43	0.24	0.38
Fe	0.11	0.15	0.14	0.21
Fe/C (10^3)	0.2749	0.3720	0.3539	0.5197
O/C	0.0364	0.0395	0.0299	0.0436
S/C	0.0347	0.0312	0.0447	0.0283
V/C (10^3)	0.7671	0.4079	1.1086	0.6240
Ni/C (10^3)	0.4994	1.0148	0.5773	0.8949

Source: Trejo et al. 2009. *Energy Fuels* 23(1): 429–439. With permission.

hindrance by the alkyl chains. In this case, the cores are free to pile up and form larger and stacked structures in solid state. Fraction B exhibits only amorphous structures with spacing around 0.354 nm. Nevertheless, graphite structures are not present, either in the edge or in the interior of the samples. Fractionation allowed us to observe more graphite-like structures in fraction A along with amorphous structures at the edge compared with fraction B. Fraction A has less alkyl chains and stacking is more evident in solid state, as shown in Figure 7.24a. Lighter fractions, included fraction B, have more alkyl chains, which make them more soluble in the mixture of solvents compared with fraction A. It is necessary the presence of alkyl chains to solubilize fraction B in the first mixture of solvents (toluene/*n*-heptane 67/33 vol%). For this reason, when separating fraction B, amorphous structures are more abundant due to alkyl chains impeding the stacking of asphaltenes at long range. Sharma et al. (2002) demonstrated with model compounds that stacking is easily carried out when alkyl chains are not present and well-ordered structures are formed. On the other hand, when having long alkyl chains, the structures are disordered and amorphous.

As can be seen, fractionation can make more evident the changes in structure of fractionated asphaltenes. This is an important parameter when studying the behavior of asphaltenes during upgrading processes, such as hydrotreating.

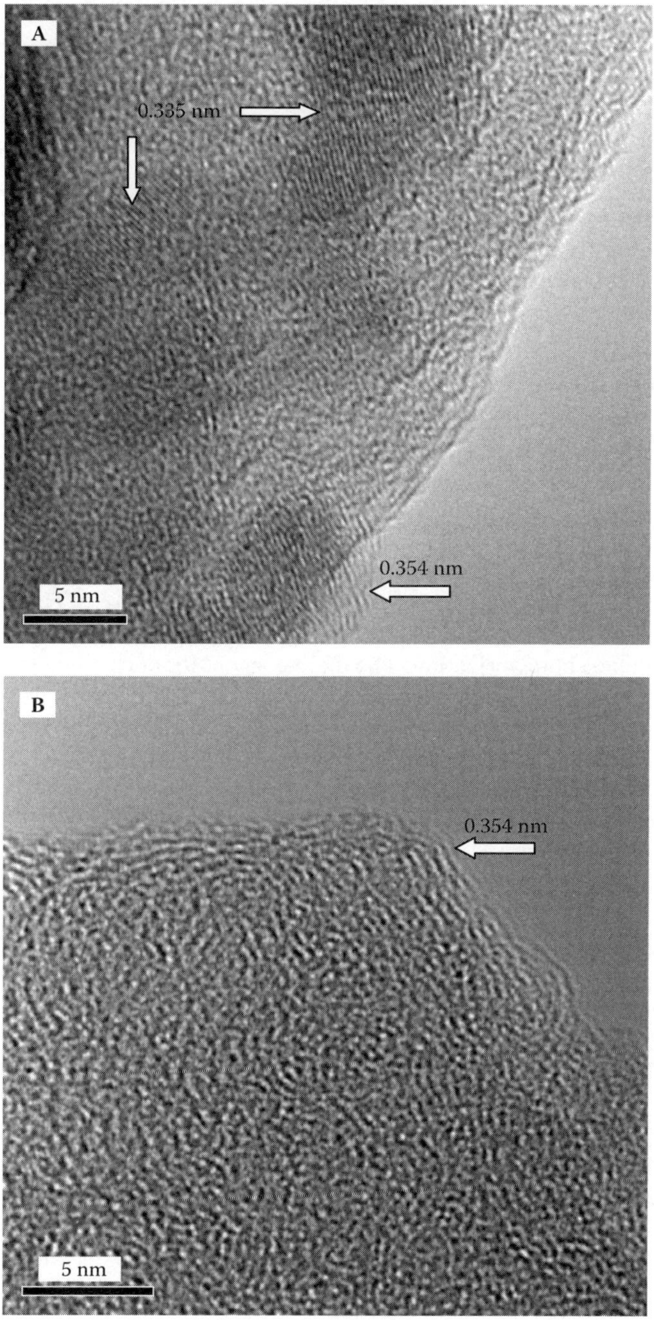

FIGURE 7.24 Transmission electron microscopy (TEM) images of: (A) Fraction A and (B) Fraction B from asphaltenes hydrotreated at 400°C, 100 kg/cm^2, and LHSV of 1 h^{-1}.

A better representation of asphaltenes is obtained when separation in groups having similar chemical structure is carried out by fractionation. Thus, we can be assured that their properties are not influenced by other heavier or lighter molecules.

7.5 CONCLUDING REMARKS

One of the most important purposes of fractionation of crude oil is to separate different fractions of a given crude, which makes it easier to characterize at the purest possible way to elucidate its physical and chemical properties. Asphaltenes are key components in this complex hydrocarbon mixture and fractionation is commonly used to separate them according to molecular weight. Fractionation has the objective of gathering information about solubility of asphaltenes, which is useful to characterize each fraction. For this reason, special emphasis is needed to find a relationship between the molecular size and average structural parameters of asphaltenes.

Characterization of asphaltenes and fractions thereof can be very useful in the petroleum refining industry specially focused on the development of heavy oils hydroprocessing catalysts and in kinetics studies of catalytic hydroprocessing reactions involving hydrocracking of asphaltenes as well. Thus, to design suitable hydroprocessing catalysts for heavy oils it is very important to perform deep asphaltene characterization. The main concept for developing suitable catalysts for heavy oil hydroprocessing depends on the particle size, pore size, and its distribution. Fractionation can give valuable information about the changes of molecular weight and size of asphaltenes during upgrading processing, so that better catalysts can be designed based on these parameters along with an optimum balance between acid and active sites.

Fractionation also helps to understand the way in which different components of petroleum are distributed in the whole sample. In this regard, SARA analysis gives information about said distribution. Taking into account the characterization data of asphaltenes and the whole crude or residue, it is possible to draw new processes suitable to manage the feedstocks available today and in the near future.

REFERENCES

Acevedo, S., Escobar, G., Ranaudo, M.A., Pinate, J., Amorin, A., Diaz, M., and Silva, P. 1997. Observations about the structure and dispersion of petroleum asphaltenes aggregates obtained from dialysis fractionation and characterization. *Energy Fuels* 11 (4): 774–778.

Ali, F.A., Ghaloum, N., and Hauser, A. 2006. Structure representation of asphaltene GPC fractions derived from Kuwaiti residual oils. *Energy Fuels* 20 (1): 231–238.

Ali, M.F., Perzanowski, H., Bukhari, A., and Al-Haji, A.A. 1993. Nickel and vanadyl porphyrins in Saudi Arabian crude oils. *Energy Fuels* 7 (2): 179–184.

Al-Muhareb, E., Morgan, T.J., Herod, A.A., and Kandiyoti, R. 2007. Characterization of petroleum asphaltenes by size exclusion chromatography, UV-fluorescence and mass spectrometry. *Petrol. Sci. Tech.* 25 (1–2): 8192.

Andersen, S.I. and Birdi, K.S. 1991. Aggregation of asphaltenes as determined by calorimetry. *J. Colloid Interface Sci.* 142 (2): 497–502.

Andersen, S.I., Keul, A., and Stenby, E. 1997. Variation in composition of subfractions of petroleum asphaltenes. *Petrol. Sci. Tech.* 15 (7–8): 611–645.

Andersen, S.I., Jensen, J.O., and Speight, J.G. 2005. X-ray diffraction of subfractions of petroleum asphaltenes. *Energy Fuels* 19 (6): 2371–2377.

ASTM D 86. Standard Test Method for Distillation of Petroleum Products at Atmospheric Pressure.

ASTM D 1160. Standard Test Method for Distillation of Petroleum Products at Reduced Pressure.

ASTM D 2887. A Standard Test Method for Boiling Range Distribution of Petroleum Fractions by Gas Chromatography.

ASTM D 2892. Standard Test Method for Distillation of Crude Petroleum (15-Theoretical Plate Column).

ASTM D 3710. Standard Test Method for Boiling Range Distribution of Gasoline and Gasoline Fractions by Gas Chromatography.

ASTM D 4124. Standard Test Methods for Separation of Asphalt into Four Fractions.

ASTM D 5236. Standard Test Method for Distillation of Heavy Hydrocarbon Mixtures (Vacuum Potstill Method).

ASTM D 5307. Standard Test Method for Determination of Boiling Range Distribution of Crude Petroleum by Gas Chromatography.

ASTM D 6352. Standard Test Method for Boiling Range Distribution of Petroleum Distillates in Boiling Range from 174 to 700°C by Gas Chromatography.

ASTM D 7169. Standard Test Method for Boiling Point Distribution of Samples with Residues, such as Crude Oils and Atmospheric and Vacuum Residues by High Temperature Gas Chromatography.

Baba, H., Matsuyama, A., and Kokubun, H. 1964. Three types of acid-base interaction in an aprotic solvent as revealed by electronic absorption spectroscopy. *J. Chem. Phys.* 41 (3): 895–896.

Bartholdy, J., Lauridsen, R., Mejlholm, M., and Andersen, S.I. 2001. Effect of hydrotreatment on product sludge stability. *Energy Fuels* 15 (5): 1059–1062.

Buch, L., Groenzin, H., Buenrostro-Gonzalez, E., Andersen, S.I., Lira-Galeana, C., and Mullins, O.C. 2003. Molecular size of asphaltene fractions obtained from residuum hydrotreatment. *Fuel* 82 (9): 1075–1084.

Buckley, J.S. 1996. Microscopic investigation of the onset of asphaltene precipitation. *Fuel Sci. Tech. Int.* 14 (1–2): 55–74.

Buckley, J.S. 1999. Predicting the onset of asphaltene precipitation from refractive index measurements. *Energy Fuels* 13 (2): 328–332.

Buenrostro-González, E., Groenzin, H., Lira-Galeana, C., and Mullins, O.C. 2001. The overriding chemical principles that define asphaltenes. *Energy Fuels* 15 (4): 972–978.

Chung, K.H., Xu, C., Hu, Y., and Wang, R. 1997. Supercritical fluid extraction reveals resid properties. *Oil Gas J.* 95 (3): 66–69.

Domin, M., Herod, A.A., Kandiyoti, R., Larsen, J.W., Lazaro, M.-J, Li, S., and Rahimi, P. 1999. Large molecular mass materials in coal-derived liquids by ^{252}Cf-Plasma and matrix-assisted laser desorption mass spectrometry. *Energy Fuels* 13 (3): 552–557.

Douda, J., Llanos, M.E., Álvarez, R., and Navarrete, J. 2004. Structure of Maya asphaltene-resin complexes through the analysis of Soxhlet extracted fractions. *Energy Fuels* 18 (3): 736–742.

Dunning, H.N., Moore, J.W., Bieber, H., and Williams, R.B. 1960. Porphyrin, nickel, vanadium, and nitrogen in petroleum. *J. Chem. Eng. Data* 5 (4): 546–549.

Dwiggins, C.W. 1965. A small angle X-ray scattering study of the colloidal nature of petroleum. *J. Phys. Chem.* 69 (10): 3500–3506.

Espinosa-Peña, M., Figueroa-Gómez, Y., and Jiménez-Cruz, F. 2004. Simulated distillation yield curves in heavy crude oils: A comparison of precision between ASTM D-5307 and ASTM D-2892 physical distillation. *Energy Fuels* 18 (6): 1832–1840.

Ferrer, E.G. and Baran, E.J. 1991. Electronic and photoelectron spectra of vanadyl(IV) tetraphenylporphyrin. *J. Electr. Spectrosc. Related Phenom.* 57 (2): 189–197.

Fish, R.H. and Komlenic, J.J. 1984. Molecular characterization and profile identifications of vanadyl compounds in heavy crude petroleums by liquid chromatography/graphite furnace atomic absorption spectrometry. *Anal. Chem.* 56 (3): 510–517.

Fish, R.H., Komlenic, J.J., and Wins, B.K. 1984. Characterization and comparison of vanadyl and nickel compounds in heavy crude petroleums and asphaltenes by reverse-phase and size-exclusion liquid chromatography/graphite furnace atomic absorption spectrometry. *Anal. Chem.* 56 (13): 2452–2460.

Galiasso, R., García, J., Caprioli, L., Pazos, J.M., and Soto, A. 1985. Reactions of porphyrinic and non-porphyrinic molecules during hydrodemetallization of heavy crude oils. *Am. Chem. Soc., Div. Petr. Chem.–Prepr.* 30 (1): 50–61.

Goulon, J., Retournard, A., Friant, P., Goulon-Ginet, C., Berthe, C., Muller, J.-F., Poncet, J.-L., Guilard, R., Escalier, J.-C., and Neff, B. 1984. Structural characterization by x-ray absorption spectroscopy (EXAFS/XANES) of the vanadium chemical environment in Boscan asphaltenes. *J. Chem. Soc., Dalton Trans.* 6: 1095–1103.

Gouterman, M. Spectra of porphyrins. 1961. *J. Mol. Spectrosc.* 6 (1): 138–163.

Grizzle, P.L. and Sablotny, D.M. 1986. Automated liquid chromatographic compound class group-type separation of crude oils and bitumens using chemically bonded silica-NH_2. *Anal. Chem.* 58 (12): 2389–2396.

Groenzin, H. and Mullins, O.C. 2000. Molecular size and structure of asphaltenes from various sources. *Energy Fuels* 14 (3): 677–684.

Gutiérrez, L.B., Ranaudo, M.A., Méndez, B., and Acevedo, S. 2001. Fractionation of asphaltene by complex formation with *p*-nitrophenol. A method for structural studies and stability of asphaltene colloids. *Energy Fuels* 15 (3): 624–628.

Hall, G. and Herron, S.P. 1979. Size characterization of petroleum asphaltenes and maltenes. *Am. Chem. Soc., Div. Pet. Chem.–Prepr.* 24 (4): 924–934.

Hansen, C.M. 1969. The universality of the solubility parameter. *Ind. Eng. Chem. Prod. Res. Dev.* 8 (1): 2–11.

Jewell, D.M., Weber, J.H., Bunger, J.W., Plancher, H., and Latham, D.R. 1972. Ion-exchange, coordination, and adsorption chromatographic separation of heavy-end petroleum distillates. *Anal. Chem.* 44 (8): 1391–1395.

Kaminski, T.J., Fogler, H.S., Wolf, N., Wattana, P., and Mairal, A. 2000. Classification of asphaltenes via fractionation and the effect of heteroatom content on dissolution kinetics. *Energy Fuels* 14 (1): 25–30.

Kharrat, A.M., Zacharia, J., Cherian, V.J., and Anyatonwu, A. 2007. Issues with comparing SARA methodologies. *Energy Fuels* 21 (6): 3618–3621.

Lee, A.K., Murray, A.M., and Reynolds, J.G. 1995. Metallopetroporphyrins as process indicators: Separation of petroporphyrins in green river oil shale pyrolysis products. *Fuel Sci. Tech. Int.* 13 (8): 1081–1097.

Li, W., Morgan, T.J., Herod, A.A., and Kandiyoti, R. 2004. Thin-layer chromatography of pitch and a petroleum vacuum residue. Relation between mobility and molecular size shown by size-exclusion chromatography. *J. Chromatog. A* 1024 (1–2): 227–243.

Liao, Z. and Geng, A. 2002. Characterization of nC7-soluble fractions of the products from mild oxidation of asphaltenes. *Org. Geochem.* 33 (12): 1477–1476.

Márquez, N., Gall, C., Tudares, C., Paredes, J., and De la Cruz, C. 1989. Isolation and spectroscopic characterization of metalloporphyrins from Venezuelan crude oils. *Am. Chem. Soc., Div. Petr. Chem.–Prepr.* 34 (2): 292–296.

Márquez, N., Ysambertt, F., and De La Cruz, C. 1999. Three analytical methods to isolate and characterize vanadium and nickel porphyrins from heavy crude oil. *Analytica Chim. Acta* 395 (3): 343–349.

McKay, J.F., Amend, P.J., Cogswell, T.E., Harnsberger, R.B., Erickson, R.B., and Latham, D.R. 1978. Analytical chemistry of liquid fuel sources. *ACS Advances. Chemistry Series* 170, American Chemical Society, Washington, D.C., 128–142.

Miller, J.T., Fisher, R.B., Thiyagarajan, P., Winans, R.E., and Hunt, J.E. 1998. Subfractionation and characterization of Mayan asphaltenes. *Energy Fuels* 12 (6): 1290–1298.

Miller, J.T., Fisher, R.B., van der Eerden, A.M.J., and Koningsberger, D.C. 1999. Structural determination by XAFS spectroscopy of non-porphyrin nickel and vanadium in Maya residuum, hydrocracked residuum, and toluene-insoluble solid. *Energy Fuels* 13 (3): 719–727.

Mushrush, G.M. and Speight, J.G. 1998. Instability and incompatibility of petroleum products, in *Petroleum Chemistry and Refining*, Speight, J.G., Ed., CRC Press: Boca Raton, FL, 199–242.

Nali, M., Fabbi, M., and Scilingo, A. 1997. A systematic preparative method for petroporphyrin purification. *Petrol. Sci. Tech.* 15 (3–4): 307–322.

Pelet, R., Behar, F., and Monin, J.C. 1986. Resins and asphaltenes in the generation and migration of petroleum. *Org. Geochem.* 10 (1–3): 481–498.

Porte, G., Zhou, H., and Lazzeri, V. 2003. Reversible description of asphaltene colloidal association and precipitation. *Langmuir* 19 (1): 40–47.

Radke, M., Willsch, H., and Welte, D.H. 1984. Class separation of aromatic compounds in rock extracts and fossil fuels by liquid chromatography. *Anal. Chem.* 56 (13): 2538–2546.

Rankel, L.A. 1981. Reactions of metalloporphyrins and petroporphyrins with H_2S and H_2. *Am. Chem. Soc., Div. Petrol. Chem.–Prepr.* 26 (3): 689–698.

Reynolds, J.G. and Biggs, W.R. 1985. Analysis of residuum demetalation by size exclusion chromatography with element specific detection. *Am. Chem. Soc., Div. Petrol. Chem.–Prepr.* 30 (4): 679–686.

Robert, E.C., Merdrignac, I., Rebours, B., Harlé, V., Kressmann, S., and Colyar, J. 2003. Contribution of analytical tools for understanding of sediment formation: application to H-oil process. *Petrol. Sci. Tech.* 21 (3–4): 615–627.

Semple, K.M., Cyr, N., Fedorak, P.M., and Westlake, D.W.S. 1990. Characterization of asphaltenes from Cold Lake heavy oil: Variations in chemical structure and composition with molecular size. *Can. J. Chem.* 68 (7): 1092–1099.

Sharma, A., Groenzin, H., Tomita, A., and Mullins, O.C. 2002. Probing order in asphaltenes and aromatic ring system by HRTEM. *Energy Fuels* 16 (2): 490–496.

Spencer, W.A., Galobardes, J.F., Curtis, M.A., and Rogers, L.B. 1982. Chromatographic studies of vanadium compounds from Boscan crude oil. *Sep. Sci. Tech.* 17 (6): 797–819.

Spiecker, P.M., Gawrys, K.L., and Kilpatrick, P.K. 2003. Aggregation and solubility behavior of asphaltenes and their subfractions. *J. Colloid Interface Sci.* 267 (1): 178–193.

Suatoni, J.C. and Swab, R.E. 1975. Rapid hydrocarbon group type analysis by high performance liquid chromatography. *J. Chromatog. Sci.* 13 (8): 361–366.

Szewczyk, V., Behar, F., Behar, E., and Scarsella, M. 1996. Evidence of the physicochemical polydispersity of asphaltenes. *Oil Gas Sci. Tech.–Rev. IFP* 51 (4): 575–590.

Tanaka, R., Sato, E., Hunt, J.E., Winans, R.E., Sato, S., and Takanohashi, T. 2004. Characterization of asphaltene aggregates using x-ray diffraction and small-angle x-ray scattering. *Energy Fuels* 18 (4): 1118–1125.

Tojima, M., Suhara, S., Imamura, M., and Furuta, A. 1998. Effect of heavy asphaltene on stability of residual oil. *Catalysis Today* 43 (3–4): 347–351.

Trejo, F. and Ancheyta, J. 2007. Characterization of asphaltene fractions from hydrotreated Maya crude oil. *Ind. Eng. Chem. Res.* 46 (23): 7571–7579.

Trejo, F., Centeno, G., and Ancheyta, J. 2004. Precipitation, fractionation and characterization of asphaltenes from heavy and light crude oils. *Fuel* 83 (16): 2169–2175.

Trejo, F., Ancheyta, J., and Rana, M.S. 2009. Structural characterization of asphaltenes obtained from hydroprocessed crude oils by SEM and TEM. *Energy Fuels* 23 (1): 429–439.

Vale, M.G.R., Silva, M.M., Damin, I.C.F., Sanches Filho, P.J., and Welz, B. 2008. Determination of volatile and non-volatile nickel and vanadium compounds in crude oil using electrothermal atomic absorption spectrometry after oil fractionation into saturates, aromatics, resins and asphaltenes. *Talanta* 74 (5): 1385–1391.

Wiehe, I.A. 1992. A solvent-resid phase diagram for tracking resid conversion. *Ind. Eng. Chem. Res.* 31 (2): 530–536.

Wood, J., Kung, J., Kingston, D., Kotlyar, L., Sparks, B., and McCracken, T. 2008. Canadian crudes: A comparative study of SARA fractions from a modified HPLC separation technique. *Oil Gas Sci. Tech.–Rev. IFP* 63 (1): 151–163.

Xu, H. and Lesage, S. 1992. Separation of vanadyl and nickel petroporphyrins on an aminopropyl column by high-performance liquid chromatography. *J. Chromatog.* 607 (1): 139–144.

Xu, Y., Koga, Y., and Strausz, O.P. 1994. Characterization of Athabasca asphaltenes by small-angle x-ray scattering. *Fuel* 74 (7): 960–964.

Yang, C., Du, F., Zheng, H., and Chung, K.H. 2005. Hydroconversion characteristics and kinetics of residue narrow fractions. *Fuel* 84 (6): 675–684.

Yang, G. and Wang, R.A. 1999. The supercritical fluid extractive fractionation and the characterization of heavy oils and petroleum residua. *J. Petrol. Sci. Eng.* 22 (1–3): 47–52.

Yang, M.G. and Eser, S. 1999. Fractionation and molecular analysis of a vacuum residue asphaltenes. *Am. Chem. Soc., Div. Fuel Chem.* 44 (4): 768–771.

Yen, T.F., Wu, W.H., and Chilingar, G.V. 1984. Study of the structure of petroleum asphaltenes and related substances by infrared spectroscopy. *Energy Sources* 7 (3): 203–225.

Ysambertt, F, Márquez, N., Rangel, B., Bauza, R., and De la Cruz, C. 1995. Isolation and characterization of metalloporphyrins from a heavy crude oil by Soxhlet adsorption chromatography and HPLC-SEC. *Sep. Sci. Tech.* 30 (12): 2539–2550.

Zhao, S., Xu, Z., Xu, C., Chung, K.H., and Wang, R. 2005. Systematic characterization of petroleum residua based on SFEF. *Fuel* 84 (6): 635–645.

Zhu, Y., Wang, X., and Liu, Z. 1998. Analysis of vanadium components in vacuum residue of Chinese crude and its impact on HDM process. *Am. Chem. Soc., Div. Petr. Chem.–Prepr.* 43 (1): 146–150.

Index

A

B

C

D

G

I

J

K

L

M

N

T

U

V

W

X

Z